Phytochemicals and Chronic Diseases Prevention

Phytochemicals and Chronic Diseases Prevention

Editors

Haixia Yang
Jianjun Deng

Basel • Beijing • Wuhan • Barcelona • Belgrade • Novi Sad • Cluj • Manchester

Editors

Haixia Yang
China Agricultural University
Beijing
China

Jianjun Deng
Chinses Academy of Agricultural Sciences
Beijing
China

Editorial Office
MDPI
St. Alban-Anlage 66
4052 Basel, Switzerland

This is a reprint of articles from the Special Issue published online in the open access journal *Nutrients* (ISSN 2072-6643) (available at: https://www.mdpi.com/journal/nutrients/special_issues/phytochemicals_chronic_disease).

For citation purposes, cite each article independently as indicated on the article page online and as indicated below:

Lastname, A.A.; Lastname, B.B. Article Title. *Journal Name* **Year**, *Volume Number*, Page Range.

ISBN 978-3-7258-1245-5 (Hbk)
ISBN 978-3-7258-1246-2 (PDF)
doi.org/10.3390/books978-3-7258-1246-2

Contents

About the Editors

Haixia Yang

Haixia Yang is a full professor at the College of Food Science and Nutritional Engineering, China Agricultural University. Her research mainly focuses on basic applied research regarding the evaluation of food nutrition, lipid metabolism and its health effects, and the development and utilization of food by-products. Haixia Yang received her PhD in Food Science College from China Agricultural University in 2011. She studied as a postdoc research fellow at the University of Massachusetts Amherst and Harvard Medical School. She has published more than 100 papers in journals such as *Sci. Transl. Med.*, *PNAS* and *Gut Microbes*. She is currently the executive editor of *Food Innovation and Advances*, and an Editorial Board Member of *Nutrients and Protein & Peptide Letters*.

Jianjun Deng

Jianjun Deng obtained his PhD in food biotechnology from China Agricultural University. He completed his post-doctoral training at the University of Hong Kong and Harvard Medical School. He is currently a professor at the Institute of Vegetable and Flower at the Chinese Academy of Agricultural Sciences. His research interests include natural products and the prevention and control of chronic diseases, food chemistry and nutrition, fatty acid metabolism and cancer. He has published more than 150 academic papers in journals such as *PNAS*, *Nucleic Acids Res.*, *Trends Food Sci. and Technol.*, among others. He serves as the Associate Editor for *eFood* and *Food Frontiers*, and is a Guest Editor for *Nutrients*, *Antioxidants* and *Phytomedicine*.

Preface

This Special Issue explores the relationship between phytochemicals and the prevention of chronic disease.

The articles included in this Special Issue explore the mechanisms through which phytochemicals ameliorate chronic diseases in clinical and animal studies, such as the regulation of blood glucose levels and lipid accumulation, alongside enhancements in the extraction, purification efficiency, and bioavailability of phytochemicals. Therefore, this Special Issue provides a comprehensive overview of the relationship between phytochemicals and the prevention of chronic disease.

We are grateful to the authors of the articles published in this Special Issue, who hail from countries such as Korea, Hungary, China, Thailand, Romania, Poland, and Japan.

We hope that this Special Issue will inform and entertain our readers.

Haixia Yang and Jianjun Deng
Editors

nutrients

MDPI

Editorial

Phytochemicals in Chronic Disease Prevention

Jing Sun [1], Shuwei Luo [1], Jianjun Deng [2] and Haixia Yang [1,*]

[1] College of Food Science and Nutritional Engineering, China Agricultural University, Beijing 100083, China
[2] State Key Laboratory of Vegetable Biobreeding, Institute of Vegetables and Flowers, Chinese Academy of Agricultural Sciences, Beijing 100081, China; dengjianjun@nwu.edu.cn
* Correspondence: hyang@cau.edu.cn

Chronic diseases, also known as noncommunicable diseases (NCD), are characterized by long durations and a slow progression of the associated medical conditions. According to the World Health Organization, they are responsible for 74% of annual deaths worldwide and have become a global pandemic with high incidence rates [1]. The development of chronic diseases is influenced by several factors, including physical activity, dietary patterns, tobacco and alcohol consumption, and environmental factors. For example, refined sugar intake and ultra-processed food products are now recognized as the most important risk factors for the development of NCD. Consequently, interventions intended to not only treat but also prevent the progression of chronic diseases early in life have become crucial [2,3]. Importantly, diet plays a vital role in promoting health and preventing chronic diseases. Phytochemicals, such as polyphenols, terpenoids, carotenoids, and alkaloids, are relatively small molecular compounds isolated from edible plants that exhibit various pharmacological activities in relation to chronic ailments such as obesity, diabetes, and cardiovascular diseases [4,5]. Therefore, the investigation of the relationship between phytochemicals and chronic disease prevention has emerged as an important topic [6,7]. This Special Issue, "Phytochemicals and Prevention of Chronic Diseases", features a series of high-quality research articles that explore the isolation, identification, and bioactivities of phytochemicals, as well as the underlying molecular mechanisms involved in chronic diseases through antioxidation, neuroprotection, and the modulation of gut microbiota.

Clinical trials have shown that natural phytochemicals can improve blood glucose levels and lipid accumulation. Lycopene, a lipophilic unsaturated carotenoid, has been found to significantly restore blood glucose levels in 152 type 2 diabetes patients based on publications from the past 5 years. This effect is attributed to the accumulation of leptin in plasma, a reduction in oxidized low-density lipoprotein cholesterol, and the amelioration of oxidative stress [8]. Furthermore, flaxseed, soy, and red clover, rich in proteins and isoflavones, have been shown to improve serum lipids by inhibiting total cholesterol and levels of low- and high-density lipoprotein cholesterol. This beneficial effect has demonstrated potential in preventing cardiovascular diseases in postmenopausal women [9]. Moreover, an increasing trend in childhood obesity has been observed worldwide, highlighting the need for the regulation of dietary food intake. Statistical analysis based on Korean preschoolers (aged 3–5 years) revealed an inverse correlation between the quartiles of dietary phytochemical intake and overweight/obesity, particularly among boys who consume high amounts of meat, milk, and dairy products [10]. Overall, these findings emphasize the importance of natural phytochemicals in improving blood glucose levels, lipid profiles, and obesity management, specifically in relation to preventing cardiovascular diseases and promoting healthy dietary habits. Meanwhile, public health policymakers should pay attention to these results so that further policies and strategies for the management of overweight/obesity can be proposed.

Meanwhile, the consumption of phytochemicals, such as ginsenoside Rh4, zerumbone, rosmarinic acid, and schisandrin B, has been widely recognized for its potential in preventing and treating chronic diseases in animal models. Ginsenoside Rh4, a terpenoid

Citation: Sun, J.; Luo, S.; Deng, J.; Yang, H. Phytochemicals in Chronic Disease Prevention. *Nutrients* **2023**, *15*, 4933. https://doi.org/10.3390/nu15234933

Received: 9 October 2023
Revised: 19 November 2023
Accepted: 20 November 2023
Published: 27 November 2023

compound derived from ginseng, is composed of triterpene aglycones and glycosides. In a mouse model of esophageal squamous cell carcinoma, promising inhibitory effects on cancer cell metastasis were induced by Ginsenoside Rh4 administration through the regulation of the Wnt/β-catenin pathway. Additionally, Ginsenoside Rh4 demonstrated the ability to inhibit the migration and invasion of gastric cancer cells in vivo by reversing epithelial–mesenchymal transition induced by the suppression of the TGF-β/Smad2/3 signaling axis [11,12]. Furthermore, another phytochemical compound, sesquiterpenoid zerumbone, exerted therapeutic effects on adult mouse microglia by reducing the expression of lipocalin-2 as well as pro-inflammatory cytokines (interleukin-6 and -1β and tumor necrosis factor-α) and chemokines (CCL-2 and CXCL-10). Additionally, zerumbone facilitated the polarization of macrophages into M2 phenotypes and regulated cellular redox homeostasis by activating the AMPK and Akt/GSK3β signaling pathways in vitro [13]. Regarding rosmarinic acid, it was reported to exhibit neuroprotective properties against rotenone-induced human neuroblastoma SH-SY5Y cell injury via restoring mitochondrial function. This effect was achieved through upregulating the expression of peroxisome proliferator-activated receptor gamma coactivator 1 and the phosphorylation of Akt and AMPK while suppressing the hyperphosphorylation of Abl [14]. Moreover, schisandrin B, a dibenzooctadiene lignan derived from *Schisandra chinensis*, exhibited a protective effect in a mouse model of ethanol-induced liver and brain injury. It achieved this by inhibiting inflammasome activation and collagen deposition and preventing neurological defects as supported by neurological tests. Surprisingly, schisandrin B showed a therapeutic effect rather than a preventive one in ethanol-induced liver injury [15]. Additionally, a molecular docking analysis revealed strong binding ability between 11-O-(4′-O-methylgalloyl)-bergenin, a potential anti-inflammatory compound isolated from *Saxifraga atrata*, and arachidonic acid 15-lipoxygenase, nitric oxide synthase, epidermal growth factor receptor 2, and e-selectin. This compound may exert its effects through the MAPK and NF-κB signaling pathways [16]. Overall, these findings highlight the potential of phytochemicals, such as ginsenoside Rh4, zerumbone, rosmarinic acid, schisandrin B, and 11-O-(4′-O-methylgalloyl)-bergenin, in the prevention and treatment of chronic diseases. Further studies are warranted to explore their mechanisms of action and potential clinical applications.

The exploration of bioactivities for phytochemicals has been limited due to the low extraction rate of pure phytochemicals. Multiple studies have utilized raw plant extracts to investigate the functions of these phytochemicals. The ethanol extract obtained from *Mitragyna speciosa (Korth.) Havil.* Leaves, which includes phenolic compounds, flavonoids, and alkaloids, exhibited significant inhibition of α-glucosidase and pancreatic lipase activities. Mitragynine, identified as the main alkaloid in the ethanol extract, noncompetitively inhibited α-glucosidase activity and synergistically interacted with acarbose, suggesting its potential in preventing diabetes mellitus [17]. Furthermore, the ethyl acetate-n-butanol extracts of *Allium tenuissimum* L. flowers, abundant in flavonoid content (1429.5 μg/g extract), particularly kaempferol (284.1 μg/g flavonoids), demonstrated strong α-glucosidase inhibition activities in vitro. In addition, these extracts ameliorated glycolipid metabolic disorders and inflammation in diabetic mice [18]. However, green extraction strategies, such as water/hydro extraction methods, should be encouraged instead of the use of organic solvents due to the side effects and toxicity of extracts as well as environmental pollution. A mixture of aqueous extracts derived from red grape, blackcurrant, redcurrant, rosehip, and black cherry was characterized by the presence of 1.74 mg/L trans-resveratrol and 2.37 mg/L trans-piceid. This mixture exhibited anticancer effects by inhibiting the expression levels of DNA methyltransferases (DNMT1, DNMT3A, and DNMT3B) and histone deacetylases (HDAC2 and HDAC3) in a mouse model exposed to the carcinogen 7,12-dimethylbenz(a)anthracene [19]. Recently, fruit juice consumption has become more popular than the consumption of fresh fruits. Notably, not-from-concentrate apple juice with minimal processing showed higher total phenol content (2169.03 ± 116.84 GAE mg/100 mL) compared to from-concentrate apple juice (1073.32 ± 72.33 GAE mg/100 mL). In a healthy

SD rat model, the former significantly alleviated intestinal inflammation and maintained intestinal homeostasis in the gut [20].

The exploration and purification methods of compounds in plants play a crucial role in studying the bioactivities of phytochemicals. In this regard, affinity ultrafiltration-HPLC, a high-throughput screening technique, has been employed to identify potential active ingredients with anti-inflammatory properties that can interact with cyclooxygenase-2 (COX-2), an inflammation-related isoenzyme, in the methanol/water extracts of *Saxifraga atrata* [16]. Through the application of this method, the phenolic compound 11-O-(4′-O-methylgalloyl)-bergenin was rapidly identified as the target component (purity > 99%) and demonstrated significant anti-inflammatory activities exerted via the AA metabolism, MAPK, and NF-κB signaling pathways [16]. The further development of more efficient strategies is required to deepen our understanding of the bioactivities exhibited by phytochemicals.

Due to their limited solubility and stability, phytochemicals suffer from low bioavailability within the body. Therefore, new strategies of bioavailability improvement have been developed, such as delivery systems (nanoparticles, liposomes, and phytoliposomes) and co-administration with absorption enhancers. The development of dietary supplements in this field has garnered considerable interest. In line with this, a successful endeavor has been made to prepare a dietary capsule utilizing powdered extracts from berries and grape pomace. The resulting capsule incorporates a rich array of phenolic compounds, including chlorogenic acid (34.51%), rutin (19.13%), ferulic acid (17.03%), (+)-catechin (12.04%), and trace amounts (<10%) of other phenolic compounds. Moreover, this composition exhibits exceptional scavenging activity against the 2,2-diphenyl-1-picrylhydrazyl free radical ($48.32 \pm 0.74\%$). The particle size distribution and powder flow of the capsule meet the stringent requirements outlined by the European Pharmacopoeia, thus establishing a promising reference point for the industrial application of polyphenol-rich supplements [21].

The publications showcased in this Special Issue encompass a wide range of phytochemicals derived from various foods and plants. Collectively, these contributions significantly enhance our understanding of the bioactivities exerted by phytochemicals in the prevention of chronic diseases. However, further research efforts are required to fully harness their potential therapeutic efficacy. Advancing our comprehension of the benefits, mechanisms, and safety profiles associated with phytochemicals will undoubtedly make substantial advancements in human health and propel the development of the food, health food, and pharmaceutical industries.

Author Contributions: Writing—original draft preparation, J.S. and H.Y.; writing—review and editing, J.S. and H.Y.; visualization, S.L. and J.D.; supervision, J.D.; project administration, H.Y.; funding acquisition, H.Y. All authors have read and agreed to the published version of the manuscript.

Funding: This research was supported by National Natural Science Foundation of China (21978229 and 21676212).

Acknowledgments: The editors like to thank all participating authors.

Conflicts of Interest: The authors declare no conflict of interest.

References

1. Noncommunicable Diseases (NCD). Available online: https://www.who.int/gho/ncd/mortality_morbidity/en/ (accessed on 16 September 2023).
2. Li, Y.P.; Schoufour, J.; Wang, D.D.; Dhana, K.; Pan, A.; Liu, X.R.; Song, M.Y.; Liu, G.; Shin, H.J.; Sun, Q.; et al. Healthy lifestyle and life expectancy free of cancer, cardiovascular disease, and type 2 diabetes: Prospective cohort study. *BMJ* **2020**, *368*, l6669. [CrossRef] [PubMed]
3. Zhang, C.; Deng, J.; Liu, D.; Tuo, X.; Xiao, L.; Lai, B.; Yao, Q.; Liu, J.; Yang, H.; Wang, N. Nuciferine ameliorates hepatic steatosis in high-fat diet/streptozocin-induced diabetic mice through a PPARα/PPARγ coactivator-1α pathway. *Br. J. Pharmacol.* **2018**, *175*, 4218–4228. [CrossRef] [PubMed]

4. Gao, Q.; Feng, J.; Liu, W.; Wen, C.; Wu, Y.; Liao, Q.; Zou, L.; Sui, X.; Xie, T.; Zhang, J.; et al. Opportunities and challenges for co-delivery nanomedicines based on combination of phytochemicals with chemotherapeutic drugs in cancer treatment. *Adv. Drug Deliv. Rev.* **2022**, *188*, 114445. [CrossRef] [PubMed]
5. Ren, Z.; Yang, H.; Zhu, C.; Fan, D.; Deng, J. Dietary phytochemicals: As a potential natural source for treatment of Alzheimer's Disease. *Food Innov. Adv.* **2023**, *2*, 36–43. [CrossRef]
6. Yang, H.; Tuo, X.; Wang, L.; Tundis, R.; Portillo, M.P.; Simal-Gandara, J.; Yu, Y.; Zou, L.; Xiao, J.; Deng, J. Bioactive procyanidins from dietary sources: The relationship between bioactivity and polymerization degree. *Trends Food Sci. Technol.* **2021**, *111*, 114–127. [CrossRef]
7. Zhao, T.; Zhu, Y.; Zhao, R.; Xiong, S.; Sun, J.; Zhang, J.; Fan, D.; Deng, J.; Yang, H. Structure-activity relationship, bioactivities, molecular mechanisms, and clinical application of nuciferine on inflammation-related diseases. *Pharmacol. Res.* **2023**, *193*, 106820. [CrossRef] [PubMed]
8. Inoue, T.; Yoshida, K.; Sasaki, E.; Aizawa, K.; Kamioka, H. Effect of lycopene intake on the fasting blood glucose level: A systematic review with meta-analysis. *Nutrients* **2023**, *15*, 122. [CrossRef] [PubMed]
9. Błaszczuk, A.; Barańska, A.; Kanadys, W.; Malm, M.; Jach, M.E.; Religioni, U.; Wróbel, R.; Herda, J.; Polz-Dacewicz, M. Role of phytoestrogen-rich bioactive substances (*Linum usitatissimum* L., *Glycine max* L., *Trifolium pratense* L.) in cardiovascular disease prevention in postmenopausal women: A systematic review and meta-analysis. *Nutrients* **2022**, *14*, 2467. [CrossRef] [PubMed]
10. Han, Y.-J.; Baek, J.-H.; Jung, S.-K.; Yang, J.S.; Shin, N.-R.; Park, M.-Y. Association between the Dietary Phytochemical Index and Lower Prevalence of Obesity in Korean Preschoolers. *Nutrients* **2023**, *15*, 2439. [CrossRef] [PubMed]
11. Jiang, H.; Ma, P.; Duan, Z.; Liu, Y.; Shen, S.; Mi, Y.; Fan, D. Ginsenoside Rh4 Suppresses Metastasis of Gastric Cancer via SIX1-Dependent TGF-β/Smad2/3 Signaling Pathway. *Nutrients* **2022**, *14*, 1564. [CrossRef] [PubMed]
12. Chen, J.; Duan, Z.; Liu, Y.; Fu, R.; Zhu, C. Ginsenoside Rh4 Suppresses Metastasis of Esophageal Cancer and Expression of c-Myc via Targeting the Wnt/β-Catenin Signaling Pathway. *Nutrients* **2022**, *14*, 3042. [CrossRef] [PubMed]
13. Yeh, W.-L.; Huang, B.-R.; Chen, G.-W.; Charoensaensuk, V.; Tsai, C.-F.; Yang, L.-Y.; Lu, D.-Y.; Chen, M.-K.; Lin, C. Role of Zerumbone, a Phytochemical Sesquiterpenoid from Zingiber zerumbet Smith, in Maintaining Macrophage Polarization and Redox Homeostasis. *Nutrients* **2022**, *14*, 5402. [CrossRef] [PubMed]
14. Han, X.; Han, B.; Zhao, Y.; Li, G.; Wang, T.; He, J.; Du, W.; Cao, X.; Gan, J.; Wang, Z.; et al. Rosmarinic Acid Attenuates Rotenone-Induced Neurotoxicity in SH-SY5Y Parkinson’s Disease Cell Model through Abl Inhibition. *Nutrients* **2022**, *14*, 3508. [CrossRef] [PubMed]
15. Lam, H.Y.; Liang, T.-R.; Peng, S.-Y. Prevention of the Pro-Aggressive Effects of Ethanol-Intoxicated Mice by Schisandrin B. *Nutrients* **2023**, *15*, 1909. [CrossRef] [PubMed]
16. Li, G.; Fang, Y.; Ma, Y.; Dawa, Y.; Wang, Q.; Gan, J.; Dang, J. Screening and Isolation of Potential Anti-Inflammatory Compounds from *Saxifraga atrata* via Affinity Ultrafiltration-HPLC and Multi-Target Molecular Docking Analyses. *Nutrients* **2022**, *14*, 2405. [CrossRef] [PubMed]
17. Limcharoen, T.; Pouyfung, P.; Ngamdokmai, N.; Prasopthum, A.; Ahmad, A.R.; Wisdawati, W.; Prugsakij, W.; Warinhomhoun, S. Inhibition of α-Glucosidase and Pancreatic Lipase Properties of *Mitragyna speciosa* (Korth.) Havil. (Kratom) Leaves. *Nutrients* **2022**, *14*, 3909. [CrossRef] [PubMed]
18. Zhang, S.-S.; Hou, Y.-F.; Liu, S.-J.; Guo, S.; Ho, C.-T.; Bai, N.-S. Exploring Active Ingredients, Beneficial Effects, and Potential Mechanism of *Allium tenuissimum* L. Flower for Treating T2DM Mice Based on Network Pharmacology and Gut Microbiota. *Nutrients* **2022**, *14*, 3980. [CrossRef] [PubMed]
19. Nowrasteh, G.; Zand, A.; Raposa, L.B.; Szabó, L.; Tomesz, A.; Molnár, R.; Kiss, I.; Orsós, Z.; Gerencsér, G.; Gyöngyi, Z.; et al. Fruit Extract, Rich in Polyphenols and Flavonoids, Modifies the Expression of DNMT and HDAC Genes Involved in Epigenetic Processes. *Nutrients* **2023**, *15*, 1867. [CrossRef] [PubMed]
20. Xu, L.; Yang, S.; Wang, K.; Lu, A.; Wang, X.; Xu, Z. Impact of Clarified Apple Juices with Different Processing Methods on Gut Microbiota and Metabolomics of Rats. *Nutrients* **2022**, *14*, 3488. [CrossRef] [PubMed]
21. Frum, A.; Dobrea, C.M.; Rus, L.L.; Virchea, L.-I.; Morgovan, C.; Chis, A.A.; Arseniu, A.M.; Butuca, A.; Gligor, F.G.; Vicas, L.G.; et al. Valorization of Grape Pomace and Berries as a New and Sustainable Dietary Supplement: Development, Characterization, and Antioxidant Activity Testing. *Nutrients* **2022**, *14*, 3065. [CrossRef] [PubMed]

nutrients

MDPI

Article

Association between the Dietary Phytochemical Index and Lower Prevalence of Obesity in Korean Preschoolers

Ye-Ji Han [1], Jung-Hyun Baek [2], Seong-Kwan Jung [2], Joshua SungWoo Yang [3], Na-Rae Shin [3] and Mi-Young Park [4,*]

1 Department of Food and Nutrition, Sungshin Women's University, Seoul 01133, Republic of Korea; yeji.h1n@gmail.com
2 Department of Pediatrics, Woori Children's Hospital, Seoul 08291, Republic of Korea
3 Healthcare Development Head, R&D Center, NGeneBio Inc., Seoul 08390, Republic of Korea; sungwoo.yang@ngenebio.com (J.S.Y.); narae.shin@ngenebio.com (N.-R.S.)
4 Institute of Health and Environment, Graduate School of Public Health, Seoul National University, Seoul 08826, Republic of Korea
* Correspondence: coolmypark@snu.ac.kr

Abstract: Little is known regarding Korean preschooler dietary phytochemical index (DPIs). We used the 24 h recall data of 1196 participants aged 3–5 years from the Korea National Health and Nutrition Examination Survey to study the association between dietary food intake and obesity prevalence. The amount of dietary intake by food group was compared according to sex and DPI quartile. Multivariable-adjusted odds ratios (ORs) and 95% confidence intervals (CIs) were calculated using logistic regression models. The average total DPI and energy from phytochemical food groups were not significantly different according to sex, although boys had a higher total daily food intake. Different inclinations between DPI quartiles and amount of intake were observed in the food groups; specifically, beans showed a higher intake difference between Q1 and Q4 for boys than in the other food groups. The highest DPI quartile had a significantly lower obesity prevalence than the lowest DPI quartile in all models for boys only when obesity prevalence by weight percentile was analyzed (Model 3, OR: 0.287, 95% CI: 0.095–0.868, p for trend < 0.05). Our results suggest a high DPI could help prevent obesity in preschoolers.

Keywords: dietary phytochemical index; obesity; preschooler; child; Korea National Health and Nutrition Examination Survey (KNHANES)

Citation: Han, Y.-J.; Baek, J.-H.; Jung, S.-K.; Yang, J.S.; Shin, N.-R.; Park, M.-Y. Association between the Dietary Phytochemical Index and Lower Prevalence of Obesity in Korean Preschoolers. *Nutrients* **2023**, *15*, 2439. https://doi.org/10.3390/nu15112439

Academic Editors: Haixia Yang and Jianjun Deng

Received: 11 April 2023
Revised: 18 May 2023
Accepted: 22 May 2023
Published: 24 May 2023

1. Introduction

An increasing trend in childhood obesity and overweight has been observed worldwide [1–3]. The prevalence of obesity among children aged 2–18 years has increased from 8.6% in 2001 to 9.8% in 2017 in Korea [4]. Many studies have demonstrated childhood obesity to be related to an increased risk of several adverse health outcomes at an early age, including cardiovascular disease, fatty liver infiltration, and sleep apnea [5–7]. Moreover, controlling childhood obesity is very important since childhood obesity has a high risk of developing into adult obesity [8,9]. A follow-up study found that overweight children aged 2–5 years had >4 times greater tendency to become overweight adults [10]. Childhood obesity reportedly increases the risk of cardiovascular disease, cancer [11], and mortality [5,12,13] in adulthood.

The increase in childhood obesity in Korea seems to be the result of a decrease in vegetable intake and an increase in animal food product intake due to rapid modernization and a westernized diet. The traditional food culture of Koreans includes vegetable-based proteins, such as tofu made of soybeans [14]. However, the proportion of individuals consuming more than 500 g vegetables and fruits daily decreased by approximately 10% from 36.5% in 2011 to 26.2% in 2020 in Korea [15].

Previous studies have demonstrated that sufficient intake of vegetables, fruits, and whole grains, which are representative phytochemical food groups, helps in weight control [16–18]. Phytochemicals, which are active substances present mainly in plant foods, protect the body from oxidative stress and free radicals with antioxidant functions [19].

The dietary phytochemical index (DPI) is defined as the percentage of dietary calories derived from foods rich in phytochemicals [20]; a higher DPI is strongly associated with a lower prevalence of overweight/obesity [21–23], cardiovascular disease, metabolic syndrome [24], and cancer [20]. A meta-analysis revealed that a lower DPI decreases the risk of overweight/obesity by 19% (95% confidence interval [8]: 0.74–0.90) [21]. However, the results demonstrated significant heterogeneity according to age, sex, and obesity criteria [21].

Therefore, studying the amount of intake by food group and DPI among Koreans is necessary. Nevertheless, few studies exist on the DPI of Korean preschoolers. Moreover, studies that focused on children did not consider sex differences. This study investigated the association between the DPI and prevalence of obesity/overweight among Korean preschoolers according to sex using the 2013–2018 data from the Korea National Health and Nutrition Examination Survey (KNHANES).

2. Materials and Methods

2.1. Data Source and Study Population

We used survey data from over 6 years (2013–2018) from the KNHANES. The KNHANES is a large-scale cross-sectional survey conducted among individuals aged ≥1 year by the Korea Centers for Disease Control and Prevention (KDCA) since 1998. The nationwide cross-sectional study was conducted to identify health-related factors with a health examination, health interview, and nutritional survey among approximately 2000–3000 South Koreans per generation. The KNHANES was approved by the Institutional Review Board of the KDCA (Approval numbers: 2013-07CON-03-4C, 2013-12EXP-03-5C, and 2018-01-03-P-A). For children aged 3–5 years, written informed consent was obtained from one of the parents to use and analyze their data.

Among children aged 3–5 years (n = 1751) who participated in the survey during 2013–2018, we excluded those using the following criteria from the analyses: (1) did not participate in the 24 h recall survey (n = 61), (2) had an extreme value in the total daily energy intake (≥3000 kcal) (n = 11), (3) missing values (weight, weight percentile, height, body mass index (BMI), BMI percentile, sex, education level of the mother, education level of the father, household income, and residential area) (n = 88), and (4) had asthma, diabetes, congenital heart disease, urinary tract infection, and pneumonia (n = 395). Finally, a total of 1196 participants (boys = 623, girls= 573) were included in the study. This study was approved by the Institutional Review Board of the Seoul National University (E2108/002-002) for analyzing the secondary processing data of KDCA.

2.2. Demographic Data, Anthropometric Measurement, and Diagnosis of Obesity

General characteristics, such as age, weight, height, BMI, education level, household income, and residential area of the participants, were described using the basic variables, health survey, and screening survey data of the KNHANES. Considering that the participants of our study included 3–5-year-old preschoolers, we presented both the age in years and number of months to denote the age of the participant. The education levels of the mothers and fathers were considered in place of the participants' education level and classified into three categories (middle school or lower, high school, and university graduation or higher). Household income level was classified in quartiles (low, middle-low, middle-high, and high) by the KNHANES with different standards for quartiles provided according to the survey conducted years [25]. The residential areas were classified into two categories (urban and rural).

Health surveys, including screening surveys, were conducted by trained staff according to a standardized protocol [26]. The weight and height were measured using a standardized height and scale, with the participants taking their shoes and socks off, wearing a test gown, and standing upright on a scale. Weight percentile (%) and BMI percentile (%) were provided as categorical variables by the KNHANES, using the percentiles by the growth chart for children and adolescents in 2017 as reference [26]. The growth chart is an important indicator for evaluating the growth status of children and adolescents who are in the process of growth, unlike adults. To evaluate the growth status of children and adolescents, such as short stature, underweight, and obesity, the Korea Centers for Disease Control and Prevention and the Korean Academy of Pediatrics have established and published a growth chart every 10 years since 1967 [27].

The weight percentile (%) was divided into three groups (<5%, 5–95%, >95%), and the BMI percentile (%) was divided into four groups (<5%, 5–85%, 85–95%, >95%). We defined obesity as a percentile >95% for both the weight and BMI percentiles according to the Korean Society for the Study of Obesity criteria [27].

2.3. Nutritional Survey Data and DPI

To calculate the DPI for children aged 3–5 years in Korea, we used the data of daily nutrient intake based on the 24 h recall survey from the KNHANES. The 24 h recall survey was conducted through face-to-face interviews with trained staff. In the case of children younger than 8 years of age, one of the parents answered the dietary recall [25]. Supplementary materials were also used to enhance recall skills and collect specific data on the survey items during the survey.

The DPI was used to calculate the approximate value of phytochemical intake from diets using the intake level of plant foods [20]. It was also useful for evaluating the health effects of the overall plant foods consumed; the calculation formula was as follows:

$$\text{DPI} = \frac{Daily\ energy\ intake\ derived\ from\ phytochemical\ rich\ foods\ (\text{kcal})}{Tatal\ daily\ energy\ intake\ (\text{kcal})} \times 100$$

In this study, a modified version of the DPI for the Korean diet was used to calculate the DPI in Korean children aged 3–5 years [23,24]. The food groups included in the DPI were whole grains, beans, seeds and nuts, vegetables, mushrooms, and fruits. All the food groups were classified according to KNHANES. Beans included soybeans and products made from beans such as tofu. Grains and their products were divided into two groups based on whether they contained phytochemicals. Whole grains, such as oats, millet, buckwheat, whole wheat, barley, sorghum, corn, and brown rice, are rich in phytochemicals, whereas refined grains are not.

2.4. Statistical Analyses

Stratification variables, colony variables, and weight were all considered during the statistical analyses, as this study used data from the KNHANES, constructed through the complex sample design method. All the analyses were performed using the PROC SURVEY procedure. General characteristics of the participants according to sex were presented as mean and standard error or frequency and percentage (%).

In the analysis of intake amount by food group according to the DPI, the range of total DPI was presented as minimum–maximum, with the mean DPI of each quartile according to sex. Energy from the phytochemical food groups (kcal/day) and dietary intake amounts by the food group were presented as mean and standard error, according to the sex and DPI quartile. Differences between the sex in the total DPI, energy from the phytochemical food groups, and amount of intake by food groups were also shown by using the PROC SURVEYREG procedure to calculate p values.

Weight (kg), BMI (kg/m^2), and daily energy intake (kcal/day) were presented as the mean and standard error by the DPI quartile for each sex. The p values were calculated using the PROC SURVEYREG procedure. Multivariable logistic regression analyses were used to calculate the odds ratios (ORs) and 95% CIs to analyze the association between DPI and obesity. Three covariate models, except for the crude model, were evaluated by adjusting for the confounding factors as follows: Model 1 was adjusted for energy intake,

Model 2 was adjusted for age, and Model 3 was adjusted for energy intake and age. The p for trend in the DPI quartiles using linear regression analysis was evaluated using the median value of the category as a continuous variable. All data analyses were performed using the Statistical Analysis System version 9.4 software (SAS Institute, Cary, NC, USA). A p-value < 0.05 was considered statistically significant.

3. Results

3.1. General Characteristics

The general characteristics of the participants stratified by sex are presented in Table 1. The average age of the participants was 4.02 years (53.75 months), and no statistical difference was observed between the sexes. Boys had a higher average height and weight than girls ($p < 0.01$); significant difference between the height according to sex was shown among participants aged 4 and 5 years old ($p < 0.01$). Weight was different only in the participants aged 3 years old when stratified by age and sex. The average daily energy intake of the participants was 1370.65 kcal/day, and boys had a significantly higher energy intake than girls ($p < 0.001$). No statistical difference in the education levels of the mother and father, household income, and residential area was observed according to sex.

Table 1. General characteristics stratified by sex among the participants.

	Total (n = 1196)	Boys (n = 623)	Girls (n = 573)	*p* Value *
Age (months)	53.75 ± 0.32	54.05 ± 1.16	53.42 ± 0.47	0.366
Age (years)	4.02 ± 0.02	4.04 ± 0.09	3.99 ± 0.04	0.360
3	386 (32.30)	203 (32.60)	183 (31.90)	0.088
4	392 (32.80)	193 (31.00)	199 (34.70)	
5	418 (34.90)	227 (36.40)	191 (33.31)	
Energy Intake (kcal/day)	1370.65 ± 14.97	1453.30 ± 48.78	1276.06 ± 20.30	<0.001
Height (cm)	105.75 ± 0.23	106.4 ± 0.82	105.0 ± 0.33	0.004
Aged 3	98.63 ± 0.60	98.86 ± 0.84	98.36 ± 0.35	0.307
Aged 4	105.86 ± 0.58	106.48 ± 0.75	105.27 ± 0.29	0.008
Aged 5	112.41 ± 0.24	113.04 ± 0.85	111.55 ± 0.37	0.002
Weight (kg)	17.59± 0.11	17.88 ± 0.37	17.26 ± 0.15	0.003
Aged 3	15.22 ± 0.36	15.47 ± 0.40	14.93 ± 0.16	0.024
Aged 4	17.52 ± 0.36	17.74 ± 0.46	17.31 ± 0.18	0.125
Aged 5	19.91 ± 0.16	20.15 ± 0.57	19.59 ± 0.24	0.088
BMI percentile (%)				
<5%	114 (9.53)	61 (9.79)	53 (9.25)	0.843
5–85%	934 (78.09)	486 (78.01)	448 (78.18)	
85–95%	77 (6.44)	37 (5.94)	40 (6.98)	
>95%	71 (5.94)	39 (6.26)	32 (5.58)	
Weight percentile (%)				
<5%	90 (7.53)	53 (8.51)	37 (6.46)	0.404
5–95%	1037 (86.71)	534 (85.71)	503 (87.78)	
>95%	69 (5.77)	36 (5.78)	33 (5.76)	

Table 1. *Cont.*

	Total (n = 1196)	Boys (n = 623)	Girls (n = 573)	p Value *
Education Level of Mother				
≤Middle school	35 (4.34)	18 (4.88)	17 (3.76)	0.717
≤High school	280 (28.66)	139 (28.03)	141 (29.34)	
≤College	700 (67.00)	361 (67.10))	339 (66.90)	
Education Level of Father				
≤Middle school	18 (2.39)	10 (2.13)	8 (2.69)	0.474
≤High school	208 (26.73)	102 (24.97)	106 (28.73)	
≤College	567 (70.88)	302 (72.90)	265 (68.58)	
Household income				
Low	82 (7.16)	41 (6.79)	41 (7.58)	0.883
Middle-low	377 (32.05)	194 (31.44)	183 (32.76)	
Middle-high	421 (34.81)	218 (35.76)	203 (33.73)	
High	316 (25.97)	170 (26.01)	146 (25.93)	
Residential area				
Urban	1018 (86.83)	533 (87.50)	485 (86.06)	0.422
Rural	178 (13.17)	90 (12.50)	88 (13.94)	

BMI, body mass index. All the values are presented as mean ± standard deviation or n (%). * p values were calculated using the PROC SURVEYREG.

3.2. Association between the DPI and Dietary Intake

Table 2 compares the energy intake of the food group across the quartiles for sex and DPI. The mean ± standard error of total DPI according to sex was 15.43 ± 1.07 and 16.49 ± 0.45 for boys and girls, respectively. Boys had a significantly higher amount of total daily food intake (g/day) than girls; however, the total DPI and energy from the phytochemical food groups (kcal/day) were not significantly different from those of girls. The intake of refined grain, beans, milk and dairy, meat, and sugars was significantly higher in boys than in girls. Upon dietary intake analysis according to the DPI by sex, a higher DPI was associated with the amount of daily food intake in most of the food groups. Beans demonstrated a higher difference in intake between Q1 and Q4 in boys than the other food groups. Eggs, fish, and shellfish showed no significant differences in the amount of daily intake between sexes. Daily intake of meat in boys increased according to the DPI quartile ($p < 0.01$); however, no significant difference was observed in girls. No statistical difference was observed in the intake of milk, dairy, and sugars across the DPI quartiles in boys; in contrast, they significantly increased across the DPI quartiles in girls.

3.3. Association between DPI and Obesity

Table 3 shows the ORs for obesity according to the DPI by sex. Obesity was defined using weight and BMI percentiles. Different results were observed according to sex. Although no significant differences were observed in the weight, BMI, and daily energy intake across the DPI quartile in each sex, an association between the DPI and obesity was observed in boys when analyzing the prevalence of obesity by the weight percentile. Boys in the highest DPI quartile showed a significantly lower prevalence of obesity than those in the lowest DPI quartile in all the models (Model 3, OR: 0.287, 95% CI: 0.095–0.868, p for trend <0.05). In contrast, no statistically significant association was observed between the DPI and obesity prevalence, according to the BMI percentile in boys. No significant association was observed between the DPI and obesity prevalence in girls.

Table 2. Comparison of the dietary intake according to the dietary phytochemical index quartile by sex.

	Total (n = 1196)			Boys (n = 623)					Girls (n = 573)				
	Boys (n = 623)	Girls (n = 573)	p * Value	Q1 (n = 155)	Q2 (n = 156)	Q3 (n = 156)	Q4 (n = 156)	p Value	Q1 (n = 143)	Q2 (n = 143)	Q3 (n = 144)	Q4 (n = 143)	p Value
Total DPI	15.43 ± 1.07	16.49 ± 0.45	0.085	5.12 ± 0.25 (0–8.71) #	11.30 ± 0.13 (0–8.71)	17.09 ± 0.16 (13.94–20.53)	28.33 ± 0.69 (20.5–59.44)	<0.001	5.71 ± 0.22 (0.01–9.22)	11.88 ± 0.13 (9.32–14.76)	17.93 ± 0.17 (14.79–21.72)	29.64 ± 0.70 (21.74–51.45)	<0.001
Energy from Phytochemical Food groups (kcal/day) †	225.6 ± 15.44	208.0 ± 6.08	0.060	68.84 ± 3.92	159.04 ± 4.67	255.36 ± 8.08	386.80 ± 15.06	<0.001	73.21 ± 3.61	150.05 ± 5.06	212.55 ± 6.56	350.28 ± 12.09	<0.001
Amount of Intake by Food groups (g/day)													
Total	1059 ± 40.30	940.2 ± 16.53	0.005	874.5 ± 96.67	1008 ± 95.75	1158 ± 97.84	1197 ± 43.73	<0.001	826.1 ± 98.47	929.2 ± 97.40	940.6 ± 97.13	1056 ± 43.22	0.001
Phytochemical food groups													
Whole grains	18.43 ± 1.10	17.60 ± 1.07	0.579	7.36 ± 1.28	12.18 ± 1.02	22.53 ± 2.27	31.91 ± 2.98	<0.001	4.88 ± 0.48	13.03 ± 1.51	21.20 ± 1.91	30.55 ± 2.77	<0.001
Fruits	196.8 ± 10.69	180.6 ± 9.29	0.243	36.51 ± 4.96	132.3 ± 10.44	231.2 ± 15.72	389.1 ± 28.79	<0.001	55.11 ± 6.15	127.1 ± 10.42	183.7 ± 14.72	346.0 ± 27.33	<0.001
Beans	26.78 ± 2.92	18.56 ± 2.58	0.036	5.17 ± 0.79	17.67 ± 3.14	30.77 ± 6.99	53.77 ± 8.32	<0.001	5.29 ± 1.31	12.18 ± 2.03	15.34 ± 1.94	40.02 ± 8.99	<0.001
Seeds and Nuts	1.73 ± 0.20	2.26 ± 0.39	0.212	0.62 ± 0.11	1.74 ± 0.28	1.98 ± 0.53	2.59 ± 0.48	<0.001	0.40 ± 0.06	1.41 ± 0.31	2.06 ± 0.47	5.02 ± 1.36	<0.001
Vegetables	95.88 ± 3.38	96.08 ± 3.68	0.966	69.49 ± 5.90	92.21 ± 6.21	112.36 ± 7.33	109.62 ± 7.87	<0.001	71.39 ± 5.33	99.73 ± 7.87	106.99 ± 7.63	105.51 ± 8.06	0.001
Mushrooms	4.06 ± 0.56	3.73 ± 0.49	0.660	2.15 ± 0.42	3.34 ± 0.56	4.16 ± 0.83	6.60 ± 1.84	0.022	2.08 ± 0.59	4.01 ± 1.30	5.32 ± 0.99	3.52 ± 0.73	0.015
Non-phytochemical food groups													
Refined grains	191.2 ± 3.99	172.0 ± 4.00	0.001	196.35 ± 7.60	214.31 ± 8.16	192.11 ± 7.64	160.68 ± 7.38	<0.001	190.14 ± 9.49	186.00 ± 8.07	163.43 ± 5.92	149.59 ± 7.35	<0.001
Milk and Dairy	244.3 ± 10.19	218.2 ± 8.23	0.039	263.9 ± 22.17	258.2 ± 18.86	250.0 ± 17.43	204.6 ± 16.16	0.059	234.5 ± 18.94	254.7 ± 13.80	214.1 ± 18.09	171.4 ± 12.72	<0.001
Meats	63.91 ± 3.62	48.71 ± 2.54	<0.001	67.71 ± 6.73	58.06 ± 4.90	83.02 ± 11.08	47.64 ± 4.19	0.002	50.98 ± 4.34	54.15 ± 5.52	45.27 ± 4.91	44.49 ± 3.68	0.395
Eggs	30.51 ± 1.87	26.02 ± 1.61	0.061	33.39 ± 3.57	32.72 ± 4.57	30.11 ± 3.51	25.76 ± 3.22	0.392	32.50 ± 4.10	26.67 ± 2.76	20.95 ± 2.29	24.03 ± 3.10	0.084
Fish and Shellfish	42.43 ± 2.91	39.10 ± 2.97	0.393	33.19 ± 4.33	39.11 ± 4.70	49.61 ± 5.92	48.03 ± 6.52	0.067	39.43 ± 7.14	41.58 ± 4.99	38.80 ± 4.74	36.69 ± 6.71	0.945
Sugars	11.11 ± 0.96	7.53 ± 0.72	0.003	13.92 ± 2.49	12.16 ± 2.15	10.20 ± 1.68	8.10 ± 1.50	0.162	11.38 ± 2.03	6.91 ± 1.22	7.45 ± 1.63	4.64 ± 0.65	0.005

DPI, dietary phytochemical index; Q, quartile. All the values are presented as mean ± standard deviation. * p values were calculated using PROC SURVEYREG. # The range of DPI is presented as (minimum–maximum). † Includes whole grains, beans, seeds, nuts, vegetables, mushrooms, and fruits.

Table 3. Odds ratio (95% confidence intervals) for obesity according to the dietary phytochemical index quartile stratified by sex.

	Boys (n = 623)					Girls (n = 573)				
	Q1 (n = 155)	**Q2 (n = 156)**	**Q3 (n = 156)**	**Q4 (n = 156)**	***p* Value/ *p* for Trend**	**Q1 (n = 143)**	**Q2 (n = 143)**	**Q3 (n = 144)**	**Q4 (n = 143)**	***p* Value/ *p* for Trend**
Weight (kg)	18.3 ± 90.68	17.98 ± 0.65	17.66 ± 0.69	17.51 ± 0.26	0.168 *	17.13 ± 0.65	17.48 ± 0.68	17.36 ± 0.64	17.07 ± 0.26	0.694
Energy (kcal/day)	1389 ± 97.6	1479 ± 96.9	1533 ± 97.5	1412 ± 40.4	0.061	1296 ± 95.4	1320 ± 90.7	1236 ± 86.0	1252 ± 35.9	0.380
Weight percentile #										
Crude	1(ref)	0.790 † (0.311–2.009)	0.461 (0.153–13.395)	0.286 (0.094–0.866)	0.017 †	1(ref)	2.128 (0.788–5.745)	1.033 (0.327–3.259)	0.956 (0.313–2.918)	0.468
Model 1	1(ref)	0.743 (0.286–1.928)	0.410 (0.141–1.192)	0.282 (0.094–0.850)	0.014	1(ref)	2.125 (0.782–5.774)	1.084 (0.337–3.493)	0.991 (0.320–3.072)	0.540
Model 2	1(ref)	0.826 (0.324–2.106)	0.467 (0.154–1.414)	0.293 (0.096–0.890)	0.018	1(ref)	2.128 (0.789–5.738)	1.032 (0.327–3.259)	0.955 (0.313–2.918)	0.468
Model 3	1(ref)	0.769 (0.295–2.009)	0.414 (0.142–1.212)	0.287 (0.095–0.868)	0.016	1(ref)	2.119 (0.783–5.730)	1.083 (0.336–3.489)	0.989 (0.319–3.064)	0.539
BMI percentile										
Crude	1(ref)	0.906 (0.371–2.21)	0.473 (0.157–1.423)	0.459 (0.148–1.143)	0.114	1(ref)	2.870 (0.964–8.541)	1.877 (0.578–6.099)	1.271 (0.360–4.486)	0.807
Model 1	1(ref)	0.850 (0.342–20,115)	0.419 (0.142–1.221)	0.454 (0.148–1.392)	0.102	1(ref)	2.883 (0.961–8.650)	2.016 (0.618–6.580)	1.341 (0.374–4.813)	0.917
Model 2	1(ref)	0.928 (0.382–2.255)	0.476 (0.158–1.438)	0.466 (0.151–1.437)	0.115	1(ref)	2.899 (0.966–8.698)	1.898 (0.581–6.201)	1.271 (0.358–4.510)	0.806
Model 3	1(ref)	0.861 (0.348–2.129)	0.419 (0.142–1.232)	0.458 (0.151–1.385)	0.101	1(ref)	2.962 (0.980–8.955)	2.047 (0.621–6.746)	1.354 (0.373–4.912)	0.920

DPI, dietary phytochemical index; Q, quartile; BMI, body mass index; BMI percentile, body mass index percentile. Model 1, adjusted for energy intake; Model 2, adjusted for age; Model 3, adjusted for energy intake and age; all the values are presented as mean ± standard deviation or odds ratio (95% confidence intervals). * Values are presented as *p*-values calculated using PROC SURVEYREG. # Values are presented as *p* for trends, calculated by logistic regression analysis using the PROC SURVEY procedure. † The weight and BMI percentiles were defined according to the Korean Society for the Study of Obesity criteria; we analyzed the odds ratio for obesity as having ≥95 percentiles.

4. Discussion

This study was conducted to determine the relationship between DPI and obesity in children aged 3–5 years in Korea based on their sex. The prevalence of obesity was more than 70% lower in the group Q4 with low DPI in boys than in the group Q1 (Table 3).

Childhood obesity has a high risk of developing into adult obesity; therefore, controlling childhood obesity is very important [8,9]. According to a study that proactively tracked children's BMI status until adolescence, 83% of 4-year-old children with obesity were overweight or obese in adolescence, and only 17% returned to normal weight [28]. Dietary factors were identified as significant predictors of the weight of preschool children [28], characterized by high consumption of energy-dense snack foods and meals in overweight and children with obesity and insufficient fruit and vegetable intake [29,30]. Food preferences and eating habits formed during preschool do not change and can result in obesity in adults [31,32]; thus, this study provides important basic data emphasizing the importance of phytochemical-rich meals for children during this period.

Many studies have highlighted the health benefits of consuming foods rich in phytochemicals. In particular, studies on obesity have focused on the metabolic function of specific phytochemical components; for example, a study exists on obesity associated with reducing fat and body fat through mechanisms such as regulating fat metabolism of polyphenols [33,34], inhibiting fat cell proliferation, increasing fatty acid oxidation of resveratrol [35], and reducing fat accumulation in the body of flavonoids [19]. However, quantifying phytochemicals in food sources is impractical in terms of cost and time because individuals consume food composed of food ingredients, not individual nutrients.

The phytochemical index (PI), proposed by McCarty, defines [20] the energy obtained from plant foods as a percentage of the total daily energy intake divided by the total daily energy intake. Foods rich in phytochemicals include whole grains, fruits, vegetables, nuts, and beans; studies on the association between PI and diseases related to the intake of such foods have been consistently published [36–38]. A meta-analysis summarized the relationship between DPI and the risk of overweight/obesity and demonstrated that a high phytochemical index was associated with a reduced risk of overweight/obesity [21]. In a cross-sectional study by Vincent et al. [39], significant adverse correlations were found between the DPI scores, BMI, waist circumference, waist circumference–hip ratio, and body fat rate in 54 American adults aged 18–30 years. Even in a cross-sectional study in Iran [40], the risk of abdominal obesity was reduced, regardless of the age, sex, and total energy intake at higher DPIs in relation to the DPI and metabolic syndrome. A study using KNHANES 2008–2018 data on Korean women [22] demonstrated that the prevalence of obesity and abdominal obesity decreased in women who consumed increased amounts of phytochemical-rich foods.

The correlation between DPI and obesity has mainly been studied in adults, and few papers targeting children, especially preschoolers, exist. A cross-sectional study of 356 school-aged children (7–10 years old) in Iran demonstrated a negative relationship between phytochemical intake and obesity [23]. Overweight/obesity odds were 0.47 (0.25–0.87, p for trend = 0.02) for DPI Q1(14.25 ± 4.13) and Q4 (61.52 ± 16.47). A study of children and adolescents aged 6–18 years similarly revealed significantly decreased trends in BMI, hip circumference, and neck circumference for overweight and obese students based on DPI quartiles [41]. Both studies were consistent with our findings that higher phytochemical intake reduced the risk of obesity.

Our study was the first to analyze the relationship between DPI and obesity in preschoolers stratified by sex. The association was analyzed separately by sex, and significant differences were observed in the daily food intake and calories between boys and girls. Boys consumed more calories per day (kcal/day) than girls did ($p < 0.085$). Sex had no substantial impact on either weight or BMI. Our findings demonstrated that DPI quartiles and obesity prevalence only inversely correlated in boys (p for trend <0.05). Boys had a higher intake of beans than girls. Moreover, the consumption of meat, milk, and dairy products, the non-phytochemicals food group, was high. Despite being significant sources of protein, they did not exhibit a linear pattern, according to DPI. Thus, we only focused on the anti-obesity effect of beans as a phytochemical-rich food.

Many studies have demonstrated that obesity and metabolic syndrome increase as soy intake decreases in women [42,43]. As isoflavones have a positive effect on women's abdominal obesity reduction due to structure similarity to that of women's sex hormones [44,45], many studies have recommended increased protein intake through beans. As children aged 3–5 years old, who were included in this study, do not yet possess physiological differences according to sex hormones, the anti-obesity effect in boys was presumably associated with high bean intake and clear differences according to the DPIs.

The prevalence of obesity was significantly lower in boys and the high DPI group. There was no significant difference in the participant's weight, BMI, and total energy intake; the average intake of milk, dairy, and meat groups was higher than that of girls; and soybean intake was higher among the food groups belonging to DPI. Therefore, to prevent obesity from spreading in children and adulthood, it is important to focus on the composition of the food group consumed rather than the calories. Gunther et al. [45] found that increased protein intake from dairy products was associated with BMI in the first 12 months and increased body fat at 7 years of age. Other studies have also considered dairy consumption in growing children and adults to protect against overweight and obesity [46,47]. A study of children 2–3 years old in Australia [48] found that there were associations between microbiota composition and dairy- and plant-based foods such as fruits, vegetables, beans, pulses, and nuts, indicating a link to microorganisms when analyzed based on food intake rather than nutrient intake. Therefore, in children's nutritional research, it is recommended to consider the intake of food groups first because melas are made up of food, not nutrients. The results of this study, considering the relationship between the phytochemical intake and obesity, indicated the DPI is considered a useful tool, and investigating the composition of the DPI and microbiome in the future could be a good strategy.

This study had several limitations. First, as this was a cross-sectional study using KNHANES data, in addition to the effect of DPIs on obesity development, the inverse correlation between obesity and food intake cannot be excluded. Other environmental factors such as genetics, breastfeeding, and lifestyle, which may have affected the relationship between DPI and obesity prevalence, were also not considered. However, the participants of this study were in the process of controlling microbial communities, unlike infants and toddlers, in whom breastfeeding has a significant health impact [49]. Second, the 24 h recall used to collect dietary data had a recall bias. However, in this study, as the average value of the group was evaluated, errors that could have been caused by recall bias were balanced. Third, this study comprised a sample of preschool-aged children within a limited age range in Korea. The generalizability of the findings of other regions of children with different dietary habits remain to be elucidated. Nevertheless, this is the first study to analyze the DPI of Korean children according to sex differences. Few studies have been conducted on DPI in preschoolers. We did not limit the diagnostic criteria for obesity to BMI percentiles, as we also used weight percentiles. Therefore, our study could serve as a steppingstone for establishing the relationship between children's DPIs and obesity in the future, which will also aid in nutrition education and formulating policies for parents.

5. Conclusions

In conclusion, in Korean boys aged 3–5 years, a direct correlation between high DPI and low obesity prevalence was discovered. Research should be conducted concerning the nutrition food group intake to corroborate these findings and clarify why the sex of the preschoolers' results varied in outcomes. Future studies are warranted to investigate the prevalence of obesity to modifications in microbiome composition attributed to the consumption of phytochemical-rich foods.

Author Contributions: Conceptualization, M.-Y.P. and J.-H.B.; methodology, M.-Y.P., J.-H.B. and S.-K.J.; formal analysis, Y.-J.H. and M.-Y.P.; investigation, N.-R.S. and J.S.Y.; writing—original draft preparation, M.-Y.P. and Y.-J.H.; writing—review and editing, M.-Y.P.; funding acquisition, M.-Y.P. All authors have read and agreed to the published version of the manuscript.

Funding: This research was funded by the National Research Foundation of Korea (NRF) (grant number NRF-2020R1I1A1A01072285).

Institutional Review Board Statement: This study was conducted according to the guidelines of the Declaration of Helsinki and approved by the Institutional Review Board of the Seoul National University (E2108/002-002).

Informed Consent Statement: Informed consent was obtained from all subjects involved in the study.

Data Availability Statement: Not applicable.

Acknowledgments: The Korea National Health and Nutrition Examination Survey provided the data for this study. (KNHANES). We owe a debt of gratitude to all KNHANES participants. We also value the contributions of the team, which includes doctors, medical technicians, interviewers for KNHANES' health or nutritional components, and expert committees that have provided technical help.

Conflicts of Interest: The funders had no role in the design of the study; in the collection, analyses, or interpretation of data; in the writing of the manuscript; or in the decision to publish the results.

References

1. Ogden, C.L.; Carroll, M.D.; Lawman, H.G.; Fryar, C.D.; Kruszon-Moran, D.; Kit, B.K.; Flegal, K.M. Trends in Obesity Prevalence among Children and Adolescents in the United States, 1988–1994 Through 2013–2014. *JAMA* **2016**, *315*, 2292–2299. [CrossRef] [PubMed]
2. Hu, K.; Staiano, A.E. Trends in Obesity Prevalence among Children and Adolescents Aged 2 to 19 Years in the US From 2011 to 2020. *JAMA Pediatr.* **2022**, *176*, 1037–1039. [CrossRef] [PubMed]
3. World Health Organization. Obesity and Overweight. Available online: https://www.who.int/news-room/fact-sheets/detail/obesity-and-overweight (accessed on 9 September 2022).
4. Kim, J.H.; Moon, J.S. Secular trends in pediatric overweight and obesity in Korea. *J. Obes. Metab. Syndr.* **2020**, *29*, 12. [CrossRef]
5. Kostas, K.; Alexandra, S.; Kyriaki, K. Childhood obesity and its associations with morbidity and mortality in adult life. *Diabetes Complet.* **2018**, *2*, 1–12. [CrossRef]
6. Cote, A.T.; Harris, K.C.; Panagiotopoulos, C.; Sandor, G.G.S.; Devlin, A.M. Childhood Obesity and Cardiovascular Dysfunction. *J. Am. Coll. Cardiol.* **2013**, *62*, 1309–1319. [CrossRef]
7. Vos, M.B.; Abrams, S.H.; Barlow, S.E.; Caprio, S.; Daniels, S.R.; Kohli, R.; Mouzaki, M.; Sathya, P.; Schwimmer, J.B.; Sundaram, S.S.; et al. NASPGHAN Clinical Practice Guideline for the Diagnosis and Treatment of Nonalcoholic Fatty Liver Disease in Children: Recommendations from the Expert Committee on NAFLD (ECON) and the North American Society of Pediatric Gastroenterology, Hepatology and Nutrition (NASPGHAN). *J. Pediatr. Gastroenterol. Nutr.* **2017**, *64*, 319–334. [CrossRef] [PubMed]
8. Jochum, F.; Abdellatif, M.; Adel, A.; Alhammadi, A.; Alnemri, A.; Alohali, E.; AlSarraf, K.; Al Said, K.; Elzalabany, M.; Isa, H.M.A.; et al. Burden of Early Life Obesity and Its Relationship with Protein Intake in Infancy: The Middle East Expert Consensus. *Pediatr. Gastroenterol. Hepatol. Nutr.* **2022**, *25*, 93–108. [CrossRef] [PubMed]
9. Chang, L.; Neu, J. Early Factors Leading to Later Obesity: Interactions of the Microbiome, Epigenome, and Nutrition. *Curr. Probl. Pediatr. Adolesc. Health Care* **2015**, *45*, 134–142. [CrossRef]
10. Freedman, D.S.; Khan, L.K.; Serdula, M.K.; Dietz, W.H.; Srinivasan, S.R.; Berenson, G.S. The Relation of Childhood BMI to Adult Adiposity: The Bogalusa Heart Study. *Pediatrics* **2005**, *115*, 22–27. [CrossRef]
11. Weihrauch-Blüher, S.; Schwarz, P.; Klusmann, J.-H. Childhood obesity: Increased risk for cardiometabolic disease and cancer in adulthood. *Metabolism* **2019**, *92*, 147–152. [CrossRef]
12. Must, A.; Phillips, S.M.; Naumova, E.N. Occurrence and Timing of Childhood Overweight and Mortality: Findings from the Third Harvard Growth Study. *J. Pediatr.* **2012**, *160*, 743–750. [CrossRef] [PubMed]
13. Horesh, A.; Tsur, A.M.; Bardugo, A.; Twig, G. Adolescent and Childhood Obesity and Excess Morbidity and Mortality in Young Adulthood—A Systematic Review. *Curr. Obes. Rep.* **2021**, *10*, 301–310. [CrossRef] [PubMed]
14. Lee, H.-S.; Cho, Y.-H.; Park, J.; Shin, H.-R.; Sung, M.-K. Dietary Intake of Phytonutrients in Relation to Fruit and Vegetable Consumption in Korea. *J. Acad. Nutr. Diet.* **2013**, *113*, 1194–1199. [CrossRef] [PubMed]
15. Division of Health and Nutrition Survey and Analysis. K.D.C.A. National Health Statistics. 2020. Available online: http://knhanes.kdca.go.kr/ (accessed on 17 August 2022).
16. He, K.; Hu, F.B.; Colditz, G.A.; Manson, J.E.; Willett, W.C.; Liu, S. Changes in intake of fruits and vegetables in relation to risk of obesity and weight gain among middle-aged women. *Int. J. Obes. Relat. Metab. Disord.* **2004**, *28*, 1569–1574. [CrossRef] [PubMed]

17. Liu, S.; Willett, W.C.; Manson, J.E.; Hu, F.B.; Rosner, B.; Colditz, G. Relation between changes in intakes of dietary fiber and grain products and changes in weight and development of obesity among middle-aged women. *Am. J. Clin. Nutr.* **2003**, *78*, 920–927. [CrossRef]
18. Serdula, M.K.; Byers, T.; Mokdad, A.H.; Simoes, E.; Mendlein, J.M.; Coates, R.J. The Association between Fruit and Vegetable Intake and Chronic Disease Risk Factors. *Epidemiology* **1996**, *7*, 161–165. [CrossRef]
19. Holubková, A.; Penesová, A.; Sturdik, E.; Mosovska, S.; Mikusova, L. Phytochemicals with potential effects in metabolic syndrome prevention and therapy. *Acta Chim. Slovaca* **2012**, *5*, 186. [CrossRef]
20. McCarty, M.F. Proposal for a dietary "phytochemical index". *Med. Hypotheses* **2004**, *63*, 813–817. [CrossRef]
21. Wei, C.; Liu, L.; Liu, R.; Dai, W.; Cui, W.; Li, D. Association between the Phytochemical Index and Overweight/Obesity: A meta-Analysis. *Nutrients* **2022**, *14*, 1429. [CrossRef]
22. Im, J.; Kim, M.; Park, K. Association between the Phytochemical Index and Lower Prevalence of Obesity/Abdominal Obesity in Korean Adults. *Nutrients* **2020**, *12*, 2312. [CrossRef]
23. Eslami, O.; Khoshgoo, M.; Shidfar, F. Dietary phytochemical index and overweight/obesity in children: A cross-sectional study. *BMC Res. Notes* **2020**, *13*, 132. [CrossRef] [PubMed]
24. Kim, M.; Park, K. Association between phytochemical index and metabolic syndrome. *Nutr. Res. Pract.* **2020**, *14*, 252–261. [CrossRef] [PubMed]
25. Korea Disease Control and Prevention Agency. *Guidelines for Examination and Inspection (2016–2018)*; Osong: Cheongju-si, Republic of Korea, 2016.
26. Korea Disease Control and Prevention Agency. 2017 Korean National Growth Charts for Children and Adolescents. Available online: https://knhanes.kdca.go.kr/knhanes/sub08/sub08_01.do (accessed on 17 August 2022).
27. Kim, J.H.; Yun, S.; Hwang, S.S.; Shim, J.O.; Chae, H.W.; Lee, Y.J.; Lee, J.H.; Kim, S.C.; Lim, D.; Yang, S.W.; et al. The 2017 Korean National Growth Charts for children and adolescents: Development, improvement, and prospects. *Korean J. Pediatr.* **2018**, *61*, 135–149. [CrossRef] [PubMed]
28. Geserick, M.; Vogel, M.; Gausche, R.; Lipek, T.; Spielau, U.; Keller, E.; Pfäffle, R.; Kiess, W.; Körner, A. Acceleration of BMI in Early Childhood and Risk of Sustained Obesity. *N. Engl. J. Med.* **2018**, *379*, 1303–1312. [CrossRef]
29. Bere, E.; van Lenthe, F.; Klepp, K.-I.; Brug, J. Why do parents' education level and income affect the amount of fruits and vegetables adolescents eat? *Eur. J. Public Health* **2008**, *18*, 611–615. [CrossRef]
30. Gerritsen, S.; Renker-Darby, A.; Harré, S.; Rees, D.; Raroa, D.A.; Eickstaedt, M.; Sushil, Z.; Allan, K.; Bartos, A.E.; Waterlander, W.E. Improving low fruit and vegetable intake in children: Findings from a system dynamics, community group model building study. *PLoS ONE* **2019**, *14*, e0221107. [CrossRef]
31. Baker, J.L.; Olsen, L.W.; Sørensen, T.I.A. Childhood Body-Mass Index and the Risk of Coronary Heart Disease in Adulthood. *N. Engl. J. Med.* **2007**, *357*, 2329–2337. [CrossRef]
32. Dietz, W.H. Health Consequences of Obesity in Youth: Childhood Predictors of Adult Disease. *Pediatrics* **1998**, *101*, 518–525. [CrossRef]
33. Lewandowska, U.; Szewczyk, K.; Hrabec, E.; Janecka, A.; Gorlach, S. Overview of Metabolism and Bioavailability Enhancement of Polyphenols. *J. Agric. Food Chem.* **2013**, *61*, 12183–12199. [CrossRef]
34. Shimoda, H.; Tanaka, J.; Kikuchi, M.; Fukuda, T.; Ito, H.; Hatano, T.; Yoshida, T. Effect of Polyphenol-Rich Extract from Walnut on Diet-Induced Hypertriglyceridemia in Mice via Enhancement of Fatty Acid Oxidation in the Liver. *J. Agric. Food Chem.* **2009**, *57*, 1786–1792. [CrossRef]
35. Aguirre, L.; Fernández-Quintela, A.; Arias, N.; Portillo, M.P. Resveratrol: Anti-Obesity Mechanisms of Action. *Molecules* **2014**, *19*, 18632–18655. [CrossRef]
36. Bahadoran, Z.; Golzarand, M.; Mirmiran, P.; Saadati, N.; Azizi, F. The association of dietary phytochemical index and cardiometabolic risk factors in adults: Tehran Lipid and Glucose Study. *J. Hum. Nutr. Diet.* **2013**, *26*, 145–153. [CrossRef] [PubMed]
37. Golzarand, M.; Mirmiran, P.; Bahadoran, Z.; Alamdari, S.; Azizi, F. Dietary phytochemical index and subsequent changes of lipid profile: A 3-year follow-up in Tehran Lipid and Glucose Study in Iran. *ARYA Atheroscler.* **2014**, *10*, 203–210. [PubMed]
38. Bahadoran, Z.; Mirmiran, P.; Tohidi, M.; Azizi, F. Dietary phytochemical index and the risk of insulin resistance and β-cell dysfunction: A prospective approach in Tehran lipid and glucose study. *Int. J. Food Sci. Nutr.* **2015**, *66*, 950–955. [CrossRef]
39. Vincent, H.K.; Bourguignon, C.M.; Taylor, A.G. Relationship of the dietary phytochemical index to weight gain, oxidative stress and inflammation in overweight young adults. *J. Hum. Nutr. Diet.* **2010**, *23*, 20–29. [CrossRef]
40. Vasmehjani, A.A.; Darabi, Z.; Nadjarzadeh, A.; Mirzaei, M.; Hosseinzadeh, M. The relation between dietary phytochemical index and metabolic syndrome and its components in a large sample of Iranian adults: A population-based study. *BMC Public Health* **2021**, *21*, 1587. [CrossRef]
41. Azizi-Soleiman, F.; Khoshhali, M.; Heidari-Beni, M.; Qorbani, M.; Pourmirzaei, M.A.; Kelishadi, R. Higher dietary phytochemical index is associated with anthropometric indices in children and adolescents: The weight disorders survey of the CASPIAN-IV study. *Int. J. Vitam. Nutr. Res.* **2020**, *91*, 531–538. [CrossRef] [PubMed]
42. Alizadeh, M.; Gharaaghaji, R.; Gargari, B.P. The effects of legumes on metabolic features, insulin resistance and hepatic function tests in women with central obesity: A randomized controlled trial. *Int. J. Prev. Med.* **2014**, *5*, 710–720.

43. Mu, Y.; Kou, T.; Wei, B.; Lu, X.; Liu, J.; Tian, H.; Zhang, W.; Liu, B.; Li, H.; Cui, W.; et al. Soy Products Ameliorate Obesity-Related Anthropometric Indicators in Overweight or Obese Asian and Non-Menopausal Women: A Meta-Analysis of Randomized Controlled Trials. *Nutrients* **2019**, *11*, 2790. [CrossRef]
44. Lee, J.-H.; Heo, J.-M.; Park, Y.-S.; Park, H.M. Survey on the Consumption of the Phytoestrogen Isoflavone in Postmenopausal Korean Women. *J. Korean Soc. Menopaus.* **2012**, *18*, 163–173. [CrossRef]
45. Günther, A.L.; Remer, T.; Kroke, A.; Buyken, A.E. Early protein intake and later obesity risk: Which protein sources at which time points throughout infancy and childhood are important for body mass index and body fat percentage at 7 y of age? *Am. J. Clin. Nutr.* **2007**, *86*, 1765–1772. [CrossRef] [PubMed]
46. Marette, A.; Picard-Deland, E. Yogurt consumption and impact on health: Focus on children and cardiometabolic risk. *Am. J. Clin. Nutr.* **2014**, *99*, 1243S–1247S. [CrossRef] [PubMed]
47. Tremblay, A.; Doyon, C.; Sanchez, M. Impact of yogurt on appetite control, energy balance, and body composition. *Nutr. Rev.* **2015**, *73*, 23–27. [CrossRef]
48. Smith-Brown, P.; Morrison, M.; Krause, L.; Davies, P.S.W. Dairy and plant based food intakes are associated with altered faecal microbiota in 2 to 3 year old Australian children. *Sci. Rep.* **2016**, *6*, 32385. [CrossRef] [PubMed]
49. Kumbhare, S.V.; Patangia, D.V.; Patil, R.H.; Shouche, Y.S.; Patil, N.P. Factors influencing the gut microbiome in children: From infancy to childhood. *J. Biosci.* **2019**, *44*, 49. [CrossRef]

nutrients

Article

Prevention of the Pro-Aggressive Effects of Ethanol-Intoxicated Mice by Schisandrin B

Ho Yin Pekkle Lam [1], Ting-Ruei Liang [2] and Shih-Yi Peng [1,*]

1 Department of Biochemistry, School of Medicine, Tzu Chi University, Hualien 970, Taiwan; pekklelavabo@mail.tcu.edu.tw
2 PhD Program in Pharmacology and Toxicology, Tzu Chi University, Hualien 970, Taiwan
* Correspondence: pengsy@mail.tcu.edu.tw

Abstract: Excessive alcohol consumption can lead to serious health complications, with liver and neurological complications being the most important. In Western nations, alcoholic liver disease accounts for 50% of mortality from end-stage liver disease and is the second most common cause of liver transplants. In addition to direct damage, hepatic encephalopathy may also arise from alcohol consumption. However, effective treatment for liver disease, as well as neurological injury, is still lacking today; therefore, finding an efficacious alternative is urgently needed. In the current study, the preventive and therapeutic effects of Schisandrin B (Sch B) against ethanol-induced liver and brain injuries were investigated. By using two treatment models, our findings indicated that Sch B can effectively prevent and ameliorate alcoholic liver diseases, such as resolving liver injuries, lipid deposition, inflammasome activation, and fibrosis. Moreover, Sch B reverses brain damage and improves the neurological function of ethanol-treated mice. Therefore, Sch B may serve as a potential treatment option for liver diseases, as well as subsequential brain injuries. Furthermore, Sch B may be useful in preventive drug therapy against alcohol-related diseases.

Keywords: alcoholic liver disease; liver fibrosis; hepatic encephalopathy; Schisandrin B

Citation: Lam, H.Y.P.; Liang, T.-R.; Peng, S.-Y. Prevention of the Pro-Aggressive Effects of Ethanol-Intoxicated Mice by Schisandrin B. *Nutrients* **2023**, *15*, 1909. https://doi.org/10.3390/nu15081909

Academic Editors: Jianjun Deng and Haixia Yang

Received: 7 March 2023
Revised: 6 April 2023
Accepted: 12 April 2023
Published: 15 April 2023

1. Introduction

Excessive and/or chronic alcohol consumption can lead to serious health complications, affecting almost every organ in the body. Alcoholism is highly related to gastrointestinal bleeding [1,2], pancreatitis [3], alcoholic liver disease [4,5], neurologic disorders [6], type II diabetes mellitus [7], and cancers [8–10]. In Western nations, alcoholic liver disease accounts for 50% of mortality from end-stage liver disease and is the second most common cause of liver transplants [11]. While the liver is the major organ responsible for the processing of alcohol [12], excessive alcohol may destroy liver cells, leading to inflammation and ultimately fibrosis. While early and slight liver problems may be treated with lifestyle modifications, effective treatment for chronic and late-stage liver problems is still lacking today [13]. In conjunction with the direct damage induced by alcohol, the brain can be affected by the damaged liver [14]. Hepatic encephalopathy is a central nervous system disorder that occurs when the liver does not function properly and, as a result, unprocessed toxins build up and travel to the brain [15]. Hepatic encephalopathy is usually associated with poor quality of life and a high mortality rate [15], so finding an effective therapy for alcohol-induced liver disease, as well as its associated encephalopathy, is urgently needed.

It has been reported that Schisandrin B (Sch B), an extract of *Schisandra chinensis*, protects against different causes of liver injuries [16–20]. Sch B has also been shown to improve damage to different organs, mainly through its anti-inflammatory [16,19,21], antioxidative [16,22], and anti-fibrotic [17–19] properties. Currently, Sch B has been shown to protect against acute alcoholic cardiomyopathy through downregulation of autophagy [23]. Sch B also enhances antioxidant status in the mitochondria of multiple tissues in mice with chronic ethanol feeding [24]. However, there seems to be a lack of focus on Sch B in ethanol-induced liver injuries.

Moreover, despite many studies focused on therapeutics for alcohol-induced liver injuries, only a few studies have examined brain damage caused by alcohol consumption. In this study, we examined the beneficial effects of Sch B on alcohol-induced liver and brain damage.

2. Materials and Methods

2.1. Drugs

Schisandrin B (Sch B; Chengdu Alfa Biotechnology, Chengdu, China; Figure 1) was dissolved in olive oil to achieve a concentration of 20 mg/kg. Ethanol was diluted with distilled water to yield the following concentrations: 30% (*v/v*), 40% (*v/v*), and 50% (*v/v*).

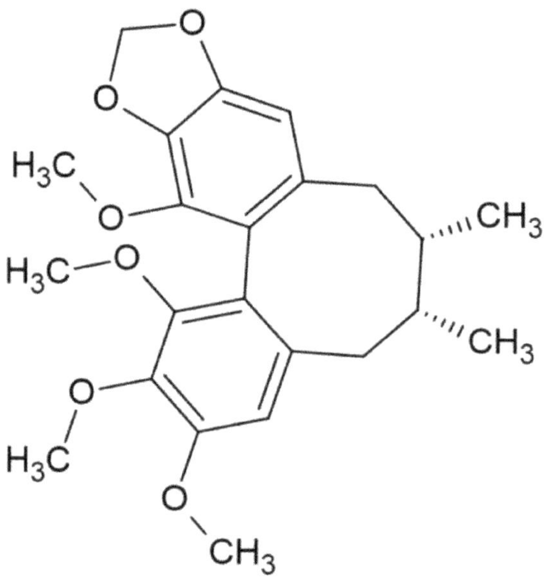

Figure 1. The chemical structure of Schisandrin B.

2.2. Animals

Eight-week-old male BALB/c mice were obtained from the National Laboratory Animal Center, Taipei, Taiwan. Mice were housed in the animal center at Tzu Chi University and maintained in 23 °C ± 1 °C with a 12 h light/dark cycle and 40–60% humidity conditions, along with an ad libitum diet. All experimental procedures involving animals were approved by the Institutional Animal Care and Use Committee (IACUC) of Tzu Chi University (No. 109066).

2.3. Animal Treatment

Two treatment models (preventive model and corrective model) were employed in this study to investigate the effect of Sch B. In the first model (preventive model), 16 mice were randomly divided into 4 groups—control, Sch B-treated control, ethanol-treated, and ethanol + Sch B-treated—with each group containing 4 mice. Ethanol was orally administered to the mice as follows: 35% ethanol for 7 days, 40% ethanol for the following 7 days, and a 52% ethanol binge on the 15th day. Sch B (20 mg/kg) was orally administered 1 h after ethanol administration for the same duration. Mice were subjected to neurofunctional tests before the administration of a new ethanol dose. Mice were sacrificed one day after the last treatment with 12 h fasting. In the second treatment model (corrective model), 16 mice were randomly divided into 4 groups—control, Sch B-treated control, ethanol-treated, and ethanol + Sch B-treated—with each group containing 4 mice. Mice were first administered ethanol in the way mentioned above. One day after the 52% ethanol binge, mice were treated with 20 mg/kg Sch B by oral gavage for 14 continuous days. Mice were sacrificed by exsanguination without anesthesia one day after the last Sch B treatment (day 30) with 12 h fasting. The method for ethanol treatment was employed as described previously [25–27]. Treatment concentrations of Sch B (20 mg/kg) were based on literature references [28,29]. No mice died or exhibited Sch B-related adverse effects during the experiment.

2.4. Hematoxylin & Eosin (H&E) and Sirius Red Staining

Livers and brains were fixed in 10% neutral buffered formalin and dehydrated in a series of alcohols. After immersion, tissue blocks were cut into sections approximately 5–8 micrometers thick and stained with hematoxylin and eosin (H&E) or picro-Sirius red, as described previously [19].

2.5. Histopathological Examination and Scoring

Histopathological liver changes were evaluated as follows: 0, no injuries; 1, mild injuries with cytoplasmic vacuolization and slight pyknosis; 2, moderate injuries with extensive pyknosis and loss of cellular borders; and 3, severe injuries with disintegration of hepatocytes, tissue congestion, and cellular infiltration. Brain sections were scored as follows: 0, no evidence of injuries; 1, dispersed pyknosis; 2, slight necrosis or apoptosis; and 3, multiple or extensive necrosis or apoptosis [30]. At least 10 random fields were scored in each liver section, and five random fields in each brain section were scored. The severity of liver fibrosis was also evaluated by Ishak scoring of the picro-Sirius red-stained slides [31].

2.6. Western Blotting

Total protein was extracted from tissues, resolved by sodium dodecyl sulfate (SDS)-polyacrylamide gel electrophoresis (PAGE), and transferred to a polyvinylidene difluoride (PVDF) membrane (EMD Millipore, Danvers, MA, USA). After blocking, the membranes were incubated overnight with antibodies against α-tubulin (GeneTex, Irvine, CA, USA), type I collagen (COLA-1; ABclonal, Woburn, MA, USA), type III collagen (COLA-3; ABclonal), fibronectin (GeneTex), NOD-, LRR-, and pyrin domain-containing protein 3 (NLRP3; Proteintech, Rosemont, IL, USA), caspase-1 (Proteintech), interleukin-1β (IL-1β; Cell Signaling Technology, Danvers, MA, USA), interleukin-18 (IL-18; Proteintech), and gasdermin D (GSDMD; Santa Cruz Biotechnology, Dallas, TX, USA). The membranes were then incubated with horseradish peroxidase (HRP)-conjugated secondary antibodies (EMD Millipore) for 1 h. Protein bands were visualized with enhanced chemiluminescence (ECL) reagents (EMD Millipore) and quantified using densitometry with Image J software (v1.46). The band intensities of the proteins of interest were normalized by that of α-tubulin or the uncleaved band.

2.7. Collection of Serum, Cerebrospinal Fluid (CSF), and Liver Lysate

A cardiac puncture was performed to obtain blood samples from the mice. Samples were allowed to clot for 30 min, followed by centrifugation for 15 min at 600× g to collect serum. Cerebrospinal fluid (CSF) was collected by injecting ice-cold sterile phosphate-buffered saline (PBS) into the cranial cavity and cerebral ventricles. The washing solution was collected and centrifuged at 600× g for 10 min and then later at 10,000× g for 30 min. The obtained supernatant was stored at −80 °C until use. Liver lysate was obtained by homogenizing a weighted fraction of the liver at 4 °C in sterile PBS using a tissue homogenizer. Homogenates were centrifuged at 1500× g at 4 °C for 15 min, and supernatants were stored at −80 °C.

2.8. Measurement of Biochemical Parameters

Alanine transaminase (ALT), aspartate aminotransferase (AST), total cholesterol, triacylglycerides, and albumin were analyzed in serum and CSF using a Hitachi 7080 chemistry analyzer (Hitachi Ltd., Tokyo, Japan). Liver lysate was analyzed for total cholesterol and triacylglycerides using a cholesterol liquicolor kit (HUMAN Diagnostics Worldwide, Wiesbaden, Germany) and a triglycerides liquicolor kit (HUMAN Diagnostics Worldwide). Levels of transforming growth factor-beta (TGF-β) in the serum and liver lysate were measured using a mouse TGF-β sandwich enzyme-linked immunosorbent assay (ELISA) kit (Thermo Fisher Scientific, Waltham, MA, USA).

2.9. Neurofunctional Tests

Mice were subjected to neurofunctional tests before the beginning of the experiment, before administration of a new ethanol dose, and every 7 days during Sch B treatment. The wire hang test was performed by hanging a mouse upside down on a wire mesh. The time when the mouse fell was documented within 120 s. The beam walk test was performed by placing a mouse on a beam of 1 cm × 70 cm and 50 cm high above the platform. The time for the mouse to walk across the beam was recorded within 120 s. Clasping scores were allocated by hanging a mouse's tail for 10 s. A score was recorded as follows: 0, the mouse showed normal escape extension; 1, the hindlimb of the mouse withdrew but did not touch the abdomen; 2, the hindlimb of the mouse withdrew and touched the abdomen but was not clasped; 3, the mouse showed immobility, with hindlimbs clasped and touching the abdomen. The hot plate test was performed by placing a mouse on a hot plate at 55 °C, and the time for their first paw reaction was documented within 15 s. The vertical pole test was performed by placing a mouse at the top of a 50 cm vertical pole. The time for the mouse to turn around and climb down was documented within 120 s. Each mouse was tested three times in each experiment.

2.10. Statistical Analysis

All results in this study were analyzed using GraphPad Prism 6.01 software (GraphPad Software, San Diego, CA, USA) and shown as mean ± standard deviation (SD), unless stated otherwise. Statistical analysis was conducted using one-way analysis of variance (ANOVA), and pairwise differences were evaluated using a post hoc test.

3. Results

3.1. Schisandrin B Prevents Liver Injuries in Mice Consuming Ethanol

To explore the protective effect of Sch B against ethanol intoxication, mice were treated with ethanol, as well as Sch B, on the same day (Figure 2A). Histopathological examination showed that ethanol consumption significantly altered hepatic cord arrangements, which was accompanied by congestion and ballooning degeneration (Figure 2B,C). Additionally, ethanol consumption significantly induced serum ALT and AST levels (Figure 2D,E), suggesting liver injury. Excessive ethanol consumption has been shown to alter lipid metabolism in the liver, leading to the accumulation of hepatic lipids and steatosis [5]; total cholesterol and triacylglyceride levels were measured in both the serum and liver. The results revealed that ethanol consumption, although it did not significantly increase total cholesterol and triacylglyceride levels in the serum, significantly increased their levels in the liver (Figure 2F–I). Sch B treatment, on the other hand, showed significant protection against ethanol-induced liver injuries, as shown by improvements in liver histology (Figure 2B,C), as well as ALT and AST levels (Figure 2D,E). Total cholesterol and triacylglyceride levels were also decreased in both serum and liver compared to ethanol-treated mice (Figure 2F–I). In the experiment, Sch B treatment alone had no adverse effect on the mice's livers. Therefore, the use of Sch B can prevent liver damage as well as lipid accumulation caused by ethanol consumption.

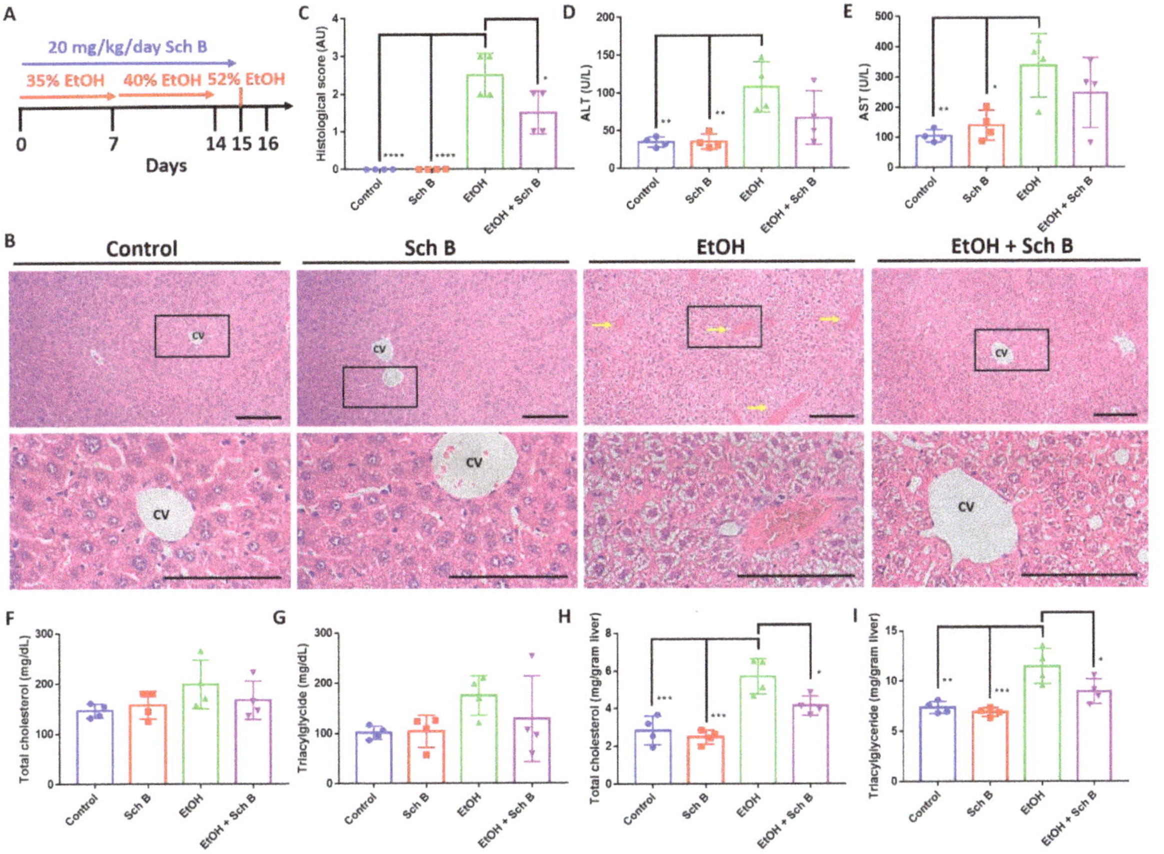

Figure 2. Schisandrin B prevents liver injuries in mice consuming ethanol. (**A**) Experimental design scheme. (**B**) Representative H&E-stained histological images showing the liver pathology of the control, Sch B-treated, ethanol-challenged, and ethanol-challenged + Sch B-treated mice. Congestion (yellow arrows) and ballooning degeneration were seen in the ethanol-treated mice, which improved after Sch B treatment. The scale bar corresponds to 200 µm. CV, central vein. (**C**) Histological scores based on the histological slides from (**B**). Serum levels of (**D**) ALT, (**E**) AST, (**F**) total cholesterol, and (**G**) triacylglycerides. Hepatic levels of (**H**) total cholesterol and (**I**) triacylglycerides. $n = 4$ mice. Results are expressed as mean ± SD. * $p < 0.05$, ** $p < 0.01$, *** $p < 0.001$, and **** $p < 0.0001$.

3.2. Schisandrin B Prevents Activation of Liver Inflammasomes in Mice Consuming Ethanol

Ethanol intake has been linked to liver fibrosis [4,5]; therefore, Sirius red staining was performed to analyze collagen deposition in the liver. The results suggest that ethanol consumption, although it did not cause fibrosis, significantly increased collagen deposits in the liver (Figure 3A). Levels of TGF-β, a powerful profibrotic cytokine, were measured in the serum and liver of ethanol-fed mice, with levels found to be significantly increased in the liver after ethanol consumption (Supplementary Figure S1A,B). Western blotting also confirmed that ethanol increases type III collagen expression as well as fibronectin expression in the liver (Figure 3B). Inflammasome-induced inflammation has been shown to be positively correlated with fibrosis [32], so markers of inflammasomes were analyzed. Inflammasome components, including NLRP3 and caspase-1, as well as their effectors, interleukin (IL)-1β and IL-18, significantly increased due to ethanol consumption compared to the control and Sch B groups (Figure 3C). GSDMD, a pyroptotic marker that can be initiated by NLRP3 signaling [33], was also significantly increased in expression by ethanol, suggesting the presence of pyroptosis (Figure 3C).

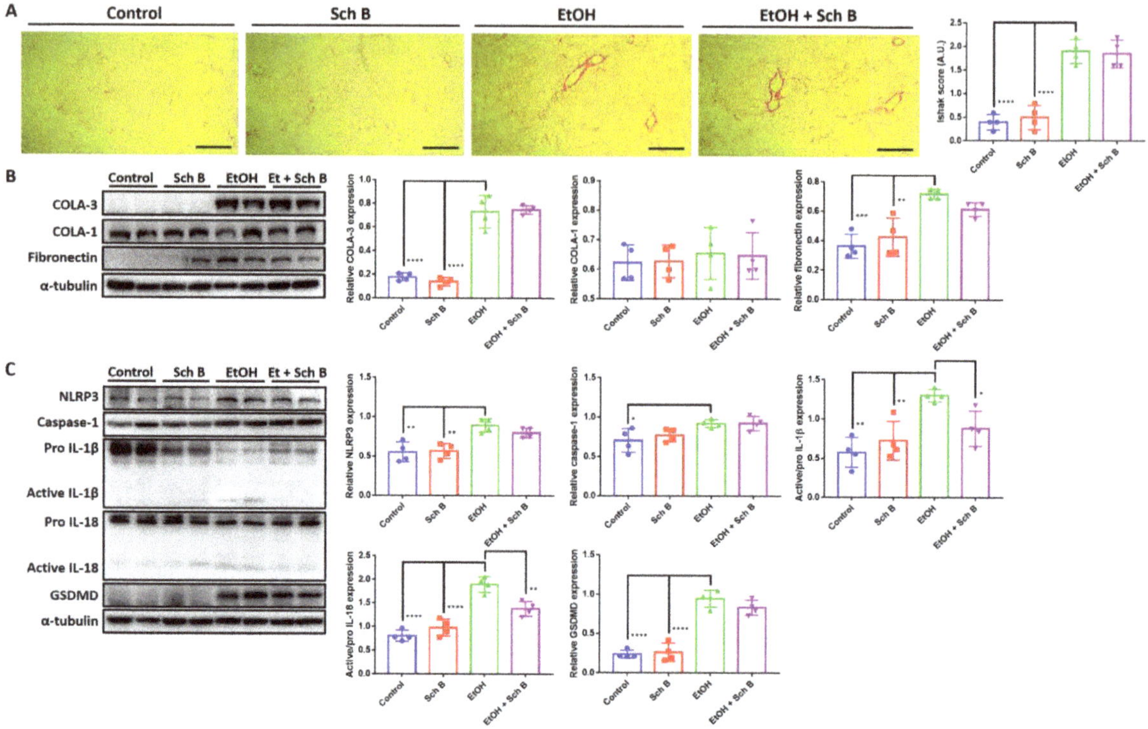

Figure 3. Schisandrin B prevents activation of liver inflammasomes in mice consuming ethanol. (**A**) Representative Sirius red-stained histological images showing collagen deposition (red color) in the livers of control, Sch B-treated, ethanol-challenged, and ethanol-challenged + Sch B-treated mice. Scale bar corresponds to 200 μm. Levels of fibrosis were quantified by Ishak score based on the Sirius red-stained slides. (**B**) Representative Western blot images and relative densitometric bar graphs of (**B**) fibrotic markers and (**C**) inflammasome components. Expression levels are standardized by α-tubulin or the pro-form protein levels as appropriate. $n = 4$ mice. Results are expressed as mean ± SD. * $p < 0.05$, ** $p < 0.01$, *** $p < 0.001$, and **** $p < 0.0001$.

Sch B treatment, when used during chronic ethanol consumption, did not prevent collagen deposition in the liver (Figure 3A). ELISA and Western blotting also showed similar results, with only fibronectin expression slightly decreased compared to the ethanol-treated group (Figure 3B and Supplementary Figure S1B). In the experiment, Sch B treatment blocked ethanol-induced inflammasome activation (Figure 3C). However, Sch B failed to inhibit pyroptosis in this treatment model, as suggested by the unchanged level of GSDMD. These results showed that Sch B, when used along with alcohol consumption, can prevent ethanol-induced inflammasome activation, but not fibrosis.

3.3. *Schisandrin B Prevents Neurological Function Defects in Mice Consuming Ethanol*

Ethanol consumption has always been shown to cause neurological defects [6]. To explore whether Sch B can prevent ethanol-caused neurological impairments, mice were subjected to different neurofunctional tests. As indicated, ethanol intake significantly decreased the mice's weight (Figure 4A) and altered the mice's neurological performances (Figure 4B–F). Sch B treatment was effective at improving some aspects of neurological performance, such as the hot plate test, ledge beam test, wire hang test, and vertical pole test. However, Sch B treatment alone did not alter the neurological functions of the mice compared to the control group.

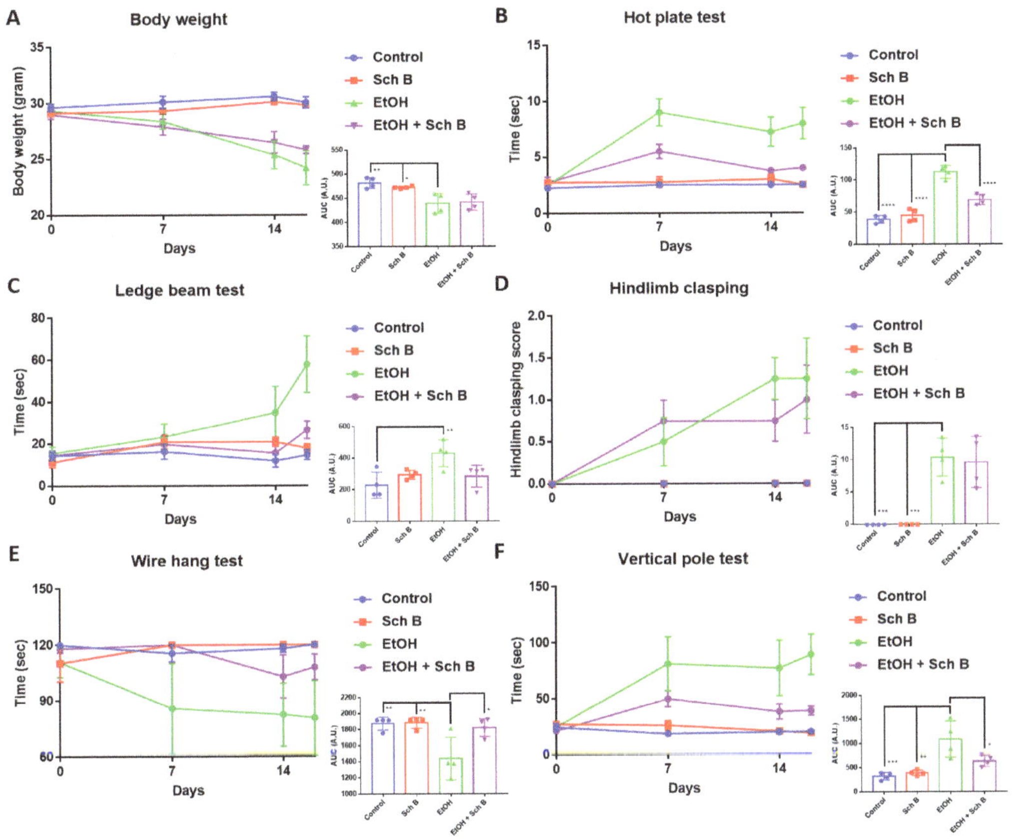

Figure 4. Schisandrin B prevents neurological function defects in mice consuming ethanol. (**A**) Mouse body weight. Assessment of neurological functions in mice by the (**B**) hot plate, (**C**) ledge beam, (**D**) hindlimb clasping, (**E**) wire hang, and (**F**) vertical pole tests. $n = 4$ mice. Results are expressed as mean ± standard error mean (SEM), and area under the curve (AUC) values are expressed as mean ± SD. * $p < 0.05$, ** $p < 0.01$, *** $p < 0.001$, and **** $p < 0.0001$.

3.4. Schisandrin B Prevents Brain Injuries in Mice Consuming Ethanol

As Sch B treatment improved neurological performance, we therefore evaluated the degree of brain injuries in ethanol-intoxicated mice, as well as the protection offered by Sch B. Histological analysis revealed numerous apoptotic neurons and hemorrhages in the brains of ethanol-drinking mice (Figure 5A,B). Ethanol consumption also increased CSF LDH levels (Figure 5C). Albumin synthesis occurs in the liver but not in the CNS, so albumin present in CSF is usually derived from plasma [28,34]. The ratio of CSF to serum albumin was used to evaluate blood-brain barrier (BBB) damage. The results showed that ethanol intake caused a significant increase in CSF albumin and CSF to serum albumin ratio, suggesting BBB damage (Figure 5D,F). To be noted, the decrease in serum albumin in ethanol-treated mice also suggests liver injury, as albumin synthesis mainly takes place in the liver (Figure 5D). Although Sch B treatment reduced the number of apoptotic neurons (Figure 5A,B), it did not affect the level of LDH (Figure 5C), which may suggest the occurrence of persistent injuries. On the contrary, Sch B decreased the level of CSF albumin, as well as the albumin ratio (Figure 5D,F), suggesting that the blood-brain barrier (BBB) was spared from damage. These results therefore suggest that Sch B, in addition to ethanol-induced liver injuries, protects against ethanol-induced neuropathy.

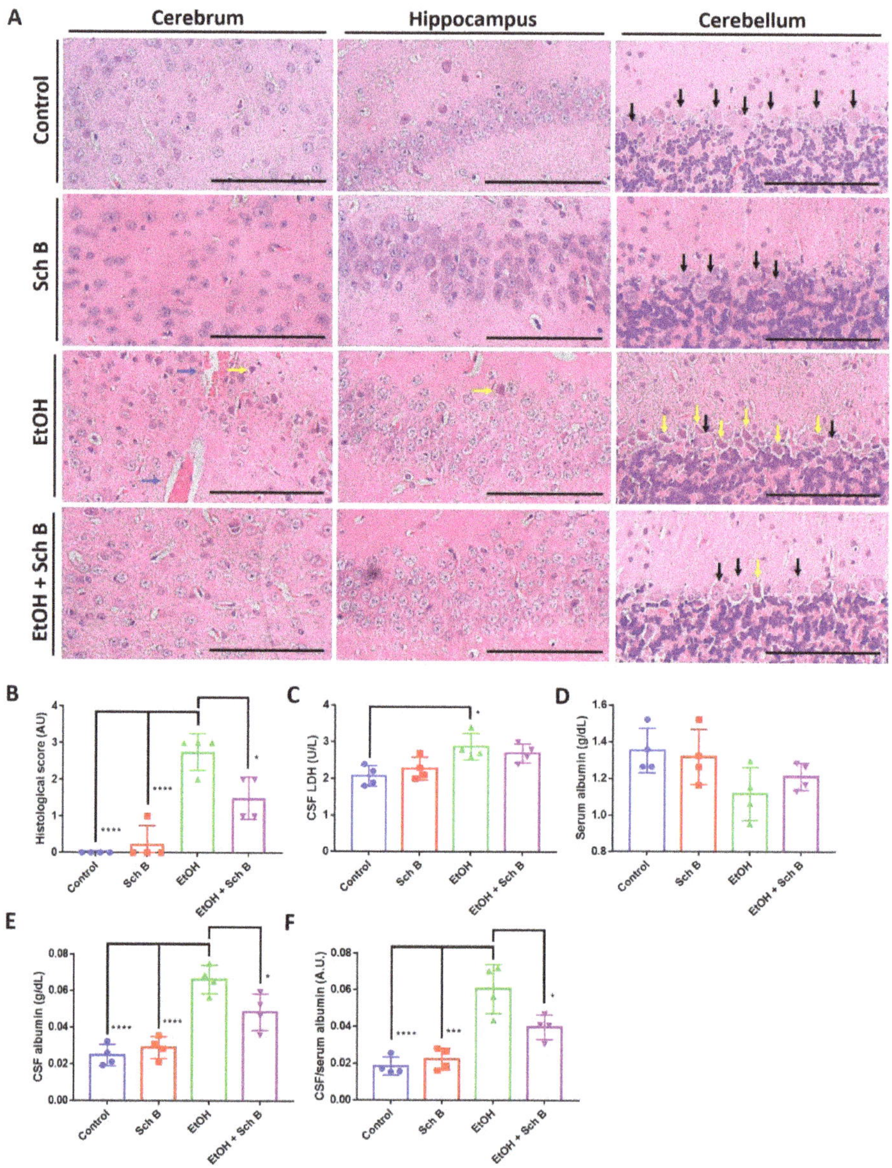

Figure 5. Schisandrin B prevents brain injuries in mice consuming ethanol. (**A**) Representative H&E-stained histological images showing the brain pathology of the mice. Congestion (blue arrows) and neuronal cell death (yellow arrows) were seen in ethanol-treated mice, which improved after Sch B treatment. Black arrows represent normal Purkinje fibers. Scale bar corresponds to 200 μm. (**B**) Histological scores based on the histological slides from (**A**). (**C**) LDH concentration in CSF. Levels of (**D**) serum albumin, (**E**) CSF albumin, and (**F**) CSF-to-serum albumin ratio. $n = 4$ mice. Results are expressed as mean ± SD. * $p < 0.05$, *** $p < 0.001$, and **** $p < 0.0001$.

3.5. Schisandrin B Alleviates Ethanol-Induced Liver Injuries in Mice

Next, we wanted to determine whether Sch B can reverse already-caused ethanol-induced liver injuries. Therefore, mice were first treated with ethanol to induce injuries in the liver and brain and then with Sch B following chronological sequence (Figure 6A). Similar to what was observed, ethanol treatment caused significant liver injuries, including congestion and ballooning degeneration, which was reversed by Sch B treatment (Figure 6B,C). This improvement was accompanied by decreased serum ALT and AST levels (Figure 6 D,E) compared with ethanol-treated mice. Sch B treatment also beneficially lowered total cholesterol and triacylglycerides in both serum and liver. Furthermore, by comparing the ethanol-fed group with the treatment group, we can observe that ethanol-induced injuries persisted, suggesting that there was no natural reversion once the ethanol was removed. These results therefore suggest that Sch B can alleviate ethanol-induced liver injuries.

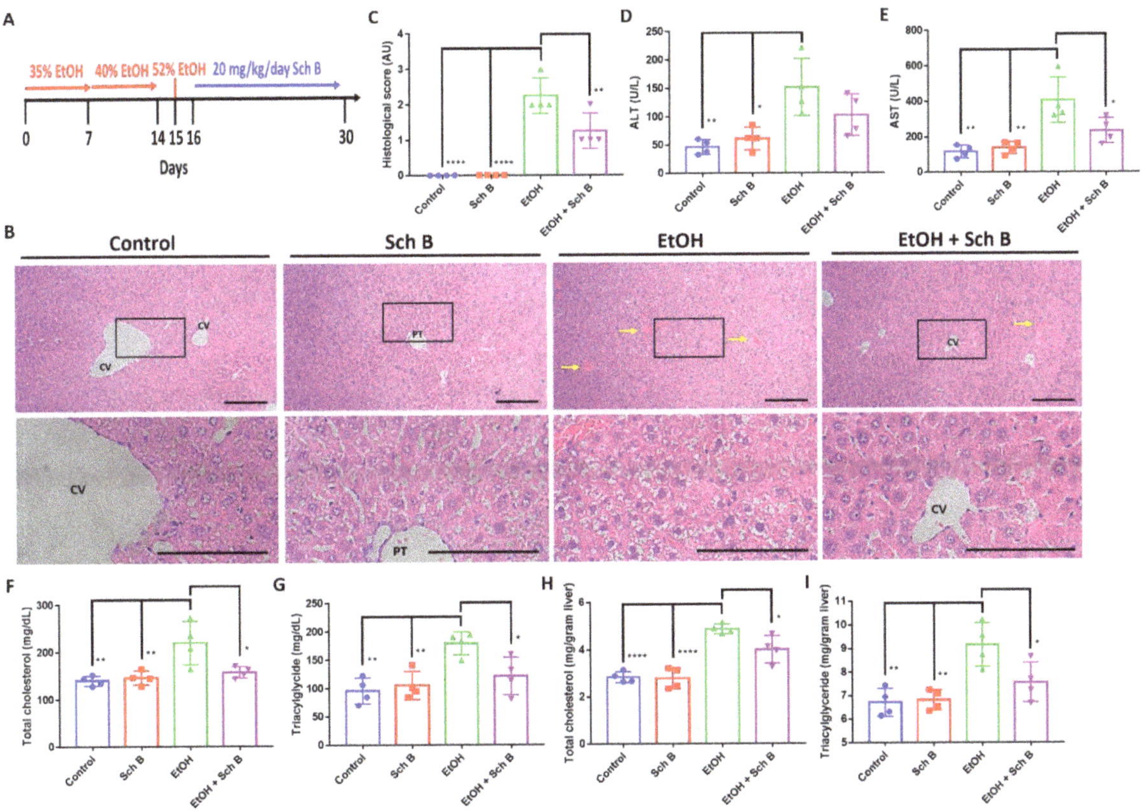

Figure 6. Schisandrin B alleviates ethanol-induced liver injuries in mice. (**A**) Experimental design scheme. (**B**) Representative H&E-stained histological images showing the liver pathology of the mice. Congestion (yellow arrows) and ballooning degeneration were seen in ethanol-treated mice, which were much improved by Sch B treatment. Scale bar corresponds to 200 μm. CV, central vein; PT, portal triad. (**C**) Histological scores based on the histological slides from (**B**). Serum levels of (**D**) ALT, (**E**) AST, (**F**) total cholesterol, and (**G**) triacylglycerides. Hepatic levels of (**H**) total cholesterol and (**I**) triacylglycerides. $n = 4$ mice. Results are expressed as mean ± SD. * $p < 0.05$, ** $p < 0.01$, and **** $p < 0.0001$.

3.6. Schisandrin B Alleviates Ethanol-Induced Liver Fibrosis, Inflammasome Activation, and Pyroptosis

Additionally, Sch B treatment significantly reduced collagen deposition (Figure 7A) along with TGF-β (Supplementary Figure S1D), collagen III, and fibronectin (Figure 7B) expression in the liver of ethanol-treated mice. Ethanol-induced inflammasome activation, as well as pyroptosis, was also suppressed by Sch B treatment (Figure 7C). Therefore, Sch B may be promising in preventing ethanol-induced liver inflammation and fibrosis.

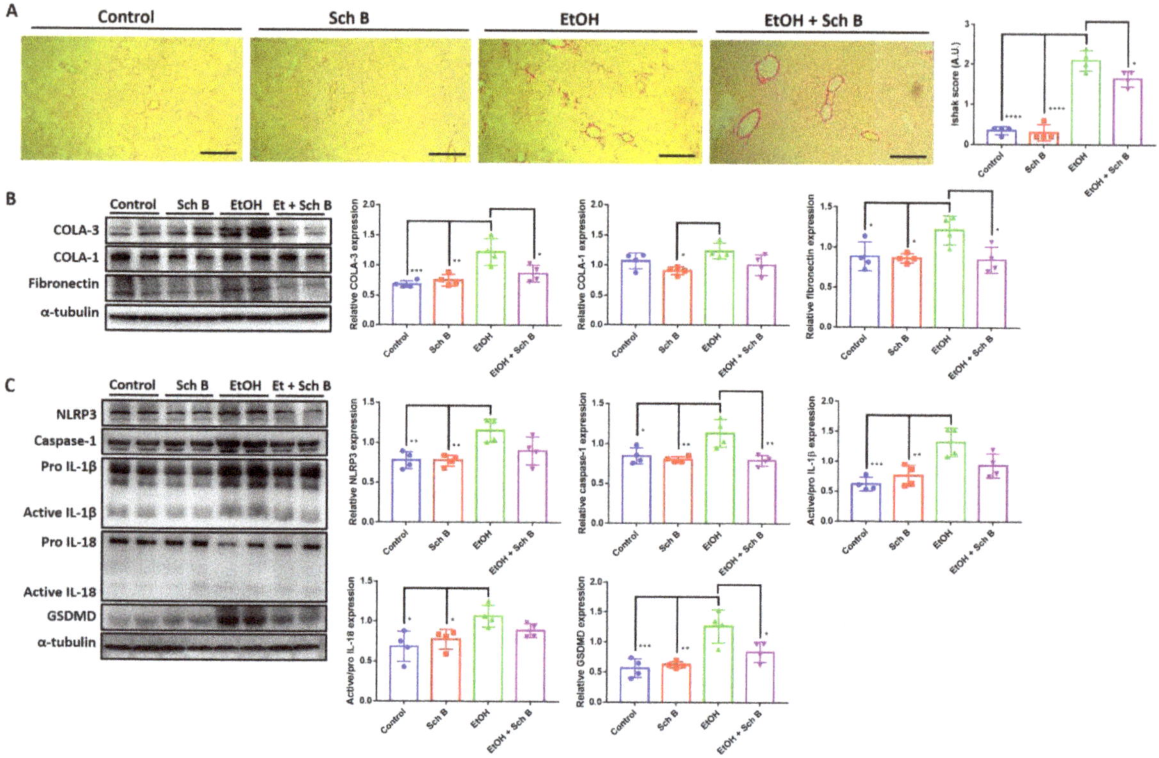

Figure 7. Schisandrin B alleviates ethanol-induced liver inflammasome activation and fibrosis. (**A**) Representative Sirius red-stained histological images showing collagen deposition (red color) in the mouse liver. Levels of fibrosis were quantified by Ishak score based on the Sirius red-stained slides. (**B**) Representative Western blot images and relative densitometric bar graphs of (**B**) fibrotic markers and (**C**) inflammasome components. Expression levels are standardized by α-tubulin or the pro-form protein levels as appropriate. $n = 4$ mice. Results are expressed as mean ± SD. * $p < 0.05$, ** $p < 0.01$, *** $p < 0.001$, and **** $p < 0.0001$.

3.7. Schisandrin B Repairs Ethanol-Induced Neurofunctional Defects by Alleviation of Brain Injuries in Mice

Sch B treatment slightly but not statistically significantly increased the mice's weight compared to ethanol-treated mice (Figure 8A). Except for the hot plate test, which showed only limited beneficial improvements, Sch B-treated mice showed beneficial improvements in the other neurological tests (Figure 8B–F), suggesting that Sch B can partially repair ethanol-induced neurofunctional defects in mice.

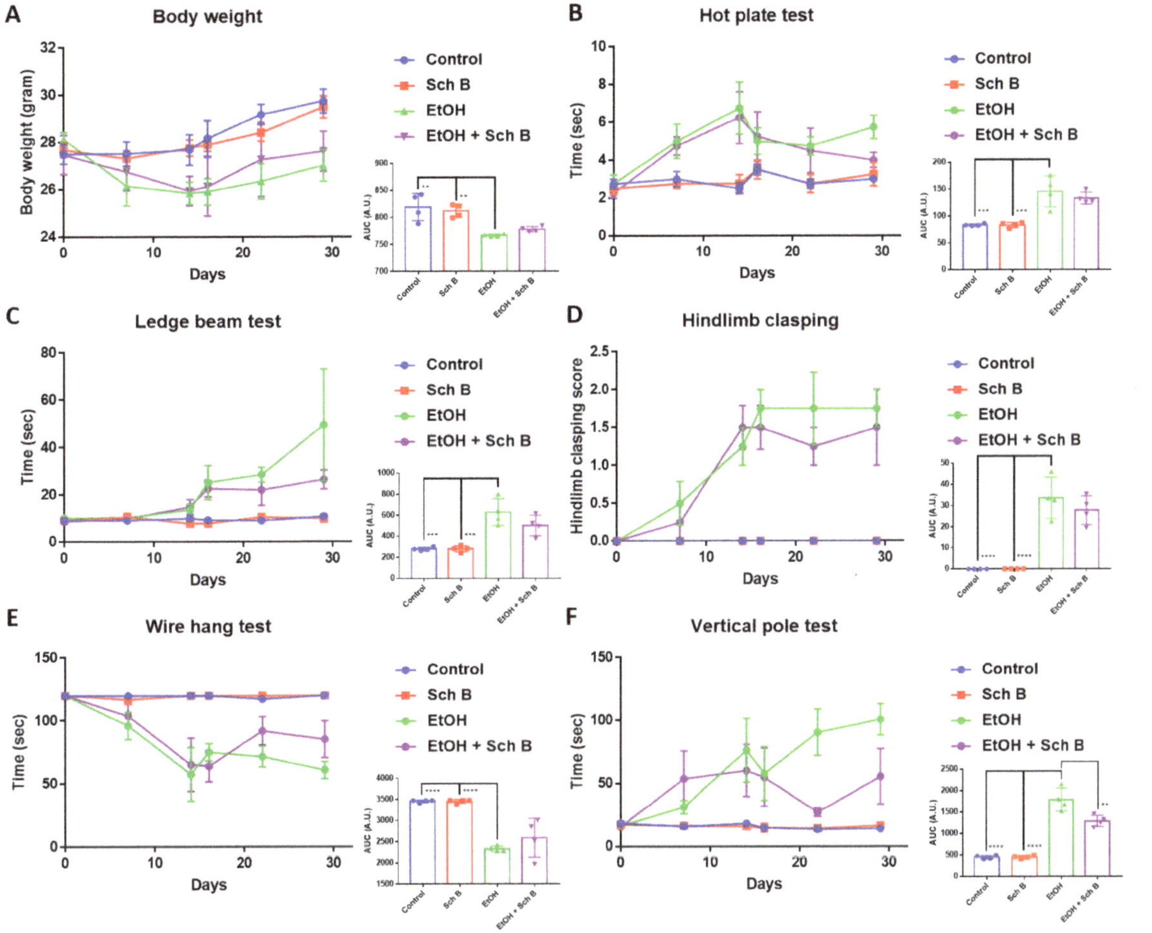

Figure 8. Schisandrin B repairs ethanol-induced neurofunctional defects. (**A**) Mouse body weight. Assessment of neurological functions in mice using the (**B**) hot plate, (**C**) ledge beam, (**D**) hindlimb clasping, (**E**) wire hang, and (**F**) vertical pole tests. $n = 4$ mice. Results are expressed as mean ± standard error mean (SEM), and area under the curve (AUC) values are expressed as mean ± SD. ** $p < 0.01$, *** $p < 0.001$, and **** $p < 0.0001$.

Finally, Sch B treatment significantly reduced ethanol-induced neuron cell death (Figure 9A,B) and improved reduced CSF LDH levels (Figure 9C). BBB damage was also significantly repaired by Sch B, as suggested by the reduced CSF-to-serum albumin ratio compared with the ethanol-treated group (Figure 9D–F).

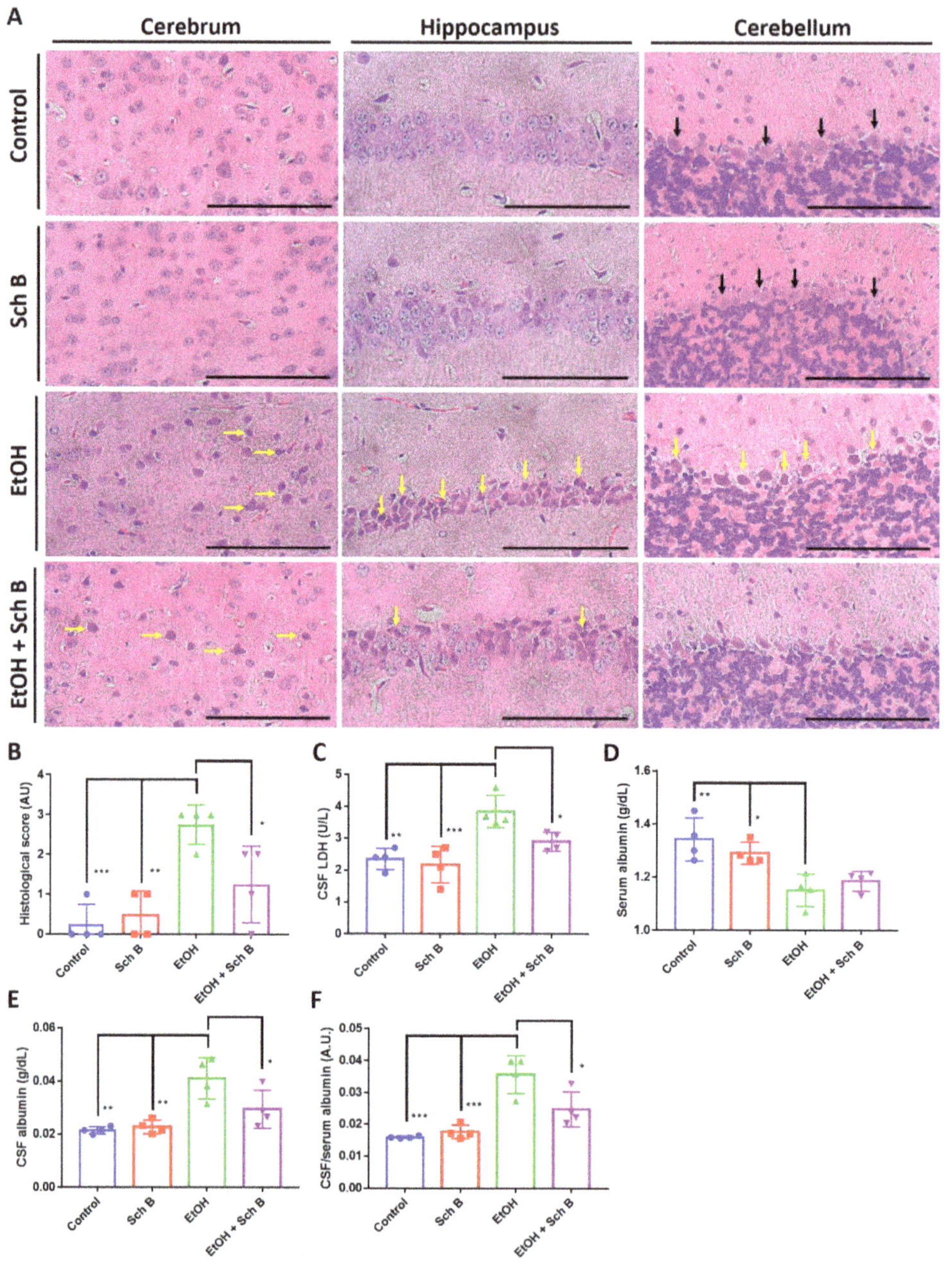

Figure 9. Schisandrin B alleviates ethanol-induced brain injuries in mice. (**A**) Representative H&E-stained histological images showing the brain pathology of mice. Neuronal cell death (yellow arrows) was seen in ethanol-treated mice, which improved after Sch B treatment. Black arrows represent normal Purkinje fibers. Scale bar corresponds to 200 μm. (**B**) Histological scores based on the histological slides from (**A**). Concentrations of (**C**) CSF LDH, (**D**) serum albumin, (**E**) CSF albumin, and (**F**) CSF-to-serum albumin ratio. $n = 4$ mice. Results are expressed as mean ± SD. * $p < 0.05$, ** $p < 0.01$, and *** $p < 0.001$.

4. Discussion

Chronic and excessive ethanol consumption may lead to intoxication and the development of chronic diseases, with liver disease being the most common and predominant [4,5]. Furthermore, excessive ethanol consumption can lead to structural and functional abnormalities in the brain [6]. In this study, we employed two treatment models to show that Sch B can be effective in preventing and treating ethanol-induced liver and brain injuries.

In the preventive treatment model, we explored the preventive effect of Sch B on ethanol-induced injuries by feeding the mice with both ethanol and Sch B on the same day (Figure 2A). The results suggested that Sch B was able to prevent liver damage (Figure 2B–E) as well as lipid deposition (Figure 2F–I). Inflammasomes are one of the key factors in liver fibrogenesis [32]. In the wake of sensing danger signals, NLRP3 activates caspase-1, thereby activating and releasing the inflammatory cytokines IL-1β and IL-18 [35]. In addition, it initiates a form of programmed cell death called pyroptosis, which is driven by a protein called GSDMD [36]. Once the cell dies, it activates Kupffer cells, which then release TGF-β and activate hepatic stellate cells (HSCs) [32,37]. The activated HSCs activate the downstream TGF-β/SMAD signaling pathway and cause fibrosis [38]. In addition to TGF-β, IL-1β also plays a role in a similar way [32]. While Sch B has some effect on inflammasome activation (Figure 2C), it cannot prevent collagen deposition in the liver (Figure 3A,B).

Since ethanol consumption also induces neurological defects [6,39], we next focused on the mice's neurological behavior as well as brain pathology. Several neurological tests were employed in this study: the hot plate test was used to measure thermal nociception in the mice, the ledged beam test was used to assess sensorimotor deficits in the mice, the hindlimb clasping score was used as an indicator of the severity of motor dysfunction, the wire hang test was used to evaluate motor functions in the mice, and the pole test was employed to assess basal ganglia-related movement in the mice [40]. Our results suggest that Sch B treatment can beneficially prevent ethanol-induced neurological defects (Figure 4). In contrast to our study, in a mouse model with *Angiostrongylus cantonensis*-induced meningoencephalitis, Sch B by itself failed to repair the mice's neurological functions; it was only effective if used together with the standard parasitic treatment drug albendazole [40]. This could be because using Sch B alone could not kill the parasites; therefore, even if Sch B ameliorated brain inflammation, neurological functions did not improve, as the injuries were persistent [40]. Another study using a mouse model with *Schistosoma manosni* (*S. mansoni*)-associated neurological deflects showed another therapeutic outcome from Sch B [41]. *S. manosni* is a parasitic worm that causes liver fibrosis, and liver fibrosis always leads to subsequent neurological damage [42]. In that study, the use of Sch B alone could already improve mice's performance on neurological tests [41]. As Sch B resolved liver injuries and fibrosis in *S. mansoni*-infected mice, it led to less severe brain injuries, thereby improving the mice's neurological functions. Along with neurological functions, improvements in brain pathology (Figure 5A–C) and BBB damage (Figure 5D–F) were also observed. Previously, the beneficial effects of Sch B have been suggested in several CNS diseases, such as cerebral ischemia [43,44], cerebral artery occlusion [44], subarachnoid hemorrhage [45], chemical-induced brain injury [28], and parasite-induced brain injury [40]. These studies have suggested that Sch B may target inflammasome activation to ameliorate brain inflammation and injuries. Therefore, it is presumed that Sch B may not only hinder inflammasome activation to prevent ethanol-induced neurological injuries, but this preventive effect may also come as a result of the resolution of liver damage.

Our second model was employed to investigate whether Sch B can be used to treat ethanol-induced injuries that have already been caused (corrective model). Surprisingly, Sch B worked better in treating than preventing ethanol-induced liver injuries (Figure 6). This treatment effect on the liver has also been observed in chemical-induced [28] and infection-related liver injuries [19]. Sch B treatment also successfully inhibited inflammasome activation as well as collagen deposition (Figure 7). Straightforwardly, unlike the first model, there was no consistent ethanol damage; therefore, Sch B can successfully perform its therapeutic functions. Sch B also improved neurological functions in ethanol-intoxicated mice (Figure 8), which was associated with improvements in neurological pathology and less BBB disruption (Figure 9).

In conclusion, our results showed that Sch B ameliorates liver injuries, lipid deposition, inflammasome activation, and fibrosis. Moreover, Sch B reverses brain damage and improves the neurological function of ethanol-treated mice. While effective treatment for liver injury is still lacking today, Sch B may serve as a potential treatment option for liver diseases as well as subsequential brain injuries. Furthermore, Sch B may be useful in preventive drug therapy against liver injuries.

Supplementary Materials: The following supporting information can be downloaded at https://www.mdpi.com/article/10.3390/nu15081909/s1: Figure S1: The use of Sch B inhibits ethanol-induced TGF-β levels in the liver. (A) Serum TGF-β levels in mice from the preventive model. (B) Hepatic TGF-β levels in mice from the preventive model. (C) Serum TGF-β levels in mice from the corrective model. (D) Hepatic TGF-β levels in mice from the corrective model. * $p < 0.05$, *** $p < 0.001$.

Author Contributions: Conceptualization, H.Y.P.L.; methodology, H.Y.P.L.; validation, H.Y.P.L.; formal analysis, H.Y.P.L. and T.-R.L.; investigation, H.Y.P.L. and T.-R.L.; writing—original draft preparation, H.Y.P.L.; writing—review and editing, H.Y.P.L.; funding acquisition, S.-Y.P. All authors have read and agreed to the published version of the manuscript.

Funding: This research was funded by the Ministry of Science and Technology of Taiwan (Grant number: MOST-111-2320-B-320-007) and the Tzu Chi Foundation, Taiwan (Grant number: TCU-B01).

Institutional Review Board Statement: The animal study protocol was approved by the Institutional Animal Care and Use Committee (IACUC) of Tzu Chi University (No. 109066).

Informed Consent Statement: Not applicable.

Data Availability Statement: The datasets generated and analyzed during the current study are available from the corresponding authors upon reasonable request.

Conflicts of Interest: The authors declare no conflict of interest. The funders had no role in the design of the study; in the collection, analyses, or interpretation of the data; in the writing of the manuscript; or in the decision to publish the results.

References

1. Kärkkäinen, J.M.; Miilunpohja, S.; Rantanen, T.; Koskela, J.M.; Jyrkkä, J.; Hartikainen, J.; Paajanen, H. Alcohol Abuse Increases Rebleeding Risk and Mortality in Patients with Non-variceal Upper Gastrointestinal Bleeding. *Dig. Dis. Sci.* **2015**, *60*, 3707–3715. [CrossRef]
2. MacMath, T.L. Alcohol and gastrointestinal bleeding. *Emerg. Med. Clin. N. Am.* **1990**, *8*, 859–872. [CrossRef]
3. Clemens, D.L.; Schneider, K.J.; Arkfeld, C.K.; Grode, J.R.; Wells, M.A.; Singh, S. Alcoholic pancreatitis: New insights into the pathogenesis and treatment. *World J. Gastrointest. Pathophysiol.* **2016**, *7*, 48–58. [CrossRef] [PubMed]
4. Osna, N.A.; Donohue, T.M., Jr.; Kharbanda, K.K. Alcoholic Liver Disease: Pathogenesis and Current Management. *Alcohol Res.* **2017**, *38*, 147–161.
5. Lieber, C.S. Alcoholic fatty liver: Its pathogenesis and mechanism of progression to inflammation and fibrosis. *Alcohol* **2004**, *34*, 9–19. [CrossRef]
6. Pervin, Z.; Stephen, J.M. Effect of alcohol on the central nervous system to develop neurological disorder: Pathophysiological and lifestyle modulation can be potential therapeutic options for alcohol-induced neurotoxication. *AIMS Neurosci.* **2021**, *8*, 390–413. [CrossRef] [PubMed]
7. Kim, S.J.; Kim, D.J. Alcoholism and diabetes mellitus. *Diabetes Metab. J.* **2012**, *36*, 108–115. [CrossRef] [PubMed]
8. Pelucchi, C.; Tramacere, I.; Boffetta, P.; Negri, E.; La Vecchia, C. Alcohol consumption and cancer risk. *Nutr. Cancer* **2011**, *63*, 983–990. [CrossRef]
9. Connor, J. Alcohol consumption as a cause of cancer. *Addiction* **2017**, *112*, 222–228. [CrossRef]
10. Weerasinghe, A.; Schoueri-Mychasiw, N.; Vallance, K.; Stockwell, T.; Hammond, D.; McGavock, J.; Greenfield, T.K.; Paradis, C.; Hobin, E. Improving Knowledge that Alcohol Can Cause Cancer is Associated with Consumer Support for Alcohol Policies: Findings from a Real-World Alcohol Labelling Study. *Int. J. Environ. Res. Public Health* **2020**, *17*, 398. [CrossRef]
11. Varma, V.; Webb, K.; Mirza, D.F. Liver transplantation for alcoholic liver disease. *World J. Gastroenterol.* **2010**, *16*, 4377–4393. [CrossRef]
12. Zakhari, S. Overview: How is alcohol metabolized by the body? *Alcohol Res. Health* **2006**, *29*, 245–254.

13. Tan, Z.; Sun, H.; Xue, T.; Gan, C.; Liu, H.; Xie, Y.; Yao, Y.; Ye, T. Liver Fibrosis: Therapeutic Targets and Advances in Drug Therapy. *Front. Cell Dev. Biol.* **2021**, *9*, 730176. [CrossRef]
14. D'Mello, C.; Swain, M.G. Liver-brain interactions in inflammatory liver diseases: Implications for fatigue and mood disorders. *Brain Behav. Immun.* **2014**, *35*, 9–20. [CrossRef] [PubMed]
15. Ferenci, P. Hepatic encephalopathy. *Gastroenterol. Rep.* **2017**, *5*, 138–147. [CrossRef] [PubMed]
16. Leong, P.K.; Ko, K.M. Schisandrin B: A Double-Edged Sword in Nonalcoholic Fatty Liver Disease. *Oxid. Med. Cell Longev.* **2016**, *2016*, 6171658. [CrossRef] [PubMed]
17. Chen, Q.; Zhang, H.; Cao, Y.; Li, Y.; Sun, S.; Zhang, J.; Zhang, G. Schisandrin B attenuates CCl(4)-induced liver fibrosis in rats by regulation of Nrf2-ARE and TGF-β/Smad signaling pathways. *Drug. Des. Dev. Ther.* **2017**, *11*, 2179–2191. [CrossRef] [PubMed]
18. Cuiqiong, W.; Chao, X.; Xinling, F.; Yinyan, J. Schisandrin B suppresses liver fibrosis in rats by targeting miR-101-5p through the TGF-β signaling pathway. *Artif. Cells Nanomed. Biotechnol.* **2020**, *48*, 473–478. [CrossRef]
19. Lam, H.Y.P.; Liang, T.-R.; Peng, S.-Y. Ameliorative effects of Schisandrin B on Schistosoma mansoni-induced hepatic fibrosis in vivo. *PLOS Negl. Trop. Dis.* **2021**, *15*, e0009554. [CrossRef] [PubMed]
20. Yang, X.; Wang, S.; Mu, Y.; Zheng, Y. Schisandrin B inhibits cell proliferation and induces apoptosis in human cholangiocarcinoma cells. *Oncol. Rep.* **2016**, *36*, 1799–1806. [CrossRef] [PubMed]
21. Liu, N.; Zheng, J.X.; Zhuang, Y.S.; Zhou, Z.K.; Zhao, J.H.; Yang, L. Anti-Inflammatory Effects of Schisandrin B on LPS-Stimulated BV2 Microglia via Activating PPAR-γ. *Inflammation* **2017**, *40*, 1006–1011. [CrossRef]
22. Xin, D.Q.; Hu, Z.M.; Huo, H.J.; Yang, X.J.; Han, D.; Xing, W.H.; Zhao, Y.; Qiu, Q.H. Schisandrin B attenuates the inflammatory response, oxidative stress and apoptosis induced by traumatic spinal cord injury via inhibition of p53 signaling in adult rats. *Mol. Med. Rep.* **2017**, *16*, 533–538. [CrossRef]
23. Tao, Y.; Zhou, H.; Huang, L.; Xu, X.; Huang, Y.; Ma, L.; Li, L.; Yao, X.; Zhang, R.; Zhang, Y.; et al. Schisandrin B Protects against Acute Ethanol-Induced Cardiac Injury by Downregulating Autophagy via the NOX4/ROS Pathway. *Pharmacology* **2021**, *106*, 177–188. [CrossRef] [PubMed]
24. Lam, P.Y.; Chiu, P.Y.; Leung, H.Y.; Chen, N.; Leong, P.K.; Ko, K.M. Schisandrin B co-treatment ameliorates the impairment on mitochondrial antioxidant status in various tissues of long-term ethanol treated rats. *Fitoterapia* **2010**, *81*, 1239–1245. [CrossRef]
25. Zhou, T.; Zhang, Y.-J.; Xu, D.-P.; Wang, F.; Zhou, Y.; Zheng, J.; Li, Y.; Zhang, J.-J.; Li, H.-B. Protective Effects of Lemon Juice on Alcohol-Induced Liver Injury in Mice. *BioMed Res. Int.* **2017**, *2017*, 7463571. [CrossRef] [PubMed]
26. Bertola, A.; Mathews, S.; Ki, S.H.; Wang, H.; Gao, B. Mouse model of chronic and binge ethanol feeding (the NIAAA model). *Nat. Protoc.* **2013**, *8*, 627–637. [CrossRef] [PubMed]
27. Zhang, Y.J.; Zhou, T.; Wang, F.; Zhou, Y.; Li, Y.; Zhang, J.J.; Zheng, J.; Xu, D.P.; Li, H.B. The Effects of Syzygium samarangense, Passiflora edulis and Solanum muricatum on Alcohol-Induced Liver Injury. *Int. J. Mol. Sci.* **2016**, *17*, 1616. [CrossRef]
28. Lam, H.Y.P.; Hung, M.Y.; Liang, T.R.; Peng, S.Y. An In-vivo Study into the Effects of Schisandrin B in the Liver, Spleen, Kidney, and Brain of Acute Thioacetamide-intoxicated Mice. *Iran. J. Pharm. Res.* **2021**, *20*, 300–314. [CrossRef]
29. Zhu, S.; Wang, Y.; Chen, M.; Jin, J.; Qiu, Y.; Huang, M.; Huang, Z. Protective Effect of Schisandrin B Against Cyclosporine A-Induced Nephrotoxicity In Vitro and In Vivo. *Am. J. Chin. Med.* **2012**, *40*, 551–566. [CrossRef] [PubMed]
30. Sheldon, R.A.; Jiang, X.; Francisco, C.; Christen, S.; Vexler, Z.S.; Täuber, M.G.; Ferriero, D.M. Manipulation of Antioxidant Pathways in Neonatal Murine Brain. *Pediatr. Res.* **2004**, *56*, 656–662. [CrossRef] [PubMed]
31. Arjmand, A.; Tsipouras, M.G.; Tzallas, A.T.; Forlano, R.; Manousou, P.; Giannakeas, N. Quantification of Liver Fibrosis—A Comparative Study. *Appl. Sci.* **2020**, *10*, 447. [CrossRef]
32. Zhang, W.-J.; Chen, S.-J.; Zhou, S.-C.; Wu, S.-Z.; Wang, H. Inflammasomes and Fibrosis. *Front. Immunol.* **2021**, *12*, 643149. [CrossRef] [PubMed]
33. He, W.-T.; Wan, H.; Hu, L.; Chen, P.; Wang, X.; Huang, Z.; Yang, Z.-H.; Zhong, C.-Q.; Han, J. Gasdermin D is an executor of pyroptosis and required for interleukin-1β secretion. *Cell Res.* **2015**, *25*, 1285–1298. [CrossRef] [PubMed]
34. Tibbling, G.; Link, H.; Ohman, S. Principles of albumin and IgG analyses in neurological disorders. I. Establishment of reference values. *Scand. J. Clin. Lab. Invest.* **1977**, *37*, 385–390. [CrossRef] [PubMed]
35. Swanson, K.V.; Deng, M.; Ting, J.P.Y. The NLRP3 inflammasome: Molecular activation and regulation to therapeutics. *Nat. Rev. Immunol.* **2019**, *19*, 477–489. [CrossRef] [PubMed]
36. Kovacs, S.B.; Miao, E.A. Gasdermins: Effectors of Pyroptosis. *Trends Cell Biol.* **2017**, *27*, 673–684. [CrossRef]
37. Higashi, T.; Friedman, S.L.; Hoshida, Y. Hepatic stellate cells as key target in liver fibrosis. *Adv. Drug. Deliv. Rev.* **2017**, *121*, 27–42. [CrossRef]
38. Xu, F.; Liu, C.; Zhou, D.; Zhang, L. TGF-β/SMAD Pathway and Its Regulation in Hepatic Fibrosis. *J. Histochem. Cytochem.* **2016**, *64*, 157–167. [CrossRef]
39. Poznanski, P.; Lesniak, A.; Korostynski, M.; Sacharczuk, M. Ethanol consumption following mild traumatic brain injury is related to blood-brain barrier permeability. *Addict. Biol.* **2020**, *25*, e12683. [CrossRef]
40. Lam, H.Y.P.; Liang, T.R.; Jiang, S.J.; Peng, S.Y. Albendazole-Schisandrin B Co-Therapy on Angiostrongylus cantonensis-Induced Meningoencephalitis in Mice. *Biomolecules* **2020**, *10*, 1001. [CrossRef]
41. Lam, H.Y.P.; Cheng, P.C.; Peng, S.Y. Resolution of systemic complications in Schistosoma mansoni-infected mice by concomitant treatment with praziquantel and Schisandrin B. *Int. J. Parasitol.* **2022**, *52*, 275–284. [CrossRef] [PubMed]
42. Jones, E.A.; Weissenborn, K. Neurology and the liver. *Neurol. Med.* **1997**, *63*, 279–293. [CrossRef] [PubMed]

43. Lee, T.; Jung, C.; Lee, D.-H. Neuroprotective effects of Schisandrin B against transient focal cerebral ischemia in Sprague-Dawley rats. *Food Chem. Toxicol.* **2012**, *50*, 4239–4245. [CrossRef] [PubMed]
44. Fan, X.; Elkin, K.; Shi, Y.; Zhang, Z.; Cheng, Y.; Gu, J.; Liang, J.; Wang, C.; Ji, X. Schisandrin B improves cerebral ischemia and reduces reperfusion injury in rats through TLR4/NF-κB signaling pathway inhibition. *Neurol. Res.* **2020**, *42*, 693–702. [CrossRef] [PubMed]
45. Chen, S.; Ding, Y.-H.; Shi, S.-S.; Tu, X.-K. Schisandrin B Inhibits NLRP3 Inflammasome Pathway and Attenuates Early Brain Injury in Rats of Subarachnoid Hemorrhage. *Chin. J. Integr. Med.* **2022**, *28*, 594–602. [CrossRef] [PubMed]

nutrients

MDPI

Article

Fruit Extract, Rich in Polyphenols and Flavonoids, Modifies the Expression of *DNMT* and *HDAC* Genes Involved in Epigenetic Processes

Ghodratollah Nowrasteh [1], Afshin Zand [1], László Bence Raposa [2], László Szabó [1], András Tomesz [1], Richárd Molnár [1], István Kiss [1], Zsuzsa Orsós [1], Gellért Gerencsér [1], Zoltán Gyöngyi [1,*] and Tímea Varjas [1]

[1] Department of Public Health Medicine, Medical School, University of Pécs, 7624 Pécs, Hungary; ghodratollah.nowrasteh@medcarehospital.com (G.N.); af.zand@gmail.com (A.Z.); laszlo.szabo.pte@gmail.com (L.S.); tomeszandras@gmail.com (A.T.); richard.molnar.pte@gmail.com (R.M.); istvan.kiss@aok.pte.hu (I.K.); zsuzsa.orsos@aok.pte.hu (Z.O.); gellert.gerencser@gmail.com (G.G.); vtimi_68@yahoo.com (T.V.)

[2] Faculty of Health Sciences, University of Pécs, 7621 Pécs, Hungary; bence.raposa@etk.pte.hu

* Correspondence: zoltan.gyongyi@aok.pte.hu

Abstract: Recently, the field of epigenetics has been intensively studied in relation to nutrition. In our study, the gene expression patterns of histone deacetylases (*HDAC*s), which regulate the stability of histone proteins, and DNA methyltransferases (*DNMT*s), which regulate DNA methylation, were determined in mice. The animals were fed a human-equivalent dose of the aqueous extract of fruit seeds and peels, which is rich in flavonoids and polyphenols, for 28 days and then exposed to the carcinogen 7,12-dimethylbenz(a)anthracene (DMBA). The concentrations of trans-resveratrol and trans-piceid were determined in the consumed extract by HPLC and were 1.74 mg/L (SD 0.13 mg/L) and 2.37 mg/L (SD 0.32 mg/L), respectively, which corresponds to the consumption of 0.2–1 L of red wine, the main dietary source of resveratrol, in humans daily. Subsequently, 24 h after DMBA exposure, the expression patterns of the *HDAC* and *DNMT* genes in the liver and kidneys were determined by qRT-PCR. The DMBA-induced expression of the tested genes *HDAC1*, *HDAC2*, *DNMT1*, *DNMT3A* and *DNMT3B* was reduced in most cases by the extract. It has already been shown that inhibition of the *DNMT* and *HDAC* genes may delay cancer development and tumour progression. We hypothesise that the extract studied may exert chemopreventive effects.

Keywords: *DNMT*s; *HDAC*s; flavonoids; fruit extract; CD1 mice; qRT-PCR; gene expression

Citation: Nowrasteh, G.; Zand, A.; Raposa, L.B.; Szabó, L.; Tomesz, A.; Molnár, R.; Kiss, I.; Orsós, Z.; Gerencsér, G.; Gyöngyi, Z.; et al. Fruit Extract, Rich in Polyphenols and Flavonoids, Modifies the Expression of *DNMT* and *HDAC* Genes Involved in Epigenetic Processes. *Nutrients* **2023**, *15*, 1867. https://doi.org/10.3390/nu15081867

Academic Editors: Haixia Yang and Jianjun Deng

Received: 11 March 2023
Revised: 10 April 2023
Accepted: 11 April 2023
Published: 13 April 2023

1. Introduction

Malignant diseases are associated with high mortality rate, and identifying novel methods of reducing the development and progression of these diseases is of high interest. Recently, epigenetic changes caused by environmental factors, including nutrition, have been intensively studied [1]. Substances in food can alter the normal methylation patterns of DNA, resulting in either the abnormal inactivation or activation of genes, both of which can lead to cancer progression.

The methylation status of DNA, especially within promoter regions, can have notable effects on both the incidence and progression of many types of malignant diseases. Aberrant methylation patterns have been observed in almost all neoplasms, suggesting that DNA methylation may serve as an important molecular marker for cancer prevention, prognosis, and therapies [1–3]. Histone acetyltransferases (*HAT*s) and histone deacetylases (*HDAC*s) play essential roles in the epigenetic regulation of gene expression by determining the acetylation status of histone proteins, which determines whether chromatin has a "relaxed" or "condensed" structure. Acetylation results in condensed structure and less active genes [4]. HDAC inhibitors (HDACis) exhibit antitumor effects by activating cell cycle arrest, inducing apoptosis and autophagy, inhibiting angiogenesis, and increasing the generation of reactive oxygen species to induce oxidative stress, which all contribute to cancer cell death [5,6].

Nutrition has been shown to mediate epigenetic mechanisms, and recent studies have demonstrated that both the quantity and quality of food intake are associated with aging and cancer incidence and prognosis. Moreover, nutrition is thought to be the most influential of all external environmental factors due to its ability to affect the transcriptional activity and expression of specific genes [1]. Natural products have received increasing attention in recent years as novel anticancer agents, and interest in the potential chemopreventive and therapeutic properties of food-derived compounds, including plant polyphenols, has increased tremendously. Several members of phytochemicals are able to inhibit tumour growth as well as carcinogenesis at many points, and are also potential natural therapeutic agents. They may inhibit the activation of several signalling pathways, such as *PI3K/AKT*, *MAPK/ERK* and *JAK/STAT3*, thereby ultimately inhibiting tumour growth and metastasis, and may also stimulate ROS-mediated apoptosis and autophagy [7–10]. Polyphenols have a full range of anti-tumour effects, including, in addition to the above, inhibition of angiogenesis and activation of the immune system [11].

Plant polyphenols can be found in many fruits and vegetables, including soy, turmeric, grapes, celery, apples, onions, parsley, bell peppers, green tea, and black pepper, and have been shown to possess chemopreventive activity [1]. Natural compounds found in plants have demonstrated potential HDACi activity, including trans-resveratrol found characteristically in grapes, red wine, blueberries, and peanuts as main active ingredients [6]. HDACis of both natural and synthetic origins are currently being evaluated in clinical trials to determine their antitumor efficacies and potential side effects. 7,12-Dimethylbenz[a]anthracene (DMBA) is a known carcinogen that is used as a general compound in in vitro studies focusing not only on tumour models, but also on epigenetic studies [12–15]. To study the effect of the carcinogenic compound DMBA on early gene expression modification, a 24 h treatment seems to be optimal in an animal model, as many genes involved in the cell regulation respond by this time point [16,17].

Our study aimed to explore whether the frequent consumption of aqueous extract of a fruit seed and peel extract rich in phytochemicals influences the expression levels of *DNMT* and *HDAC* genes, which are responsible for the epigenetic changes that affect tumour formation and prevention. Specifically, we tested whether an extract derived from fruit seeds, lyophilized shells, and dried red grapes, blackberries, blackcurrants, redcurrants, and rosehips under the brand name Fruit Café (ProVitamix, Budapest, Hungary), which contains 100–200 mg/portion/day of polyphenols/flavonoids, could prevent the early stages of tumour formation following carcinogenic DMBA exposure by modulating *DNMT* and *HDAC* mRNA levels. Detection of trans-resveratrol and its glucoside trans-piceid was also performed. Our study showed that the tested polyphenol-rich extract, which provides dietary levels of polyphenols, can effectively reduce the expression of *DNMT* and *HDAC* genes.

2. Materials and Methods

2.1. HPLC Measurements

Chemicals and reagents: The *trans*-resveratrol standard (99%) was purchased from Sigma-Aldrich Co. (Budapest, Hungary), the *trans*-piceid standard was purchased from Herbstandard Inc. (Chesterfield, MO, USA), acetic acid (96%) was obtained from Riedel-de Haën GmbH & Co. (Seelze, Germany) and methanol (HPLC grade) was acquired from Scharlau Chemie S.A. (Barcelona, Spain).

Standard solutions and sample preparation: The standards were dissolved in a small portion of ethanol and filled up with the eluent. The fruit extract samples were prepared in a coffee machine with tap water as it is described in "Treatments of animals" section to obtain 100 mL of solution per portion, filtrated by using Millex syringe filter (Millipore Kft., Budapest, Hungary) and injected. All standard solutions and fruit extract samples were stored in the dark place at 5 °C to avoid oxidative degradation and isomerisation of *trans*-resveratrol and *trans*-piceid to *cis*-configuration (Figure 1).

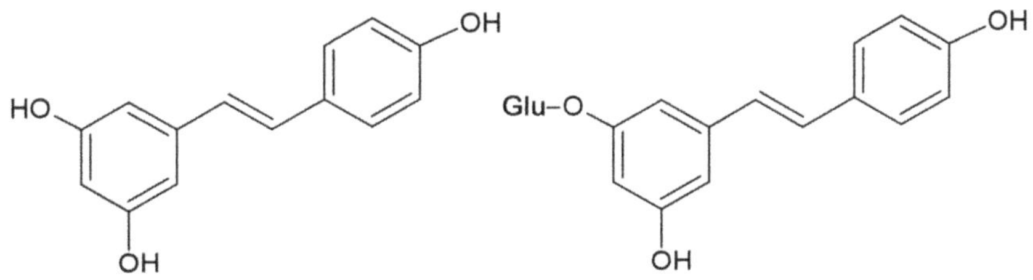

Trans-resveratrol
***trans*-3,5,4′-trihydroxystilbene**

Piceid
***trans*-resveratrol-3-O-β-glucoside**

Figure 1. Structure of *trans*-resveratrol and *trans*-piceid.

HPLC instrumentation and conditions: The system used for high-performance liquid chromatography (HPLC) consisted of a Gynkotek M 580 GT pump, a Rheodyne 8125 (20 μL loop) injector and a Gynkotek M 340S UV diode array detector (Gynkotek GmbH, Germering, Germany). A 250 × 4.6 mm column packed with 6 μm particle size C18 material has been used for the separations. A Chromeleon data management software (Dionex Corp., Sunnyvale, CA, USA) was used for the control of the equipment and for data evaluation. Quantization was carried out using peak areas method. A multistep gradient method was applied using methanol–water–acetic acid (10:90:1 v/v) mixture as solvent A and methanol–water–acetic acid (90:10:1 v/v) mixture as solvent B at a flow rate of 1.5 cm^3 min^{-1}. The gradient profile was 0.0–18.0 min from 0% to 40% B; 18.0–25.0 min from 40% to 100% B; 25.0–27.0 min 100% B. Chromatographic separations were monitored at 306 nm. Chromatographic peaks were identified by comparing retentions and UV spectra and MS spectra of the samples with those of the standard compounds. Quantisation was carried out by external standardisation.

MS instrumentation and conditions: MS (mass spectrometry) analysis was performed using a Finnigan AQA (Thermoquest, San Jose, CA, USA) mass spectrometer. The auxiliary and the curtain gas were nitrogen at the flow rate of 600 L/h. For HPLC-MS analysis, we used ESI (electrospray ionization) source, the probe temperature was 250 °C, the probe voltage was 3.5 kV. Spectra were recorded by 1.2 scan/s in the negative ion mode between m/z 10 and 700. The scan filter on the quadrupole analyser was 10 and 20 V. Finnigan Xcalibur (version: XCALI-97006) was used to acquire the mass spectra of the compounds.

2.2. Animals

Male CD1 mice (Charles River Laboratories International, Budapest, Hungary) were housed separately in polycarbonate cages and maintained in a 12:12 h light–dark cycle, provided with water and standard pellets ad libitum throughout the experiment. Four groups of 6-week-old (18 ± 2 g) male CD1 mice were used for this study, with six animals in each group.

2.3. Treatment of Animals

The negative control group (Control group) was provided with standard feed and tap water ad libitum. The positive control group (DMBA group) was provided with standard feed and tap water ad libitum, and 20 mg/kg body weight (bw) DMBA (Sigma-Aldrich, Budapest, Hungary), dissolved 0.2 mL corn oil, was administered intraperitoneally 24 h before cervical dislocation. The third group (FC group) received drinking water containing a human-equivalent dose of the fruit seed and peel extract for 28 days, in addition to ad libitum access to standard feed. The fourth group (FC + DMBA group) received drinking water containing a human-equivalent dose of the fruit seed and peel extract for 28 days, in addition to ad libitum access to standard feed, and 20 mg/kg bw DMBA, dissolved in

0.2 mL corn oil, was administered intraperitoneally 24 h before cervical dislocation (Table 1). The animals in first and third groups received 0.2 mL corn oil vehicle intraperitoneally. The fruit seed and peel extract contained lyophilized seeds, the skins of red grape, blackcurrant, redcurrant, rosehip, and black cherry, and fructose [13,14,18] commercialized under the brand name Fruit Café (ProVitamix, Budapest, Hungary). The extract was prepared in a regular high-pressure coffee from two components. One component is the seed extract and the other one is peel extract. A total of 60 doses from 1000 g of seed extract and 700 g of peel extract can be produced, and 8 mL of the 100 mL solution was dissolved in tap water to a final volume of 40 mL, which generated a daily dose for 12 mice, equivalent to the human dose [19].

Table 1. Treatment program.

		Fluid Consumption (for 28 Days)		**Carcinogenic Exposure (on Day 27) ****	
Group 1	Control	Tap water	ad libitum	Vehicle	
Group 2	DMBA	Tap water	ad libitum	DMBA	20 mg/kg bw
Group 3	FC	Fruit seed and peel extract	Human-equivalent dose *	Vehicle	
Group 4	FC + DMBA	Fruit seed and peel extract	Human-equivalent dose *	DMBA	20 mg/kg bw

The experimental animals received ad libitum water after full consumption of the polyphenolic fruit seed and peel extract (FC; *) and purified corn oil as the vehicle or DMBA (7,12-dimethylbenz(a)anthracene) dissolved in purified corn oil was administered intraperitoneally 24 h prior to cervical dislocation and dissection (**).

Animals were autopsied after euthanasia was performed by cervical dislocation. All animals received humane care, and the experimental protocol received ethical approval.

2.4. Total RNA Isolation

During the autopsy, the liver and kidneys were collected from every animal and total mRNA was isolated using TRIzol reagent (MRTR118-20, NucleotestBio, Budapest, Hungary) according to the manufacturer's protocol.

Tissue samples were homogenised with TRIzol reagent (1 mL per 100 mg tissue) using a Janke and Kunkel Ultra Turrax T25 stirrer (IKA-Werke, Staufen, Germany), and 100 μL of chloroform (Merck, Budapest, Hungary) was added to the TRIzol lysate and mixed thoroughly by shaking. After 5 min at room temperature, the samples were centrifuged at 12,000× *g* for 15 min at 4 °C, and the upper aqueous phase containing RNA was collected in a new Eppendorf tube.

Then, 250 μL of isopropanol (Merck, Budapest, Hungary) was added to the sample, mixed and kept at room temperature for 10 min before centrifugation at 12,000× *g* at 4 °C for 10 min. The supernatant was discarded and the pellet was washed with 70% ethanol (Merck, Budapest, Hungary) and then centrifuged at 7500× *g* at 4 °C for 5 min. The pellet was air-dried and RNase-free water (Merck, Budapest, Hungary) was added. Finally, the RNA was quantified with a MaestroNano (Maestrogen, Hsinchu City, Taiwan) spectrophotometer.

2.5. qRT-PCR

The assays were performed on the Roche 480 instrument (Roche, Budapest, Hungary) using the KAPA SYBR FAST One-Step qRT-PCR Master Mix Kit (Sigma-Aldrich, Budapest, Hungary). The amplifications were performed in 20 μL reaction volume, mixing 5 μL RNA target (100 ng) and 15 μL of master mix with forward and reverse primers (10 μL KAPA SYBR FASTqPCR Master Mix, 0.4 μL KAPA RT Mix, 0.4 μL dUTP, 0.4 μL primer (200 nM), 3.8 μL sterile double distilled water). The reactions were performed using the following thermal profile: 42 °C for 5 min for the reverse transcription step, a hot-start denaturation step of 95 °C for 3 min, followed by 45 cycles of 95 °C for 10 s (denaturation), 60 °C for 20 s (annealing/extension). The fluorogenic signal emitted during the annealing/extension step was read and analysed using the software. Immediately after amplification, a melting curve protocol was generated by increasing each cycle by 0.5 °C for 80 cycles of 10 s each, starting from the target temperature (55.0 °C).

The primary sequences of the housekeeping gene used as an internal control, hypoxanthine phosphoribosyltransferase 1 (*HPRT1*), and the genes of interest, *DNMT1*, *DNMT3A*, *DNMT3B*, *HDAC2*, *HDAC3* and *HDAC8*, are listed in Table 2. Primers were designed using Primer Express™ software (Applied Biosystems, Budapest, Hungary) and synthesised by Integrated DNA Technologies (Bio-Sciences, Budapest, Hungary). The results were analysed using the relative quantification method ($2^{-\Delta\Delta CT}$).

Table 2. PCR primers.

Gene Name	Forward Primer	Reverse Primer
DNA methyltransferase 1 (*DNMT1*)	F-AAGAATGGTGTTGTCTACCGAC	R-CATCCAGGTTGCTCCCCTTG
DNA methyltransferase 3A (*DNMT3A*)	F-GAGGGAACTGAGACCCCAC	R-CTGGAAGGTGAGTCTTGGCA
DNA methyltransferase 3B (*DNMT3B*)	F-AGCGGGTATGAGGAGTGCAT	R-GGGAGCATCCTTCGTGTCTG
Histone deacetylase 2 (*HDAC2*)	F-GGAGGAGGCTACACAATCCG	R-TCTGGAGTGTTCTGGTTTGTCA
Histone deacetylase 3 (*HDAC3*)	F-GCCAAGACCGTGGCGTATT	R-GTCCAGCTCCATAGTGGAAGT
Histone deacetylase 8 (*HDAC8*)	F-ACTATTGCCGGAGATCCAATGT	R-CCTCCTAAAATCAGAGTTGCCAG
Hypoxanthine phosphoribo-syltransferase 1 (*HPRT1*)	F-TCAGTCAACGGGGGACATAAA	R-GGGGCTGTACTGCTTAACCAG

Sequences of primers used for qRT-PCR are listed.

2.6. Statistical Analysis

The Kolmogorov–Smirnov test was used to verify the normality of the results. Means between groups were compared using analysis of variance (ANOVA). Statistical analyses were performed using IBM SPSS Statistics for Windows, Version 26.0 (Armonk, NY, USA). Significance was established at $p < 0.05$.

3. Results

After analysis of the fruit extract by HPLC, the calculated amount of trans-resveratrol (Figure 2) was 1.74 mg/L (SD 0.13) and that of trans-piceid (Figure 3) was 2.37 mg/L (SD 0.32). Therefore, the mean daily intake of trans-resveratrol was 5.8 μg, while the mean daily intake of trans-piceid was 7.9 μg for each mouse, respectively. Using the metabolic multiplier between humans and mice (12.3×), this would result in a human consuming 366.76 micrograms of trans-resveratrol and 499.55 micrograms of trans-piceid daily.

We tested the efficacy of a fruit seed and peel extract to attenuate *HDAC* and *DNMT* gene expression following the administration of the carcinogen DMBA in an animal model.

We measured the relative gene expression of *HDAC2*, *HDAC3*, and *HDAC8* in both liver and kidney isolates. DMBA administration significantly increased the relative expression of *HDAC2* in the liver of the DMBA group twofold compared to the Control group; however, the fruit seed and peel extract completely diminished the effects of DMBA (Figure 4). The relative gene expression pattern of *HDAC3* in the liver was similar to that observed for *HDAC2*; however, gene expression of both genes was reduced threefold in the FC group compared with that in the Control group. No significant changes were observed for *HDAC8* expression. A similar trend was observed in the kidney for all three genes (Figure 4).

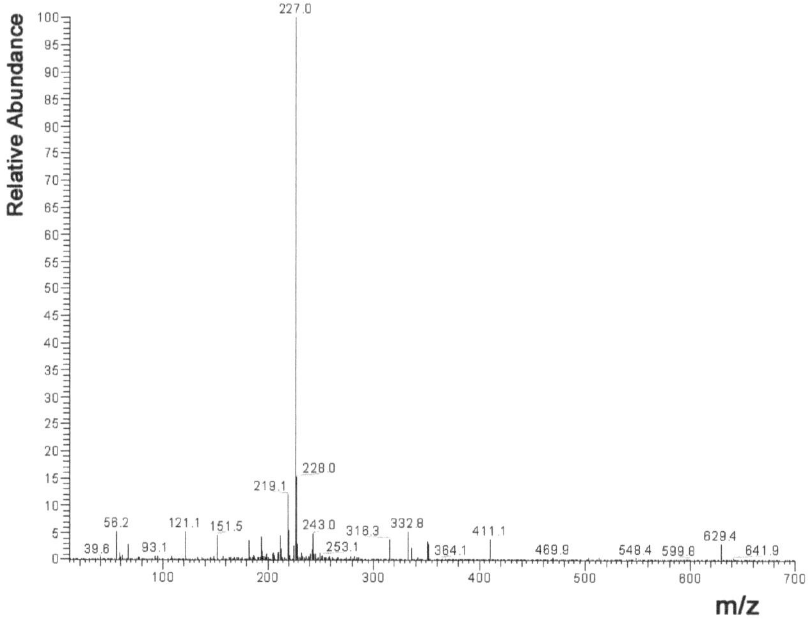

Figure 2. ESI negative ion MS spectra of trans-resveratrol. The negative polarised quasi-molecular ion derived from trans-resveratrol was detected at 227.0 m/z (ESI = electrospray ionization, MS = mass spectrometry).

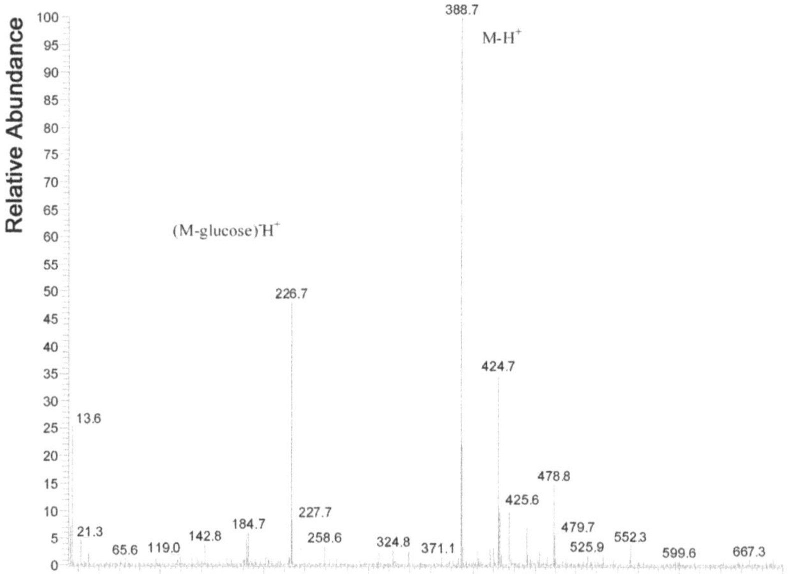

Figure 3. ESI negative ion MS spectra of quasi-molecular trans-piceid at 388.7 m/z. The peak of trans-resveratrol, which was partially presented in native form and generated through the fragmentation of piceid precursor as well, was detected at 226.7 m/z (ESI = electrospray ionization, MS = mass spectrometry).

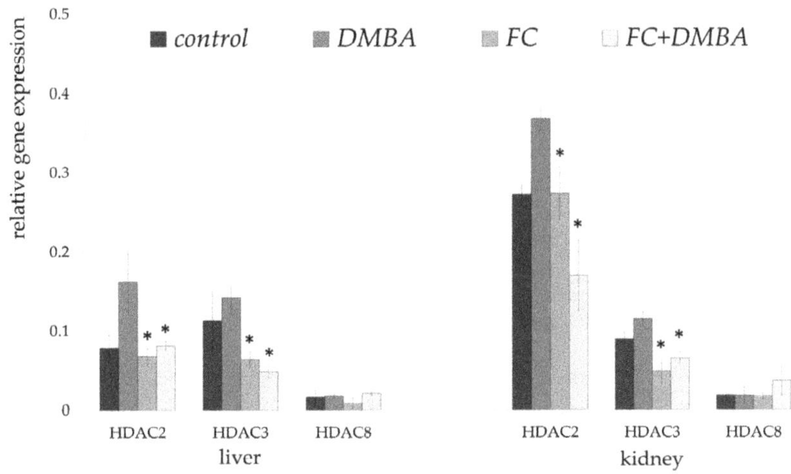

Figure 4. Relative gene expression of *HDAC* genes. The relative gene expression levels of histone deacetylases at the mRNA level in the liver and kidney of experimental animals relative to the mRNA levels of *HPRT1* (DMBA: 7,12-dimethylbenz(a)anthracene, HPcFE: fruit seed and peel extract); * significantly different from DMBA-treated mice ($p < 0.05$). Error bars represent the standard deviation.

In the liver, fruit and peel extract significantly reduced the expression of the genes tested not only in those elevated by DMBA, but also in the control groups. However, the most significant reduction occurred in the FC + DMBA groups. FC induced a more than tenfold, 3.5-fold and 3-fold decrease in the expression of *DNMT1*, *DNMT3A* and *DNMT3B* genes, respectively, compared to the DMBA groups.

In the kidneys, DMBA exposure significantly increased the gene expression of *DNMT3A* and *DNMT3B* compared with the control. The fruit seed and peel extract reduced this change in gene expression but less emphatically than in the liver (Figure 5).

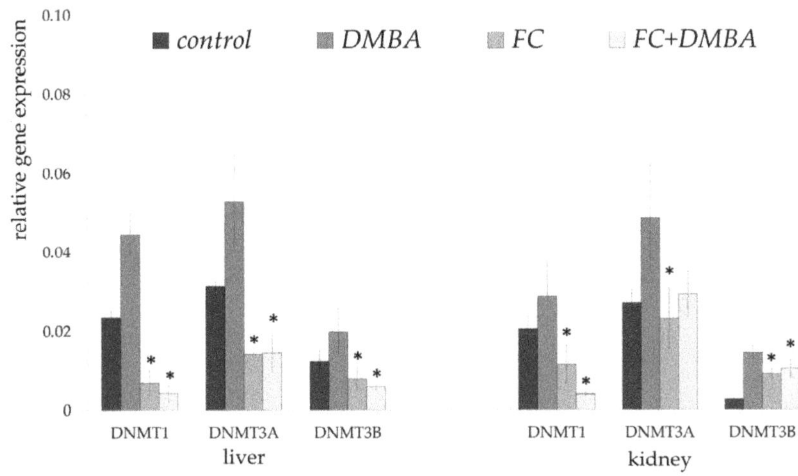

Figure 5. Relative gene expression of *DNMT* genes. Relative gene expression of DNA methyltransferases at the mRNA level in the liver and kidney of experimental animals compared with the mRNA level of *HPRT1* (DMBA: 7,12-dimethylbenz(a)anthracene, HPcFE: Fruit seed and peel extract); * significantly different from DMBA-treated mice ($p < 0.05$). Error bars represent the standard deviation.

4. Discussion

In our animal model, DMBA was used as a carcinogenic compound, and we analysed whether pre-treatment with a commercially available fruit seed and peel extract exerted chemopreventive and cellular regulatory function in an animal model. Our results revealed, in general, that the fruit and peel extract, which is rich in polyphenols and flavonoids, protected against DMBA-induced alterations in gene expression in the liver and kidneys. DMBA treatment significantly increased the expression of most examined *DNMT* and *HDAC* genes, whereas pre-treatment with the fruit seed and peel extract decreased the expression of these genes, even relative to the Control group.

Based on our HPLC measurements, the recommended daily intake of the fruit and peel extract contains trans-resveratrol and trans-piceid in amounts equivalent to the mean consumption in the adult Spanish population. This amount corresponds to 0.2 L of red wine for trans-resveratrol and 1 L of red wine for trans-piceid. This population in the lowest quartile consumes only a fraction of the mean [20]. Considering that resveratrol is mainly found in red wine, consumption of similar products tested could significantly increase resveratrol intake in the population without red wine consumption. The HPLC measurement also indicated that the animals received amounts of resveratrol close to the mean human intake. Similar animal studies carried out by us are very rare. The studies focus mainly on tolerance, toxicity, renal and hepatic damage and the effects of high doses in general [21].

Histone acetylation, which is regulated by *HDAC* activity [4], has been shown to modulate gene expression, and the post-transcriptional modifications of acetylation and methylation may play roles in cancer development by regulating the expression of tumour suppressor genes and oncogenes. Our results demonstrate that the fruit seed and peel extract significantly decreased *HDAC2* and *HDAC3* expression in both the liver and kidney compared with the Control group, which consumed only standard laboratory pellets, and with the DMBA-treated group. However, no significant changes in *HDAC8* expression were observed in either the liver or kidney.

There may be a special case where DMBA combined with fruit and peel extract results in lower expression than fruit and peel extract alone. Such examples are shown in Figure 4 for *HDAC3* in the liver and *HDAC2* in the kidney. Similar observations have been made by other researchers, but this may be a gene- and tissue-dependent issue, and not a universal one. For example, Singhal and colleagues investigated the effect of oestrogen modification on DMBA carcinogenesis in the liver of ovariectomised rats. They determined that the expression of *CYP1A1* and *CYP1B1* was higher in the co-administration of oestrogen and DMBA than in the DMBA group. The expression of *NQO1* was increased by DMBA, but this increase was more modest in the presence of DMBA and oestrogen together. For *GSTM3*, DMBA and oestrogen resulted in lower expression than DMBA or oestrogen alone [20].

Mirza et al. described results, similar to our findings, for the relative gene expression levels of *DNMT1*, *DNMT3A*, and *DNMT3B* measured by RT-PCR in breast cancer cells treated with multiple natural chemopreventive substances, such as EGCG, genistein, withaferin A, curcumin, resveratrol, and guggulsterone. These natural substances significantly decreased *DNMT* gene expression, suggesting potential chemopreventive activities. *DNMT*s are recognized as major enzymes involved in the somatic inheritance of DNA methylation, playing crucial roles in epigenome maintenance. The upregulation of the DNMT1 expression may induce the dissociation of p21WAF1 from PCNA, which may make p21WAF1 more vulnerable to ubiquitination and proteasomal degradation [22]. Supplementation with the fruit seed and peel extract in our experiment resulted in relatively low gene expression levels of *DNMT1* and *DMNT3A* in both the kidneys and liver compared with the Control group and prevented increased gene expression following DMBA exposure, whereas *DNMT3B* only showed lower levels in the FC groups compared to the DMBA groups.

The chemopreventive effects of fruit seed and peel extracts, which contain polyphenols and flavonoids, have been demonstrated in several studies. Flavonoids, such as anthocyanin and luteolin, are abundant in many fruits and vegetables. Ursolic acid, a natural triterpenoid found in blueberries and cranberries, activates the nuclear factor-erythroid factor 2–related factor 2 (*Nrf2*) pathway and decreases the enzymatic activities of DNMTs and HDACs, leading to chemopreventive and antitumor activities [23–25].

Various studies have demonstrated that DNA methylation, which is regulated by the histone acetylation status, results in gene silencing in breast cancer cells. Several hypermethylated genes have been associated with gene silencing, such as the cell cycle inhibitor genes *CDKN2A* (*p16*) and *RASSF1A*, and the DNA repair gene *BRCA1*. The combination of sulforaphane (SFN) with the main polyphenol found in green tea, epigallocatechin-3-gallate (EGCG), can reactivate oestrogen receptor (ER)-α expression in ER-α-negative breast cancer cell lines, restoring sensitivity to antioestrogen treatment [26,27]. Therefore, the inhibition of HDAC and DNMT activities may provide new treatment options for patients with ER-α-negative breast cancer [23]. According to a recent study, the carcinogenic polycyclic aromatic hydrocarbons, such as benzofluorene and benzo (a) pyrene (BaP), can significantly increase the enzymatic levels of DNMT1l [28].

Recently, black raspberry-derived anthocyanins, a subclass of flavonoids, have been demonstrated to suppress the enzymatic activities and protein levels of DNMT1 and DNMT3B in multiple colon cancer cell lines, including HCT116, Caco-2, and SW480 cells leading to demethylation. The demethylation of promoters for cyclin-dependent kinase inhibitor 2A (*CDKN2A*) and secreted frizzled-related protein 2 (*SFRP2*), in addition to *SFRP5* and Wnt inhibitory factor 1 (*WIF1*), which are upstream of the *Wnt* signaling pathway, results in the increased mRNA expression of these genes. The mRNA expression genes downstream of the *Wnt* pathway, including *β-catenin* and *c-Myc*, also decreased, resulting in reduced cell proliferation and increased apoptosis [29]. Resveratrol, an active ingredient found in red grapes, peanuts, and berries, has been linked to health and disease prevention due to reported cardioprotective, antioxidative, anti-inflammatory, and anticancer activities [26]. Resveratrol decreases *DNMT1* and *DNMT3B* expression levels and modulates the aberrant expression profiles of microRNAs (miRNAs) in cancer cell lines and tumour tissues [30].

Trans-resveratrol can reduce the occurrence of Ras association domain family 1 isoform A (*RASSF1A*) hypermethylation by inhibiting *DNMT* activity in women with increased breast cancer risk [31,32]. Pterostilbene is a dimethyl ether derivative of resveratrol found in blueberries, grapes, and herbs, such as *Pterocarpus indicus*. By suppressing the activities of Sirtuin 1 (*SIRT1*) and *DNMT*s, resveratrol and pterostilbene can synergistically inhibit cell viability, induce apoptosis, and arrest the cell cycle in triple-negative breast cancer cells, inhibiting telomerase activity and histone *H2AX* expression [3,5]. In summary, DNMT and HDAC inhibitors target several pathways that contribute to cancer development [33].

Overall, it can be assumed that the increase in carcinogenic DMBA-induced expression of *HDAC1*, *HDAC2*, *DNMT1*, *DNMT3A* and *DNMT3B* genes tested was reduced in most cases by the fruit peel and seed extract tested. This finding is significant because normal dietary conditions were modelled in our studies. However, it provides the amount that already has a significant protective effect against many diseases [34,35].

5. Conclusions

The results of this study indicated that fruit and peel extract administered to mice at human-equivalent doses protected against the DMBA-induced upregulation of *DNMT* and *HDAC* genes, which was likely due to the chemopreventive effects of the various compounds found in fruit seed and peel extract, which is rich in polyphenols and flavonoids. Consumption of one serving of the fruit seeds and peel extracts tested does not exceed the regular daily intake of polyphenols, trans-resveratrol and trans-piceid. Although vast amounts of data are available in the literature associated with specific gene expression, relatively few research groups have examined the epigenetic effects of natural extracts in normal dietary intake level in animal models. Our results suggest that an extract derived from fruit peels and seeds provides effective protection against early gene expression changes of genes involved in epigenetic processes induced by the carcinogen DMBA.

Supplementary Materials: The following supporting information can be downloaded at: https://www.mdpi.com/article/10.3390/nu15081867/s1, Dataset: Original measurement data and calculations.

Author Contributions: Conceptualization, G.N., A.Z. and T.V.; methodology, G.N., A.Z. and T.V.; validation, L.B.R.; formal analysis, I.K. and Z.O.; investigation, G.N., A.Z., L.B.R., L.S., A.T. and R.M.; resources, T.V.; data curation, T.V.; writing—original draft preparation, G.N., A.Z. and T.V.; writing—review and editing, T.V. and Z.G.; visualization, Z.O. and G.G.; supervision, T.V.; project administration, L.B.R.; funding acquisition, T.V. All authors have read and agreed to the published version of the manuscript.

Funding: This research was funded by University of Pécs, EFOP 3.6.1-16.2016.00004. The APC was funded by University of Pécs.

Institutional Review Board Statement: The animal study protocol was approved by the Animal Welfare Committee of the University of Pécs issued the ethical license No. BA02/2000-79/2017.

Informed Consent Statement: Not applicable.

Data Availability Statement: The data presented in this study are available in this article and in the Supplementary Materials.

Acknowledgments: The authors would like to thank Csilla Egyed for editorial support with the manuscript and Zsuzsanna Brunnerné Bayer and Mónika Herczeg for laboratory work.

Conflicts of Interest: The authors declare no conflict of interest.

References

1. Daniel, M.; Tollefsbol, T.O. Epigenetic linkage of aging, cancer and nutrition. *J. Exp. Biol.* **2015**, *218*, 59–70. [CrossRef] [PubMed]
2. Aldawsari, F.S.; Aguayo-Ortiz, R.; Kapilashrami, K.; Yoo, J.; Luo, M.; Medina-Franco, J.L.; Velázquez-Martínez, C.A. Resveratrol-salicylate derivatives as selective DNMT3 inhibitors and anticancer agents. *J. Enzym. Inhib. Med. Chem.* **2016**, *31*, 695–703. [CrossRef] [PubMed]
3. Huang, Z.; Huang, Q.; Ji, L.; Wang, Y.; Qi, X.; Liu, L.; Liu, Z.; Lu, L. Epigenetic regulation of active Chinese herbal components for cancer prevention and treatment: A follow-up review. *Pharmacol. Res.* **2016**, *114*, 1–12. [CrossRef] [PubMed]
4. Shen, L.; Orillion, A.; Pili, R. Histone deacetylase inhibitors as immunomodulators in cancer therapeutics. *Epigenomic* **2016**, *8*, 415–428. [CrossRef]
5. Kala, R.; Shah, H.N.; Martin, S.L.; Tollefsbol, T.O. Epigenetic-based combinatorial resveratrol and pterostilbene alters DNA damage response by affecting SIRT1 and DNMT enzyme expression, including SIRT1-dependent γ-H2AX and telomerase regulation in triple-negative breast cancer. *BMC Cancer* **2015**, *15*, 672. [CrossRef]
6. Singh, A.K.; Bishayee, A.; Pandey, A.K. Targeting histone deacetylases with natural and synthetic agents: An emerging anticancer strategy. *Nutrients* **2018**, *10*, 731. [CrossRef]
7. Arimoto-Kobayashi, S.; Sasaki, K.; Hida, R.; Miyake, N.; Fujii, N.; Saiki, Y.; Daimaru, K.; Nakashima, H.; Kubo, T.; Kiura, K. Chemopreventive effects and anti-tumorigenic mechanisms of 2, 6-dimethoxy-1, 4-benzoquinone, a constituent of Vitis coignetiae Pulliat (crimson glory vine, known as yamabudo in Japan), toward 4-(methylnitrosamino)-1-(3-pyridyl)-1-butanone (NNK)-induced lung tumorigenesis in A/J mice. *Food Chem. Toxicol.* **2021**, *154*, 112319. [CrossRef]
8. Kim, H.G.; Sung, N.Y.; Kim, J.H.; Cho, J.Y. In vitro anti-cancer effects of beauvericin through inhibition of actin polymerization and Src phosphorylation. *Phytomedicine* **2023**, *109*, 154573. [CrossRef]
9. Huang, C.F.; Hung, T.W.; Yang, S.F.; Tsai, Y.L.; Yang, J.T.; Lin, C.L.; Hsieh, Y.H. Asiatic acid from centella asiatica exert anti-invasive ability in human renal cancer cells by modulation of ERK/p38MAPK-mediated MMP15 expression. *Phytomedicine* **2022**, *100*, 154036. [CrossRef]
10. Peng, Y.; Zhou, T.; Wang, S.; Bahetjan, Y.; Li, X.; Yang, X. Dehydrocostus lactone inhibits the proliferation of esophageal cancer cells in vivo and in vitro through ROS-mediated apoptosis and autophagy. *Food Chem. Toxicol.* **2022**, *170*, 113453. [CrossRef]
11. Zhou, Y.; Yu, Y.; Lv, H.; Zhang, H.; Liang, T.; Zhou, G.; Huang, L.; Tian, Y.; Liang, W. Apigenin in cancer therapy: From mechanism of action to nano-therapeutic agent. *Food Chem. Toxicol.* **2022**, *168*, 113385. [CrossRef] [PubMed]
12. Yang, A.Y.; Lee, J.H.; Shu, L.; Zhang, C.; Su, Z.-Y.; Lu, Y.; Huang, M.-T.; Ramirez, C.; Pung, D.; Huang, Y.; et al. Genome-wide analysis of DNA methylation in UVB-and DMBA/TPA-induced mouse skin cancer models. *Life Sci.* **2014**, *113*, 45–54. [CrossRef] [PubMed]
13. Tomesz, A.; Szabo, L.; Molnar, R.; Deutsch, A.; Darago, R.; Raposa, B.L.; Nowrasteh, G.; Varjas, T.; Nemeth, B.; Orsos, Z.; et al. Changes in miR-124-1, miR-212, miR-132, miR-134, and miR-155 expression patterns after 7, 12-dimethylbenz (a) anthracene treatment in CBA/Ca mice. *Cells* **2022**, *11*, 1020. [CrossRef]

14. Molnar, R.; Szabo, L.; Tomesz, A.; Deutsch, A.; Darago, R.; Raposa, B.L.; Nowrasteh, G.; Varjas, T.; Nemeth, B.; Orsos, Z.; et al. The chemopreventive effects of polyphenols and coffee, based upon a DMBA mouse model with microRNA and mTOR gene expression biomarkers. *Cells* **2022**, *11*, 1300. [CrossRef]
15. Chen, Y.K.; Hsue, S.S.; Lin, L.M. Survivin expression is regulated by an epigenetic mechanism for DMBA-induced hamster buccal-pouch squamous-cell carcinomas. *Arch. Oral Biol.* **2005**, *50*, 593–598. [CrossRef]
16. Singhal, R.; Shankar, K.; Badger, T.M.; Ronis, M.J. Estrogenic status modulates aryl hydrocarbon receptor—Mediated hepatic gene expression and carcinogenicity. *Carcinogenesis* **2008**, *29*, 227–236. [CrossRef] [PubMed]
17. Larsen, M.C.; Almeldin, A.; Tong, T.; Rondelli, C.M.; Maguire, M.; Jaskula-Sztul, R.; Jefcoate, C.R. Cytochrome P4501B1 in bone marrow is co-expressed with key markers of mesenchymal stem cells. BMS2 cell line models PAH disruption of bone marrow niche development functions. *Toxicol. Appl. Pharm.* **2020**, *401*, 115111. [CrossRef]
18. Szabó, L.; Kiss, I.; Orsós, Z.; Ember, I. Effect of Nano-Fruit-Café on the expression of oncogenes and tumor suppressor genes. *Anticancer Res.* **2008**, *28*, 3274.
19. Reagan-Shaw, S.; Nihal, M.; Ahmad, N. Dose translation from animal to human studies revisited. *FASEB J.* **2007**, *22*, 659–661. [CrossRef]
20. Zamora-Ros, R.; Andres-Lacueva, C.; Lamuela-Raventós, R.M.; Berenguer, T.; Jakszyn, P.; Martínez, C.; Sánchez, M.J.; Navarro, C.; Chirlaque, M.D.; Tormo, M.-J.; et al. Concentrations of resveratrol and derivatives in foods and estimation of dietary intake in a Spanish population: European Prospective Investigation into Cancer and Nutrition (EPIC)-Spain cohort. *Brit. J. Nutr.* **2008**, *100*, 188–196. [CrossRef]
21. Shaito, A.; Posadino, A.M.; Younes, N.; Hasan, H.; Halabi, S.; Alhababi, D.; Al-Mohannadi, A.; Abdel-Rahman, M.W.; Eid, A.H.; Nasrallah, G.K.; et al. Potential adverse effects of resveratrol: A literature review. *Int. J. Mol. Sci.* **2020**, *21*, 2084. [CrossRef] [PubMed]
22. Mirza, S.; Sharma, G.; Parshad, R.; Gupta, S.D.; Pandya, P.; Ralhan, R. Expression of DNA methyltransferases in breast cancer patients and to analyze the effect of natural compounds on DNA methyltransferases and associated proteins. *J. Breast Cancer* **2013**, *16*, 23–31. [CrossRef] [PubMed]
23. Fan, J.; Yin, W.-J.; Lu, J.-S.; Wang, L.; Wu, J.; Wu, F.-Y.; Di, G.-H.; Shen, Z.-Z.; Shao, Z.-M. ERα negative breast cancer cells restore response to endocrine therapy by combination treatment with both HDAC inhibitor and DNMT inhibitor. *J. Cancer Res. Clin.* **2008**, *134*, 883–890. [CrossRef]
24. Kim, H.; Ramirez, C.N.; Su, Z.-Y.; Kong, A.-N.T. Epigenetic modifications of triterpenoid ursolic acid in activating Nrf2 and blocking cellular transformation of mouse epidermal cells. *J. Nutr. Biochem.* **2016**, *33*, 54–62. [CrossRef]
25. Kuo, H.-C.D.; Wu, R.; Li, S.; Yang, A.Y.; Kong, A.-N. Anthocyanin Delphinidin Prevents Neoplastic Transformation of Mouse Skin JB6 P+ Cells: Epigenetic Re-activation of Nrf2-ARE Pathway. *AAPS J.* **2019**, *21*, 83. [CrossRef] [PubMed]
26. Kumar, A.; Dhar, S.; Rimando, A.M.; Lage, J.M.; Lewin, J.R.; Zhang, X.; Levenson, A.S. Epigenetic potential of resveratrol and analogs in preclinical models of prostate cancer. *Ann. N. Y. Acad. Sci.* **2015**, *1348*, 1–9. [CrossRef] [PubMed]
27. Basse, C.; Arock, M. The increasing roles of epigenetics in breast cancer: Implications for pathogenicity, biomarkers, prevention, and treatment. *Int. J. Cancer* **2015**, *137*, 2785–2794. [CrossRef]
28. Olguín-Reyes, S.R.; Hernández-Ojeda, S.L.; Govezensky, T.; Camacho-Carranza, R.; Espinosa-Aguirre, J.J. Sub-acute exposure effect of selected polycyclic aromatic hydrocarbons on protein levels of epigenetic modifiers in non-cancerous hepatic model. *Biomed. Res.* **2018**, *29*, 342. [CrossRef]
29. Wang, L.-S.; Kuo, C.-T.; Cho, S.-J.; Seguin, C.; Siddiqui, J.; Stoner, K.; Weng, Y.-I.; Huang, T.H.-M.; Tichelaar, J.; Yearsley, M.; et al. Black raspberry-derived anthocyanins demethylate tumor suppressor genes through the inhibition of DNMT1 and DNMT3B in colon cancer cells. *Nutr. Cancer* **2013**, *65*, 118–125. [CrossRef]
30. Qin, W.; Zhang, K.; Clarke, K.; Weiland, T.; Sauter, E.R. Methylation and miRNA effects of resveratrol on mammary tumors vs. normal tissue. *Nutr. Cancer* **2014**, *66*, 270–277. [CrossRef]
31. Gerhauser, C. Cancer chemoprevention and nutri-epigenetics: State of the art and future challenges. *Top. Curr. Chem.* **2013**, *329*, 73–132. [CrossRef] [PubMed]
32. Zhu, W.; Qin, W.; Zhang, K.; Rottinghaus, G.E.; Chen, Y.-C.; Kliethermes, B.; Sauter, E.R. Trans-resveratrol alters mammary promoter hypermethylation in women at increased risk for breast cancer. *Nutr. Cancer* **2012**, *64*, 393–400. [CrossRef] [PubMed]
33. Yuan, Z.; Chen, S.; Gao, C.; Dai, Q.; Zhang, C.; Sun, Q.; Lin, J.-S.; Guo, C.; Chen, Y.; Jiang, Y. Development of a versatile DNMT and HDAC inhibitor C02S modulating multiple cancer hallmarks for breast cancer therapy. *Bioorg. Chem.* **2019**, *87*, 200–208. [CrossRef]
34. Del Bo, C.; Bernardi, S.; Marino, M.; Porrini, M.; Tucci, M.; Guglielmetti, S.; Cherubini, A.; Carrieri, B.; Kirkup, B.; Kroon, P.; et al. Systematic review on polyphenol intake and health outcomes: Is there sufficient evidence to define a health-promoting polyphenol-rich dietary pattern? *Nutrients* **2019**, *11*, 1355. [CrossRef] [PubMed]
35. El Khawand, T.; Courtois, A.; Valls, J.; Richard, T.; Krisa, S. A review of dietary stilbenes: Sources and bioavailability. *Phytochem. Rev.* **2018**, *17*, 1007–1029. [CrossRef]

nutrients

MDPI

Article

Role of Zerumbone, a Phytochemical Sesquiterpenoid from *Zingiber zerumbet Smith*, in Maintaining Macrophage Polarization and Redox Homeostasis

Wei-Lan Yeh [1,2,†], Bor-Ren Huang [3,4,†], Guan-Wei Chen [2], Vichuda Charoensaensuk [5,6], Cheng-Fang Tsai [7], Liang-Yo Yang [6,8], Dah-Yuu Lu [5,9], Mao-Kai Chen [10] and Chingju Lin [6,*]

1 Department of Biochemistry, School of Medicine, China Medical University, Taichung 40402, Taiwan
2 Institute of Translational Medicine and New Drug Development, China Medical University, Taichung 40402, Taiwan
3 Department of Neurosurgery, Taichung Tzu Chi Hospital, Buddhist Tzu Chi Medical Foundation, Taichung 427213, Taiwan
4 School of Medicine, Tzu Chi University, Hualien 97004, Taiwan
5 Department of Pharmacology, School of Medicine, College of Medicine, China Medical University, Taichung 40402, Taiwan
6 Department of Physiology, School of Medicine, China Medical University, Taichung 40402, Taiwan
7 Department of Medical Laboratory Science and Biotechnology, Asia University, Taichung 41354, Taiwan
8 Laboratory for Neural Repair, China Medical University Hospital, Taichung 404327, Taiwan
9 Department of Photonics and Communication Engineering, Asia University, Taichung 41354, Taiwan
10 Department of Anesthesiology, China Medical University Hospital, Taichung 404327, Taiwan
* Correspondence: clin33@mail.cmu.edu.tw
† These authors contributed equally to this work.

Citation: Yeh, W.-L.; Huang, B.-R.; Chen, G.-W.; Charoensaensuk, V.; Tsai, C.-F.; Yang, L.-Y.; Lu, D.-Y.; Chen, M.-K.; Lin, C. Role of Zerumbone, a Phytochemical Sesquiterpenoid from *Zingiber zerumbet Smith*, in Maintaining Macrophage Polarization and Redox Homeostasis. *Nutrients* **2022**, *14*, 5402. https://doi.org/10.3390/nu14245402

Academic Editors: Haixia Yang and Jianjun Deng

Received: 2 November 2022
Accepted: 15 December 2022
Published: 19 December 2022

Abstract: Macrophages and microglia are highly versatile cells that can be polarized into M1 and M2 phenotypes in response to diverse environmental stimuli, thus exhibiting different biological functions. In the central nervous system, activated resident macrophages and microglial cells trigger the production of proinflammatory mediators that contribute to neurodegenerative diseases and psychiatric disorders. Therefore, modulating the activation of macrophages and microglia by optimizing the inflammatory environment is beneficial for disease management. Several naturally occurring compounds have been reported to have anti-inflammatory and neuroprotective properties. Zerumbone is a phytochemical sesquiterpenoid and also a cyclic ketone isolated from Zingiber zerumbet Smith. In this study, we found that zerumbone effectively reduced the expression of lipocalin-2 in macrophages and microglial cell lines. Lipocalin-2, also known as neutrophil gelatinase-associated lipocalin (NGAL), has been characterized as an adipokine/cytokine implicated in inflammation. Moreover, supplement with zerumbone inhibited reactive oxygen species production. Phagocytic activity was decreased following the zerumbone supplement. In addition, the zerumbone supplement remarkably reduced the production of M1-polarization-associated chemokines CXC10 and CCL-2, as well as M1-polarization-associated cytokines interleukin (IL)-6, IL-1β, and tumor necrosis factor-α. Furthermore, the expression of inducible nitric oxide synthase (iNOS) and cyclooxygenase-2 and the production of NO were attenuated in macrophages and microglial cells supplemented with zerumbone. Notably, we discovered that zerumbone effectively promoted the production of the endogenous antioxidants heme oxygenase-1, glutamate–cysteine ligase modifier subunit, glutamate–cysteine ligase catalytic subunit, and NAD(P)H quinone oxidoreductase-1 and remarkably enhanced IL-10, a marker of M2 macrophage polarization. Endogenous antioxidant production and M2 macrophage polarization were increased through activation of the AMPK/Akt and Akt/GSK3 signaling pathways. In summary, this study demonstrated the protective role of zerumbone in maintaining M1 and M2 polarization homeostasis by decreasing inflammatory responses and enhancing the production of endogenous antioxidants in both macrophages and microglia cells. This study suggests that zerumbone can be used as a potential therapeutic drug for the supplement of neuroinflammatory diseases.

Keywords: zerumbone; microglial cells; macrophage polarization; neuroinflammation; redox homeostasis

1. Introduction

Macrophages are key immune cells that maintain homeostasis and defense during periods of both good health and disease by regulating the onset and resolution of inflammation [1]. Tissue macrophages reside in almost every part of the human body, including the brain. Resident microglia in the central nervous system (CNS) are local phagocytic cells that mediate immunological and inflammatory reactions in response to pattern-associated molecular patterns (PAMPs) and damage-associated molecular patterns (DAMPs) in the environment [2]. Several environmental factors alter the phenotype of macrophages, thereby affecting their functions. Macrophages can be polarized into M1-like and M2-like phenotypes [3]. Typically, microglia in healthy brain tissues are in a quiescent M2 phenotype, and these microglia are critical for the maintenance of neuron–microglia interactions and neuronal development [4]. M1 macrophages are activated by bacterial lipopolysaccharide (LPS) and proinflammatory cytokines, including tumor necrosis factor (TNF) and interferon (IFN). Activated M1 macrophages overexpress inducible nitric oxide synthase (iNOS), nitric oxide (NO), and reactive oxygen species (ROS) and upregulate proinflammatory mediators such as interleukin (IL)-1β, IL-6, and TNF-α [5–7]. ROS, which are produced in response to oxidative and environmental stress, cause the activation of microglia [2]. Activated M1 microglia produce various proinflammatory mediators and free radicals that inhibit brain repair and regeneration, leading to neuroinflammation, neurodegenerative diseases, and psychiatric disorders [8]. Moreover, M1 macrophages produce several chemokines, such as the C–X–C motif chemokine ligand (CXCL)-10 [9] and C–C motif chemokine ligand (CCL)-2 [10], which trigger the activation of type 1 T helper (Th1) response, accelerate phagocytic activity, and promote inflammation [6]. By contrast, M2 macrophages are activated in response to the Th2 response. Upregulation of arginase-1 (Arg-1) and IL-10 in M2 macrophages promotes cell proliferation, tissue repair, and anti-inflammatory cytokines, thereby counteracting the inflammation triggered by activated M1 macrophages [11]. M2 microglia increase brain repair and regeneration by promoting phagocytosis, producing endogenous trophic factors, and alleviating brain inflammation [1]. However, disruption in the homeostasis of M1 versus M2 phenotypes results in the development of several diseases, including obesity, atherosclerosis, and insulin resistance [12].

Generally, the proportion of M1 and M2 macrophages is tightly controlled in healthy tissues [5]. Substantial evidence has been obtained indicating that the modulation of macrophage polarization plays a crucial role in the pathology of several diseases, including obesity [13], atherosclerosis [14], and cancers [15]. According to a study by Jiang et al., spinal cord injury (SCI) induced the expression of M1 phenotypic markers (CD86, iNOS, IL-6, and TNF-α) and decreased the expression of M2 phenotypic markers (CD206, IL-10, and Arg-1) [16]. In addition, treatment with substance P improved recovery from SCI by inducing the production of endogenous anti-inflammatory mediators [16]. Our previous study showed that treatment with paliperidone effectively decreased the expression of an M2 phenotype marker (CD206) while increasing that of an M1 phenotype marker (CD80), resulting in the inhibition of glioblastoma and suggesting that regulation of macrophage polarization is a potential treatment strategy for certain diseases [17]. Notably, our recent findings indicate that inhibiting lipocalin-2 expression in macrophages and microglial cells may be a novel strategy for the treatment of neuroinflammation and neurodegenerative diseases [18]. Lipocalin-2 has been characterized as an adipokine/cytokine and was found to be associated with several cellular processes, including cell survival, death, differentiation, invasion, migration, inflammatory response, iron homeostasis, insulin resistance, and tissue regeneration [19]. An increased level of lipocalin-2 expression was correlated with acute and chronic liver injury [20]. Following acute inflammation, the liver overexpresses lipocalin-2, triggering inflammatory cell infiltration for phagocytosis and ensuring homeostasis [21]. One study discovered that lipocalin-2 promoted microglial M1 polarization, resulting in impairment of cognitive function and motor behavior due to neuroinflammation [22]. A recent study reported that the neutralization of lipocalin-2 diminishes the severity of brain injury caused by ischemia reperfusion [23]. Moreover, lipocalin-2-deficient mice exhibited a weaker M1 phenotype with an increase in the strength of the M2 phenotype, suggesting that lipocalin-2 plays a critical role in microglial polarization [22].

Zerumbone is a dietary compound presented in a variety of natural foods. It is naturally occurring in plants of the *Zingiberaceae* and *Curcuma* families, notably *Zingiber zerumbet* Smith, and features as a monocyclic sesquiterpene phytochemical [24]. Zerumbone has been reported to possess diverse biological activities, including activities against microbes, osteoporosis, prostatic hyperplasia, and polycystic ovary syndrome [24,25]. The safety, cytotoxicity, and chemopreventive potential of zerumbone have been reported [26–28]. Zerumbone was also reported to possess anti-inflammatory effects against acute and chronic inflammation of granulomatous tissue in mice [29]. In addition, oral administration of zerumbone did not result in any clinical abnormalities or other adverse effects in one study [30]. Zerumbone has been discovered to possess anti-inflammatory and antioxidant activities in various inflammation-related diseases [31,32]. Additionally, zerumbone was found to be beneficial for the treatment of learning and memory impairment in an animal model [33]. A recent study suggested that zerumbone decreased proinflammatory cytokine expression, β-amyloid production, and behavioral deficits in APP/PS1 transgenic mice [34]. However, the effects (and underlying mechanisms) of zerumbone on lipocalin 2 expression and macrophage polarization, as well as the generation of endogenous antioxidant enzymes and anti-inflammatory proteins in the CNS, remain poorly understood.

This study aimed to elucidate the regulatory effects of zerumbone on homeostasis and M1/M2 macrophage polarization by considering lipocalin-2 expression, oxidation/antioxidation effects, and inflammatory/anti-inflammatory effects in macrophages and microglial cells. Furthermore, this study investigated the effect of zerumbone on the activity of endogenous antioxidants and anti-inflammatory proteins in macrophages and microglial cells. In summary, this study suggests that zerumbone may be a potential supplement for inflammatory diseases and neurodegenerative diseases.

2. Materials and Methods

2.1. Materials

Primary antibodies against GSK3α/β, β-actin, and phosphor-Akt^{Ser743} were obtained from Santa Cruz Biotechnology (Santa Cruz, CA, USA). Antibodies against phosphor-AMPK^{Thr172} and phosphor-GSK3α/β$^{Ser21/Ser9}$ were purchased from Cell Signaling Technology. Anti-iNOS antibody (610431) was acquired from BD Transduction Laboratories (Lexington, KY, USA). Cyclooxygenase (COX)-2 polyclonal antibody (aa 570–598) was purchased from Cayman Chemicals (Ann Arbor, MI, USA). Heme oxygenase (HO)-1 polyclonal antibody was obtained from Enzo Life Sciences Inc. (Farmingdale, NY, USA). Antibodies against glyceraldehyde-3-phosphate dehydrogenase (GAPDH; GCLC, GCLM, and NQO1) were acquired from Abcam (Cambridge, MA, USA).

2.2. Cell Culture

In a humidified incubator containing 5% CO_2 and 95% air at 37 °C, mouse macrophages RAW264.7 cells were cultured in high glucose Dulbecco's modified Eagle's medium (DMEM), 10% fetal bovine serum (FBS), and 100 U/mL penicillin/streptomycin. The adult mouse microglia (IMG) was obtained from the Harvard School of Public Health (Boston, MA, USA). IMG cells expressing a microglial-specific marker represent brain microglia features morphologically and functionally. The IMG cells were cultured in DMEM with low glucose content (1 g/L), 10% FBS, and 100 U/mL penicillin/streptomycin.

2.3. Western Blotting Analysis

The cells were lysed on ice for 30 min with radioimmunoprecipitation assay buffer containing a protease inhibitor cocktail. The supernatant was collected after centrifugation, and proteins in the supernatant were separated using sodium dodecyl sulfate–polyacrylamide gel electrophoresis. The blots were transferred onto polyvinylidene fluoride membranes. After being blocked with nonfat milk, the membranes were probed with primary antibodies and secondary antibodies. Proteins were visualized through enhanced chemiluminescence using Kodak X-OMAT LS film (Eastman Kodak, Rochester, NY, USA). The densitometric values were quantified by ImageJ software.

2.4. NO Assay

The NO assay method is described in our previous publication [35]. Briefly, culture supernatant containing nitrite was reacted for 10 min with 0.1% NED solution and 1% sulfanilamide in 5% phosphoric acid avoiding light. NO was quantified by measuring the amount of nitrite under OD 520 nm using a microplate reader.

2.5. Quantitative Real-Time Polymerase Chain Reaction (PCR)

mRNA levels were detected using quantitative real-time PCR, and total RNA was extracted using TRIzol reagent (Invitrogen, Carlsbad, CA, USA). An amount of 2 μg of total RNA was used for reverse transcription (RT) by using an RT Kit (Invitrogen, Carlsbad, CA, USA). SYBR Green Master Mixes (Applied Biosystems, Waltham, MA, USA) was used for conducting PCR. To calculate the transcripts cycle (denoted CT), the threshold was set within the linear phase of gene amplification.

2.6. Phagocytosis Assay

The phagocytosis assay method was performed in accordance with the method in our previous study [35]. The cells were seeded onto culture dishes and grown at 37 °C and 5% CO_2. After drug treatment, the medium was replaced with medium containing carboxylate-modified polystyrene fluorescent yellow–green latex beads (YG beads; Cat#L4655; Sigma Aldrich, St. Louis, MA, USA), and the cells were incubated at 37 °C. The cells were trypsinized after several washes to remove the noninternalized beads, and their phagocytic activity was quantified using flow cytometry.

2.7. Statistical Analysis

GraphPad Prism 6.0 (Graph Pad Software, San Diego, CA, USA) was used for statistical analysis. Values are presented as the mean ± standard error of the mean (SEM). Significance of the differences between the groups was analyzed by Student's *t*-test. One-way analysis of variance (ANOVA) with the Bonferroni post hoc test was used for comparisons of more than two groups. A $p < 0.05$ was considered significant.

3. Results

3.1. Zerumbone Lowers the Expression of Lipocalin-2 in Macrophages and Microglial Cells

RAW264.7 mouse macrophages (Figure S1) and IMG adult mouse microglia (Figure S2) were supplemented with zerumbone (1, 5, or 10 μM), and no toxicity was then observed. As shown in Figure 1, supplement with zerumbone alone did not affect the expression of lipocalin-2 in either cell model. Application of LPS resulted in significantly increased lipocalin-2 expression in both the macrophages (Figure 1A) and microglia (Figure 1B). Moreover, the zerumbone supplement effectively decreased LPS-stimulated lipocalin-2 expression in a concentration-dependent manner for macrophages (Figure 1A) and microglia (Figure 1B), with a 40% and 75% reduction under the maximum concentration of zerumbone.

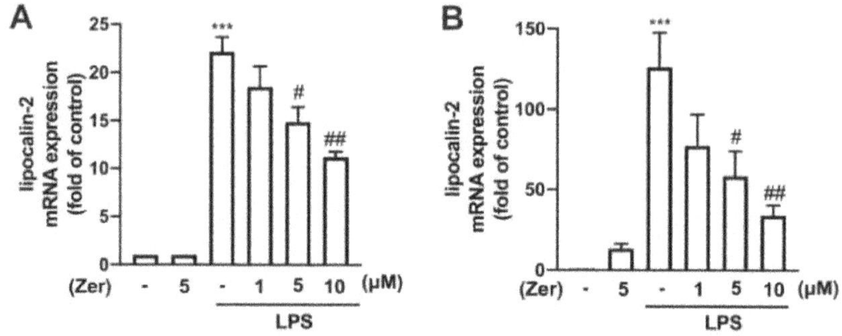

Figure 1. Inhibitory effects of zerumbone on the expression of lipocalin-2 in macrophages and microglia. RAW264.7 macrophages (**A**) and IMG adult mouse microglial cell lines (**B**) were supplemented with different concentrations of zerumbone (1, 5, or 10 μM) for 30 min and administered with lipopolysaccharide (LPS; 50 ng/mL) for 6 h. Lipocalin-2 mRNA expression levels were determined using real-time PCR and normalized to β-actin. Data are presented as the mean ± standard error of the mean (SEM) (n = 3 or 4). *** $p < 0.005$ compared with the control group. # $p < 0.05$, ## $p < 0.01$ compared with the LPS alone group.

3.2. Supplement with Zerumbone Decreases H_2O_2, ROO•, and HO• Production in Microglial Cells

The microglial cells were treated with either hydrogen peroxide (H_2O_2), 2, 2′-azobis (2-amidinopropane) hydrochloride (AAPH), or iron (II) plus H_2O_2 to stimulate the production of various ROS. Then, the effects of zerumbone on ROS production were determined. As illustrated in Figure 2, supplement with zerumbone alone did not influence ROS production. H_2O_2, AAPH, and iron resulted in ROS levels in microglial cells that were approximately four- to six-fold higher than those in the control group. Notably, zerumbone decreased H_2O_2 production in a concentration-dependent manner (Figure 2A). Supplement with zerumbone further decreased AAPH-induced peroxyl radical (ROO•) production (Figure 2B). Moreover, hydroxyl radical (HO•) production stimulated by iron (II) and H_2O_2 following the zerumbone supplement was markedly decreased in a concentration-dependent manner (Figure 2C). This study suggests that supplements with zerumbone concentration-dependently inhibited H_2O_2, ROO•, and HO• production in microglial cells.

3.3. Inhibitory Effect of Zerumbone against Phagocytic Activity in Microglial Cells

We further investigated the effect of zerumbone on phagocytosis in microglial cells. The results revealed that the nonphagocytic populations were remarkably smaller in the LPS-activated microglial cells than in the non-LPS-activated microglial cells (Figure 3). However, the phagocytic populations that engulfed two or more beads were larger. Furthermore, supplement with zerumbone alone did not change the ability of phagocytosis of microglial cells either in one or in two or more beads (Figure 3A upper-left panel, B). Notably, 1 μM zerumbone slightly decreased LPS-stimulated phagocytosis in microglial cells (Figure 3A upper-right panel, B). In addition, zerumbone at higher concentrations (5 and 10 μM) effectively decreased the phagocytic populations engulfing two or more beads in the LPS-stimulated microglial cells (Figure 3A lower panel, B). These results confirm that supplement with zerumbone alone did not affect the ability of phagocytosis. Moreover, zerumbone effectively inhibited LPS-stimulated microglial phagocytosis.

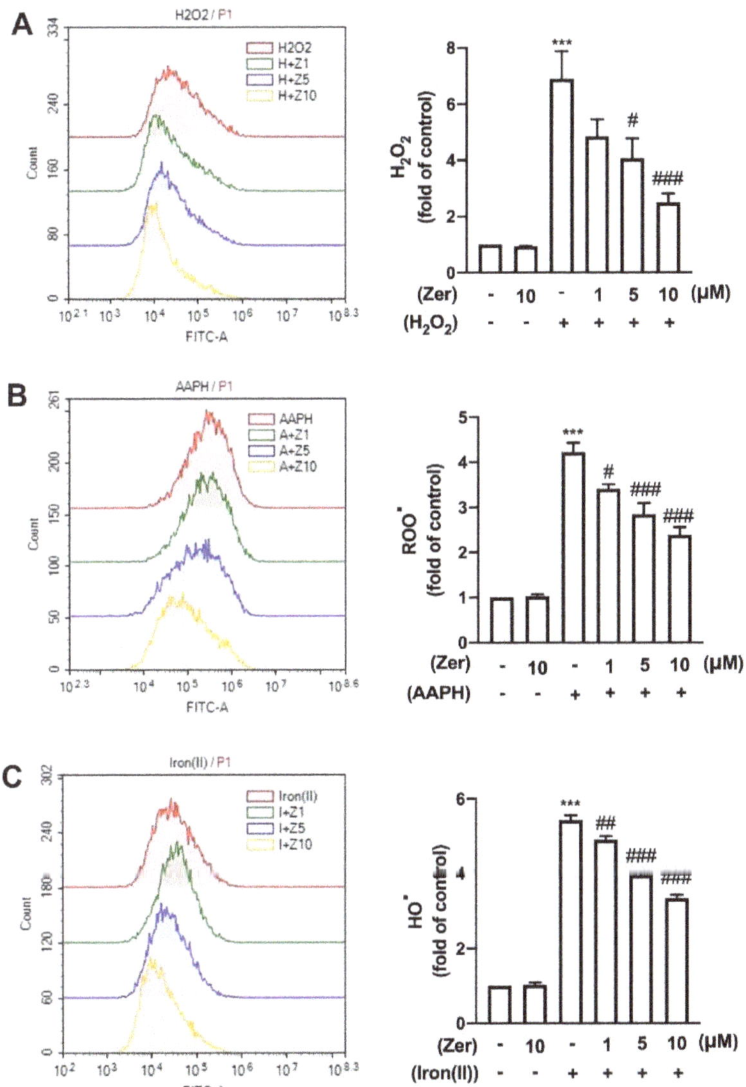

Figure 2. Effects of zerumbone on ROS production in microglia. IMG cells were supplemented with various concentrations of zerumbone (1, 5, or 10 μM) for 30 min, followed by 5 mM H_2O_2 (**A**), 5 mM AAPH (**B**), or 1 mM iron (II) with 0.5 mM H_2O_2 (**C**) for another 90 min. The intensity of dichlorofluorescein (DCF) fluorescence was detected through flow cytometry after 40 min of incubation with 10 μM dichloro-dihydro-fluorescein diacetate (DCFH-DA). Quantitative data are represented as the mean ± SEM ($n = 4$). *** $p < 0.005$ compared with the control group. # $p < 0.05$, ## $p < 0.01$, ### $p < 0.005$ compared with the treatment group alone.

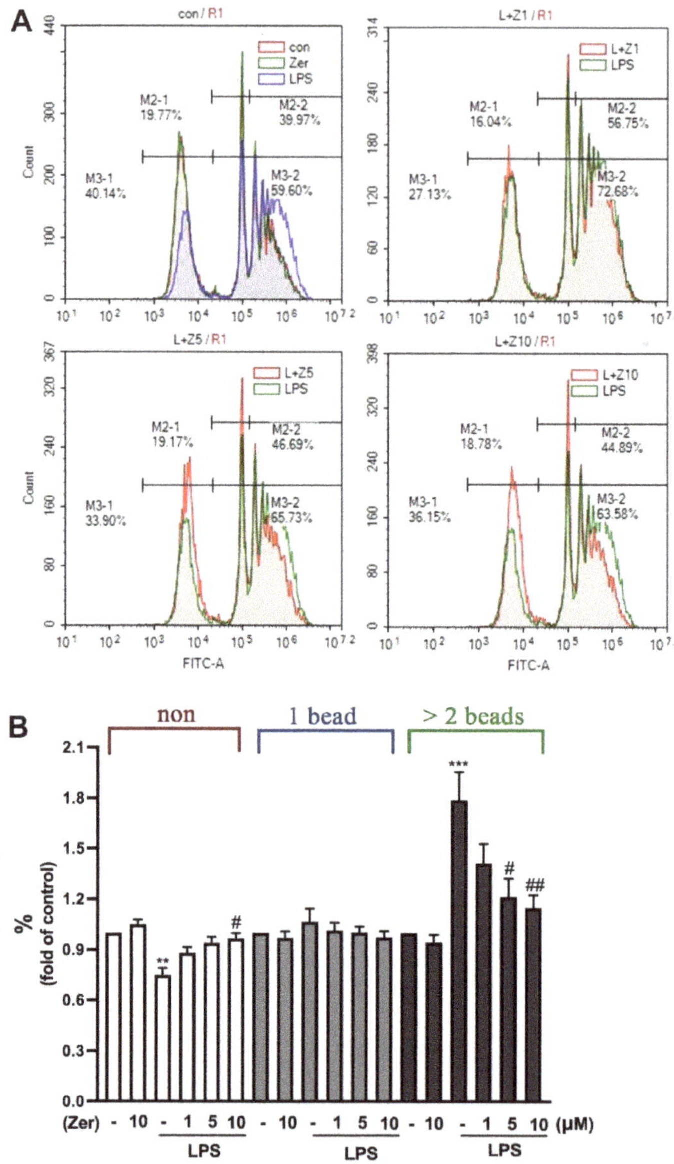

Figure 3. Effect of zerumbone on phagocytic ability in microglia. (**A**) IMG cells were presupplemented with different concentrations of zerumbone (1, 5, or 10 μM) for 30 min and LPS (50 ng/mL) for another 24 h. After incubation of the cells with 1 μm fluorescent YG beads for 1 h at 37 °C, the intensity of the beads was analyzed using flow cytometry. The quantitative results shown in (**B**) are the mean ± SEM (n = 4): non, no bead was uptaken by cell; 1 bead, cell uptake 1 bead; >2 beads, cell uptake more than 2 beads. ** $p < 0.01$, *** $p < 0.005$ compared with the control group. # $p < 0.05$, ## $p < 0.01$ compared with the LPS alone group.

3.4. Zerumbone Reduces the Expression of Proinflammatory Mediators and M1-Macrophage Polarization Markers in Macrophages and Microglial Cells

We investigated the effect of zerumbone on the LPS-stimulated expression of proinflammatory mediators associated with M1-like macrophage/microglia polarization. The mRNA expression of CXCL-10 and CCL-2 was elevated in the mouse macrophages (Figure 4A,B) and IMG cells (Figure 4C,D) following LPS stimulation. Furthermore, zerumbone concentration-dependently decreased the LPS-induced increased expression of CXCL-10 (Figure 4A,C) and CCL-2 (Figure 4B,D) in both cell models.

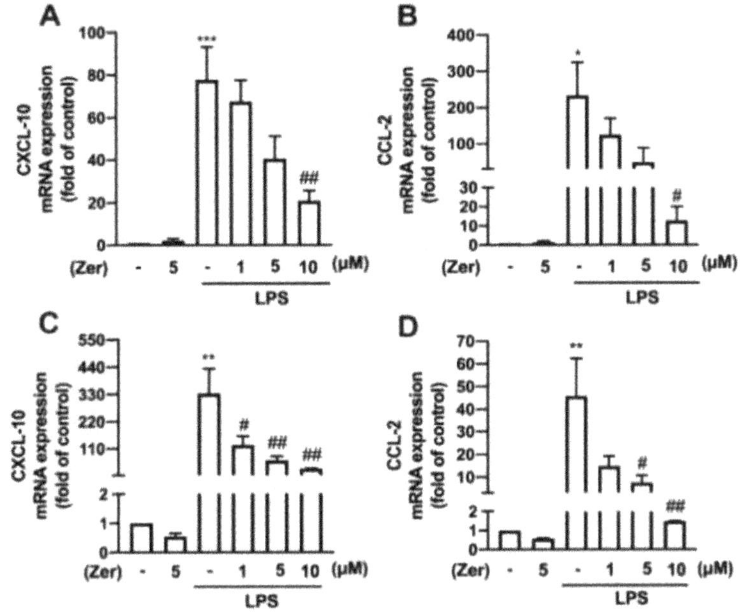

Figure 4. The expression of proinflammatory mediators in macrophages and microglia in response to zerumbone. RAW264.7 (**A**,**B**) and IMG (**C**,**D**) cells were supplemented with different concentrations of zerumbone (1, 5, or 10 μM) for 30 min and then activated by LPS (50 ng/mL) for another 6 h. CXCL-10 (**A**,**C**) and CCL-2 (**B**,**D**) mRNA expression was analyzed using real-time PCR and normalized to β-actin. Data are presented as the mean ± SEM ($n = 3$). * $p < 0.05$, ** $p < 0.01$, *** $p < 0.005$ compared with the control group. # $p < 0.05$, ## $p < 0.01$ compared with the LPS alone.

Additionally, zerumbone considerably and concentration-dependently reduced the LPS-induced upregulation of M1 polarization markers such as IL-6 (Figure 5A), IL-1β (Figure 5B), and TNF-α (Figure 5C) in macrophages. We further observed similar inhibitory effects of zerumbone on LPS-induced IL-6 (Figure 6A), IL-1β (Figure 6B), and TNF-α (Figure 6C) in microglial cells. Moreover, supplement with zerumbone attenuated LPS-stimulated expression of iNOS (Figure 5D,E) and COX-2 (Figure 5D,F) proteins in macrophages in a concentration-dependent manner. Zerumbone further inhibited LPS-induced NO production in macrophages (Figure 5G). Furthermore, zerumbone effectively reduced the mRNA expression of iNOS (Figure 6D) and COX-2 (Figure 6E) induced by LPS. Supplement with zerumbone attenuated the expression of iNOS (Figure 6F,G) and COX-2 (Figure 6F,H) proteins induced by LPS, as well as NO production, dose-dependently (Figure 6I). We did not observe any change in the expression of proinflammatory mediators in macrophages (Figure 4A,B and Figure 5) or microglia (Figure 4C,D and Figure 6) supplemented with zerumbone alone. The results suggest that supplement with zerumbone reversed LPS-activated macrophage and microglia polarization toward the M1 phenotype.

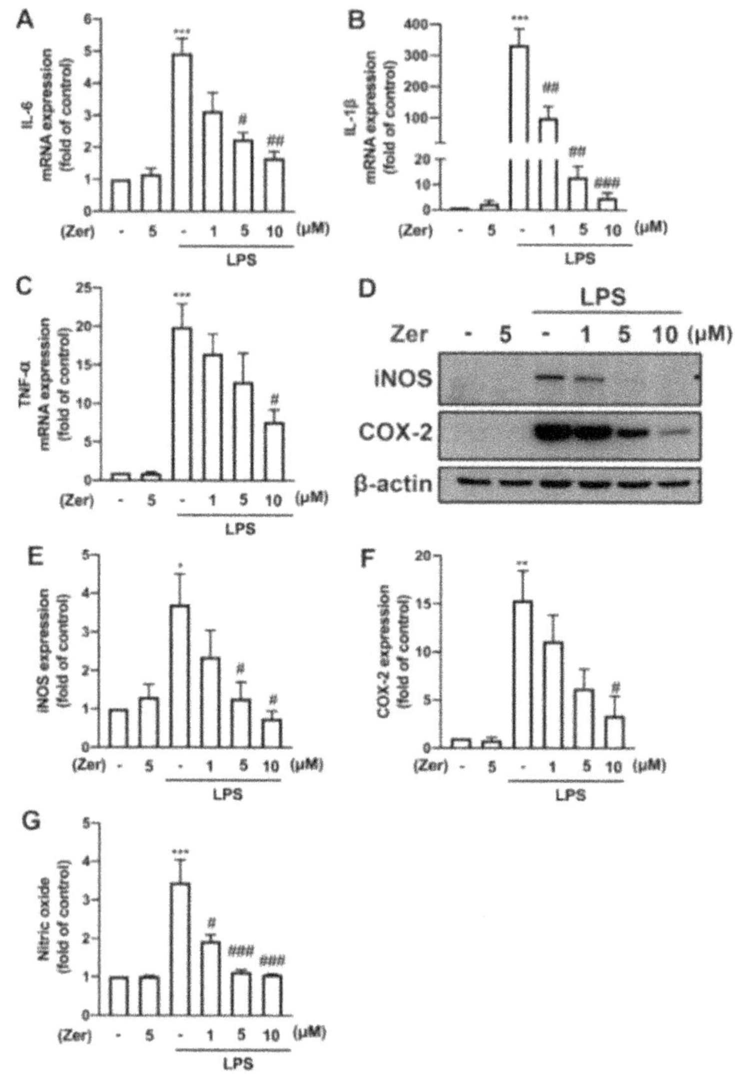

Figure 5. The expression of proinflammatory mediators in response to zerumbone in macrophages. RAW264.7 cells were supplemented with different concentrations of zerumbone (1, 5, or 10 μM) for 30 min and then stimulated with LPS (50 ng/mL) for another 6 h (**A**–**C**) or 24 h (**D**–**G**). Expressions of IL-6 (**A**), IL-1β (**B**), and TNF-α (**C**) mRNA were analyzed using real-time PCR and normalized to β-actin. (**D**) iNOS and COX-2 protein expressions were analyzed using Western blotting. Quantitative results are shown in (**E**,**F**). (**G**) The cultural supernatant was harvested for measuring NO production by NO assay. Each bar represents the mean ± SEM (n = 3 or 4). * $p < 0.05$, ** $p < 0.01$, *** $p < 0.005$ compared with the control group. # $p < 0.05$, ## $p < 0.01$, ### $p < 0.005$ compared with the LPS alone group.

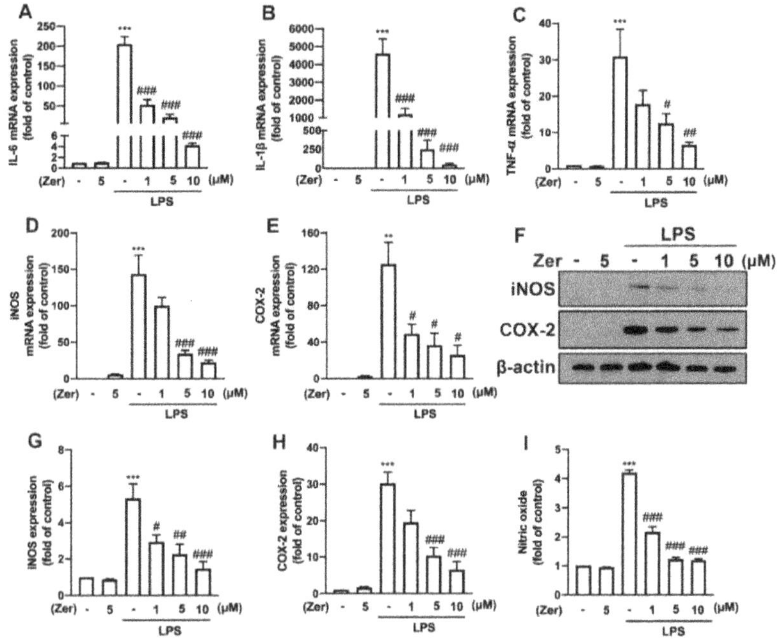

Figure 6. The expression of proinflammatory mediators in response to zerumbone in microglia. IMG cells were supplemented with different concentrations of zerumbone (1, 5, or 10 μM) for 30 min and administered with LPS (50 ng/mL) for another 6 h (**A–C**) or 24 h (**D–G**). IL-6 (**A**), IL-1β (**B**), TNF-α (**C**), iNOS (**D**), and COX-2 (**E**) mRNA expressions were analyzed using real-time PCR and normalized to β-actin. (**F**) iNOS and COX-2 protein expressions were analyzed using Western blotting. Quantitative results are shown in (**G**,**H**). (**I**) The cultural supernatant was harvested for measuring NO production by NO assay. Each bar represents the mean ± SEM (n = 3 or 4). ** $p < 0.01$, *** $p < 0.005$ compared with the control group. # $p < 0.05$, ## $p < 0.01$, ### $p < 0.005$ compared with the LPS alone group.

3.5. Zerumbone Promotes Endogenous Antioxidant Production and IL-10 Expression in Microglial Cells

Several naturally occurring compounds stimulate the production of endogenous antioxidants—such as heme oxygenase (HO)-1, glutamate–cysteine ligase modifier subunit (GCLM), glutamate–cysteine ligase catalytic subunit (GCLC), and NAD(P)H quinone oxidoreductase-1 (NQO1)—and promotes microglial polarization toward M2-like phenotypes that are beneficial for maintaining cellular redox homeostasis and are anti-inflammatory [18,36,37]. This study showed that supplement with zerumbone remarkably promoted the expression of the endogenous antioxidant proteins HO-1 (Figure 7A,B), GCLM (Figure 7A,C), GCLC (Figure 7A,D), and NQO1 (Figure 7A,E) in microglial cells. Moreover, the mRNA expression of HO-1 (Figure 7F), GCLM (Figure 7G), GCLC (Figure 7H), and NQO1 (Figure 7I) was upregulated following zerumbone supplement in microglial cells. Moreover, as shown in Figure 7J, zerumbone increased the expression of the M2 phenotype marker IL-10 in a dose-dependent manner. These data suggest that the antineuroinflammatory properties of zerumbone were modulated by the production of endogenous antioxidants.

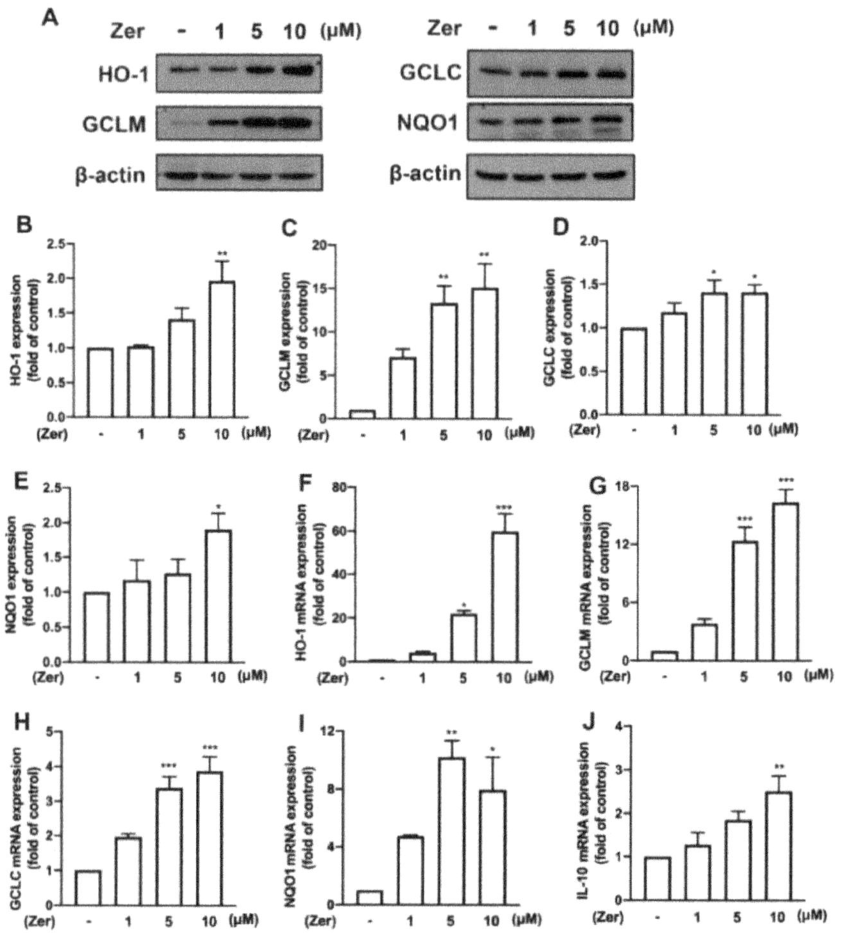

Figure 7. The expression of endogenous antioxidants and anti-inflammatory cytokines in response to zerumbone in microglia. (**A**) Different concentrations of zerumbone (1, 5, or 10 μM) were supplemented on IMG cells for 24 h. Protein expressions of HO-1, GCLM, GCLC, and NQO1 were evaluated using Western blotting. The quantitative results of HO-1 (**B**), GCLM (**C**), GCLC (**D**), and NQO1 (**E**) were determined by using ImageJ. Different concentrations of zerumbone (1, 5, or 10 μM) were supplemented on IMG cells for 6 h. HO-1 (**F**), GCLM (**G**), GCLC (**H**), NQO1 (**I**), and IL-10 (**J**) mRNA expressions were quantified using real-time PCR. Each bar represents the mean ± SEM (n = 3 or 4). * $p < 0.05$, ** $p < 0.01$, *** $p < 0.001$ compared with the control group.

3.6. AMPK and Akt/GSK3 Signaling Pathways Mediate Zerumbone-Stimulated Production of Endogenous Antioxidants in Microglial Cells

Studies have shown that zerumbone activates AMPK signaling pathways and the downstream target of AMPK, acetyl-CoA carboxylase (ACC), contributing to a protective role in high-glucose-stimulated renal tubular cells [38] and high-fat-diet-induced obesity in mice [39]. Furthermore, zerumbone attenuated inflammatory responses in mice with acute lung injury [40] and macrophages [26] by modulating the Akt pathway. In the present study, the zerumbone supplement enhanced activation of the AMPK (Figure 8A) and Akt (Figure 8B) signaling pathways, as well as their downstream targets ACC (Figure 8A) and GSK3 (Figure 8B). In addition, supplement with an AMPK inhibitor (compound C) suppressed the expression of HO-1 (Figure 8C,D), GCLM (Figure 8C,E), and GCLC (Figure 8C) proteins in microglial cells supplemented with zerumbone. We further confirmed the involvement of the AMPK/Akt signaling pathways in the protective effects of zerumbone. As illustrated in Figure 8, the zerumbone-induced expression of endogenous antioxidant genes such as HO-1, GCLC, GCLM, and NQO1 (Figure 8F–I), as well as M2 phenotype marker IL-10 (Figure 8J), was inhibited by compound C, Akt inhibitor, and GSK3 inhibitor (SB21). These findings indicate that zerumbone promoted endogenous antioxidants and polarization toward M2 phenotypes by mediating the AMPK and Akt/GSK3 pathways in microglial cells.

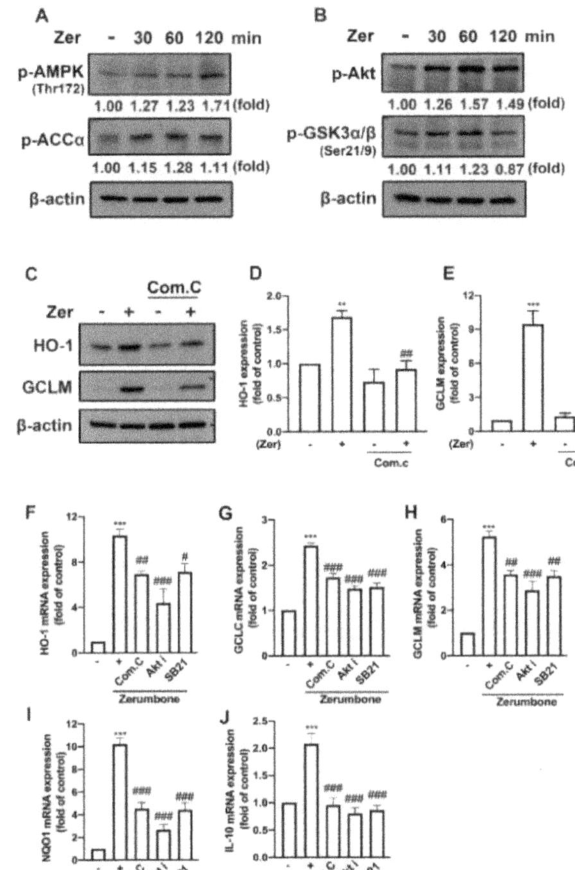

Figure 8. Zerumbone-induced endogenous antioxidant expression is mediated through AMPK and Akt/GSK3 signaling pathways. IMG cells were supplemented with zerumbone (10 μM) for 30, 60, or 120 min. The phosphorylation of AMPK and ACCα (**A**) and of Akt and GSK3α/β (**B**) were examined by Western blotting. AMPK inhibitor compound C (15 μM) was administered 30 min before supplemented with zerumbone (10 μM) for another 24 h. (**C**) HO-1 and GCLM protein expressions were detected by Western blotting, with quantitative data shown in (**D**) and (**E**). Compound C, Akt inhibitor (10 μM), or SB 216763 (SB21; 20 μM) were administered 30 min before supplemented with zerumbone for another 6 h. HO-1 (**F**), GCLC (**G**), GCLM (**H**), NQO1 (**I**), and IL-10 (**J**) mRNA expressions were determined using real-time PCR and normalized to β-actin. The quantitative results in bar graphs represent the mean ± SEM (n = 3 or 4). ** $p < 0.01$, *** $p < 0.005$ compared with the control group. # $p < 0.05$, ## $p < 0.01$, ### $p < 0.005$ compared with zerumbone alone.

4. Discussion

Zerumbone has been reported to modulate oxidative stress in several cancers, including breast cancer [18], nonsmall-cell lung cancer [36], and colon cancer [37]. Furthermore, it has been reported that zerumbone enhances the radiosensitivity and chemosensitivity of these malignancies. Notably, zerumbone has been reported to induce oxidative stress and cell apoptosis in protozoan parasites [38]. A recent report indicated that zerumbone protects against zearalenone-induced hepatotoxicity in mice by activating endogenous antioxidants, including glutathione and superoxide dismutase [39]. Moreover, zerumbone treatment also protected against oxidative stress in a mouse model of acute liver injury [40]. A few studies have reported that zerumbone exerts biological effects on the CNS. Treatment with zerumbone reversed scopolamine-induced memory impairments in rats [33] and social memory in triple transgenic Alzheimer's disease (AD) mouse models [41]. In addition, the cotreatment of zerumbone with polyunsaturated fatty acids attenuated oxidative stress in the brain by increasing antioxidants and neurotrophins [42]. A recent study suggested that in APP/PS1 transgenic mice, zerumbone decreased the expression of proinflammatory cytokines, lessened the amount of β-amyloid accumulation, and reduced behavioral deficits, partly due to the production of IL-10 by activated microglia [34]. The present study confirmed the antioxidative and anti-inflammatory properties of zerumbone by demonstrating that zerumbone effectively reduced inflammation and oxidative stress in macrophages and microglia without any toxicity being incurred.

Production of lipocalin-2 may act as a signal during oxidative stress and inflammation. Lipocalin-2 is produced in the CNS in response to acute-phase brain injury, further triggering the inflammation-related chemokine CXCL10, which promotes the migration of astrocytes to injury sites [43,44]. Neutralization of lipocalin-2-attenuated neurological deficits and cerebral infarction by diminishing the expression of M1 macrophage polarization in the brain in a stroke-reperfusion injury mouse model [45]. Additionally, lipocalin-2-deficient mice had less hyperalgesia, M1 macrophage polarization, and macrophage inflammatory protein 2 production in response to complete Freund's adjuvant. [46]. The level of lipocalin-2 was found to be increased in patients with AD [47] and Parkinson's disease (PD) [48]; the patient's pathophysiology was also aggravated along the lipocalin-2 levels. Moreover, a recent study considered lipocalin-2 to be a promising therapeutic target in the management of dementia [49]. In our previous study, we demonstrated that management of lipocalin-2 reduced astrocyte activation and improved cognitive functions, social avoidance, and anxiety-like behaviors [50]. Notably, one study [23] and our recent study [51] have reported that lipocalin-2 may be a regulator of M1 and M2 macrophage polarization. The present study demonstrated the role of zerumbone in decreasing the expression of lipocalin-2 and thereby improving the inflammatory response and tissue homeostasis in activated microglia and macrophages.

Macrophages respond to external and endogenous stimuli by switching their phenotypes to resolve inflammation and maintain immune defense and homeostasis [1]. One study demonstrated that in an experimental PD mouse model, an increase in M1 macrophages in the peripheral immune system triggered the expression of proinflammatory mediators such as iNOS, IL-1β, and TNF-α in brain, leading to neuronal cell death [52]. Notably, by depleting peripheral M1 macrophages and promoting M2 macrophages, T cell infiltration to the brain was reduced, thereby reducing brain inflammation, neuronal cell death, and behavioral deficits [52]. In experimental autoimmune encephalomyelitis (EAE), activated M1 microglia upregulated CCL-2, which facilitated the recruitment of circulating monocytes to the injured sites, as well as TNF and iNOS, which contributed to inflammation [53]. Overproduction of ROS in macrophages may trigger necrosis, which leads to the production of proinflammatory mediators and aggravates inflammation [54]. One study revealed that mitochondrial ROS generated by activated macrophages stimulated the expression of IL-1β, TNF-α, and CCL-2, thereby increasing the risk of developing high-fat-induced insulin resistance and atherosclerosis [55]. Moreover, increased levels of ROS were shown to trigger inflammation and cause apoptotic death of microglia under oxygen–glucose deprivation conditions [56]. Notably, NADPH oxidase (NOX)-1-derived ROS increased the expression of lipocalin-2 in colon epithelial cells stimulated with TNF-α and IL-17 [57]. Accordingly, NOX-1-knockout mice exhibited reduced expression of lipocalin-2 along with improved colon condition [57]. Treatment with zerumbone was found to diminish ROS production and protect cells from high-glucose-stimulated pancreatic β cells [58]. The present study supported previous studies that zerumbone could protect microglia and macrophages by regulating the production of ROS and expression of proinflammatory mediators.

Macrophages require ROS during the uptake and clearance of dying cellular parts. ROS control the phagocytic activity of macrophages [59]. Increased intracellular production of ROS was reported to enhance the phagocytosis of macrophages [60]. Phagocytosis by microglia is associated with neurodegenerative diseases [61]. The roles of microglia and phagocytosis in different stages of neurodegenerative disorders remain unknown. Phagocytosis by activated microglia can be beneficial in terms of the clearance of Aβ in AD [62]. However, microglia may be detrimental to the pathophysiology of AD because they stimulate neurotoxins [61]. Similarly, a study suggested that there is a delicate balance between activated microglial damage of myelin-generating cells and activated microglial repair and support of neurogenesis in multiple sclerosis [63]. Thus, maintaining macrophage homeostasis and phagocytic activity is beneficial for disease management. Notably, lipocalin-2-deficient mice exhibited lower phagocytic activity than wild-type mice [64]. A recent study suggested that lipocalin-2 regulated myelin phagocytosis in an ischemic stroke mouse model [65]. The present study demonstrated the regulatory effects of zerumbone on proinflammatory-stimulus-triggered lipocalin-2 expression and microglial phagocytic activity. On the other hand, zerumbone is also reported to exert anti-inflammatory effects through pathways other than lipocalin-2, such as the Akt-NFkB pathway and NLRP3 inflammasome [40,66], indicating that zerumbone may augment inhibitory effects on M1/M2 polarization, cytokine production, and ROS formation.

The production of IL-10 is facilitated by the protein expression of M2 phenotypes, which suppress inflammation and restore homeostasis [67]. Treatment with recombinant IL-10 considerably decreased M1 macrophage polarization in LPS-activated microglia [68]. Moreover, IL-10-deficient mice exhibited a decreased inflammatory response and persistent ischemia, suggesting the role of IL-10 in attenuating local inflammatory responses [68]. IL-10 overexpression was found to be beneficial for the treatment of several neurodegenerative diseases—including SCI [16], EAE [69], and AD [70]—by reducing the expression of proinflammatory mediators and improving neurological functions. Evidence was found that IL-10 signaling is correlated with the expression of HO-1 [71]. The induction of endogenous antioxidants, such as glutathione-S-transferases, regulates inflammatory responses [72]. Our previous studies have demonstrated that treatment with naturally occurring compounds—quercetin [51], paeonol [73], fisetin [74], and caffeic acid phenethyl ester [75]—induces the expression of HO-1 and promotes the polarization of macrophages toward the M2 phenotype that inhibits proinflammatory responses in microglia. In addition, zerumbone has been reported to upregulate the expression of HO-1 and γ-glutamyl cysteine ligase in human keratinocyte cells [76]. Zerumbone enhances GSK3β phosphorylation in meningioma cells [75,77]. A recent study reported that zerumbone supplement activates the PI3/Akt signaling pathway and upregulates the expression of endogenous antioxidants against hepatotoxicity [39]. Additionally, stimulation of endogenous antioxidants is regulated by activation of the AMPK/Akt signaling pathway [78,79]. The present study demonstrated that zerumbone significantly stimulated the expression of endogenous antioxidants and M2 macrophage markers involving AMPK/Akt and Akt/GSK3β signaling pathways, resulting in zerumbone having antioxidant and protective roles in microglia and macrophages.

The limitation of this study includes that no result was obtained from in vivo experimental models. If additional animal models were carried out, we could provide substantial information considering the effectiveness of zerumbone and improve our understanding of zerumbone under systemic conditions. From our established in vivo model, we found that LPS-induced inflammation provokes IL-6 and TNF-α production in mouse brain microglia [80]. LPS also causes impaired motor balance and coordination function in mice [36]. Moreover, LPS injection also induces microglia to change their normal ramified morphology into an activated hypertrophic form [81]. In this study, we attempted to focus on the effects of zerumbone on M1/M2 polarization and the ability of phagocytosis, and we chose macrophages and microglia cell lines to clarify the effect of zerumbone.

5. Conclusions

This study demonstrated the potential role of zerumbone in reducing the expression of lipocalin-2- and M1-associated inflammatory responses, including the overexpression of proinflammatory cytokines (IL-1β, IL-6, and TNF-α), chemokines (CCL-2 and CXCL-10), iNOS, NO, and COX-2 in activated microglia and macrophages. Supplement with zerumbone was discovered to effectively reduce ROS production stimulated by H_2O_2, ROO•, and HO•. The phagocytic activity of microglial cells triggered by proinflammatory stimuli was also lower in cells subjected to zerumbone supplement. The expression of IL-10 in both microglia and macrophages was increased following supplementation with zerumbone. Notably, we discovered that zerumbone increased the expression of HO-1, GCLM, GCLC, and NQO1 by regulating the AMPK and Akt/GSK3β signaling pathways. This study suggests that zerumbone could be a potential supplement for inflammatory diseases in both the CNS and peripheral systems due to its ability to regulate cellular redox homeostasis and macrophage polarization.

Supplementary Materials: The following supporting information can be downloaded at: https://www.mdpi.com/article/10.3390/nu14245402/s1, Figure S1: effects of zerumbone on viability of macrophages, Figure S2: effects of zerumbone on viability of microglia.

Author Contributions: Conceptualization, W.-L.Y., B.-R.H. and C.L.; methodology, W.-L.Y., G.-W.C., B.-R.H. and C.L.; resources, W.-L.Y., B.-R.H., D.-Y.L., C.-F.T., L.-Y.Y., M.-K.C. and C.L.; data curation, G.-W.C. and B.-R.H.; writing—original draft preparation, W.-L.Y., V.C. and C.L.; writing—review and editing, W.-L.Y., B.-R.H., D.-Y.L., C.-F.T. and C.L.; supervision, W.-L.Y. and C.L.; funding acquisition, W.-L.Y., B.-R.H. and C.L. All authors have read and agreed to the published version of the manuscript.

Funding: This work is supported in part by grants from the Ministry of Science and Technology (MOST 111-2320-B-039-028-, MOST 110-2320-B-039-030-MY3, and MOST 110-2320-B-039-011), China Medical University (CMU110-S-39), China Medical University Hospital (DMR-110-120), and Taichung Tzu Chi Hospital (TTCRD109-17 and TTCRD110-04).

Institutional Review Board Statement: Not applicable.

Informed Consent Statement: The study did not involve humans.

Data Availability Statement: Data are available from the corresponding author on reasonable request.

Conflicts of Interest: The authors declare no conflict of interest.

References

1. Jha, M.K.; Lee, W.H.; Suk, K. Functional polarization of neuroglia: Implications in neuroinflammation and neurological disorders. *Biochem. Pharmacol.* **2016**, *103*, 1–16. [CrossRef] [PubMed]
2. Colton, C.A. Heterogeneity of microglial activation in the innate immune response in the brain. *J. Neuroimmune Pharmacol.* **2009**, *4*, 399–418. [CrossRef] [PubMed]
3. Yunna, C.; Mengru, H.; Lei, W.; Weidong, C. Macrophage M1/M2 polarization. *Eur. J. Pharmacol.* **2020**, *877*, 173090. [CrossRef] [PubMed]
4. Dey, A.; Allen, J.; Hankey-Giblin, P.A. Ontogeny and polarization of macrophages in inflammation: Blood monocytes versus tissue macrophages. *Front. Immunol.* **2015**, *5*, 683. [CrossRef]
5. Bashir, S.; Sharma, Y.; Elahi, A.; Khan, F. Macrophage polarization: The link between inflammation and related diseases. *Inflamm. Res.* **2016**, *65*, 1–11. [CrossRef] [PubMed]
6. Atri, C.; Guerfali, F.Z.; Laouini, D. Role of human macrophage polarization in inflammation during infectious diseases. *Int. J. Mol. Sci.* **2018**, *19*, 1801. [CrossRef]

7. Labonte, A.C.; Tosello-Trampont, A.C.; Hahn, Y.S. The role of macrophage polarization in infectious and inflammatory diseases. *Mol. Cells* **2014**, *37*, 275–285. [CrossRef]
8. Nakagawa, Y.; Chiba, K. Diversity and plasticity of microglial cells in psychiatric and neurological disorders. *Pharmacol. Ther.* **2015**, *154*, 21–35. [CrossRef]
9. Xuan, W.; Qu, Q.; Zheng, B.; Xiong, S.; Fan, G.H. The chemotaxis of M1 and M2 macrophages is regulated by different chemokines. *J. Leukoc. Biol.* **2015**, *97*, 61–69. [CrossRef]
10. Sierra-Filardi, E.; Nieto, C.; Domínguez-Soto, A.; Barroso, R.; Sánchez-Mateos, P.; Puig-Kroger, A.; López-Bravo, M.; Joven, J.; Ardavín, C.; Rodríguez-Fernández, J.L.; et al. CCL2 shapes macrophage polarization by GM-CSF and M-CSF: Identification of CCL2/CCR2-dependent gene expression profile. *J. Immunol.* **2014**, *192*, 3858–3867. [CrossRef]
11. Mills, C.D.; Ley, K. M1 and M2 macrophages: The chicken and the egg of immunity. *J. Innate Immun.* **2014**, *6*, 716–726. [CrossRef] [PubMed]
12. West, M. Dead adipocytes and metabolic dysfunction: Recent progress. *Curr. Opin. Endocrinol. Diabetes Obes.* **2009**, *16*, 178–182. [CrossRef] [PubMed]
13. Ito, A.; Suganami, T.; Yamauchi, A.; Degawa-Yamauchi, M.; Tanaka, M.; Kouyama, R.; Kobayashi, Y.; Nitta, N.; Yasuda, K.; Hirata, Y.; et al. Role of CC chemokine receptor 2 in bone marrow cells in the recruitment of macrophages into obese adipose tissue. *J. Biol. Chem.* **2008**, *283*, 35715–35723. [CrossRef] [PubMed]
14. Lin, P.; Ji, H.H.; Li, Y.J.; Guo, S.D. Macrophage plasticity and atherosclerosis therapy. *Front. Mol. Biosci.* **2021**, *8*, 679797. [CrossRef]
15. Zhou, B.; Yang, Y.; Li, C. SIRT1 inhibits hepatocellular carcinoma metastasis by promoting M1 macrophage polarization via NF-κb pathway. *OncoTargets Ther.* **2019**, *12*, 2519–2529. [CrossRef] [PubMed]
16. Jiang, M.H.; Chung, E.; Chi, G.F.; Ahn, W.; Lim, J.E.; Hong, H.S.; Kim, D.W.; Choi, H.; Kim, J.; Son, Y. Substance P induces M2-type macrophages after spinal cord injury. *Neuroreport* **2012**, *23*, 786–792. [CrossRef] [PubMed]
17. Liu, Y.S.; Huang, B.R.; Lin, C.J.; Shen, C.K.; Lai, S.W.; Chen, C.W.; Lin, H.J.; Lin, C.H.; Hsieh, Y.C.; Lu, D.Y. Paliperidone inhibits glioblastoma growth in mouse brain tumor model and reduces PD-L1 expression. *Cancers* **2021**, *13*, 4357. [CrossRef]
18. Tsai, C.F.; Chen, G.W.; Chen, Y.C.; Shen, C.K.; Lu, D.Y.; Yang, L.Y.; Chen, J.H.; Yeh, W.L. Regulatory effects of quercetin on M1/M2 macrophage polarization and oxidative/antioxidative balance. *Nutrients* **2021**, *14*, 67. [CrossRef]
19. Lee, S.; Jha, M.K.; Suk, K. Lipocalin-2 in the inflammatory activation of brain astrocytes. *Crit. Rev. Immunol.* **2015**, *35*, 77–84. [CrossRef]
20. Borkham-Kamphorst, E.; Drews, F.; Weiskirchen, R. Induction of lipocalin-2 expression in acute and chronic experimental liver injury moderated by pro-inflammatory cytokines interleukin-1β through nuclear factor-κb activation. *Liver Int.* **2011**, *31*, 656–665. [CrossRef]
21. Asimakopoulou, A.; Weiskirchen, S.; Weiskirchen, R. Lipocalin 2 (LCN2) expression in hepatic malfunction and therapy. *Front. Physiol.* **2016**, *7*, 430. [CrossRef] [PubMed]
22. Jang, E.; Lee, S.; Kim, J.H.; Kim, J.H.; Seo, J.W.; Lee, W.H.; Mori, K.; Nakao, K.; Suk, K. Secreted protein lipocalin-2 promotes microglial M1 polarization. *FASEB J.* **2013**, *27*, 1176–1190. [CrossRef] [PubMed]
23. Wang, G.; Weng, Y.C.; Chiang, I.C.; Huang, Y.T.; Liao, Y.C.; Chen, Y.C.; Kao, C.Y.; Liu, Y.L.; Lee, T.H.; Chou, W.H. Neutralization of lipocalin-2 diminishes stroke-reperfusion injury. *Int. J. Mol. Sci.* **2020**, *21*, 6253. [CrossRef] [PubMed]
24. Singh, Y.P.; Girisa, S.; Banik, K.; Ghosh, S.; Swathi, P.; Deka, M.; Padmavathi, G.; Kotoky, J.; Sethi, G.; Fan, L. Potential application of zerumbone in the prevention and therapy of chronic human diseases. *J. Funct. Foods* **2019**, *53*, 248–258. [CrossRef]
25. Girisa, S.; Shabnam, B.; Monisha, J.; Fan, L.; Halim, C.E.; Arfuso, F.; Ahn, K.S.; Sethi, G.; Kunnumakkara, A.B. Potential of zerumbone as an anti-cancer agent. *Molecules* **2019**, *24*, 734. [CrossRef]
26. Haque, M.A.; Jantan, I.; Harikrishnan, H. Zerumbone suppresses the activation of inflammatory mediators in LPS-stimulated U937 macrophages through Myd88-dependent NF-κb/MAPK/PI3K-Akt signaling pathways. *Int. Immunopharmacol.* **2018**, *55*, 312–322. [CrossRef] [PubMed]
27. Moreira da Silva, T.; Pinheiro, C.D.; Puccinelli Orlandi, P.; Pinheiro, C.C.; Soares Pontes, G. Zerumbone from *Zingiber zerumbet* (L.) Smith: A potential prophylactic and therapeutic agent against the cariogenic bacterium *Streptococcus mutans*. *BMC Complement. Altern. Med.* **2018**, *18*, 301. [CrossRef]
28. Abdelwahab, S.I.; Abdul, A.B.; Devi, N.; Taha, M.M.; Al-zubairi, A.S.; Mohan, S.; Mariod, A.A. Regression of cervical intraepithelial neoplasia by zerumbone in female Balb/c mice prenatally exposed to diethylstilboestrol: Involvement of mitochondria-regulated apoptosis. *Exp. Toxicol. Pathol.* **2010**, *62*, 461–469. [CrossRef]
29. Sulaiman, M.R.; Perimal, E.K.; Akhtar, M.N.; Mohamad, A.S.; Khalid, M.H.; Tasrip, N.A.; Mokhtar, F.; Zakaria, Z.A.; Lajis, N.H.; Israf, D.A. Anti-inflammatory effect of zerumbone on acute and chronic inflammation models in mice. *Fitoterapia* **2010**, *81*, 855–858. [CrossRef]
30. Rahman, H.S.; Rasedee, A.; Othman, H.H.; Chartrand, M.S.; Namvar, F.; Yeap, S.K.; Abdul Samad, N.; Andas, R.J.; Muhammad Nadzri, N.; Anasamy, T.; et al. Acute toxicity study of zerumbone-loaded nanostructured lipid carrier on BALB/c mice model. *BioMed Res. Int.* **2014**, *2014*, 563930. [CrossRef]
31. Kiyama, R. Nutritional implications of ginger: Chemistry, biological activities and signaling pathways. *J. Nutr. Biochem.* **2020**, *86*, 108486. [CrossRef] [PubMed]
32. Sut, S.; Maggi, F.; Nicoletti, M.; Baldan, V.; Dall Acqua, S. New drugs from old natural compounds: Scarcely investigated sesquiterpenes as new possible therapeutic agents. *Curr. Med. Chem.* **2018**, *25*, 1241–1258. [CrossRef] [PubMed]

33. Jafarian, S.; Ling, K.H.; Hassan, Z.; Perimal-Lewis, L.; Sulaiman, M.R.; Perimal, E.K. Effect of zerumbone on scopolamine-induced memory impairment and anxiety-like behaviours in rats. *Alzheimer's Dement. Transl. Res. Clin. Interv.* **2019**, *5*, 637–643. [CrossRef] [PubMed]
34. Li, L.; Wu, X.H.; Zhao, X.J.; Xu, L.; Pan, C.L.; Zhang, Z.Y. Zerumbone ameliorates behavioral impairments and neuropathology in transgenic APP/PS1 mice by suppressing MAPK signaling. *J. Neuroinflammation* **2020**, *17*, 61. [CrossRef]
35. Wu, L.H.; Huang, B.R.; Lai, S.W.; Lin, C.; Lin, H.Y.; Yang, L.Y.; Lu, D.Y. SIRT1 activation by minocycline on regulation of microglial polarization homeostasis. *Aging* **2020**, *12*, 17990–18007. [CrossRef]
36. Chuang, J.-Y.; Chang, P.-C.; Shen, Y.-C.; Lin, C.; Tsai, C.-F.; Chen, J.-H.; Yeh, W.-L.; Wu, L.-H.; Lin, H.-Y.; Liu, Y.-S.; et al. Regulatory effects of fisetin on microglial activation. *Molecules* **2014**, *19*, 8820–8839. [CrossRef]
37. Tsai, C.-F.; Kuo, Y.-H.; Yeh, W.-L.; Wu, C.Y.-J.; Lin, H.-Y.; Lai, S.-W.; Liu, Y.-S.; Wu, L.-H.; Lu, J.-K.; Lu, D.-Y. Regulatory effects of caffeic acid phenethyl ester on neuroinflammation in microglial cells. *Int. J. Mol. Sci.* **2015**, *16*, 5572–5589. [CrossRef]
38. Shrikanth, C.B.; Chilkunda, N.D. Zerumbone ameliorates high glucose-induced reduction in AMP-activated protein kinase phosphorylation in tubular kidney cells. *J. Agric. Food Chem.* **2017**, *65*, 9208–9216. [CrossRef]
39. Ahn, J.; Lee, H.; Jung, C.H.; Choi, W.H.; Ha, T.Y. Zerumbone ameliorates high-fat diet-induced adiposity by restoring AMPK-regulated lipogenesis and microRNA-146b/SIRT1-mediated adipogenesis. *Oncotarget* **2017**, *8*, 36984–36995. [CrossRef]
40. Ho, Y.C.; Lee, S.S.; Yang, M.L.; Huang-Liu, R.; Lee, C.Y.; Li, Y.C.; Kuan, Y.H. Zerumbone reduced the inflammatory response of acute lung injury in endotoxin-treated mice via Akt-NFκB pathway. *Chem. Biol. Interact.* **2017**, *271*, 9–14. [CrossRef]
41. Li, J.; Wang, L.; Sun, Y.; Wang, Z.; Qian, Y.; Duraisamy, V.; Antary, T.M.A. Zerumbone-induced reactive oxygen species-mediated oxidative stress re-sensitizes breast cancer cells to paclitaxel. *Biotechnol. Appl. Biochem.* 2022; *ahead of print.*
42. Hu, Z.; Zeng, Q.; Zhang, B.; Liu, H.; Wang, W. Promotion of p53 expression and reactive oxidative stress production is involved in zerumbone-induced cisplatin sensitization of non-small cell lung cancer cells. *Biochimie* **2014**, *107 Pt B*, 257–262. [CrossRef] [PubMed]
43. Deorukhkar, A.; Ahuja, N.; Mercado, A.L.; Diagaradjane, P.; Raju, U.; Patel, N.; Mohindra, P.; Diep, N.; Guha, S.; Krishnan, S. Zerumbone increases oxidative stress in a thiol-dependent ROS-independent manner to increase DNA damage and sensitize colorectal cancer cells to radiation. *Cancer Med.* **2015**, *4*, 278–292. [CrossRef] [PubMed]
44. Mukherjee, D.; Singh, C.B.; Dey, S.; Mandal, S.; Ghosh, J.; Mallick, S.; Hussain, A.; Swapana, N.; Ross, S.A.; Pal, C. Induction of apoptosis by zerumbone isolated from *Zingiber zerumbet* (L.) smith in protozoan parasite *Leishmania donovani* due to oxidative stress. *Braz. J. Infect. Dis. Off. Publ. Braz. Soc. Infect. Dis.* **2016**, *20*, 48–55. [CrossRef] [PubMed]
45. AbuZahra, H.M.; Rajendran, P.; Ismail, M.B. Zerumbone exhibit protective effect against zearalenone induced toxicity via ameliorating inflammation and oxidative stress induced apoptosis. *Antioxidants* **2021**, *10*, 1593. [CrossRef] [PubMed]
46. Wang, M.; Niu, J.; Ou, L.; Deng, B.; Wang, Y.; Li, S. Zerumbone protects against Carbon Tetrachloride (CCL(4))-induced acute liver injury in mice via inhibiting oxidative stress and the inflammatory response: Involving the TLR4/NF-κB/COX-2 pathway. *Molecules* **2019**, *24*, 1964. [CrossRef]
47. Nguyen, E.T.; Selmanovic, D.; Maltry, M.; Morano, R.; Franco-Villanueva, A.; Estrada, C.M.; Solomon, M.B. Endocrine stress responsivity and social memory in 3xTg-AD female and male mice: A tale of two experiments. *Horm. Behav.* **2020**, *126*, 104852. [CrossRef]
48. Uppin, V.; Acharya, P.; Bettadaiah Bheemanakere, K.; Talahalli, R.R. Hyperlipidemia downregulate brain antioxidant defense enzymes and neurotrophins in rats: Assessment of the modulatory potential of EPA+DHA and zerumbone. *Mol. Nutr. Food Res.* **2020**, *64*, 2000381. [CrossRef] [PubMed]
49. Shin, H.J.; Jeong, E.A.; Lee, J.Y.; An, H.S.; Jang, H.M.; Ahn, Y.J.; Lee, J.; Kim, K.E.; Roh, G.S. Lipocalin-2 deficiency reduces oxidative stress and neuroinflammation and results in attenuation of kainic acid-induced hippocampal cell death. *Antioxidants* **2021**, *10*, 100. [CrossRef] [PubMed]
50. Lee, S.; Kim, J.-H.; Kim, J.-H.; Seo, J.-W.; Han, H.-S.; Lee, W.-H.; Mori, K.; Nakao, K.; Barasch, J.; Suk, K. Lipocalin-2 is a chemokine inducer in the central nervous system: Role of chemokine ligand 10 (CXCL10) in lipocalin-2-induced cell migration. *J. Biol. Chem.* **2011**, *286*, 43855–43870.
51. Jha, M.K.; Jeon, S.; Jin, M.; Ock, J.; Kim, J.H.; Lee, W.H.; Suk, K. The pivotal role played by lipocalin-2 in chronic inflammatory pain. *Exp. Neurol.* **2014**, *254*, 41–53. [CrossRef]
52. Naudé, P.J.; Nyakas, C.; Eiden, L.E.; Ait-Ali, D.; van der Heide, R.; Engelborghs, S.; Luiten, P.G.; De Deyn, P.P.; den Boer, J.A.; Eisel, U.L. Lipocalin 2: Novel component of proinflammatory signaling in Alzheimer's disease. *FASEB J.* **2012**, *26*, 2811–2823. [CrossRef] [PubMed]
53. Kim, B.W.; Jeong, K.H.; Kim, J.H.; Jin, M.; Kim, J.H.; Lee, M.G.; Choi, D.K.; Won, S.Y.; McLean, C.; Jeon, M.T.; et al. Pathogenic upregulation of glial lipocalin-2 in the parkinsonian dopaminergic system. *J. Neurosci.* **2016**, *36*, 5608–5622. [CrossRef] [PubMed]
54. Lim, D.; Jeong, J.H.; Song, J. Lipocalin 2 regulates iron homeostasis, neuroinflammation, and insulin resistance in the brains of patients with dementia: Evidence from the current literature. *CNS Neurosci. Ther.* **2021**, *27*, 883–894. [CrossRef] [PubMed]
55. Chen, Y.-H.; Xie, S.-Y.; Chen, C.-W.; Lu, D.-Y. Electroacupuncture improves repeated social defeat stress-elicited social avoidance and anxiety-like behaviors by reducing lipocalin-2 in the hippocampus. *Mol. Brain* **2021**, *14*, 150. [CrossRef]
56. Shen, H.; Wang, Z.; Huang, A.; Zhu, D.; Sun, P.; Duan, Y. Lipocalin 2 is a regulator during macrophage polarization induced by soluble worm antigens. *Front. Cell. Infect. Microbiol.* **2021**, *11*, 747135. [CrossRef]

57. Yan, A.; Zhang, Y.; Lin, J.; Song, L.; Wang, X.; Liu, Z. Partial depletion of peripheral M1 macrophages reverses motor deficits in MPTP-treated mouse by suppressing neuroinflammation and dopaminergic neurodegeneration. *Front. Aging Neurosci.* **2018**, *10*, 160. [CrossRef]
58. Starossom, S.C.; Mascanfroni, I.D.; Imitola, J.; Cao, L.; Raddassi, K.; Hernandez, S.F.; Bassil, R.; Croci, D.O.; Cerliani, J.P.; Delacour, D.; et al. Galectin-1 deactivates classically activated microglia and protects from inflammation-induced neurodegeneration. *Immunity* **2012**, *37*, 249–263. [CrossRef]
59. Tan, H.-Y.; Wang, N.; Li, S.; Hong, M.; Wang, X.; Feng, Y. The reactive oxygen species in macrophage polarization: Reflecting its dual role in progression and treatment of human diseases. *Oxidative Med. Cell. Longev.* **2016**, *2016*, 2795090. [CrossRef]
60. Ming, X.F.; Rajapakse, A.G.; Yepuri, G.; Xiong, Y.; Carvas, J.M.; Ruffieux, J.; Scerri, I.; Wu, Z.; Popp, K.; Li, J.; et al. Arginase II promotes macrophage inflammatory responses through mitochondrial reactive oxygen species, contributing to insulin resistance and atherogenesis. *J. Am. Heart Assoc.* **2012**, *1*, e000992. [CrossRef]
61. Yao, Z.; Liu, N.; Zhu, X.; Wang, L.; Zhao, Y.; Liu, Q.; Gao, C.; Li, J. Subanesthetic isoflurane abates ROS-activated MAPK/NF-κB signaling to repress ischemia-induced microglia inflammation and brain injury. *Aging* **2020**, *12*, 26121–26139. [CrossRef]
62. Makhezer, N.; Ben Khemis, M.; Liu, D.; Khichane, Y.; Marzaioli, V.; Tlili, A.; Mojallali, M.; Pintard, C.; Letteron, P.; Hurtado-Nedelec, M.; et al. NOX1-derived ROS drive the expression of lipocalin-2 in colonic epithelial cells in inflammatory conditions. *Mucosal Immunol.* **2019**, *12*, 117–131. [CrossRef] [PubMed]
63. Wang, C.; Zou, S.; Cui, Z.; Guo, P.; Meng, Q.; Shi, X.; Gao, Y.; Yang, G.; Han, Z. Zerumbone protects INS-1 rat pancreatic beta cells from high glucose-induced apoptosis through generation of reactive oxygen species. *Biochem. Biophys. Res. Commun.* **2015**, *460*, 205–209. [CrossRef] [PubMed]
64. Rendra, E.; Riabov, V.; Mossel, D.M.; Sevastyanova, T.; Harmsen, M.C.; Kzhyshkowska, J. Reactive oxygen species (ROS) in macrophage activation and function in diabetes. *Immunobiology* **2019**, *224*, 242–253. [CrossRef] [PubMed]
65. Lo, H.M.; Chen, C.L.; Yang, C.M.; Wu, P.H.; Tsou, C.J.; Chiang, K.W.; Wu, W.B. The carotenoid lutein enhances matrix metalloproteinase-9 production and phagocytosis through intracellular ROS generation and ERK1/2, p38 MAPK, and RARβ activation in murine macrophages. *J. Leukoc. Biol.* **2013**, *93*, 723–735. [CrossRef]
66. Su, C.C.; Wang, S.C.; Chen, I.C.; Chiu, F.Y.; Liu, P.L.; Huang, C.H.; Huang, K.H.; Fang, S.H.; Cheng, W.C.; Huang, S.P.; et al. Zerumbone suppresses the LPS-induced inflammatory response and represses activation of the NLRP3 inflammasome in macrophages. *Front. Pharmacol.* **2021**, *12*, 652860. [CrossRef]
67. Neumann, H.; Kotter, M.R.; Franklin, R.J. Debris clearance by microglia: An essential link between degeneration and regeneration. *Brain* **2009**, *132*, 288–295. [CrossRef]
68. Anwar, S.; Rivest, S. Alzheimer's disease: Microglia targets and their modulation to promote amyloid phagocytosis and mitigate neuroinflammation. *Expert Opin. Ther. Targets* **2020**, *24*, 331–344. [CrossRef]
69. Goldmann, T.; Prinz, M. Role of microglia in CNS autoimmunity. *Clin. Dev. Immunol.* **2013**, *2013*, 208093. [CrossRef]
70. Du, H.; Liang, L.; Li, J.; Xiong, Q.; Yu, X.; Yu, H. Lipocalin-2 alleviates LPS-induced inflammation through alteration of macrophage properties. *J. Inflamm. Res.* **2021**, *14*, 4189–4203. [CrossRef]
71. Wan, T.; Zhu, W.; Zhao, Y.; Zhang, X.; Ye, R.; Zuo, M.; Xu, P.; Huang, Z.; Zhang, C.; Xie, Y.; et al. Astrocytic phagocytosis contributes to demyelination after focal cortical ischemia in mice. *Nat. Commun.* **2022**, *13*, 1134. [CrossRef]
72. Cherry, J.D.; Olschowka, J.A.; O'Banion, M.K. Neuroinflammation and M2 microglia: The good, the bad, and the inflamed. *J. Neuroinflammation* **2014**, *11*, 98. [CrossRef] [PubMed]
73. Pérez-de Puig, I.; Miró, F.; Salas-Perdomo, A.; Bonfill-Teixidor, E.; Ferrer-Ferrer, M.; Márquez-Kisinousky, L.; Planas, A.M. IL-10 deficiency exacerbates the brain inflammatory response to permanent ischemia without preventing resolution of the lesion. *J. Cereb. Blood Flow Metab.* **2013**, *33*, 1955–1966. [CrossRef] [PubMed]
74. Yang, J.; Jiang, Z.; Fitzgerald, D.C.; Ma, C.; Yu, S.; Li, H.; Zhao, Z.; Li, Y.; Ciric, B.; Curtis, M.; et al. Adult neural stem cells expressing IL-10 confer potent immunomodulation and remyelination in experimental autoimmune encephalitis. *J. Clin. Investig.* **2009**, *119*, 3678–3691. [CrossRef] [PubMed]
75. Kiyota, T.; Ingraham, K.L.; Swan, R.J.; Jacobsen, M.T.; Andrews, S.J.; Ikezu, T. AAV serotype 2/1-mediated gene delivery of anti-inflammatory interleukin-10 enhances neurogenesis and cognitive function in APP+PS1 mice. *Gene Ther.* **2012**, *19*, 724–733. [CrossRef] [PubMed]
76. Ricchetti, G.A.; Williams, L.M.; Foxwell, B.M. Heme oxygenase 1 expression induced by IL-10 requires STAT-3 and phosphoinositol-3 kinase and is inhibited by lipopolysaccharide. *J. Leukoc. Biol.* **2004**, *76*, 719–726. [CrossRef] [PubMed]
77. Simpson, D.S.A.; Oliver, P.L. ROS generation in microglia: Understanding oxidative stress and inflammation in neurodegenerative disease. *Antioxidants* **2020**, *9*, 743. [CrossRef] [PubMed]
78. Lin, C.; Lin, H.-Y.; Chen, J.-H.; Tseng, W.-P.; Ko, P.-Y.; Liu, Y.-S.; Yeh, W.-L.; Lu, D.-Y. Effects of paeonol on anti-neuroinflammatory responses in microglial cells. *Int. J. Mol. Sci.* **2015**, *16*, 8844–8860. [CrossRef] [PubMed]
79. Yang, H.-L.; Lee, C.-L.; Korivi, M.; Liao, J.-W.; Rajendran, P.; Wu, J.-J.; Hseu, Y.-C. Zerumbone protects human skin keratinocytes against UVA-irradiated damages through Nrf2 induction. *Biochem. Pharmacol.* **2018**, *148*, 130–146. [CrossRef] [PubMed]

80. Huang, B.R.; Chang, P.C.; Yeh, W.L.; Lee, C.H.; Tsai, C.F.; Lin, C.; Lin, H.Y.; Liu, Y.S.; Wu, C.S.; Hsu, H.C.; et al. Anti-neuroinflammatory effects of the calcium channel blocker nicardipine on microglial cells: Implications for neuroprotection. *PLoS ONE* **2014**, *9*, e91167. [CrossRef]
81. Lin, H.Y.; Huang, B.R.; Yeh, W.L.; Lee, C.H.; Huang, S.S.; Lai, C.H.; Lin, H.; Lu, D.Y. Antineuroinflammatory effects of lycopene via activation of adenosine monophosphate-activated protein kinase-α1/heme oxygenase-1 pathways. *Neurobiol. Aging* **2014**, *35*, 191–202. [CrossRef]

nutrients

MDPI

Article

Exploring Active Ingredients, Beneficial Effects, and Potential Mechanism of *Allium tenuissimum* L. Flower for Treating T2DM Mice Based on Network Pharmacology and Gut Microbiota

Shan-Shan Zhang [1], Yu-Fei Hou [2], Shao-Jing Liu [3], Sen Guo [2], Chi-Tang Ho [4] and Nai-Sheng Bai [2,*]

1 Department of Pharmaceutical Engineering, College of Chemical Engineering, Northwest University, Xi'an 710069, China
2 Department of Food Science, College of Food Science and Technology, Northwest University, Xi'an 710069, China
3 Department of Medicinal Chemistry, College of Pharmacy, Xi'an Medical University, Xi'an 710021, China
4 Department of Food Science, Rutgers University, New Brunswick, NJ 08901, USA
* Correspondence: nsbai@nwu.edu.cn; Tel.: +029-88305208

Abstract: Forty compounds were isolated and characterized from *A. tenuissimum* flower. Among them, twelve flavonoids showed higher α−glucosidase inhibition activities in vitro than acarbose, especially kaempferol. The molecular docking results showed that the binding of kaempferol to α−glucosidase (GAA) could reduce the hydrolysis of substrates by GAA and reduce the glucose produced by hydrolysis, thus exhibiting α−glucosidase inhibition activities. The in vivo experiment results showed that flavonoids−rich *A. tenuissimum* flower could decrease blood glucose and reduce lipid accumulation. The protein expression levels of RAC−alpha serine/threonine−protein kinase (AKT1), peroxisome proliferator activated receptor gamma (PPARG), and prostaglandin G/H synthase 2 (PTGS2) in liver tissue were increased. In addition, the Firmicutes/Bacteroidetes (F/B) ratio was increased, the level of gut probiotics *Bifidobacterium* was increased, and the levels of *Enterobacteriaceae* and *Staphylococcus* were decreased. The carbohydrate metabolism, lipid metabolism, and other pathways related to type 2 diabetes mellitus were activated. This study indicating flavonoids−rich *A. tenuissimum* flower could improve glycolipid metabolic disorders and inflammation in diabetic mice by modulating the protein expression and gut microbiota.

Keywords: *A. tenuissimum* flower flavonoids; network pharmacology; molecular docking; gut microbiota; type 2 diabetes mellitus

Citation: Zhang, S.-S.; Hou, Y.-F.; Liu, S.-J.; Guo, S.; Ho, C.-T.; Bai, N.-S. Exploring Active Ingredients, Beneficial Effects, and Potential Mechanism of *Allium tenuissimum* L. Flower for Treating T2DM Mice Based on Network Pharmacology and Gut Microbiota. *Nutrients* **2022**, *14*, 3980. https://doi.org/10.3390/nu14193980

Academic Editors: Maria Traka and Flávio Reis

Received: 24 August 2022
Accepted: 22 September 2022
Published: 25 September 2022

1. Introduction

Diabetes mellitus (DM) is a complex chronic metabolic disease. Type 2 diabetes mellitus (T2DM) accounts for more than 90% of diabetic patients. Previous studies showed that T2DM is always accompanied by glucolipid metabolism disorders. Conventional therapies such as long−term usage of insulin injection and other oral medicines might have some side effects, such as gastrointestinal discomfort and hepatic metabolic burden [1]. Chinese herbs and their active ingredients are considered potential therapeutic materials for diabetes based on the safety of natural ingredients. It has been shown that *Pueraria lobate* (kudzu), *Portulaca oleracea*, *Folium mori*, *Radix scutellariae*, and other traditional Chinese herbs are widely used for treating diabetes [2,3]. Isoflavone−rich kudzu extract could decrease blood glucose levels in diabetic Wistar rats [4]. *Portulaca oleracea* flavonoids could significantly improve liver injury and promote insulin secretion and glucose uptake in diabetic mice [5]. Although the anti−diabetic effects of flavonoids are wildly reported in animal studies, however, most individual flavonoid ingredients are not successful in treating diabetes in clinical research. On the other hand, a previous study has shown that dietary flavonoids could reduce the risk of T2DM [6].

Network pharmacology is a cross−disciplinary product of traditional pharmacology, biochemistry, and systems biology [7,8]. Network pharmacology is conducted based on multiple online platforms and related visualization software. Multi−target analysis methods are increasingly used to predict the major active ingredients in herbal medicines and their potential targets and mechanisms of action [9]. Furthermore, molecular docking is an important method for drug design and screening, which could predict the binding mode and stability of ligand and target by calculating binding energy [10]. Combining network pharmacological analysis with molecular docking could be a useful method to explain the mechanism of interaction between active ingredients and their targets.

A. tenuissimum flower (*Allium*, Liliaceae) is wildly distributed in China. *A. tenuissimum* flower with a unique aroma is a vegetable, pickled product, and condiment. Our previous study has characterized and identified volatile compounds and confirmed the key aroma compounds of the *A. tenuissimum* flower [11]. Previous studies had also focused on the nutritional analysis, antioxidant and antibacterial activities, and the inhibitory effect on α−glucosidase of the ethanol extract of *A. tenuissimum* flower [12–14]. α−Glucosidase is the key enzyme of glucose metabolism, which can hydrolyze glycosides to produce glucose and play a key role in the modulation of blood glucose levels [15]. The in vitro α−glucosidase inhibitory activity assay is widely used to evaluate the potential activity of natural compounds including flavonoids and phenolics. Studies have demonstrated that α−glucosidase inhibitors have a positive effect on T2DM animal models [16–18]. Due to the containing of multiple hydroxyl and glycosidic functional groups, flavonoids are considered potential effective α−glucosidase inhibitors. Therefore, it has been suggested that flavonoids might have a therapeutic effect on diabetes mellitus [19–21]. In this research, we will focus on the flavonoids of *A. tenuissimum* flower, their anti−diabetic activity, and potential mechanisms based on network pharmacology analysis and high−fat diet feeding plus a streptozotocin−induced mice model.

Therefore, we aimed to (1) isolate and characterize the ingredients of *A. tenuissimum* flower, (2) use network pharmacology analysis and molecular docking to screen active ingredients and potential targets, (3) perform animal experiments to explore the anti−diabetic activity of *A. tenuissimum* flower and to verify the results of network pharmacology analysis and molecular docking, (4) explore the modulation effect of *A. tenuissimum* flower on the gut microbiota of diabetic mice.

2. Materials and Methods

2.1. Chemicals and Reagents

Methanol, ethyl acetate, *n*−butanol, dichloromethane, petroleum ether, and ethanol were of analytical grade and purchased from Fuyu Fine Chemical Co., Ltd. (Tianjin, China). α−Glucosidase (from yeast), phosphate−buffered saline (PBS), and streptozotocin (STZ) were purchased from Sigma−Aldrich (St Louis, MO, USA). Na_2CO_3, 4−nitrophenyl α−D−galactopyranoside (PNPG), acarbose, and TBST buffer were obtained from Aladdin Biochemical Technology Co., Ltd. (Shanghai, China). Bicinchoninic acid (BCA) protein assay kit was purchased from Nanjing Jiancheng Bioengineering Institute (Nanjing China). Pre−stained protein marker was obtained from New Cell & Molecular Biotech Co., Ltd. (Suzhou, China). RAC−alpha serine/threonine−protein kinase (AKT1), peroxisome proliferator−activated receptor gamma (PPARG), prostaglandin G/H synthase 2 (PTGS2), β−actin, and HRP−conjugated Affinipure Goat Anti−Mouse IgG (H + L) were purchased from Proteintech Group, Inc. (Rosement, PA, USA).

2.2. *A. tenuissimum* Flower Extract Preparation

The *A. tenuissimum* flower was collected from Xingtai City (Hebei Province, China) in August 2020. The dried *A. tenuissimum* flower (10 kg) was extracted with 60 L 75% ethanol/water three times (24 h/time). Approximately 2268 g crude residue (AF) was obtained after concentration and then dispersed in water. Petroleum ether was used for degreasing and decolorizing. Ethyl acetate and *n*−butanol were used for liquid−liquid extraction sequentially, yielding about 120.97 g of ethyl acetate extract and 514.85 g of $n-$butanol extract. Sample AFr (ca. 635.8 g) included the ethyl acetate and $n-$butanol parts.

2.3. Isolation, Characterization, and Quantitation

The $n-$butanol fraction was separated into five fractions from B1 to B5 by silica gel (100−200 mesh) column chromatography eluting with the MeOH−CH_2Cl_2 system (from 1:30 to 1:1). In the subsequent separation process, both silica gel and CHP−20P column chromatography were used for the separation, and the elution system for CHP−20P column chromatography was a methanol−water system (volume ratio from 0:1 to 1:0). Thirty−one compounds were isolated from the $n-$butanol part. The ethyl acetate fraction was separated into three fractions from EA1 to EA3 by column chromatography with silica gel (200−300 mesh) eluting with the MeOH−CH_2Cl_2 system from 1:20 to 1:1. The silica gel and CHP−20P column chromatography were used for further separation, and also the recrystallization method. Finally, nine compounds were obtained from the ethyl acetate part. The separation details were shown in Figure S1. The chemical structures of isolated compounds were characterized by comparing ^{13}C NMR and ^{1}H NMR data from the literature.

High performance liquid chromatography (HPLC) analysis was conducted by using an Agilent 1260 Infinity LC (Agilent Technologies, USA) equipped with a Phenomenex Gemini 5 μm C18 column (250 × 4.60 mm). The HPLC−DAD analysis conditions were as follows: the mobile phase was composed of 0.2% phosphoric acid aqueous solution (A) and methanol (B) with a gradient elution program: 0 min, 95% A; 10 min, 85% A; 20 min, 75% A; 35 min, 65% A; 40 min, 55% A; 50 min, 35% A; 60 min, 15% A; 70 min, 5% A. The flow rate was 1.0 mL/min, and the column temperature was 25 °C. The content of flavonoids from the extract of *A. tenuissimum* flower (AF) and AFr was detected through HPLC analysis at 254 nm wavelength using the external standard method (Figure S2).

2.4. Network Pharmacology Predictive Analysis

2.4.1. Collecting and Analyzing Targets of Ingredients and T2DM

Active ingredients were screened in the Traditional Chinese Medicine Systems Pharmacology (TCMSP, https://tcmsp-e.com/tcmsp.php, accessed on 17 June 2022) database. Oral bioavailability (OB) and drug−likeness (DL) were the most important pharmacokinetic parameters of absorption, distribution, metabolism, and excretion (ADME). The active ingredients were screened by filter criteria OB $\geq$ 30% and DL $\geq$ 0.18. The compound information including Mol ID, structure, and relevant targets name was collected. Meantime, the disease−related targets were retrieved from a database of gene−disease associations DisGeNET platform (https://www.disgenet.org/, accessed on 17 June 2022) by using the "diabetes mellitus" term. The targets were normalized in the UniProt database (https://www.uniprot.org/, accessed on 19 June 2022) with "human species" and "reviewed" filter criteria. To find the common targets of active ingredients−T2DM, the screened active ingredients targets and T2DM targets were visualized in a draw Veen diagram online platform (http://bioinformatics.psb.ugent.be/webtools/Venn/, accessed on 20 June 2022).

2.4.2. Protein−Protein Interaction Network Construction and Key Target Analysis

To explore the interaction of the active ingredients' targets for treating T2DM, the common targets of active ingredients−T2DM were imported into an online platform STRING (version 11.5, https://cn.string-db.org/, accessed on 21 June 2022) to construct a protein−protein interaction (PPI) network. And the PPI network results were analyzed and visualized by Cytoscape v.3.9.0. The Centiscape 2.2 plug−in of Cytoscape was used to screen high interaction targets based on Closeness unDir threshold, Betweeness unDir threshold, and Degree unDir threshold.

2.4.3. GO and KEGG Enrichment Analysis

Gene ontology (GO) analysis and Kyoto encyclopedia of genes and genomes (KEGG) pathway enrichment analysis were conducted on the platform Metascape (http://metascape.org/gp/index.html#/main/step1, accessed on 21 June 2022). The main GO terms of potential targets were analyzed and the terms in biological process (BP), cellular component (CC), and molecular function (MF) were enriched. The KEGG pathways were analyzed and enriched. The significant enrichment terms ($p \leq 0.01$) were selected to visualize through an online platform (http://www.bioinformatics.com.cn/, accessed on 21 June 2022).

2.4.4. Ingredient-Target Protein Molecular Docking

According to the above−mentioned analysis, the key proteins were obtained. The 2D structures of active ingredients were downloaded from PubChem (https://pubchem.ncbi.nlm.nih.gov/, accessed on 22 June 2022), then converted to 3D structures in Chem3D software and saved as a docking ligand mol2 format. The 3D structures of key proteins were obtained from the PSCB PDB platform (https://www.rcsb.org/, accessed on 22 June 2022). The key proteins were removed from organic and solvent, and added hydrogen by using PyMOL software. The Autodock (v.4.2.6) software was used for molecular docking. Finally, the visualization of molecular docking results was performed by PyMOL software.

2.5. *In Vitro α−Glucosidase Inhibitory Activity Assay*

The α−glucosidase inhibitory activity assay method of *A. tenuissimum* flower flavonoids was slightly modified based on previous studies [22–24]. A 96−well plate was used for preparing the mixture consisting of 0.1 M PBS (pH 6.8), 1.0 mg/mL sample solution, and 2 U/mL α−glucosidase, which were incubated at 37 °C for 10 min. 20 μL PNPG solution (2.5 Mm, dissolved in PBS) was added to each well to start the reaction and incubated at 37 °C. Twenty minutes later, 80 μL Na_2CO_3 solution (0.2 mol/L) was added to each well to terminate the reaction. The absorbance was measured at 405 nm by Tecan Infinite M200 PRO (Tecan Trading Co. Ltd., Shanghai, China). Acarbose was used as a positive control. The inhibitory rate was calculated by the absorbance formula below:

$$\text{Inhibition\%} = [1 - (A_a - A_b) / A_c] \times 100\%$$

a. 20 μL PNPG + 80 μL Na_2CO_3 + 20 μL PBS + 20 μL sample + 20 μL enzyme;

b. 20 μL PNPG + 80 μL Na_2CO_3 + 40 μL PBS +20 μL sample;

c. 20 μL PNPG + 80 μL Na_2CO_3 + 40 μL PBS + 20 μL enzyme.

2.6. *In Vivo Animal Experiments*

Animal experiments protocols and procedures following the guidelines established by the China Science Council were approved by the Laboratory Animal Center of Xi'an Jiaotong University Health Science Center (license number: SCXK 2018−001).

Sixty male KM (Kunming) mice (22 ± 2 g) were obtained from the laboratory animal center of Xi'an Jiaotong University Health Science Center. All mice were raised at a temperature of 25 ± 2 °C, relative humidity of 40−60%, and 12 h light/dark cycles laboratory conditions with free access to water and food. The schematic diagram of the animal experiments is shown in Figure 1. After adaptive feeding for one week, they were divided into two groups. The normal group (8 mice) was provided with a normal diet. The other group was provided with a high−fat diet (HFD) containing 45% energy from fat (FBSH Biotechnology Co., Ltd., Shanghai, China). After feeding for four weeks, the HFD−fed group mice were intraperitoneally injected with a dose of 35 mg per kg·bw 0.1 mol/L streptozotocin (STZ) dissolved in pH 4.5 Citric Acid−Sodium Citrate Buffer (Sigma−Aldrich, St Louis, MO, USA) for three consecutive days, while the normal group mice received the same dose of citric acid−sodium citrate buffer.

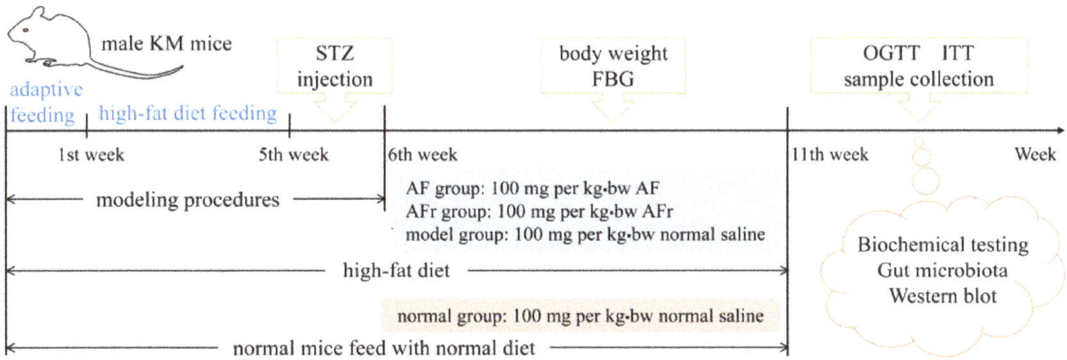

Figure 1. The schedule of animal experiment procedures. KM mice, Kunming mice; STZ, streptozotocin; FBG, fast blood glucose; OGTT, oral glucose tolerance test; ITT, insulin tolerance test. AF represents A. tenuissimum flower extract. AFr represents the mixture of ethyl acetate part and *n*-butanol part.

The mice with fasting blood glucose (FBG) $\geq$ 11.1 mmol/L were considered diabetic mice and divided into three groups for treating 35 days [3]. The AF group mice were fed with 100 mg per kg·bw *A. tenuissimum* flower extract (AF) every day. The AFr group mice were fed with 100 mg per kg·bw the mixture of ethyl acetate part and *n*−butanol part (AFr) every day. The model group and the normal group were treated with the same dose of normal saline. Mice's body weight and FBG were recorded once a week. After the last treatment, the mice fasted for 12 h to carry out an oral glucose tolerance test (OGTT) and insulin tolerance test (ITT). After the OGTT and ITT, the blood samples of all mice were collected at −80 °C. After euthanizing, liver and colon contents were collected and stored at −80 °C until analysis.

2.7. Western Blot Analysis

The protein of liver tissue was extracted and measured in concentration using bicinchoninic acid (BCA) protein assay kit for further western blot analysis. Protein samples were separated using sodium dodecyl sulfate−polyacrylamide gel electrophoresis (SDS−PAGE) and transferred to a 0.22 μm polyvinylidene fluoride (PVDF, Millipore, 6.6 × 8.5 cm) membrane. The membranes were blocked in 5% skimmed milk for 2 h at room temperature. Membranes were incubated with primary antibodies as follows overnight at 4 °C: anti−AKT1 (1:1000), anti−PTGS2 (1:1000), anti−PPARG (1:1000), and anti−β−actin (1:5000). After incubation, the membranes were washed three times and incubated at room temperature for 2 h with a secondary antibody goat anti−mouse IgG (1:5000). Then the membranes were washed three times with TBST buffer. Imaging protein bands were completed by using Tanon 5200 Multi (Tanon, Shanghai China), and ImageJ software was used for protein band grayscale analysis.

2.8. Gut Microbiota Analysis

The total DNA of colon samples was extracted according to the previous method [25]. PCR amplification of the V3−V4 region of the bacteria 16S rRNA genes was detected by 2% agarose gel electrophoresis and measured by fluorescence quantification. The Illumina Miseq platform was used to sequence. Alpha/beta diversity analysis, network analysis, species differences and marker species analysis, and functional forecasting were used to evaluate the diversity of the gut microbiota.

2.9. Statistical Analysis

Origin 2017 (OriginLab, Northampton, Massachusetts, USA) used statistical analysis. The results were analyzed by GraphPad Prism 8 and expressed as mean ± standard deviation (mean ± SD). A value of $p < 0.05$ was considered significantly different.

3. Results and Discussion

3.1. Identification and Quantitation of Compounds in A. tenuissimum Flower

A total of forty compounds numbering from AF−1 to AF−40 were obtained from *A. tenuissimum* flower. The chemical structures of these compounds were characterized by comparing their ^{13}C NMR and ^{1}H NMR data with those in the literature (Supplementary Materials). They are multiflorin A (AF−1), sonchifoliasolide G (AF−2), (-)−ent−prelacinan−7S−ol (AF−3), tritriacontane (AF−4), laricinolic acid (AF−5), di−D−fructose (AF−6), cedr−6−en −12−ol−14−oic acid (AF−7), kaempferol (AF−8), astragalin (AF−9), methyl 11,12,15−trihydroxy−13(14)−octadecenoate (AF−10), isolicoflavonol (AF−11), 8E−decaene-4,6−diyn −1−*O*−β−D−glucopyranoside (AF−12), methyl−α−D−fructofuranoside (AF−13), arbutin (AF−14), kaempferol−4′,7−dimethyl−3−*O*−glucoside (AF−15), 8*Z*−decaene−4,6−diyne−1−*O*−β−D−glucopyranoside (AF−16), (3Z) −3−hexene−1,5−diol 1−*O*−α−L−arabinopyranosyl (1→6) −β−D−glucopyranoside (AF−17), elentheroside E (AF−18), methyl−α−*D*−glucopyranoside (AF−19), burkholone (AF−20), gynostemoside C (AF−21), 5,7−dihydroxy−2−pentadecylchromen−4−one (AF−22), anisodamine (AF−23), onopornoid C (AF−24), 3−methylspongia−3,12−dien−16−one (AF−25), methyl 2,3,4,6−tetra −*O*−methyl−α−D−mannopyranosyl−(1→4)−6−*O*−acetyl−2,3−di−*O*−methyl−α−*D* −glucopyranoside (AF−26), β1−chaconine (AF−27), pestalafuranone E (AF−28), nortetillapyrone (AF−29), tricosanol (AF−30), pentacosane (AF−31), pinellactam (AF−32), kaempferol−3−*O*−β−*D*−4′′,6′′− di−(*E*)−p−coumaroyl glucoside (AF−33), kaempferol 3−*O*−β −*D*−glucopyranosyl−(1→2) −β−*D*−glucopyranosyl−(1→6)−β−*D*−glucopyranodide (AF−34), α−*D*−fructofuranose (AF−35), isorhamnetin 3−*O*−β−*D*−(6−acetyl)−galactopyranoside (AF−36), rhamnocitrin (AF−37), macasiamenol B (AF−38), afzelin (AF−39), tamarixin (AF−40). The chemical structures of all compounds were shown in Figure S3. Twelve flavonoids, eleven glycosides, five terpenes, and twelve other compounds were obtained from *A. tenuissimum* flower. The chemical structures of other flavonoids were derived from the skeleton of kaempferol (Figure 2). The quantitation of twelve flavonoids was developed by HPLC and the results are shown in Table 1. All calibration curves exhibited excellent linear regressions with the determination coefficients (R^2) ranging from 0.9992 to 1.0000. The highest content was 284.1 μg/g (kaempferol, AF−8) and the lowest content was 15.1 μg/g (rhamnocitrin, AF−37) in the AFr sample. The total content of flavonoids was 1429.5 μg/g in the AFr sample. The highest content in the AF sample was 105.1 μg/g (multiflorin A, AF−1) and the lowest content was 7.6 μg/g (rhamnocitrin, AF−37). The total content of flavonoids was 402.0 μg/g in the AF sample. Total flavonoid content was higher in the AFr sample than that in the AF sample.

Table 1. Quantitative analysis of twelve flavonoids in the AF and the AFr.

No.	Calibration Curves [1]	R_2	LOD [2] (μg/mL)	LOQ [3] (μg/mL)	Content in AFr (μg/g)	Content in AF (μg/g)
AF−1	y = 118.4x + 1441.9	0.9995	0.15	0.40	211.1	105.1
AF−8	y = 83.8x − 148.3	0.9999	0.05	0.25	284.1	59.2
AF−9	y = 185.1x − 177.3	0.9994	0.25	1.50	99.4	10.5
AF−11	y = 120.3x − 369.5	0.9997	0.50	1.25	188.8	26.8
AF−15	y = 279.6x − 226.0	0.9997	0.20	1.30	15.7	8.7
AF−33	y = 104.5x − 241.8	0.9998	0.15	0.70	196.3	54.7

Table 1. *Cont.*

No.	Calibration Curves [1]	R_2	LOD [2] (μg/mL)	LOQ [3] (μg/mL)	Content in AFr (μg/g)	Content in AF (μg/g)
AF−34	y = 352.5x − 292.2	0.9992	0.10	1.55	26.2	21.4
AF−36	y = 77.6x − 46.5	0.9997	0.45	2.75	143.7	37.3
AF−37	y = 57.2x + 121.7	0.9998	0.15	1.70	15.1	7.6
AF−38	y = 33.3x + 579.1	0.9998	0.20	1.50	171.6	21.1
AF−39	y = 304.6x + 280.8	1.0000	0.10	1.20	19.6	8.9
AF−40	y = 243.0x + 294.2	0.9995	0.05	0.50	57.9	40.7

[1] y is the peak area, x is the concentration (μg/mL). [2] LOD is the detection limit which means the lowest detectable concentration. [3] LOQ is the quantification limit which means the lowest concentration that can be quantified. AF represents *A. tenuissimum* flower extract. AFr represents the mixture of ethyl acetate part and *n*−butanol part.

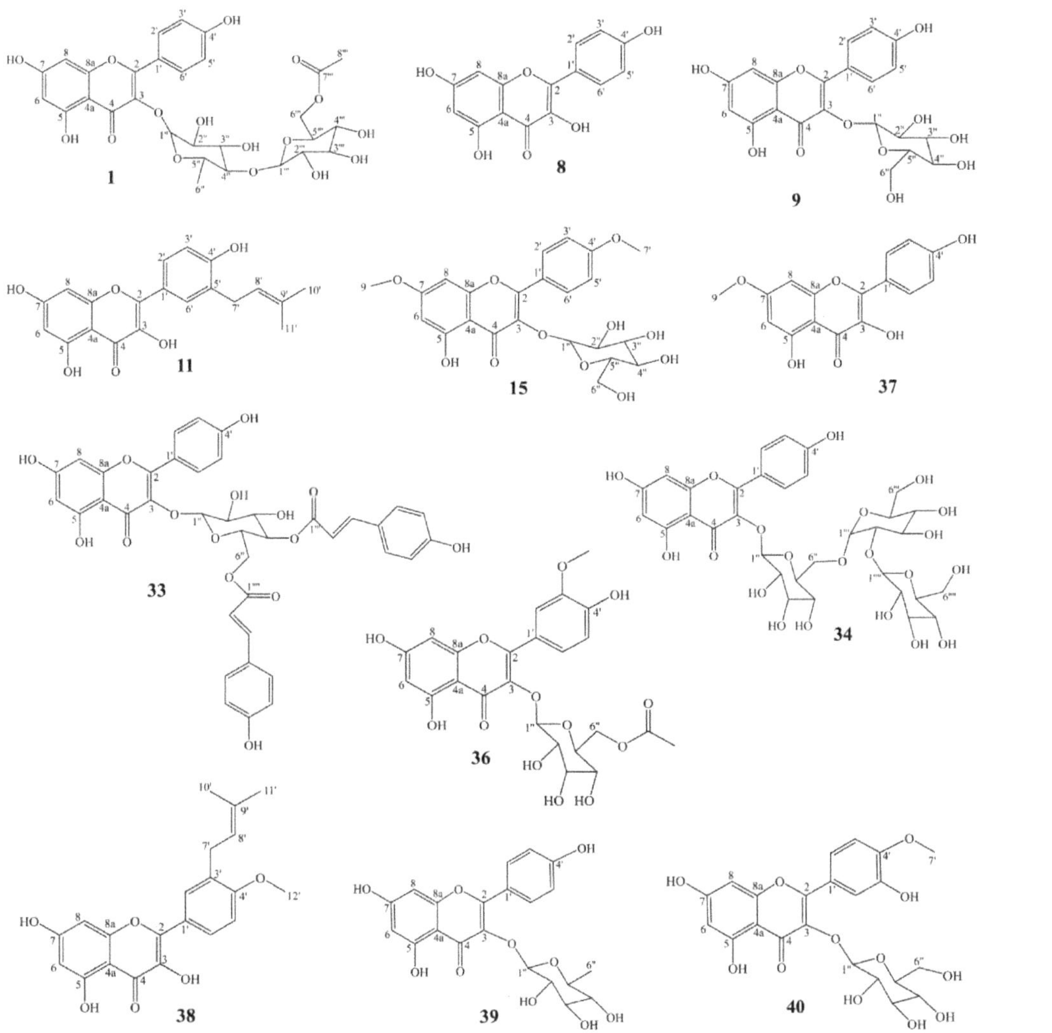

Figure 2. The structures of twelve flavonoids from *A. tenuissimum* flower.

3.2. *Network Pharmacology Analysis*

3.2.1. Potential Targets Analysis of Active Ingredients and T2DM

Three active ingredients were collected as kaempferol (MOL000422), isolicoflavonol (MOL004949), and anisodamine (MOL005409) under OB $\geq$ 30% and DL $\geq$ 0.18 filter conditions, and the active ingredients—related targets were obtained from TCMSP database. The potential targets of T2DM were collected from the DisGeNET database. The "ingredients—targets" interaction network is shown in Figure 3A. A Venn diagram was used to analyze the ingredients' targets and the disease's targets. A total of 42 ingredient-diseases common targets were collected and used for further analysis (Figure 3B). Some information on those common targets is shown in Table S1.

3.2.2. PPI Network Analysis

Protein—protein interaction (PPI) network analysis of ingredients—T2DM intersection targets was conducted by using the STRING platform (Figure 3C). The average node degree was 9.95 and the p—value was less than 1.0e-16. The STRING analysis results showed that AKT1, PPARG, PTGS2, and other proteins had high interaction. Then, the STRING analysis results were imported to Cytoscape software Centiscape 2.2 plug—in to analyze high interacting protein modules (Figure 3D). The Centiscape 2.2 analysis results showed the top three highly interacting targets were AKT1, PPARG, and PTGS2. The above results showed that AKT1 (RAC—alpha serine/threonine—protein kinase, PDB ID: 1UNQ), PPARG (peroxisome proliferator activated receptor gamma, PDB ID: 2PRG), and PTGS2 (Prostaglandin G/H synthase 2, PDB ID: 5F1A) might be good potential targets for molecular docking.

3.2.3. GO Analysis and KEGG Pathway Enrichment

Metascape platform was used to analyze the 42 common targets of active ingredients—T2DM. The visualization results showed that the potential targets function related to biological processes (BP), cell components (CC), and molecular functions (MF). A total of 582 BP GO terms, 31 CC GO terms, and 56 MF GO terms were enriched, and the significantly enriched terms ($p \leq 0.01$) were selected for analysis, respectively (Figure 3E). The results showed that the top 20 enriched BP GO terms involved response to lipopolysaccharide, negative regulation of apoptotic signaling pathway, inflammatory response, regulation of inflammatory response, and other biological processes—related terms. The significantly enriched CC GO terms included membrane raft, receptor complex, side of membrane, transcription regulator complex, and other cell components—related terms. The potential targets of MF GO terms were enriched in nuclear receptor activity, protein homodimerization activity, carboxylic acid binding, protein domain—specific binding, and other molecular function—related terms. For further potential mechanism exploring, the KEGG pathway enrichment analysis was conducted. The KEGG pathway enrichment results (Figure 3F) showed that fourteen pathways were enriched including lipid and atherosclerosis, pathways in cancer, and other related signaling pathways, suggesting the possible involvement and the potential mechanisms for *A. tenuissimum* flower in treating diabetes.

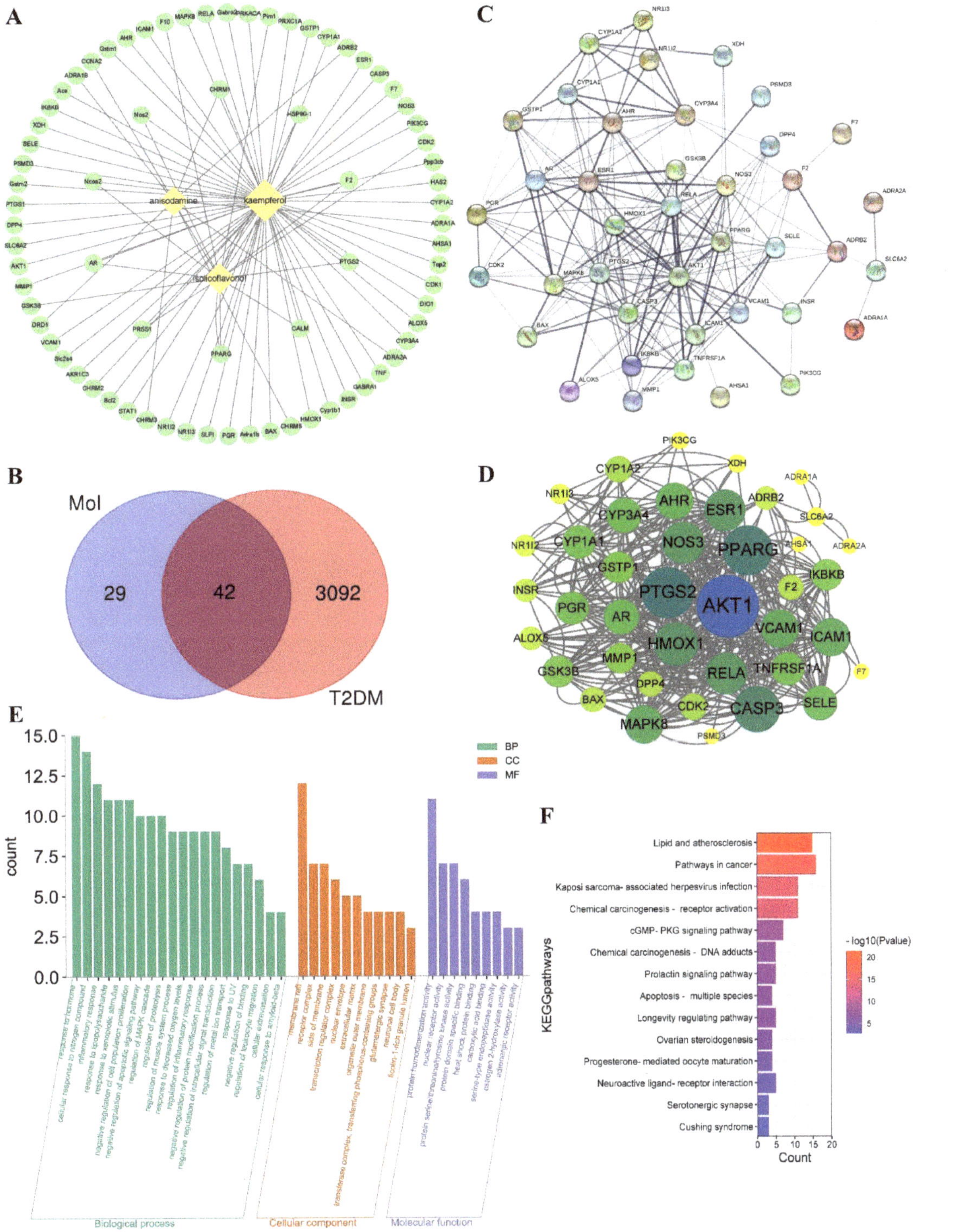

Figure 3. Network construction of ingredients−targets (**A**) and common targets Venn diagram of ingredients and T2DM (**B**), the "diamond" nodes represent the active ingredients. The "ellipse" nodes

represent the potential targets. Protein−protein interaction (PPI) network analysis of potential targets of active ingredients−T2DM by STRING platform (**C**), and the high interaction modules analyzed by Centiscape 2.2 plug−in (**D**), the node size and color represented the connectivity degree. Gene ontology (GO) analysis (**E**) and Kyoto encyclopedia of genes and genomes (KEGG) pathway enrichment analysis (**F**) of active ingredients−T2DM potential targets.

3.2.4. Molecular Docking

Kaempferol was the only flavonoid of the active ingredients screened by network pharmacology analysis. Molecular docking was used to simulate the binding of kaempferol to the potential key targets AKT1, PPARG, and PTGS2. α−Glucosidase is a class of enzymes associated with glucose metabolism, and α−glucosidase (GAA, PDB ID: 5NN4) was also used as a receptor. Autodock (v.4.2.6) software was used to conduct molecular docking. The diagrams of ligand and potential target docking results are shown in Figure 4. It showed that kaempferol could bind to ARG−41, GLU−40, and LYS−39 residues of AKT1, the binding energy was -6.42 kcal·mol^{-1}. The binding energy of kaempferol with GAA residues (ASP−185, LYS−184, GLU−192, and ARG−189) was -5.58 kcal·mol^{-1}. The residues were binding sites for kaempferol on proteins. Detailed information on ligands and potential targets is listed in Table S2.

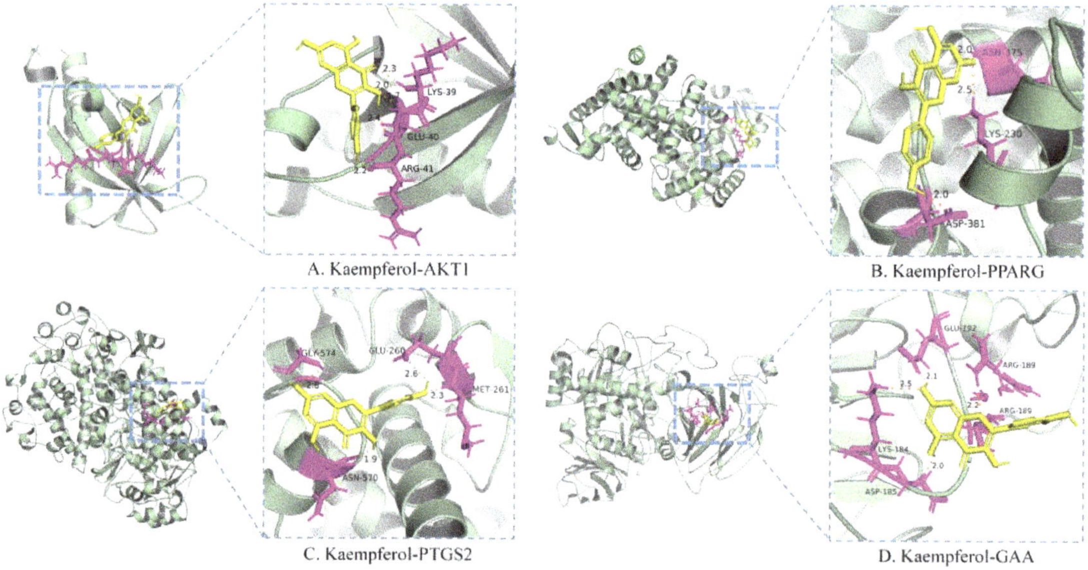

Figure 4. Molecular docking of the active ingredient and potential target by Autodock software.

3.3. *α−Glucosidase Inhibitory Activity of Flavonoids*

The results of the α−glucosidase inhibitory activity of flavonoids from *A. tenuissimum* flower were analyzed. It could be seen intuitively from the histogram in Figure 5 (More data listed in Table S3) that 12 flavonoids had superior α−glucosidase inhibitory activity than acarbose. Among them, AF−8 (kaempferol) showed superior α−glucosidase inhibitory activity (0.135 ± 0.011 mM) compared with acarbose (0.750 ± 0.002 mM). And the molecular docking results showed that kaempferol had high bonding energy with α−glucosidase (GAA), which was responsible for the high α−glucosidase inhibitory activity of kaempferol. The binding of kaempferol to GAA might replace the binding of substrate to GAA to exhibit the α−glucosidase inhibitory activity, which was consistent with the above experimental results. Other *A. tenuissimum* flower flavonoids were based on the kaempferol backbone with different substitution positions and substituents. Structurally, the C−ring phenolic hydroxyl hydrogen of AF−9 (astragalin) was substituted by the glucose group, and the experimental results showed that the α−glucosidase inhibitory activity of AF−8 was superior to that of AF−9 (0.264 ± 0.021 mM). The IC_{50} value of 0.506 ± 0.001 mM for compound AF−15 (kaempferol−4′,7−dimethyl−3−O−glucoside), in which the two hydroxyl hydrogens at the 4′ and 7 positions were substituted with methyl groups and the hydroxyl hydrogens at the 3 position was substituted with glucose groups, which had the smallest IC_{50} among the 12 flavonoids indicating that the α−glucosidase inhibitory activity of AF−15 was the lowest among these 12 flavonoids. The above results suggested that the reduction of phenolic hydroxyl groups in the B and C rings of the flavonoid backbone and the occurrence of glucosyl substitution led to the reduction of α−glucosidase inhibitory activity, which was consistent with the results of previous studies [24,26].

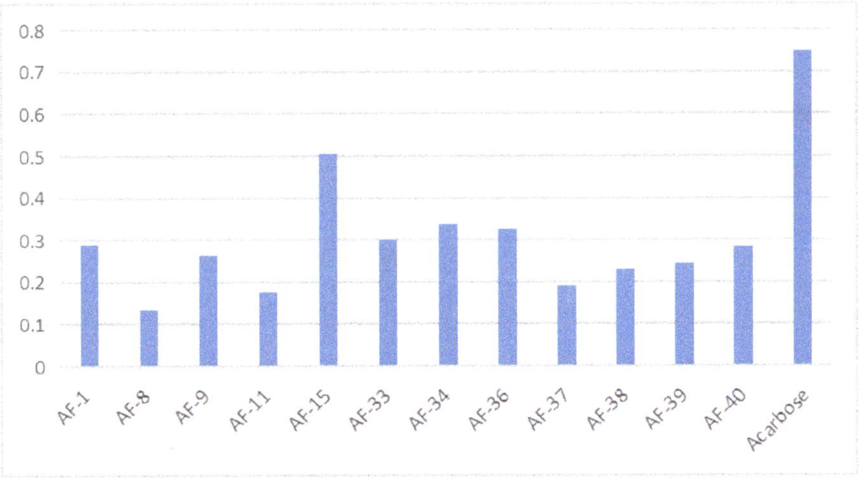

Figure 5. The result of *A. tenuissimum* flower flavonoids α−glucosidase inhibitory activity.

3.4. *In Vivo Experiments Results*

3.4.1. Effect on Body Weight, FBG, OGTT, and ITT of Diabetic Mice

The body weight, fasting blood glucose, OGTT, and ITT were measured and shown in Figure 6. After STZ injection, the body weight of the model group (34.32 ± 1.33 g) was lower than the normal group (39.57 ± 2.57 g). Compared with the normal group (5.2 ± 0.21 mmol·L^{-1}), the FBG level was significantly increased in the model group (17.32 ± 0.46 mmol·L^{-1}). The results showed that the T2DM model was successfully constructed. The OGTT and ITT were conducted after the treatment procedure. The results of OGTT showed that after 1.0 g per kg·bw glucose solution intragastrically, the blood glucose levels significantly increased in thirty minutes, and then gradually decreased. The peak serum glucose level of the model group was about 1.42 times the initial level. The peak blood glucose levels of the two treatment groups were about 1.60 times of initial levels. The ITT results showed a decreasing trend after injecting 1.0 U per kg·bw insulin solution, and the downward trends tended to be flat after 40 min. At 60 min after injection, the blood glucose in the AF group decreased by 46.0%, the blood glucose in the AFr group decreased by 52.2%, and the blood glucose in the Model group decreased by 36.2% compared to the initial values. The results showed that *A. tenuissimum* flower could decrease blood glucose, and improve diabetic mice's blood glucose metabolism. And the AFr group with higher flavonoid content showed a better effect than the AF group.

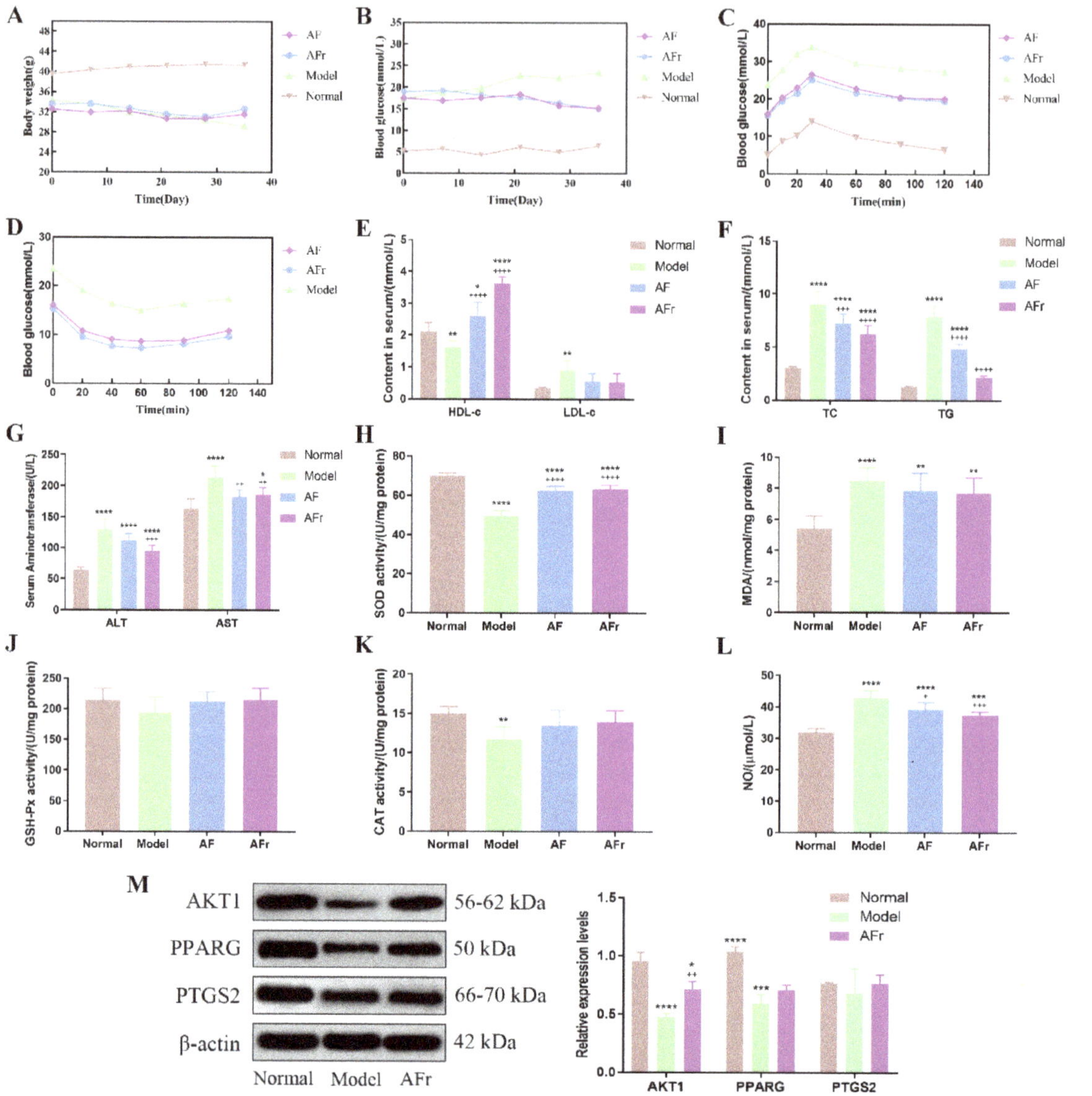

Figure 6. Effects on body weight (**A**), fasting blood glucose (FBG, **B**), oral glucose tolerance test (OGTT, **C**), insulin tolerance test (ITT, **D**), high/low−density lipoprotein cholesterol (HDL−c & LDL−c, **E**), total cholesterol & triglyceride (TC &TG, **F**), alanine aminotransferase & aspartate aminotransferase (ALT & AST, **G**), superoxide dismutase (SOD, **H**), malonaldehyde (MDA, **I**), glutathione peroxidase (GSH−Px, **J**), catalase (CAT, **K**), and nitric oxide (NO, **L**) (six mice per group). The expression levels of RAC−alpha serine/threonine−protein kinase (AKT1), peroxisome proliferator activated receptor gamma (PPARG), and prostaglandin G/H synthase 2 (PTGS2) proteins in liver tissue (**M**), data were shown as mean ± SD (n = 3). (compared to the normal group, * $p < 0.05$, ** $p < 0.01$, *** $p < 0.005$, **** $p < 0.001$; compared to the model group, + $p < 0.05$, ++ $p < 0.01$, +++ $p < 0.005$, ++++ $p < 0.001$).

3.4.2. Effect on Serum Biomarkers of Diabetic Mice

Figure 6 showed that the insulin level of the model group was significantly decreased by 43.75% ($p < 0.001$) in the normal group while increased in the treatment groups. Compared with the normal group, the high−density lipoprotein cholesterol (HDL−c) level was decreased by 24.01% ($p < 0.01$) in the model group, and the HDL−c levels of two treatment groups were significantly increased especially in the AFr group (70.49%, $p < 0.001$). The low−density lipoprotein cholesterol (LDL−c) level of the model group was increased by 174.48% ($p < 0.01$) compared with the normal group. In the two treatment groups, the LDL−c levels were decreased. The total cholesterol (TC) and triglyceride (TG) levels of the model group were significantly increased than the normal group ($p < 0.001$). The TC and TG levels of the two treatment groups were decreased than the model group, but it still higher than the normal group. The alanine aminotransferase (ALT) and aspartate aminotransferase (AST) levels of the model group were significantly increased by 101.62% and 30.75% to the normal group, respectively. The levels of superoxide dismutase (SOD), catalase (CAT), and glutathione peroxidase (GSH−Px) were decreased in the model group compared with the normal group. After administration, the levels of SOD, CAT, and GSH−Px were increased in the AF group and the AFr group. The levels of malonaldehyde (MDA) and nitric oxide (NO) were significantly up−regulated ($p < 0.001$) in the model group and decreased after treatment. The histopathological results of the liver (Figure S4) showed that the normal group liver tissue had a clear structure and normal cell morphology. The model group's liver tissue had many obvious fatty vacuoles, and the hepatocytes were enlarged and disordered. After treatment, the liver tissue injury degree was reduced, and the morphology tended to that of healthy mice. The above results showed that the AFr group with higher flavonoid content could better improve the abnormal lipid metabolism and ameliorate the oxidative stress and injury degree of the liver in diabetic mice. Flavonoid−rich *A. tenuissimum* flower could be a potential therapeutic material for diabetes.

3.4.3. Western Blot Analysis

According to the above results, the AFr group with higher content of *A. tenuissimum* flower flavonoids showed an excellent anti−diabetic effect on mice. The protein expression levels in liver tissue were measured using Western blotting. All data were normalized by β−actin. As shown in Figure 6M, compared to the normal group, the expression levels of AKT1 ($p < 0.001$) and PPARG ($p < 0.005$) were significantly decreased in the model group. The expression level of PTGS2 was decreased without a significant difference. The expression level of AKT1 in the AFr group ($p < 0.01$) was up−regulated significantly compared with the model group. Also, the expression levels of PPARG and PTGS2 in the AFr group were up−regulated. The results indicated that the administration of AFr could increase the expression of these three proteins, especially AKT1 and PPARG. Previous studies showed that AKT1 might play a role in insulin−related pathways [27], and the increased expression of AKT1 plays a role in the treatment of T2DM and prevented its complications [28]. PPARG is a transcription factor revolving around adipocyte differentiation, lipid metabolism, and inflammation [29]. Some evidence has shown that PPARG is a candidate gene for obesity, insulin resistance, and T2DM [30–32]. PTGS2 has been confirmed to play a role in treating liver injury [33]. The KEGG pathways analysis indicated that the AKT1 participated in the TNF signaling pathway, lipid and atherosclerosis pathway, and regulation of lipolysis in the adipocytes pathway. The PPARG was the key protein of the PPAR signaling pathway, which revolved around adipocyte differentiation and lipid metabolism. The PTGS2 was related to the TNF signaling pathway, metabolism pathway, regulation of lipolysis in adipocytes, VEGF signaling pathway, and chemical carcinogenesis pathway. The results indicated that *A. tenuissimum* flower flavonoids could increase AKT1, PPARG, and PTGS2 expression levels, and play a role in T2DM by activating lipid metabolism and other related pathways.

3.4.4. Gut Microbiota Analysis

As shown in Figure 7A, the curves were gradually gentle as the sequencing depth increased, which reflected the diversity and abundance of samples. Figure 7B showed a total of 23,880 operational taxonomic units (OTUs) were detected from all samples. Venn diagram showed the number of unique or common species in four groups, and the differential species were the focus of subsequent research. The Chao 1 index, Shannon index, and Simpson index were used to reflect the results of alpha diversity analysis (Figure 7C). The levels of the Chao 1 index in the two treatment groups were higher than in the normal group. The increase of the Chao 1 index indicated that the species richness increased. The Shannon index and Simpson index reflected that the two treatment groups had higher species diversity than the normal mice. The results indicated that *A. tenuissimum* flower had a beneficial effect on improving the species richness and diversity of T2DM mice's gut microbiota. The heat map (Figure 7D) visualized the differences and abundance of species composition. At the phylum level (Figure 7E), the species with higher relative abundance were *Firmicutes* (Normal: 68.97%, AF: 65.89%, AFr: 68.67%, Model: 59.44%), *Bacteroidetes* (Normal: 20.01%, AF: 22.63%, AFr: 17.45%, Model: 30.32%), *Proteobacteria*, and *Actinobacteria*, which together accounted for about over 95%. Compared with the normal group, the *Firmicutes* phylum was slightly lower, the *Bacteroidetes* phylum was higher, and the ratio of *Firmicutes/Bacteroidetes* (F/B) was lower in the model group. The F/B ratio in the AFr group (F/B = 3.93) was much higher than that in the model group (F/B = 1.96) and slightly higher than that in the normal group (F/B = 3.04) and the AF group (F/B = 2.91), which suggested that *A. tenuissimum* flower could improve the ratio of F/B in the intestine of diabetic mice and thus alleviate the symptoms of obesity. At the genus level (Figure 7F), the species with higher abundance were *Lactobacillus* (Normal: 48.07%, AF: 20.66%, AFr: 37.16%, Model: 27.63%), *Oscillospira, Bacteroides, Corynebacterium, Weissella, Ruminococcaceae,* and *Mucispirillum*. *Lactobacillus* had an important role in maintaining the health of the organism. Studies had shown that *Lactobacillus* could regulate gut microbiota, and reduce liver damage associated with T2DM [34]. It also had a regulatory effect on lipid metabolism, reduced high−fat diet−induced obesity in mice, as well as lowered blood glucose [35]. The relative abundance of *Lactobacillus* spp. was decreased in the model mice compared to normal mice. The relative abundance of *Lactobacillus* spp. in the AFr group was close to that of the normal group [36]. *Oscillospira* belongs to *Firmicutes*, which are widely found in the intestine of animals and humans and are positively associated with health [37].

To investigate the species differences among samples, beta diversity analysis was performed on all samples. Principal coordinates analysis (PCoA) and the between−group difference analysis (Figure 7G,H) were performed based on weighted unifrac. The results showed that the contribution of PCo1 was 24.7%, and PCo2 was 17.6%. The samples of the four groups were significantly different, especially between the normal group and the model group. The AFr group was closer to the normal group than the AF group. The between−group difference analysis results intuitively reflected the similarity and variability between samples. It could be seen that the gut microbiota of the normal group significantly differed from the model group, and the similarity of the gut microbiota between the two treatment groups was higher. And the distances between the two treatment groups and the normal group were closer than that between the treatment groups and the model group. This was consistent with the results of the PCoA analysis.

As shown in Figure 7I, the key species at the phylum level were *Firmicutes, Bacteroidetes, Actinobacteria, Proteobacteria, Verrucomicrobia*, and this results were generally consistent with the results of species composition analysis. The results of the LEfSe analysis were shown in Figure 7J. When LDA values were equal to 4, a total of thirteen species with significant group differences ($p < 0.5$) in the AF group and the model group were found, while there had no significant differences between the AFr group and the normal group. Among them, four species in the AF group had significantly higher relative abundance than the model group, including *Clostridia, Clostridiales, Ruminococcaceae*, and *Oscillospira*. Most *Clostridia* are not pathogenic, and only a few *Clostridia* are pathogenic. *Ruminococcaceae* plays a crucial role in metabolism. *Ruminococcaceae* is one of the most effective bacteria to decompose carbohydrates and a key bacterium to degrade resistant starch, which can stabilize the intestinal barrier. *Oscillospira* belongs to the *Firmicutes*, which widely exists in animal and human intestines and is positively related to health [37].

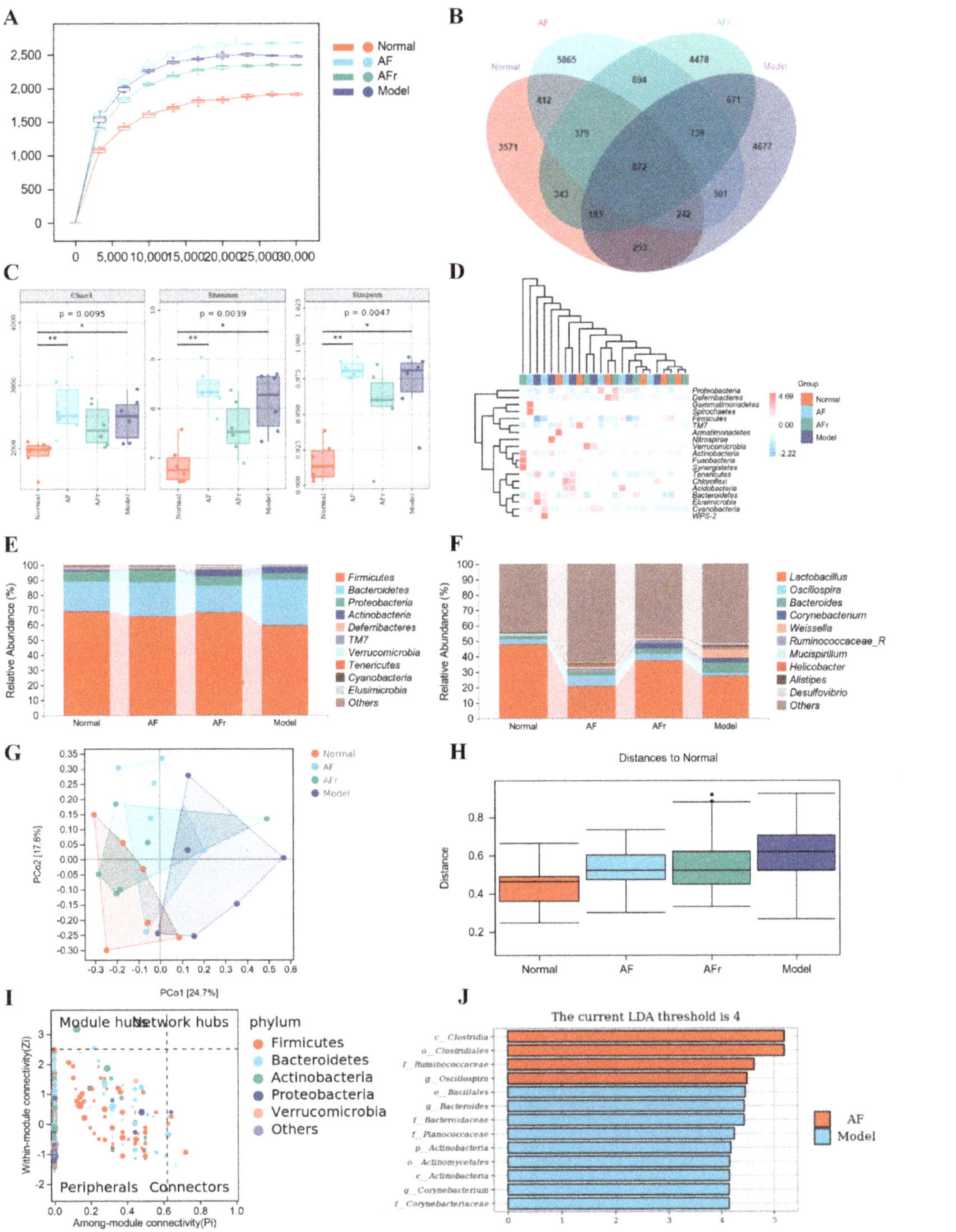

Figure 7. Effect of *A. tenuissimum* flower on gut microbiota consumption. Rarefaction curve (**A**), ASV/OTU Venn diagram (**B**), alpha diversity analysis (**C**, * $p < 0.05$, ** $p < 0.01$), the heat map of species composition at the phylum level (**D**), the gut microbiota abundance at the phylum level (**E**), and the gut microbiota abundance at the genus level (**F**), PCoA analysis (**G**), permutational multivariate analysis of variance (**H**), ZIPI−score diagram (**I**), and LEfSe analysis (**J**).

The KEGG pathway enrichment (Figure 8) results showed that many pathways had higher abundance, especially metabolism–related pathways. The KEGG pathways abundance heat map of all samples was shown in Figure S5. The metabolism–related pathways were arranged by abundance as carbohydrate metabolism, metabolism of cofactors and vitamins, amino acid metabolism, metabolism of terpenoids and polyketides, lipid metabolism, metabolism of other amino acids, energy metabolism, glycan biosynthesis and metabolism, nucleotide metabolism, xenobiotics biodegradation and metabolism, and biosynthesis of other secondary metabolites. Other pathways with high abundance include replication repair, translation, folding, sorting and degradation, membrane transport, cell motility, and other pathways. The results showed that the therapeutic effect of *A. tenuissimum* flower on T2DM was related to the modulation of metabolism-related pathways including carbohydrate metabolism, energy metabolism, lipid metabolism, and glycan biosynthesis and metabolism.

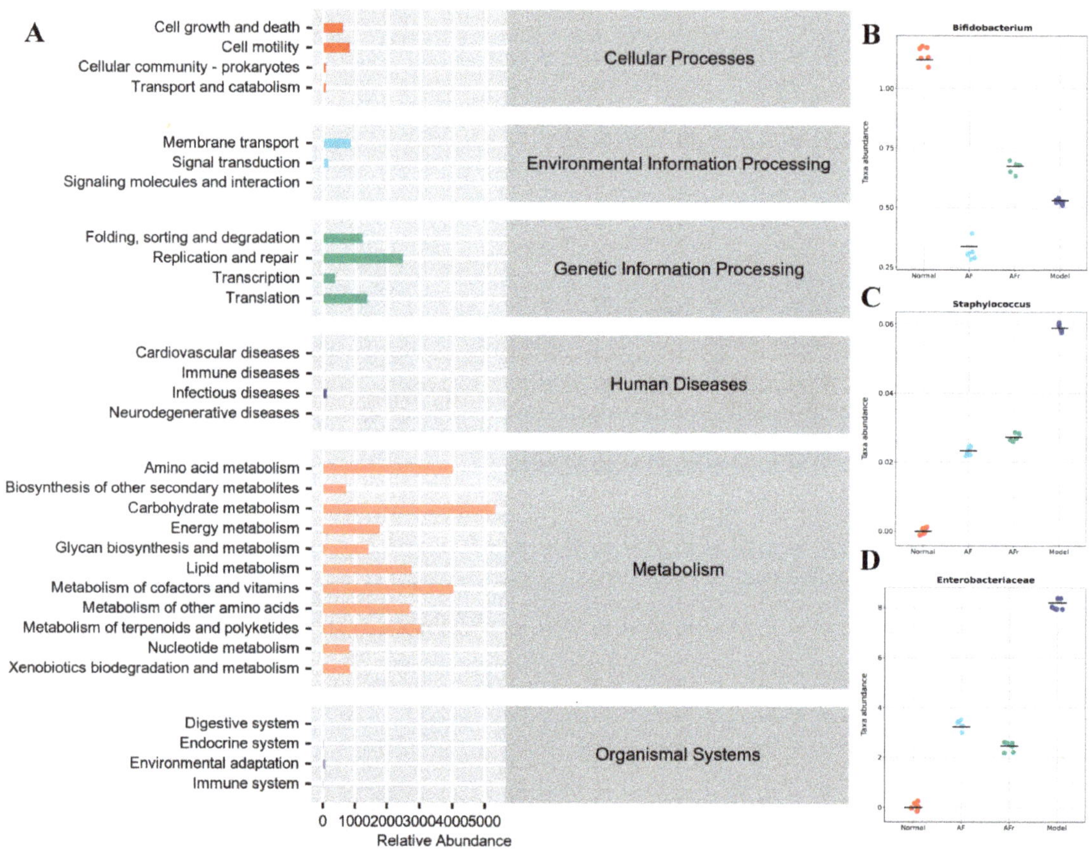

Figure 8. Metabolic pathways enrichment (**A**), and species composition difference analysis at the genus level between groups: (**B**) *Bifidobacterium*, (**C**) *Staphylococcus*, and (**D**) *Enterobacteriaceae*.

Species composition difference analysis at the genus level between groups was conducted based on the above pathways. *Enterobacteriaceae*, *Staphylococcus*, and *Bifidobacterium* had high abundance in the model group than in the other groups (Figure 8). *Staphylococcus* was a common purulent coccus that predisposes to a variety of purulent infections [38]. *Enterobacteriaceae* was usually found in fresh produce and many *Enterobacteriaceae* were known pathogens, e.g., *Salmonella* and pathogenic *E. coli* [39]. The microorganisms of the *Enterobacteriaceae* family induced inflammation and infection in vivo [40]. The results showed that the diabetic model mice had higher inflammation levels due to a significant increase of *Enterobacteriaceae* compared to the normal group. The *Bifidobacterium*, a probiotic with significant health benefits, was higher in the AFr group compared to the model group. *Bifidobacterium* could stimulate intestinal motility, purify the intestinal environment, stimulate the immune system, and improve immunity and anti−infection ability [41,42]. The experiment results showed that the abundance of *Enterobacteriaceae* and *Staphylococcus* microorganisms decreased, and the abundance of *Bifidobacterium* increased in the two treatment groups, indicating that *A. tenuissimum* flower could improve the inflammation in diabetic mice and enhance the anti-infection ability, thus alleviating the symptoms of diabetes.

4. Conclusions

To summarize, forty compounds were isolated and characterized by the *A. tenuissimum* flower. Twelve flavonoids were quantitatively analyzed as the main active ingredients of *A. tenuissimum* flower. Kaempferol had a higher content in *A. tenuissimum* flower compared with other flavonoids. Based on the network pharmacology analysis and molecular docking results, kaempferol could bind to AKT1, PPARG, PTGS2, and GAA. The expression levels of key target proteins AKT1, PPARG, and PTGS2 in liver tissue were increased in the AFr group by Western blot analysis. The flavonoids had higher α−glucosidase inhibitory activities than acarbose, especially kaempferol. The results of an animal experiment showed that *A. tenuissimum* flower could decrease blood glucose and lipid accumulation. The AFr group showed a better improvement effect in T2DM mice than the AF group. Based on quantitative analysis results, the total content of flavonoids in the AFr sample was higher than that of the AF sample. At the same dose, the AFr group mice were administrated with more flavonoids. That might be the reason for the AFr group showing an excellent alleviative effect on T2DM than that of the AF group. A dose of 100 mg per kg·bw of AF (AFr) was used on the T2DM mice model. The human equivalent dose (HED) for AF (AFr) was calculated as 8.1 mg/kg, which equates to a 486.5 mg dose of AF (AFr) for a 60 kg adult [43]. The plant *A. tenuissimum* flower is edible after being fried or pickled. The *A. tenuissimum* flower extract also could be made as a sauce and added to the daily diet as a condiment. Under the current dose, a consumption of *A. tenuissimum* flower might have potential health benefits for human and this data have some reference values for *A. tenuissimum* flower consumption. The gut microbiota analysis results showed that *A. tenuissimum* flower could modulate the species structure and abundance of diabetic mice, which might be related to the flavonoid content. The *A. tenuissimum* flower could modulate the ratio of F/B, decrease the level of *Enterobacteriaceae* and *Staphylococcus*, increase the level of *Bifidobacterium*, and activate carbohydrate metabolism, energy metabolism, lipid metabolism, glycan biosynthesis and metabolism, and other pathways related to T2DM and metabolism. It is indicated that *A. tenuissimum* flower could improve glycolipid metabolic disorders and inflammation in diabetic mice by modulating gut microbiota. *A. tenuissimum* flower is a healthy vegetable for daily consumption and also could be a potential medicinal ingredient. This research might provide support for subsequent applications of the *A. tenuissimum* flower, and will provide evidence for further study of the anti−diabetic mechanism of the *A. tenuissimum* flower and its active ingredients.

Supplementary Materials: The following supporting information can be downloaded at: https://www.mdpi.com/article/10.3390/nu14193980/s1, Figure S1. The separation process of *A. tenuissimum* flower; Figure S2. HPLC-DAD chromatograms of AF and AFr at 254 nm; Figure S3. The structure of forty compounds from *A. tenuissimum* flower; Figure S4. Histopathological results of H&E stained liver tissue; Figure S5. The heat map of the KEGG pathway abundance of all samples; Table S1. Common targets of active ingredients and T2DM; Table S2. Information of ligands and potential targets; Table S3. The α-glucosidase inhibitory activity of flavonoids from *A. tenuissimum* flower; NMR data [20,44–82].

Author Contributions: S.-S.Z.: Conceptualization, Methodology, Investigation, Visualization, Writing—original draft; Y.-F.H.: Methodology; S.-J.L.: Investigation; S.G.: Conceptualization; C.-T.H.: Writing—review & editing; N.-S.B.: Supervision, Writing—review & editing, Funding Acquisition. All authors have read and agreed to the published version of the manuscript.

Funding: This research was funded by the National Natural Science Foundation of China (No. 31570348).

Institutional Review Board Statement: The animal experiments in this study operated in strict accordance with the "Chinese Animal Ethics Standards" and "Chinese Animal Ethics Guidelines", which were approved by the Northwestern University Animal Ethics Committee.

Informed Consent Statement: Not applicable.

Data Availability Statement: All data generated or analyzed during this study are included in this published article (and its Supplementary Materials file).

Acknowledgments: The authors wish to thank the National Natural Science Foundation of China (No. 31570348) for financial support.

Conflicts of Interest: The authors declare no conflict of interest.

References

1. Zhang, Y.; Cao, Y.; Chen, J.; Qin, H.; Yang, L. A new possible mechanism by which punicalagin protects against liver injury induced by type 2 diabetes mellitus: Upregulation of autophagy via the Akt/FoxO3a signaling pathway. *J. Agric. Food Chem.* **2019**, *67*, 13948–13959. [CrossRef] [PubMed]
2. Ma, G.; Chai, X.; Hou, G.; Zhao, F.; Meng, Q. Phytochemistry, bioactivities and future prospects of mulberry leaves: A review. *Food Chem.* **2022**, *372*, 131335. [CrossRef] [PubMed]
3. Yan, X.; Zhang, Y.; Peng, Y.; Li, X. The water extract of Radix scutellariae, its total flavonoids and baicalin inhibited CYP7A1 expression, improved bile acid, and glycolipid metabolism in T2DM mice. *J. Ethnopharmacol.* **2022**, *293*, 115238. [CrossRef] [PubMed]
4. Duru, K.C.; Mukhlynina, E.A.; Moroz, G.A.; Gette, I.F.; Danilova, I.G.; Kovaleva, E.G. Anti-diabetic effect of isoflavone rich kudzu root extract in experimentally induced diabetic rats. *J. Funct. Foods* **2020**, *68*, 103922. [CrossRef]
5. Kumar, A.; Sreedharan, S.; Kashyap, A.K.; Singh, P.; Ramchiary, N. A review on bioactive phytochemicals and ethnopharmacological potential of purslane (*Portulaca oleracea* L.). *Heliyon* **2022**, *8*, e08669. [CrossRef]
6. Xiao, J. Recent advances in dietary flavonoids for management of type 2 diabetes. *Curr. Opin. Food Sci.* **2022**, *44*, 100806. [CrossRef]
7. Shi, X.Q.; Yue, S.J.; Tang, Y.P.; Chen, Y.Y.; Zhou, G.S.; Zhang, J.; Zhu, Z.H.; Liu, P.; Duan, J.A. A network pharmacology approach to investigate the blood enriching mechanism of Danggui buxue Decoction. *J. Ethnopharmacol.* **2019**, *235*, 227–242. [CrossRef]
8. Zhou, W.; Dai, Y.; Meng, J.; Wang, P.; Wu, Y.; Dai, L.; Zhang, M.; Yang, X.; Xu, S.; Sui, F.; et al. Network pharmacology integrated with molecular docking reveals the common experiment-validated antipyretic mechanism of bitter-cold herbs. *J. Ethnopharmacol.* **2021**, *274*, 114042. [CrossRef]
9. Zhou, C.; Liu, L.J.; Zhuang, J.; Wei, J.Y.; Zhang, T.T.; Gao, C.; Liu, C.; Li, H.Y.; Si, H.Z.; Sun, C.G. A systems biology-based approach to uncovering molecular mechanisms underlying effects of traditional chinese medicine qingdai in chronic myelogenous leukemia, involving integration of network pharmacology and molecular docking technology. *Med. Sci. Monit.* **2018**, *24*, 4305–4316. [CrossRef]
10. Luo, Q.; Shi, X.; Ding, J.; Ma, Z.; Chen, X.; Leng, Y.; Zhang, X.; Liu, Y. Network pharmacology integrated molecular docking reveals the Antiosteosarcoma Mechanism of Biochanin A. *Evid. Based Complement. Altern. Med.* **2019**, *2019*, 1–10.
11. Zhang, S.S.; Guo, S.; Zheng, Z.J.; Liu, S.J.; Hou, Y.F.; Ho, C.T.; Bai, N.S. Characterization of volatiles in *Allium tenuissimum* L. flower by headspace-gas chromatography-olfactometry-mass spectrometry, odor activity values, and the omission and recombination experiments. *LWT-Food Sci. Technol.* **2021**, *151*, 112144. [CrossRef]
12. Li, M.; Liu, Y.; Zhang, S.; Guo, C.; Li-gang, Y. Study on chemical constituents and antibacterial activity of ethanol extract from *Allium tenuissimum* flowers. *China Condiment* **2020**, *45*, 35–38. (In Chinese)
13. Li, M.; Ding, P.; Li, R.; Zhang, S.; Guo, C. Antioxidant properties and inhibitory effect on nitrosation of extracts from *Allium tenuissimum* flowers. *Food Res. Dev.* **2019**, *45*, 46–51. (In Chinese)
14. Li, M.; Hou, J.; Liu, Y.; Wei, L.; Guo, C. Study on antioxidation of ethanol extract from *Allium tenuissimum* L. flowers *in vitro* and its inhibition on α-glucosidase. *China Condiment* **2022**, *47*, 30–34. (In Chinese)
15. Kang, M.G.; Yi, S.H.; Lee, J.S. Production and characterization of a new α-glucosidase inhibitory peptide from *Aspergillus oryzae* N159-1. *Mycobiology* **2013**, *41*, 149–154. [CrossRef]
16. Osonoi, T.; Saito, M.; Mochizuki, K.; Fukaya, N.; Muramatsu, T.; Inoue, S.; Fuchigami, M.; Goda, T. The α-glucosidase inhibitor miglitol decreases glucose fluctuations and inflammatory cytokine gene expression in peripheral leukocytes of Japanese patients with type 2 diabetes mellitus. *Metabolism* **2010**, *59*, 1816–1822. [CrossRef]
17. Jang, J.H.; Park, J.E.; Han, J.S. Scopoletin inhibits α-glucosidase *in vitro* and alleviates postprandial hyperglycemia in mice with diabetes. *Eur. J. Pharmacol.* **2018**, *834*, 152–156. [CrossRef]

18. Hu, P.; Li, D.H.; Jia, C.C.; Liu, Q.; Wang, X.F.; Li, Z.L.; Hua, H.M. Bioactive constituents from *Vitex negundo* var. *heterophylla* and their antioxidant and α-glucosidase inhibitory activities. *J. Funct. Foods* **2017**, *35*, 236–244. [CrossRef]
19. Zhang, Y.; Xiao, G.; Sun, L.; Wang, Y.; Wang, Y.; Wang, Y. A new flavan-3-ol lactone and other constituents from *Euonymus alatus* with inhibitory activities on a-glucosidase and differentiation of 3T3-L1 cells. *Nat. Prod. Res.* **2013**, *27*, 1513–1520. [CrossRef]
20. Gou, S.H.; Liu, J.; He, M.; Qiang, Y.; Ni, J.M. Quantification and bio-assay of α-glucosidase inhibitors from the roots of *Glycyrrhiza uralensis* Fisch. *Nat. Prod. Res.* **2016**, *30*, 2130–2134. [CrossRef]
21. Sun, J.; Zhang, F.; Yang, M.; Zhang, J.; Chen, L.; Zhan, R.; Li, L.; Chen, Y. Isolation of α-glucosidase inhibitors including a new flavonol glycoside from *Dendrobium devonianum*. *Nat. Prod. Res.* **2014**, *28*, 1900–1905. [CrossRef] [PubMed]
22. Hong, D.F.; Hu, G.L.; Peng, X.R.; Wang, X.Y.; Wang, Y.B.; Al-Romaima, A.; Li, Z.R.; Qiu, M.H. Unusual ent-Kaurane Diterpenes from the coffea cultivar S288 coffee beans and molecular docking to α-glucosidase. *J. Agric. Food Chem.* **2022**, *70*, 615–625. [CrossRef] [PubMed]
23. Xie, L.; Zhang, T.; Karrar, E.; Zheng, L.; Xie, D.; Jin, J.; Chang, M.; Wang, X.; Jin, Q. Insights into an α-glucosidase inhibitory profile of 4,4-dimethylsterols by multispectral techniques and molecular docking. *J. Agric. Food Chem.* **2021**, *69*, 15252–15260. [CrossRef] [PubMed]
24. Zhang, S.S.; Zhang, N.N.; Guo, S.; Liu, S.J.; Hou, Y.F.; Li, S.; Ho, C.T.; Bai, N.S. Glycosides and flavonoids from the extract of *Pueraria thomsonii* Benth leaf alleviate type 2 diabetes in high-fat diet plus streptozotocin-induced mice by modulating the gut microbiota. *Food Funct.* **2022**, *13*, 3931–3945. [CrossRef]
25. Cheng, N.; Chen, S.; Liu, X.; Zhao, H.; Cao, W. Impact of schisandrachinensis bee pollen on nonalcoholic fatty liver disease and gut microbiota in highfat diet induced obese mice. *Nutrients* **2019**, *11*, 346. [CrossRef]
26. Han, L.; Song, J.; Yan, C.; Wang, C.; Wang, L.; Li, W.; Du, Y.; Li, Q.; Liang, T. Inhibitory activity and mechanism of calycosin and calycosin-7-O-β-D-glucoside on α-glucosidase: Spectroscopic and molecular docking analyses. *Process Biochem.* **2022**, *118*, 227–235. [CrossRef]
27. Gu, Z.; Pan, M.; Liu, J.; Yang, M.; Zhang, W.; Mai, K. Molecular cloning of AKT1 and AKT2 and their divergent responses to insulin and glucose at transcriptional level in the liver of Japanese flounder *Paralichthys olivaceus*. *Aquac. Rep.* **2022**, *23*, 101066. [CrossRef]
28. Zhang, L.; Han, L.; Ma, J.; Wu, T.; Wei, Y.; Zhao, L.; Tong, X. Exploring the synergistic and complementary effects of berberine and paeoniflorin in the treatment of type 2 diabetes mellitus by network pharmacology. *Eur. J. Pharmacol.* **2022**, *919*, 174769. [CrossRef]
29. Verma, S.; Srivastava, N.; Banerjee, M. Genetic polymorphisms in TCF7L2 and PPARG genes and susceptibility to Type 2 diabetes mellitus. *Meta Gene* **2021**, *28*, 100864. [CrossRef]
30. Lefterova, M.I.; Haakonsson, A.K.; Lazar, M.A.; Mandrup, S. PPARγ and the global map of adipogenesis and beyond. *Trends Endocrinol. Metab.* **2014**, *25*, 293–302. [CrossRef]
31. Celi, F.S.; Shuldiner, A.R. The role of peroxisome proliferator-activated receptor gamma in diabetes and obesity. *Curr. Diabetes Rep.* **2002**, *2*, 179–185. [CrossRef]
32. Horiki, M.; Ikegami, H.; Fujisawa, T.; Kawabata, Y.; Ono, M.; Nishino, M.; Shimamoto, K.; Ogihara, T. Association of Pro12Ala polymorphism of PPARγ gene with insulin resistance and related diseases. *Diabetes Res. Clin. Pract.* **2004**, *66*, S63–S67. [CrossRef] [PubMed]
33. Xu, G.; Lv, X.; Feng, Y.; Li, H.; Chen, C.; Lin, H.; Li, H.; Wang, C.; Chen, J.; Sun, J. Study on the effect of active components of *Schisandra chinensis* on liver injury and its mechanisms in mice based on network pharmacology. *Eur. J. Pharmacol.* **2021**, *910*, 174442. [CrossRef]
34. El-Baz, A.M.; Shata, A.; Hassan, H.M.; El-Sokkary, M.M.A.; Khodir, A.E. The therapeutic role of *Lactobacillus* and montelukast in combination with metformin in diabetes mellitus complications through modulation of gut microbiota and suppression of oxidative stress. *Int. Immunopharmacol.* **2021**, *96*, 107757. [CrossRef] [PubMed]
35. Wang, M.; Zhang, B.; Hu, J.; Nie, S.; Xiong, T.; Xie, M. Intervention of five strains of *Lactobacillus* on obesity in mice induced by high-fat diet. *J. Funct. Foods* **2020**, *72*, 104078. [CrossRef]
36. Wu, T.; Zhang, Y.; Li, W.; Zhao, Y.; Long, H.; Muhindo, E.M.; Liu, R.; Sui, W.; Li, Q.; Zhang, M. Lactobacillus rhamnosus LRa05 ameliorate hyperglycemia through a regulating glucagon-mediated signaling pathway and gut microbiota in type 2 diabetic mice. *J. Agric. Food Chem.* **2021**, *69*, 8797–8806. [CrossRef]
37. Konikoff, T.; Gophna, U. Oscillospira: A central, enigmatic component of the human gut microbiota. *Trends Microbiol.* **2016**, *24*, 523–524. [CrossRef]
38. Chaudhari, S.S.; Chauhan, H.C.; Sharma, K.K.; Patel, S.S.; Patel, A.C.; Mohapatra, S.K.; Srimali, M.D.; Chandel, B. Antibiotic susceptibility pattern of canine coagulase positive and coagulase negative *Staphylococcus* spp. in a hot and dry region of India. *Top. Companion Anim. Med.* **2022**, *50*, 100679. [CrossRef]
39. Kilonzo-Nthenge, A.; Liu, S.; Hashem, F.; Millner, P.; Githua, S. Prevalence of *Enterobacteriaceae* on fresh produce and food safety practices in small-acreage farms in Tennessee, USA. *J. Consum. Prot. Food Saf.* **2018**, *13*, 279–287. [CrossRef]
40. Palmer, J.D.; Mortzfeld, B.M.; Piattelli, E.; Silby, M.W.; McCormick, B.A.; Bucci, V. Microcin H47: A Class IIb microcin with potent activity against multidrug resistant *Enterobacteriaceae*. *ACS Infect. Dis.* **2020**, *6*, 672–679. [CrossRef]
41. Derrien, M.; Turroni, F.; Ventura, M.; van Sinderen, D. Insights into endogenous *Bifidobacterium* species in the human gut microbiota during adulthood. *Trends Microbiol.* **2022**, *30*, 940–947. [CrossRef] [PubMed]

42. Chen, C.; Li, T.; Chen, G.; Chen, D.; Peng, Y.; Hu, B.; Sun, Y.; Zeng, X. Commensal relationship of three *Bifidobacterial* Species leads to increase of *Bifidobacterium in Vitro* Fermentation of sialylated immunoglobulin G by human gut microbiota. *J. Agric. Food Chem.* **2020**, *68*, 9110–9119. [CrossRef] [PubMed]
43. Reagan-Shaw, S.; Nihal, M.; Ahmad, N. Dose translation from animal to human studies revisited. *FASEB J.* **2008**, *22*, 659–661. [CrossRef] [PubMed]
44. Li, M.X.; Xie, J.; Bai, X.; Du, Z.Z. Anti-aging potential, anti-tyrosinase and antibacterial activities of extracts and compounds isolated from Rosa chinensis cv. 'JinBian'. *Ind. Crops Prod.* **2021**, *159*, 113059. [CrossRef]
45. Zhang, W.Z.; Li, X.L.; Shi, L.G.; Wang, J.L.; Zhao, M.; Zhao, D.F.; Zhang, S.J. Sesquiterpene lactones from *Ixeris sonchifolia* (Bge.) Hance II. *J. Asian Nat. Prod. Res.* **2008**, *10*, 1087–1091. [CrossRef] [PubMed]
46. Nagashima, F.; Suzuki, M.; Asakawa, Y. seco-cuparane-type sesquiterpenoid from the Japanese liverwort *Jungermannia infusca*. *Phytochemistry* **2001**, *56*, 807–810. [CrossRef]
47. Parmar, V.S.; Bisht, K.S.; Malhotra, A.; Jha, A.; Errington, W.; Howarth, O.W.; Tyagi, O.D.; Stein, P.C.; Jensen, S.; Boll, P.M.; et al. A benzoic acid ester from uvaria narum. *Phytochemistry* **1995**, *38*, 951–955. [CrossRef]
48. Erb, B.; Borschberg, H.J.; Arigoni, D. The structure of laricinolic acid and its biomimetic transformation into officinalic acid. *J. Chem. Soc. Perkin Trans. 1* **2000**, *31*, 2307–2309. [CrossRef]
49. Samoylenko, V.; Rahman, M.M.; Tekwani, B.L.; Tripathi, L.M.; Wang, Y.H.; Khan, S.I.; Khan, I.A.; Miller, L.S.; Joshi, V.C.; Muhammad, I. Banisteriopsis caapi, a unique combination of MAO inhibitory and antioxidative constituents for the activities relevant to neurodegenerative disorders and Parkinson's disease. *J. Ethnopharmacol.* **2010**, *127*, 357–367. [CrossRef]
50. Paridhavi, M.; Agrawal, S.S. Isolation and characterization of flowers of *Rosa damascena*. *Asian J. Chem.* **2007**, *19*, 2751–2756.
51. Da Liu, C.; Chen, J.; Wang, J.H. A novel kaempferol triglycoside from flower buds of *Panax quinquefolium*. *Chem. Nat. Compd.* **2009**, *45*, 808–810. [CrossRef]
52. Rezende, F.M.; Ferreira, M.J.P.; Clausen, M.H.; Rossi, M.; Furlan, C.M. Acylated flavonoid glycosides are the main pigments that determine the flower colour of the brazilian native tree *Tibouchina pulchra* (Cham.) Cogn. *Molecules* **2019**, *24*, 718. [CrossRef] [PubMed]
53. Tewari, A.; Bhakuni, R.S. Terpenoid and lipid constituents from *Artemisia annua*. *Indian J. Chem.-Sect. B Org. Med. Chem.* **2003**, *42*, 1782–1785. [CrossRef]
54. Wang, C.Z.; Yu, D.Q. Lignan and acetylenic glycosides from *Aster auriculatus*. *Phytochemistry* **1998**, *48*, 711–717. [CrossRef]
55. Hyun, S.K.; Jung, H.A.; Min, B.S.; Jung, J.H.; Choi, J.S. Isolation of phenolics, nucleosides, saccharides and an alkaloid from the root of *Aralia cordata*. *Nat. Prod. Sci.* **2010**, *16*, 20–25.
56. Nycz, J.E.; Malecki, G.; Morag, M.; Nowak, G.; Ponikiewski, L.; Kusz, J.; Switlicka, A. Arbutin: Isolation, X-ray structure and computional studies. *J. Mol. Struct.* **2010**, *980*, 13–17. [CrossRef]
57. Wei-li, Y.; Jun, T.; Liu-sheng, D. Chemical Constituents of *Gymnotheca involucrata* Pei. *China J. Chinese Mater. Medica* **2001**, *26*, 43–44.
58. Zhou, Y.Z.; Ma, H.Y.; Chen, H.; Qiao, L.; Yao, Y.; Cao, J.Q.; Pei, Y.H. New acetylenic glucosides from *Carthamus tinctorius*. *Chem. Pharm. Bull.* **2006**, *54*, 1455–1456. [CrossRef]
59. Ding, Z.; Liu, Y.; Ruan, J.; Yang, S.; Yu, H.; Chen, M.; Zhang, Y.; Wang, T. Bioactive constituents from the whole plants of *Gentianella acuta* (Michx.) Hulten. *Molecules* **2017**, *22*, 1309. [CrossRef]
60. Chun-xia, X.; Yan, X.; Cai-xia, D.; Chun-mao, Y.; Lie-jun, H.; Wei, C.; Xiao-jiang, H. Chemical Constituents from Water-soluble Parts of *Baphicacanthus cusia*. *J. Chinese Med. Mater.* **2020**, *43*, 596–601.
61. Abou-Hussein, D.R.; Badr, J.M.; Youssef, D.T.A. Dragmacidoside: A new nucleoside from the red sea sponge dragmacidon coccinea. *Nat. Prod. Res.* **2014**, *28*, 1134–1141. [CrossRef] [PubMed]
62. Mori, T.; Yamashita, T.; Furihata, K.; Nagai, K.; Suzuki, K.I.; Hayakawa, Y.; Shin-ya, K. Burkholone, a new cytotoxic antibiotic against IGF-I dependent cells from *Burkholderia* sp. *J. Antibiot. (Tokyo)* **2007**, *60*, 713–716. [CrossRef] [PubMed]
63. Zhang, Z.; Zhang, W.; Ji, Y.P.; Zhao, Y.; Wang, C.G.; Hu, J.F. Gynostemosides A-E, megastigmane glycosides from *Gynostemma pentaphyllum*. *Phytochemistry* **2010**, *71*, 693–700. [CrossRef]
64. Rovirosa, J.; Sepulveda, M.; Quezada, E.; San-Martin, A. Isoepitaondiol, a diterpenoid of *Stypopodium flabelliforme* and the insecticidal activity of stypotriol, epitaondiol and derivatives. *Phytochemistry* **1992**, *31*, 2679–2681. [CrossRef]
65. Zang, E.H.; Chen, Z.W.; Zhang, C.H.; Li, M.H. Chemical constituents of *Physochlaina physaloides* (L.) G. Don (*Solanaceae*). *Biochem. Syst. Ecol.* **2021**, *98*, 104332. [CrossRef]
66. Sugimoto, S.; Yamano, Y.; Desoukey, S.Y.; Katakawa, K.; Wanas, A.S.; Otsuka, H.; Matsunami, K. Isolation of Sesquiterpene-Amino Acid Conjugates, Onopornoids A-D, and a Flavonoid Glucoside from *Onopordum alexandrinum*. *J. Nat. Prod.* **2019**, *82*, 1471–1477. [CrossRef]
67. Costa, M.; Fernández, R.; Pérez, M.; Thorsteinsdottir, M. Two new spongian diterpene analogues isolated from the marine sponge *Acanthodendrilla* sp. *Nat. Prod. Res.* **2020**, *34*, 1053–1060. [CrossRef]
68. León, E.I.; Martín, A.; Peréz-Martín, I.; Quintanal, L.M.; Suárez, E. Hydrogen atom transfer experiments provide chemical evidence for the conformational differences between C-and O-disaccharides. *European J. Org. Chem.* **2010**, *2010*, 5248–5262. [CrossRef]
69. Kitajima, J.; Komori, T.; Kawasaki, T.; Schulten, H. rolf Basic steroid saponins from aerial parts of *Fritillaria thunbergii*. *Phytochemistry* **1982**, *21*, 187–192. [CrossRef]

70. Liu, H.; Liu, S.; Guo, L.; Zhang, Y.; Cui, L.; Ding, G. New furanones from the plant endophytic fungus *Pestalotiopsis besseyi*. *Molecules* **2012**, *17*, 14015–14021. [CrossRef]
71. El-Gamal, A.A.; Al-Massarani, S.M.; Shaala, L.A.; Alahdald, A.M.; Al-Said, M.S.; Ashour, A.E.; Kumar, A.; Abdel-Kader, M.S.; Abdel-Mageed, W.M.; Youssef, D.T.A. Cytotoxic compounds from the Saudi red sea sponge *Xestospongia testudinaria*. *Mar. Drugs* **2016**, *14*, 82. [CrossRef] [PubMed]
72. Sun, F.; Zhang, L.; Tian, J.K.; Cheng, J.Y.; Xiao, P.G. Studies on Chemical Constituents of *Clematis terniflora*. *Chin. Pharm. J.* **2007**, *42*, 102–103.
73. Huang, X.A.; Yang, R.Z. A new hydroquinone diglucoside from *Lysimachia fordiana*. *Chem. Nat. Compd.* **2004**, *40*, 457–459. [CrossRef]
74. Gao, Q.; Cheng, Y. A New Lactam from *Pinellia ternata*. *Nat. Prod. Res. Dev.* **2015**, *27*, 1693–1696.
75. Yang, C.P.; Shie, P.H.; Huang, G.J.; Chien, S.C.; Kuo, Y.H. New anti-inflammatory flavonol glycosides from *Lindera akoensis* hayata. *Molecules* **2019**, *24*, 563. [CrossRef]
76. Lin, Q.G.; Jie, O.Y.; De Yun, K.; Shan, D.S. Studies on Chemical Condtituents of *Primula maximowiczii* Regel II. *Chin. Pharm. J.* **2008**, *43*, 1300–1304.
77. Zhang, Z.; Wang, D.; Zhao, Y.; Gao, H.; Hu, Y.H.; Hu, J.F. Fructose-derived carbohydrates from *Alisma orientalis*. *Nat. Prod. Res.* **2009**, *23*, 1013–1020. [CrossRef]
78. Gudej, J.; Nazaruk, J. Flavonol glycosides from the flowers of *Bellis perennis*. *Fitoterapia* **2001**, *72*, 839–840. [CrossRef]
79. Johnst, P.I.M.; Barrientos, R.E.; Simirgiotis, M.J.; Palacios, J. Characterization of Polyphenol Compounds from Endothelium-Dependent Vascular Relaxation Effect in Rat Aorta. *Molecules* **2020**, *25*, 3105.
80. Pailee, P.; Sangpetsiripan, S.; Mahidol, C.; Ruchirawat, S.; Prachyawarakorn, V. Cytotoxic and cancer chemopreventive properties of prenylated stilbenoids from *Macaranga siamensis*. *Tetrahedron* **2015**, *71*, 5562–5571. [CrossRef]
81. Duan, Y.; Hu, Y.; Yang, W.; Xiong, Y.; Du, C.; Yuan, C.; Hao, X.; Gu, W. Study on chemical constituents and α-glucosidase inhibitory activity of *Cyclocarya paliurus* in Guizhou province. *Nat. Prod. Res. Dev.* **2019**, *31*, 940–945.
82. Kavtaradze, N.; Alaniya, M.; Masullo, M.; Cerulli, A.; Piacente, S. New Flavone Glycosides from *Astragalus tanae* Endemic to Georgia. *Chem. Nat. Compd.* **2020**, *56*, 70–74. [CrossRef]

nutrients

MDPI

Article

Inhibition of α-Glucosidase and Pancreatic Lipase Properties of *Mitragyna speciosa* (Korth.) Havil. (Kratom) Leaves

Thanchanok Limcharoen [1,2], Phisit Pouyfung [2,3], Ngamrayu Ngamdokmai [1,2], Aruna Prasopthum [4,5], Aktsar Roskiana Ahmad [6], Wisdawati Wisdawati [6], Woraanong Prugsakij [7] and Sakan Warinhomhoun [1,2,*]

1 School of Medicine, Walailak University, Nakhon Si Thammarat 80160, Thailand
2 Center of Excellent in Marijuana, Hemp and Kratom, Walailak University, Nakhon Si Thammarat 80160, Thailand
3 School of Public Health, Walailak University, Nakhon Si Thammarat 80160, Thailand
4 School of Pharmacy, Walailak University, Nakhon Si Thammarat 80160, Thailand
5 Biomass and Oil Palm Center of Excellent, Walailak University, Nakhon Si Thammarat 80160, Thailand
6 Department of Pharmacognosy and Phytochemistry, Faculty of Pharmacy, Universitas of Muslim Indonesia, Makassar 90241, Indonesia
7 Department of Pharmaceutics and Industrial Pharmacy, Faculty of Pharmaceutical Sciences, Chulalongkorn University, Bangkok 10330, Thailand
* Correspondence: sakan.wa@wu.ac.th

Citation: Limcharoen, T.; Pouyfung, P.; Ngamdokmai, N.; Prasopthum, A.; Ahmad, A.R.; Wisdawati, W.; Prugsakij, W.; Warinhomhoun, S. Inhibition of α-Glucosidase and Pancreatic Lipase Properties of *Mitragyna speciosa* (Korth.) Havil. (Kratom) Leaves. *Nutrients* **2022**, *14*, 3909. https://doi.org/10.3390/nu14193909

Academic Editors: Haixia Yang and Jianjun Deng

Received: 26 August 2022
Accepted: 19 September 2022
Published: 21 September 2022

Abstract: Kratom (*Mitragyna speciosa* (Korth.) Havil.) has been used to reduce blood sugar and lipid profiles in traditional medicine, and mitragynine is a major constituent in kratom leaves. Previous data on the blood sugar and lipid-altering effects of kratom are limited. In this study, phytochemical analyses of mitragynine, 7-hydroxymitragynine, quercetin, and rutin were performed in kratom extracts. The effects on α-glucosidase and pancreatic lipase activities were investigated in kratom extracts and mitragynine. The LC-MS/MS analysis showed that the mitragynine, quercetin, and rutin contents from kratom extracts were different. The ethanol extract exhibited the highest total phenolic content (TPC), total flavonoid content (TFC), and total alkaloid content (TAC). Additionally, compared to methanol and aqueous extracts, the ethanol extract showed the strongest inhibition activity against α-glucosidase and pancreatic lipase. Compared with the anti-diabetic agent acarbose, mitragynine showed the most potent α-glucosidase inhibition, with less potent activity of pancreatic lipase inhibition. Analysis of α-glucosidase and pancreatic lipase kinetics revealed that mitragynine inhibited noncompetitive and competitive effects, respectively. Combining mitragynine with acarbose resulted in a synergistic interaction with α-glucosidase inhibition. These results have established the potential of mitragynine from kratom as a herbal supplement for the treatment and prevention of diabetes mellitus.

Keywords: *Mitragyna speciosa*; kratom; α-glucosidase; pancreatic lipase; anti-diabetes mellitus

1. Introduction

Obesity is a major risk factor that is closely related to various pathological conditions and chronic diseases worldwide and has become a major global public health problem. In 2022, the World Health Organization (WHO) reported that among adults aged 18 years and older, 39% were overweight and 13% were obese [1]. Recent studies revealed that the common underlying cause of both diabetes and obesity is related to insulin resistance, which occurs due to the stimulation of insulin production or a reduction in insulin receptors [2,3]. Common molecular targets for designing anti-obesity drugs are enzymes, which are accounted for virtually half of the small-molecule drugs available on the market. Because of their protein structures with several validated sites for drug interaction, enzymes are proved to be an appealing target for the discovery of novel therapeutic molecules [4]. Natural products are known to be remarkable sources for discovering possible therapeutic approaches for metabolic disorders, such as anti-diabetic and anti-obesity drugs. Hence, extensive experiments were made to investigate the effects of molecules from natural products through enzyme inhibition assay, particularly activities of α -Glucosidase and lipase [5–7].

α-glucosidase is a carbohydrate-hydrolyzing enzyme that is secreted from the intestinal chronic epithelium. These enzymes are essential enzymes of digestion that stimulate the breakdown of disaccharides and oligosaccharides into small, simple, and absorbable carbohydrates [8]. One of the most common alpha-glucosidase inhibitors (AGIs) drugs currently available is acarbose, which is proven to be efficient in blood glucose level stabilization. Moreover, acarbose was found to have activity against oxidative stress and endothelial dysfunction; therefore, it might lower the risk of developing cardiovascular diseases and help increase the lifespans of type-2 DM patients [9]. In addition, the dysfunctions of insulin-producing pancreatic β cells lose their functions due to the excessive accumulation of lipids in the pancreas [10,11]. Lipase is an enzyme produced by the exocrine portion of the pancreas and is released into the intestinal lumen to catalyze the hydrolytic breakdown of triacylglycerols in ingested fats, into free fatty acids and monoacylglycerols; thus, it is the most important enzyme in lipid digestion and is absorbed into circulation [12]. Orlistat, a lipase inhibitor that competes with dietary lipid molecules for enzymatic active sites, is an archetypic medication prescribed for obese patients in need of losing weight. Orlistat, with a therapeutic oral dose of 120 mg administered three times per day, accounts for nearly one-third of the reduction of dietary fat absorption via intestinal epithelium due to its inhibitory effect against lipase in the hydrolysis of triacylglycerol. This decrease in lipid absorption is shown to be without prominent effects on appetite [13]. However, long-term use of these drugs is often reported to cause some severe side effects, such as insomnia, headaches, hypoglycemia, weight gain, constipation, and renal damage [14,15]. Efforts have been directed toward discovering medicines from natural products due to their low costs, relative safety, probability of high compliance, and low incidences of undesirable side effects [16]. Based on previous studies, secondary metabolites of natural products, such as phenolics, alkaloids, terpenoids, flavonoids, and glycosides, have shown potent pancreatic lipase and α-glucosidase inhibitory activities [17–22]. The α-glucosidase inhibitory potential was also shown by plants in the Juglandaceae family [23]. However, it remains necessary to continue searching for more effective α-glucosidase and lipase inhibitors from traditional herbs.

Mitragyna speciosa (Korth.) Havil. or kratom (Rubiaceae) is an indigenous plant in Southeast Asia that grows naturally in several regions, including Thailand, Indonesia, Malaysia, Sumatra, Java, Bali, and Borneo (Figure 1) [24]. Kratom leaves have been used in local communities to treat pain, cough, fever, and diabetes; to enhance work performance; and are used as substitutes for illicit substances, mainly opioids. The leaves have been used as a traditional medicinal herb in southern Thailand, where it has become a culturally accepted stimulant drink similar to coffee and tea [25]. Previous studies revealed that *M. speciosa* contains a diverse group of secondary metabolites, such as indole alkaloids, flavonoids, triterpenoids, saponins, and glycosides [26]. More than forty of these compounds have been identified. Mitragynine is the most abundant compound available in the kratom preparation, with an estimated 2% by mass and up to two-thirds (66%) of total alkaloid content [27]. In addition, through in vitro and in vivo studies, kratom has been observed to exhibit various pharmacological properties, such as antioxidant, anti-inflammation, antibacterial, antiproliferative, and anti-analgesic properties [28,29]. Although previous studies have reported that kratom leaves exhibit activities that control diabetes and lipid profile [30–35], there is still a lack of scientific evidence related to the inhibitory effects and mechanism of enzymatic activities. Thus, in this research, we aimed to evaluate the α-glucosidase and pancreatic lipase inhibitory activities of kratom leaves. Additionally, the phytochemical profile, total alkaloid content, total phenolic content, and total flavonoid content of kratom extract were also investigated.

Figure 1. *Mitragyna speciosa* (Korth.) Havil. or Kratom.

2. Materials and Methods

2.1. Plant Material

Fresh leaves of *M. speciosa* were collected from Thasala district, Nakhon Si Thammarat Province, Thailand in February 2022 (Figure 1). The leaves were identified by the Plant Varieties Protection office, Department of Agriculture, Thailand; the voucher specimen (BK083621) was deposited.

2.2. Preparation of Kratom Extracts

The leaves of kratom (1 kg) were dried in a hot air oven at 60 °C and ground to a coarse powder (600 g). The kratom powder was extracted with various organic solvents, including methanol (MeOH), ethanol (EtOH), and water, respectively for 24 h (three times). The extracts were filtered using a Whatman No. 1 filter paper. Each filtrate was concentrated to dryness in a rotary evaporator (Büchi Labortechnik, Germany) under reduced pressure and controlled temperature (40 °C) to give the final extracts including EtOH extract (52.8 g), MeOH extract (60.5 g), and aqueous extract (57.99 g), which were stored at 4 °C in a refrigerator until further use.

2.3. Determination of Total Phenolic Content, Total Flavonoid Content, and Total Alkaloid Content

2.3.1. Total Phenolic Contents (TPC)

The total phenolic content was determined for individual extracts using the Folin–Ciocalteu method with some modifications [36]. Briefly, the 20 µL of extracts were mixed with 100 µL of tenfold diluted Folin–Ciocalteu reagent (Sigma-Aldrich, St Louis, MO, USA) and 80 µL of sodium bicarbonate (75 g/L). The mixture was incubated at room temperature for 1 h, and the absorbance was measured at 765 nm with a microplate reader (Thermo Scientific, Göteborg, Sweden). All samples were analyzed in triplicate. Gallic acid (4–30 µg/mL) was used as a positive control (Sigma-Aldrich, St Louis, MO, USA, lot number 099K0128). The results are expressed as milligrams of gallic acid equivalent (mg GAE/g).

2.3.2. Total Flavonoid Content (TFC)

The total flavonoid content method was determined by using an aluminum chloride colorimetric assay with some modifications [37]. Briefly, 50 µL of samples were mixed with 10 µL of 10% aluminum chloride (Sigma-Aldrich, St Louis, MO, USA), 1 M sodium acetate (Sigma-Aldrich, St Louis, MO, USA), and 150 μL of 95% ethanol. The mixtures were further incubated in the dark at room temperature for 40 min. The absorbance was measured at 415 nm using a microplate reader (Thermo Scientific, Göteborg, Sweden). All samples were analyzed in triplicate. A solution of quercetin (Sigma-Aldrich, St Louis, MO, USA, lot number Q4951) in a range of 10–100 μg/mL was used to prepare a standard curve for determining the total flavonoid contents. The results are expressed as milligrams of quercetin equivalent (mg QE/g).

2.3.3. Total Alkaloid Content (TAC)

Bromocresol green solution (1×10^{-4}) was prepared by heating 69.8 mg bromocresol green (Sigma-Aldrich, St Louis, MO, USA) with 3 mL of 2N NaOH and 5 mL of distilled water until completely dissolved and the solution was diluted to 1000 mL with distilled water. A phosphate buffer solution (pH 4.7) was prepared by adjusting the pH of 2 M sodium phosphate (71.6 g Na_2HPO_4 in 1 L of distilled water) to 4.7 with 0.2 M citric acid (42.02 g citric acid in 1 L of distilled water). Atropine standard (Glentham Life Sciences, Ltd., Wiltshire, United Kingdom, lot number 756VTN) solution (20–140 μg/mL) was dissolved at 1 mg in 10 mL of distilled water [38]. All samples were analyzed in triplicate. The results are expressed as milligrams of atropine equivalent (mg ATR/g).

2.4. *Liquid Chromatography Analysis of Kratom Extracts*

UHPLC model Ultimate 3000 with an LC-MS/MS model Altis Plus (Thermo Fisher Scientific, MA, USA) was used. All samples were filtrated by a nylon filter with a pore size of 0.22 μm. The scanning range was from 100 to 1700 *m/z* for MS/MS in positive mode. Separation was achieved through Thermo Hypersil GOLD C-18 (2.1 × 100 mm, 1.9 μm). The gradient mobile phase was a mixture of solvent A: 0.1% formic acid (pH 2.99), and solvent B: acetonitrile, with a flow rate of 0.5 mL/min. The gradient program was set as 0–0.5 min, 25% B; 3–4 min, 80–100% B; and 4–6.5 min, 25% B with an injection volume of 10 μL. Standard mitragynine (0.1–1 μg/mL) (Lipomed, Inc., lot number 1610.1B0.2), quercetin (0.1–1 μg/mL) (Sigma-Aldrich, St Louis, MO, USA, lot number Q4951), 7-hydroxymitragynine (0.1–1 μg/mL) (ChromaDex, Inc., lot number 00008624-00377), and rutin (0.1–1 μg/mL) (Acros Organics, lot number A0355330) were eluted at 2.08, 1.20, 3.06, and 1.22 min, respectively. The mass scan mode was the positive multiple reaction monitoring mode. The precursor ion and product ion were *m/z* 399.20 → 174.10 for mitragynine, *m/z* 415.20 → 190.10 for 7-hydroxymitragynine, *m/z* 303.05 → 229.00 for quercetin, and *m/z* 611.16 → 303.10 for rutin.

2.5. *Kinetic Study of α-Glucosidase Inhibition*

The assay was performed as described previously with some modifications [39]. The kinetic parameters (Michaelis–Menten constant; K_m and maximum velocity; V_{max}) of α-glucosidase were determined by using *p*-nitrophenol-α-D-glucopyranoside (*p*NPG) as a substrate. Briefly, 0.1 unit of α-glucosidase (Sigma-Aldrich, St Louis, MO, USA) were incubated at 37 °C with 100 μL of total reaction. After 10 min, the reactions were started by adding 50 μL *p*NPG (Sigma-Aldrich, St Louis, MO, USA, lot number 3698310) ranging from 0 to 15 mM, and the *p*-nitrophenol product was measured at 405 nm using a microplate reader (Multiskan SkyHigh, Thermo Scientific, Göteborg, Sweden). The substrate concentration rates were plotted against *p*NPG concentrations and fitted to the Michaelis–Menten equation. For the inhibition assay, 100 μL containing 0.1 unit of α-glucosidase was incubated at 37 °C with acarbose (positive control) (Sigma-Aldrich, St Louis, MO, USA, Lot number SLCF5122) or kratom extracts at various concentrations (0–100 μg/mL).

The reactions were started by adding 50 μL of 0.3 mM *p*NPG (K_m), and the *p*-nitrophenol product was measured at 405 nm using a microplate reader. The IC_{50} was determined by comparing the rate of each test inhibitor concentration to that of the vehicle control (ethanol), and the IC_{50} was calculated by plotting the remaining activity of each inhibitor concentration to the log test inhibitor concentration. The inhibition modes of pure compounds (mitragynine) and the positive control (acarbose) against the α-glucosidase enzyme were determined with at least four different concentrations of test inhibitors (mitragynine 0–0.4 mM; acarbose 0–3 mM). After co-incubation with each inhibitor for 5 min at 37 °C, the reactions were initiated by adding the *p*-NPG substrate (0–15 mM) [39]. The inhibition constant (K_i) values were determined from Lineweaver–Burk Plots.

2.6. Kinetic Study of Pancreatic Lipase Inhibition

The kinetic parameters (K_m and V_{max}) of pancreatic lipase were determined using 4-methylumbelliferyl oleate (4MUO) as the substrate (Sigma-Aldrich, St Louis, MO, USA, Lot number BCCF8781). Briefly, 0.5 units of lipase (Sigma-Aldrich, St Louis, MO, USA, Lot number SLCG8579) were incubated in 96-well plates with a total volume of 100 μL (15 mM Tris-HCl buffer, pH 8.0) at 37 °C for 10 min. The reactions were initiated by adding 50 μL of 4MUO ranging from 0 to 5 mM. The reactions were left to proceed for 10 min, and the rates of the reactions were measured by monitoring the increase in 4-methylumbelliferone (4MU), a fluorescent product (excitation, 355 nm; and emission, 460 nm). The rates at which substrate concentrations were plotted against 4MUO concentrations and fitted to the Michaelis–Menten equation. For the inhibition assay, 0.5 units of lipase were incubated with orlistat (Sigma-Aldrich, St Louis, MO, USA, Lot number 0000117290) or kratom extracts at various concentrations (0–100 μg/mL) in a 100 μL total reaction volume at 37 °C. After 10 min, the reactions were started by adding 0.35 mM 4MUO. The reaction rates were monitored by measuring the release of 4MU from 4MUO. Fluorescence from the release of 4MU was measured using a microplate reader (Synergy Mx, Agilent Technology, Santa Clara, USA) with excitation and emission wavelengths of 355 and 460 nm, respectively. The remaining activity of lipase was determined by comparing the rate of each test inhibitor concentration to the vehicle control (ethanol), and IC_{50} was calculated by plotting the remaining activity of each inhibitor concentration to the log test inhibitor concentration [39].

2.7. The Combination of Kratom Extracts and Acarbose Inhibited Enzymatic α-Glucosidase Activities

For the combination inhibition of α-glucosidase, 0.1 units of α-glucosidase was incubated with acarbose ranging from 0 to 100 μM in the presence and absence of ethanol extract (IC_{50}; 16 μg/mL), methanol extract (IC_{50}; 42 μg/mL), aqueous extract (IC_{50}; 70 μg/mL), and mitragynine (IC_{50}; 82 μg/mL) at 37 °C in a total volume of 100 μL of 15 mM Tris-HCl buffer, pH 8.0. After 10 min, the residual reactions were started by the addition of 50 μL of 0.3 mM *p*NPG. The reactions were left to proceed for 10 min. The absorption at 405 nm was then measured using a microplate reader, and IC_{50} values were calculated by GraphPad Prism version 9.3.1.

2.8. Statistical Analysis

All data were obtained from three dependent experiments and are expressed as the mean ± standard deviation (SD). Statistical analysis was performed using GraphPad Prism 9.3.1 (GraphPad Software, Inc., San Diego, CA, USA) with one-way ANOVA. Differences with * $p < 0.05$ were statically significant.

3. Results

3.1. TPC, TFC, and TAC of All Kratom Extracts

The effects of the extraction solvents (ethanol, methanol, and aqueous) on the TPC, TFC, and TAC of the kratom leaf extracts were evaluated, as shown in Table 1. The TPC was calculated from the regression equation of the calibration curve ($Y = 0.0236x$; $R^2 = 0.9995$) and expressed as mg GAE/g of samples. The TFC was also reported as mg QE/g of samples ($Y = 0.0059x$; $R^2 = 0.9923$). The content of alkaloids was measured in terms of atropine equivalents ($Y = 0.0052x$; $R^2 = 0.9980$). The results showed that the ethanol extract exhibited a higher TPC (252.92 ± 1.15 mg GAE/g), TFC (26.07 ± 0.01 mg GAE/g), and TAC

(88.04 ± 0.15 mg ATR/g) than those of the other kratom extracts. These results suggest that the ethanol extract is rich in phenolics, flavonoids, and alkaloids.

Table 1. TPC (mg GAE/g extract), TFC (mg QE/g extract), and TAC (mg ATR/g extract) of kratom leaf extracts.

Samples	TPC (mg GAE/g) ± SD	TFC (mg QE/g) ± SD	TAC (mg ATR/g) ± SD
Ethanol	252.92 ± 1.15 *	26.07 ± 0.01 *	88.04 ± 0.15 *
Methanol	159.30 ± 2.01	13.15 ± 0.09	52.82 ± 0.85
Aqueous	130.58 ± 0.68	0.82 ± 0.02	5.61 ± 0.13

Results are expressed as means ± SDs, $n = 3$. * $p < 0.05$ compared to methanol and aqueous extracts of each group.

3.2. LC-MS/MS Analysis of Mitragynine, 7-hydroxymitragynine, Quercetin, and Rutin

LC-MS/MS is a confirmed hyphenated and accurate tool for rapid analysis, and it was used for the identification of a total of the mitragynine, 7-hydroxymitragynine, and two flavonoid compounds, which were identified by comparing their retention times and mass fragmentation patterns with data obtained from previous studies [26]. Then, each individual compound was quantified by comparing its peak area with the calibration curve obtained for the corresponding standard (Figure 2). Table 2 summarizes the bioactive phytochemical, mainly indole alkaloids and flavonoids. The presence and identification of these phytochemicals correlate with the reports published in a previous report [26].

Table 2. MS/MS data of compounds identified tentatively in kratom ethanol, methanol, and aqueous leaf extracts using UHPLC and LC-MS/MS.

Identification	Calculated m/z $[M+H]^+$	Precursor ion Experimental m/z $[M+H]^+$	Major Ion in MS/MS Spectra (Key Fragment Ions)	Ethanol RT, min	Methanol RT, min	Aqueous RT, min
Mitragynine	399.2278	399.20	174.10	3.284	3.224	3.224
7-hydroxymitragynine	415.2227	415.2	190.10	3.286	3.212	3.212
Rutin	611.1602	611.16	303.10	1.209	1.175	1.175
Quercetin	303.0508	303.05	229.00	1.217	1.223	1.223

The quantitative analysis of kratom leaf extracts is shown in Table 3. Mitragynine appeared to be the major alkaloid that was found in ethanol, followed by methanol and aqueous extracts, respectively. From this result, we can calculate that the percentages of mitragynine in TAC are 66% in methanol extract, 68% in ethanol extract, and 6.9% in aqueous extract. However, 7-hydroxymitragynine was observed at less than 1 mg/g. Additionally, quercetin and rutin were found to contain the highest amounts of ethanol compared to the other kratom extracts.

Table 3. Quantitative analysis of mitragynine, quercetin, and rutin of kratom extracts.

Compounds	Amount (mg/g) ± SD		
	Ethanol Extract	Methanol Extract	Aqueous Extract
Mitragynine	58.75 ± 0.21 *	35.87 ± 1.01	3.85 ± 0.17
Quercetin	19.10 ± 0.85 *	5.90 ± 0.14	1.28 ± 0.02
Rutin	11.36 ± 0.11 *	3.19 ± 0.22	1.22 ± 0.05

Results are expressed as means ± SDs, $n = 3$. * $p < 0.05$ compared to methanol and aqueous extracts of each group.

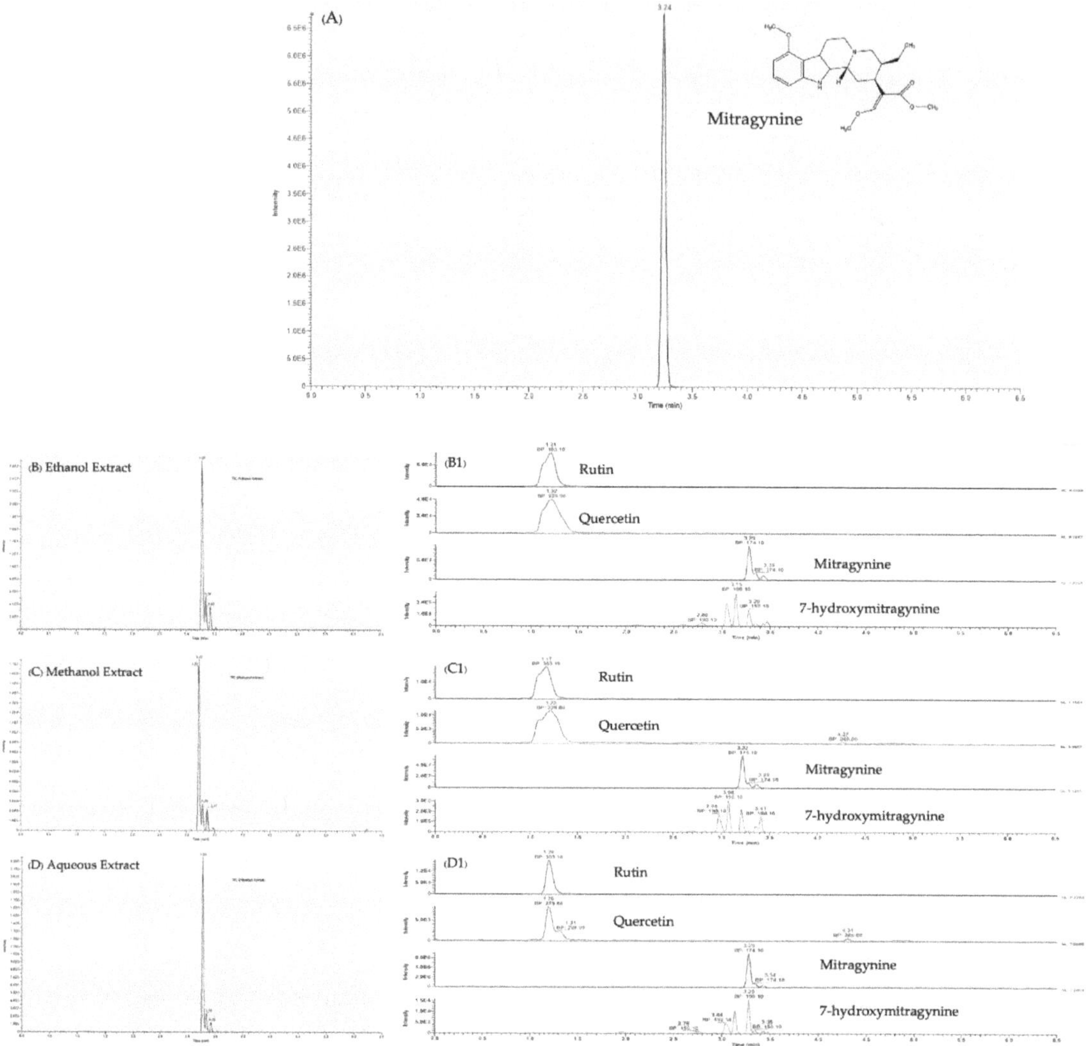

Figure 2. Total ion chromatograms (TIC) from LC-MS/MS of (**A**) the reference standard (mitragynine) concentration at 1 μg/mL, (**B**) ethanol extract, (**C**) methanol extract, and (**D**) aqueous extract from kratom leaves. The peaks of four major constituents were identified by comparison with the reference standards, their retention times, and mass fragmentation patterns (**B1–D1**) as rutin, quercetin, mitragynine, and 7-hydroxymitragynine.

3.3. α-Glucosidase Inhibition Activity

The three different solvent extracts of kratom were evaluated for their α-glucosidase inhibitory activities (Table 4). Each kratom extract was initially treated at 0–100 μg/mL. The ethanol extract showed the strongest activity (IC_{50} 15.9 ± 1.34 μg/mL), followed by methanol extract (IC_{50} 42.12 ± 1.76 μg/mL) and aqueous extract (IC_{50} 69.48 ± 2.67 μg/mL), with a potency higher than that of the drug acarbose (IC_{50} 728.20 ± 7.01 μg/mL). However, compared to acarbose as a positive control, mitragynine has lower IC_{50} (81.68 ± 1.70 μg/mL) and, thus, has higher inhibitory activity (Figure 3A,B).

Table 4. The IC_{50} values of kratom extracts and mitragynine for α-glucosidase and pancreatic lipase inhibitory activities.

Samples	α-Glucosidase		Pancreatic Lipase	
	IC_{50} (μM)	IC_{50} (μg/mL)	IC_{50} (μM)	IC_{50} (μg/mL)
Ethanol extract	-	15.90 ± 1.34 *	-	14.15 ± 1.71 *
Methanol extract	-	42.12 ± 1.76 *	-	28.38 ± 2.34 *
Aqueous extract	-	69.48 ± 2.67 *	-	41.43 ± 3.32 *
Mitragynine	205.04 ± 15.11 *	81.68 ± 1.70 *	24.9 ± 1.38 *	9.86 ± 0.45 *
Acarbose	1121.09 ± 67.01	728.20 ± 7.01	-	-
Orlistat	-	-	0.84 ± 0.10	0.42 ± 0.05

Results are expressed as means ± SDs, $n = 3$. * $p < 0.05$ compared to positive controls (acarbose and orlistat).

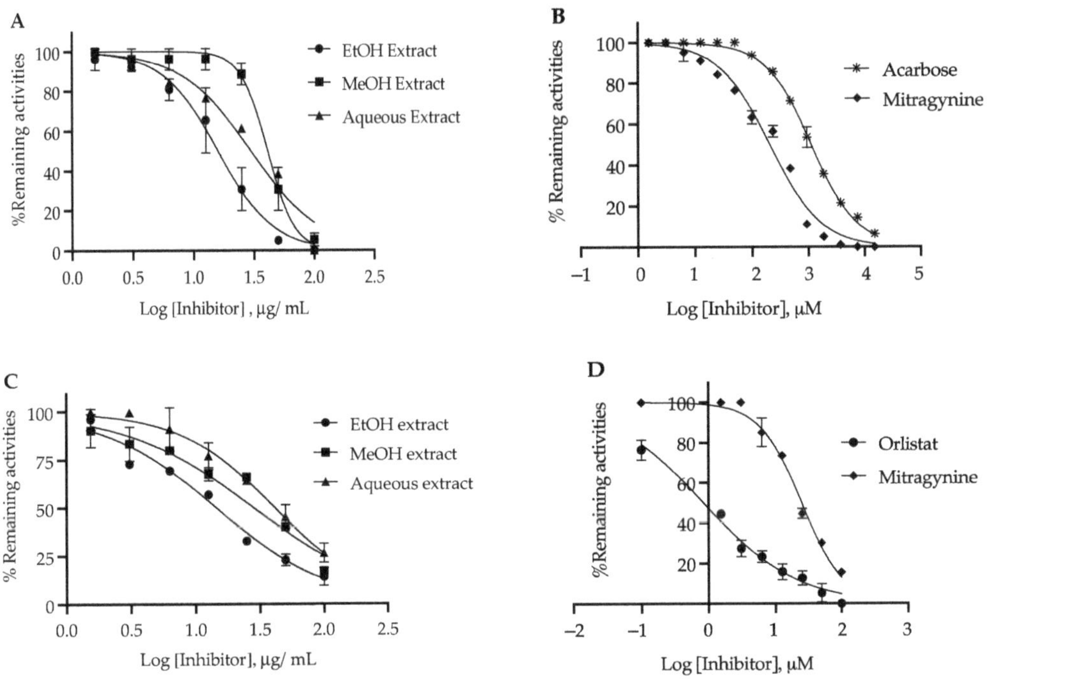

Figure 3. Inhibitory effect of the kratom extracts and mitragynine against α-glucosidase (**A**,**B**) and pancreatic lipase (**C**,**D**). Inhibition curves of acarbose (**B**) and orlistat (**D**) were used as positive controls.

The inhibition kinetics of acarbose and kratom extracts against α-glucosidase were analyzed using the Lineweaver–Burk plots of the inverted values of velocity (1/V) versus the inverted values of substrate concentration (1/[S]), which are presented in Figure 4A–D. The drug acarbose showed the intersection of the lines on the ordinate, indicative of mixed-type inhibition. The results showed that k_m 0.3 mM and V_{max} (28 μmol/min/mg) were consistent with a previous report [39]. The K_i of each mode was evaluated (Figure 4A,B). The intersection on the abscissa yielded a K_i value of acarbose (0.28 mM), indicating mixed-type inhibition. Mitragynine was present at concentrations of 100, 200, and 300 μM with a K_i value of 0.10 mM (Figure 4C,D). These results suggest that mitragynine is a noncompetitive inhibitor of this enzyme (Table 5).

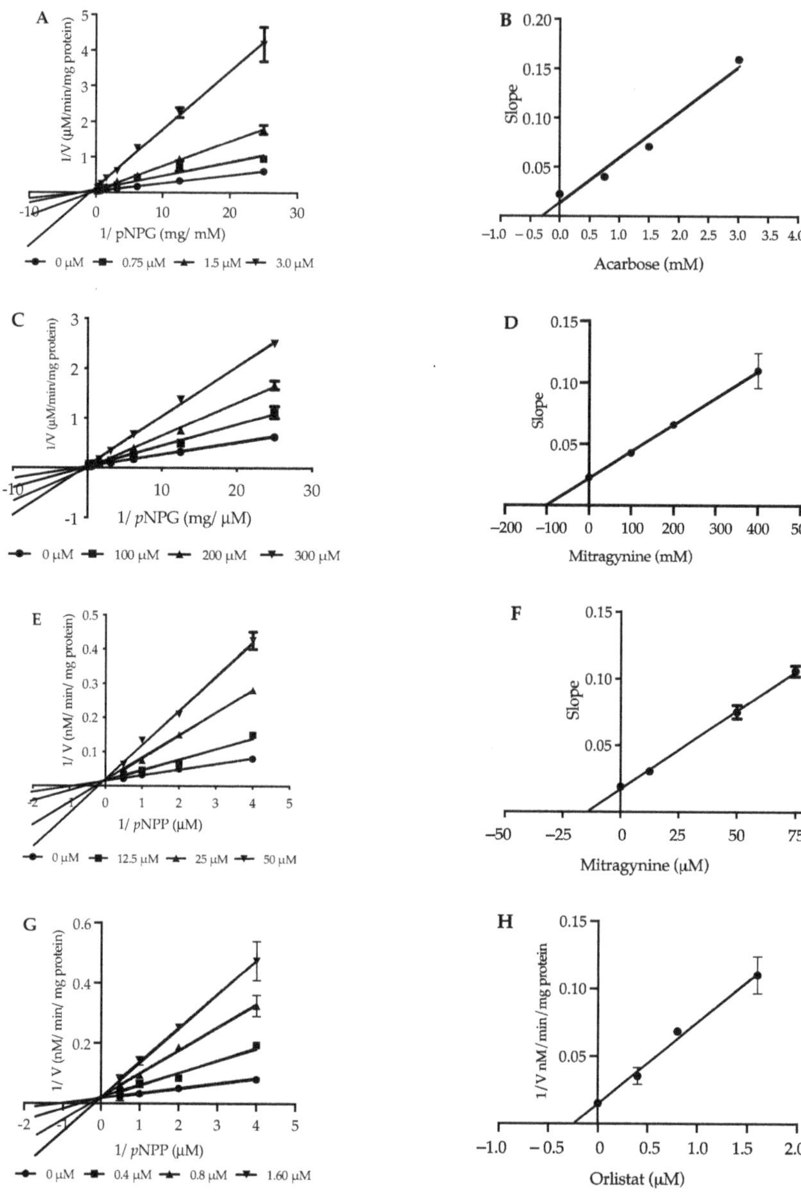

Figure 4. Lineweaver–Burk plots of α-glucosidase (**A–D**) and pancreatic lipase (**E–H**) in the presence and absence of kratom extracts, mitragynine, and positive controls (acarbose and orlistat).

Table 5. Kinetic parameters in α-glucosidase and pancreatic lipase in the presence of mitragynine and positive controls.

Inhibitors	α-Glucosidase		Pancreatic Lipase	
	K_i (mM)	Mode	K_i (µM)	Mode
Mitragynine	0.10	noncompetitive	14.94	competitive
Acarbose	0.28	mixed-type	-	-
Orlistat	-	-	0.24	competitive

3.4. Pancreatic Lipase Inhibition Activity

The kratom leaf extracts were subjected to pancreatic lipase inhibitory activity with 4MUO as the substrate evaluation. Each kratom extract was initially treated at 0–100 μg/mL (Table 4). The results revealed that the ethanol extract exhibited strong inhibition of lipase (IC_{50} 14.15 ± 1.71 μg/mL), followed by the methanol extract (IC_{50} 28.38 ± 2.34 μg/mL) and aqueous extract (IC_{50} 41.43 ± 3.32 μg/mL), but the potency was lower than that of orlistat (IC_{50} 0.42 ± 0.05 μg/mL), which was a positive control (Figure 3C,D). In addition, we found that the IC_{50} value of mitragynine was 30-fold higher than that of orlistat. In experiments on the kinetics parameter of lipase with 4MUO as the substrate, the K_m and V_{max} values were determined to be 0.45 ± 0.01 mM and 58 nmol/min/mg, respectively. The K_i values of orlistat and mitragynine were determined. The intersection on the abscissa yielded a K_i value of orlistat (0.24 ± 0.03 μM) and mitragynine (14.94 ± 0.29 μM). These observations suggested that both orlistat and mitragynine were competitive inhibitors of the lipase enzyme (Table 5), but mitragynine was shown to have weaker inhibition than that of orlistat (Figure 4E–H).

3.5. The Combined Inhibitory Effect of Kratom Extracts and Mitragynine on α-Glucosidase Activities

The mode of inhibition between acarbose and mitragynine on the α-glucosidase enzyme was as described earlier. We hypothesized that acarbose and mitragynine might have synergist effects. The inhibitory effects of kratom and its combination with α-glucosidase are shown in Figure 5. The concentration of acarbose in the range of 0–15 mM was combined with the IC_{50} values of kratom extracts and mitragynine. The results showed that the IC_{50} of kratom extracts combined with acarbose was lowered almost two-fold compared with that of acarbose alone. Interestingly, combination treatments with mitragynine at the concentrations of 200 and 300 μM decreased IC_{50} values compared with that of mixed kratom extracts. However, 100 μM of mitragynine showed less potency than that of kratom extracts. Thus, the results indicated that mitragynine in kratom is a major constituent for reducing blood glucose as well as improving the efficiency of acarbose, which is a reference standard for glucose-lowering drug.

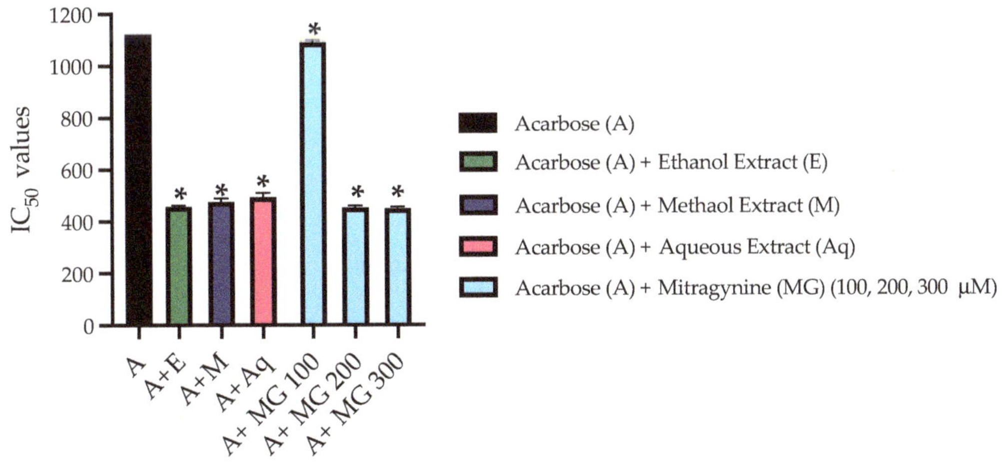

Figure 5. The IC_{50} values α-glucosidase of combinatorial kratom extracts and mitragynine with acarbose. The results are expressed as the means ± SDs, $n = 3$. * $p < 0.05$ compared to acarbose alone.

4. Discussion

M. speciosa or kratom contains more than 40 identified bioactive compounds including indole alkaloids, flavonoids, triterpenoids, saponins, and glycosides. These compounds have been described to have antioxidant, anti-inflammatory, antibacterial, antiproliferative, and analgesic activities by several studies [26,28,29,40]. In this study, the presence of phenolic, flavonoid, and alkaloid compounds in our kratom extracts were confirmed by TPC, TFC, and TAC, respectively (Table 1). The four major kratom extract constituents identified by LC-MS/MS included mitragynine, 7-hydroxymitragynine, quercetin, and rutin (Table 2), which corresponded to the previous studies [26,27]. The amounts of these four compounds appeared to be different depending on the extraction solvent used (Table 3). Ethanol, the most common extraction solvent used in maceration, could extract the highest amounts of all types of compounds, while water extracted the least amount of all the compounds regardless of the polarities of the compounds. This result could imply that the polarities of the extracting solvents directly affected the yield of extraction following the like–dissolve–like principle [41]. Additionally, these data suggest that the water-based preparation of kratom leaves as a tea or stimulant drink for use as an alternative medicine for pain relief and diabetes in different cultures, including the Thai culture [25], may not be effective, as water cannot extract many bioactive ingredients from kratom leaves.

Bioactive constituents of kratom leaves, including mitragynine, are already well-known for their analgesic activities, primarily via activation of μ-opioid receptors, in which 7-hydroxymitragynine has 16-fold higher analgesic activity than that of mitragynine. [42]. Increasing evidence indicates that kratom extract could be used to treat metabolic syndrome, i.e., controlling blood glucose and lipid profiles [30–35], even though information about the underlying molecular mechanisms or targeted molecules, in which bioactive compounds in kratom leaves exhibit these actions, is limited. There has been only one previous study that reported the potential of kratom leaf extracts in glucose-lowering effects, in which the methanolic extract of kratom leaves and the major constituent mitragynine promoted in vitro glucose uptake to muscle cells via glucose transporter-1 (GLUT-1) [31]. Thus, this study evaluated whether kratom extracts and the major constituent mitragynine could inhibit the enzymatic activities of α-glucosidase and pancreatic lipase, which are two of the most common molecular targets of anti-diabetic agents (i.e., acarbose) and anti-obesity agents (i.e., orlistat), respectively [5–7]. Inhibiting α-glucosidase results in fewer transformations of the oligosaccharides and disaccharides to glucose, which plays an important role in controlling the postprandial blood glucose levels of diabetics and keeping the blood glucose levels normal by delaying the digestion of carbohydrates and diminishing the absorption of monosaccharides [43], meanwhile, the suppression of pancreatic lipase activity reduces the breakdown of dietary triglycerides into free fatty acids and glycerol and, thus, helps lower blood triglyceride levels [12].

Among the three different kratom leaf extracts, ethanolic extract showed the strongest inhibitory effects toward both α-glucosidase and pancreatic lipase, followed by the methanolic and aqueous extracts (Table 4). It appeared that these trends of inhibition also corresponded to the amounts of the identified four major bioactive compounds that were identified in kratom extracts, which were highest in ethanolic extracts, followed by methanolic and aqueous extracts. Therefore, our results suggested that the highest inhibitory activities of the ethanolic extract were attributed to the presence of the four main compounds, mitragynine, 7-hydroxymitragynine, quercetin, and rutin, which were present in the highest amounts compared to the methanolic and aqueous extracts. Interestingly, it should be noted that all the kratom extracts outperformed the anti-diabetic agent acarbose, which is well known as an α-glucosidase inhibitor, to inhibit α-glucosidase activity [9]. Our discovery could primarily imply that kratom extracts could be used for lowering blood glucose levels by inhibiting the α-glucosidase enzyme. However, this was not a similar case for the inhibition, in which the kratom extracts appeared to inhibit pancreatic lipase activity 30-fold compared to inhibition by the well-known drug inhibitor orlistat [13]. These compounds in the kratom extracts could act synergistically for the activation or downregulation of some key pathways [44].

To further investigate whether the major constituent mitragynine mainly contributed to the inhibition of both α-glucosidase and pancreatic lipase, the purified mitragynine compound was also used in enzymatic assays and the determination of inhibition kinetics. We found that mitragynine exhibited an approximately 3.5-fold stronger inhibitory effect toward α-glucosidase than that of acarbose (Table 5). However, mitragynine appeared to exhibit weaker inhibition toward pancreatic lipase when compared to the known inhibitor drug orlistat judging by double-reciprocal plots (Figure 4A–H), mitragynine inhibited α-glucosidase in a noncompetitive mode with a K_i value of 0.10 mM, whereas the known inhibitor drug acarbose inhibited α-glucosidase by a mixed-type inhibition mode with a K_i value of 0.28 mM. In contrast, both mitragynine and the known inhibitor drug orlistat inhibited pancreatic lipase in a competitive mode, in which the K_i of mitragynine was approximately 62.5-time higher than that of orlistat (K_i = 14.94 ± 0.29 μM versus K_i = 0.24 ± 0.03 μM, respectively). This finding suggested that mitragynine exhibited relatively weak inhibition toward pancreatic lipase and could not be effective for use in obesity management via lowering blood triglyceride levels. The orlistat molecule comprises several aliphatic chains that could span over the active site of the pancreatic lipase enzyme [45]. However, the relatively more rigid molecular structure of mitragynine may not fit well into the active site of pancreatic lipase, resulting in compromised inhibitory activities. Based on molecular docking, several previous studies have predicted that flavonoids with subclasses of flavones, flavanones, and chalcones could be potential candidate compounds for the effective inhibition of pancreatic lipase [46]. Further work will be conducted to investigate whether the other bioactive compounds in kratom leaves could exhibit anti-diabetic and anti-obesity properties.

It has been hypothesized that the two inhibitors with different modes of inhibition could contribute synergistically to each other (to the inhibition of α-glucosidase). Therefore, we investigated whether this postulation was correct by performing synergistic inhibition assays. When mitragynine was added to the enzymatic reactions in which acarbose was also present, a synergistic effect could be observed as the IC_{50} value of the compound mixture was lower than that of acarbose alone. Therefore, our results have reported for the first time the potential of kratom leaf extracts and the major constituent mitragynine for use as herbal medicinal therapies for diabetes. In addition, our information could primarily provide healthcare professionals with significant notes on the potential of the use of kratom in combination with the anti-diabetic agent acarbose for more effective control of blood glucose levels in patients with diabetes and can increase the knowledge [47] of using kratom in combination with bioactive substances and medicines.

5. Conclusions

Kratom leaves are known to be rich sources of alkaloids, flavonoids, and phenolic compounds, and our study found that the ethanolic extraction of kratom leaves produced the highest amount of these compounds compared with that of methanolic and aqueous extractions. The extracted compounds from kratom leaves also exhibited inhibitory activity against α-glucosidase and pancreatic lipase, and the highest inhibition was found from the ethanolic extract. Mitragynine, one of the major alkaloid constituents in kratom leaves, was found to have stronger inhibitory activity against α-glucosidase than that of the well-known anti-diabetic drug acarbose. We deduced that mitragynine is a main component for the inhibition of α-glucosidase in kratom leaves since the combination of acarbose with mitragynine showed a higher inhibitory effect than that of acarbose combined with kratom leaf extracts. In addition to the mentioned effect against α-glucosidase, the ethanol extracts of kratom leaves and mitragynine were revealed to possess repressive activity against pancreatic lipase. Our research suggests that kratom leaves, particularly mitragynine, have promising potential for use in therapeutic and protective applications in diabetic patients as herbal supplements in conjunction with standard pharmacological approaches.

Author Contributions: Conceptualization, S.W.; methodology, S.W., P.P., N.N., T.L. and A.P.; software, P.P., N.N. and S.W.; validation, N.N., P.P., A.P. and S.W.; investigation, S.W., P.P., N.N., A.P. and T.L.; resources, P.P. and A.P.; data curation, S.W. and P.P.; writing—original draft preparation, S.W., P.P., A.P., A.R.A., W.W. and W.P.; writing—review and editing, S.W., P.P., N.N., T.L. and A.P.; supervision, S.W.; funding acquisition, S.W. All authors have read and agreed to the published version of the manuscript.

Funding: This work was supported by Walailak University and Dr. CBD Co., Ltd., under a joint research project, grant number WUSTP-11/2565.

Institutional Review Board Statement: Not applicable.

Informed Consent Statement: Not applicable.

Data Availability Statement: Data are contained within the article.

Acknowledgments: We thank the Dr. CBD Co., Ltd. and Dr. Kratom Bio Co., Ltd., for providing the research facilities. We are also thankful to Jitnapa Voranitikul and Boontariga Wongsa (Scispec, Co., Ltd., Bangkok, Thailand) for the LC-MS/MS identification.

Conflicts of Interest: The authors have no conflict of interest to declare.

References

1. Ambele, M.A.; Dhanraj, P.; Giles, R.; Pepper, M.S. Adipogenesis: A Complex Interplay of Multiple Molecular Determinants and Pathways. *Int. J. Mol. Sci.* **2020**, *21*, 4283. [CrossRef] [PubMed]
2. Mihailović, M.; Jovanović, A.J.; Uskoković, A.; Grdović, N.; Dinić, S.; Vidović, S.; Poznanović, G.; Mujić, I.; Vidaković, M. Protective Effects of the Mushroom *Lactarius deterrimus* Extract on Systemic Oxidative Stress and Pancreatic Islets in Streptozotocin-Induced Diabetic Rats. *J. Diabetes Res.* **2015**, *2015*, 576726. [CrossRef] [PubMed]
3. Mohamed, G.A.; Ibrahim, S.R.M.; Elkhayat, E.S.; El Dine, R.S. Natural anti-obesity agents. *Bull. Fac. Pharm. Cairo Univ.* **2014**, *52*, 269–284. [CrossRef]
4. Lankatillake, C.; Luo, S.; Flavel, M.; Lenon, G.B.; Gill, H.; Huynh, T.; Dias, A.D. Screening natural product extracts for potential enzyme inhibitors: Protocols, and the standardisation of the usage of blanks in α-amylase, α-glucosidase and lipase assays. *Plant Methods* **2021**, *17*, 3. [CrossRef]
5. Buchholz, T.; Melzig, M.F. Medicinal Plants Traditionally Used for Treatment of Obesity and Diabetes Mellitus—Screening for Pancreatic Lipase and α-Amylase Inhibition. *Phytother. Res.* **2016**, *30*, 260–266. [CrossRef]
6. Rajan, L.; Palaniswamy, D.; Mohankumar, S.K. Targeting obesity with plant-derived pancreatic lipase inhibitors: A comprehensive review. *Pharmacol. Res.* **2020**, *155*, 104681. [CrossRef]
7. Tundis, R.; Loizzo, M.R.; Menichini, F. Natural Products as alpha-Amylase and alpha-Glucosidase Inhibitors and their Hypoglycaemic Potential in the Treatment of Diabetes: An Update. *Mini Rev. Med. Chem.* **2010**, *10*, 315–331. [CrossRef]
8. Tomasik, P.; Horton, D. Enzymatic conversions of starch. *Adv. Carbohydr. Chem. Biochem.* **2012**, *68*, 59–436. [CrossRef]
9. DiNicolantonio, J.J.; Bhutani, J.; O'Keefe, J.H. Acarbose: Safe and Effective for Lowering Postprandial Hyperglycaemia and Improving Cardiovascular Outcomes. *Open Heart* **2015**, *2*, e000327. [CrossRef]
10. Tushuizen, M.E.; Bunck, C.M.; Pouwels, P.J.; Bontemps, S.; Van Waesberghe, J.H.T.; Schindhelm, R.K.; Mari, A.; Heine, R.J.; Diamant, M. Pancreatic Fat Content and Beta-Cell Function in Men with and without Type 2 Diabetes. *Diabetes Care* **2007**, *30*, 2916–2921. [CrossRef]
11. You, Q.; Chen, F.; Wang, X.; Jiang, Y.; Lin, S. Anti-Diabetic Activities of Phenolic Compounds in Muscadine Against Alpha-Glucosidase and Pancreatic Lipase. *LWT—Food. Sci. Technol.* **2012**, *46*, 164–168. [CrossRef]
12. Yang, M.H.; Chin, Y.-W.; Yoon, K.D.; Kim, J. Phenolic Compounds with Pancreatic Lipase Inhibitory Activity from Korean yam (*Dioscorea opposita*). *J. Enzym. Inhib. Med. Chem.* **2014**, *29*, 1–6. [CrossRef]
13. Jaradat, N.; Zaid, A.N.; Hussein, F.; Maram, Z.; Alijammal, H.; Ayesh, O. Anti-Lipase Potential of the Organic and Aqueous Extracts of Ten Traditional Edible and Medicinal Plants in Palestine; a Comparison Study with Orlistat. *Medicines* **2017**, *4*, 89. [CrossRef]
14. Harp, J.B. An Assessment of the Efficacy and Safety of Orlistat for the Long-Term Management of Obesity. *J. Nutr. Biochem.* **1998**, *9*, 516–521. [CrossRef]
15. Shai, L.J.; Masoko, P.; Mokgotho, M.P.; Magano, S.R.; Mogale, A.M.; Boaduo, N.; Eloff, J.N. Yeast Alpha Glucosidase Inhibitory and Antioxidant Activities of Six Medicinal Plants Collected in Phalaborwa, South Africa. *S. Afr. J. Bot.* **2010**, *76*, 465–470. [CrossRef]
16. Gatto, L.J.; Oliveira de Bellei, G.R.; Rech, K.S.; Moura, P.F.; Gribner, C.; Merino, Z.; Avila, S.; Fatima Gaspari Dias, J.; Miguel, O.G.; Miguel, M.D. Inhibition of α-Glucosidase, Pancreatic Lipase, and Antioxidant Property of *Myrcia hatschbachii D.* Legrand containing Gallic and Ellagic Acids. *Bol. Latinoam.* **2021**, *20*, 226–243. [CrossRef]
17. Itoh, K.; Matsukawa, T.; Murata, K.; Nishitani, R.; Yamagami, M.; Tomohiro, N.; Kajiyama, S.; Fumuro, M.; Iijima, M.; Shigeoka, S.; et al. Pancreatic Lipase Inhibitory Activity of *Citrus unshiu* Leaf Extract. *Nat. Prod. Commun.* **2019**, *14*. [CrossRef]

18. Chang, Y.; Zhang, D.; Yang, G.; Zheng, Y.; Guo, L. Screening of Anti-Lipase Components of *Artemisia argyi* Leaves Based on Spectrum-Effect Relationships and HPLC-MS/MS. *Front. Pharmacol.* **2021**, *12*, 675396. [CrossRef]
19. Bhatia, A.; Singh, B.; Arora, R.; Arora, S. In vitro Evaluation of the α-Glucosidase Inhibitory Potential of Methanolic Extracts of Traditionally used Antidiabetic Plants. *BMC Complement. Altern. Med.* **2019**, *19*, 74. [CrossRef]
20. Chelladurai, G.R.M.; Chinnachamy, C. Alpha Amylase and Alpha Glucosidase Inhibitory Effects of Aqueous Stem Extract of *Salacia oblonga* and Its GC-MS Analysis. *Braz. J. Pharm. Sci.* **2018**, *54*, e17151. [CrossRef]
21. Elya, B.; Basah, K.; Mun'im, A.; Yuliastuti, W.; Bangun, A.; Septiana, E.K. Screening of α-Glucosidase Inhibitory Activity from Some Plants of Apocynaceae, Clusiaceae, Euphorbiaceae, and Rubiaceae. *J. Biomed. Biotechnol.* **2012**, *2012*, 281078. [CrossRef] [PubMed]
22. Dechakhamphu, A.; Wongchum, N. Screening for Anti-Pancreatic Lipase Properties of 28 Traditional Thai Medicinal Herbs. *Asian Pac. J. Trop. Biomed.* **2015**, *5*, 1042–1045. [CrossRef]
23. Pop, A.; Fizeșan, I.; Vlase, L.; Rusu, M.E.; Cherfan, J.; Babota, M.; Gheldiu, A.-M.; Tomuta, I.; Popa, D.-S. Enhanced Recovery of Phenolic and Tocopherolic Compounds from Walnut (*Juglans regia* L.) Male Flowers Based on Process Optimization of Ultrasonic Assisted-Extraction: Phytochemical Profile and Biological Activities. *Antioxidants* **2021**, *10*, 607. [CrossRef] [PubMed]
24. Boffa, L.; Ghe, C.; Barge, A.; Muccioli, G.; Cravotto, G. Alkaloid Profiles and Activity in Different *Mitragyna speciosa* Strains. *Nat. Prod. Commun.* **2018**, *13*. [CrossRef]
25. Charoenratana, S.; Anukul, C.; Aramrattana, A. Attitudes Towards Kratom Use, Decriminalization and the Development of a Community-Based Kratom Control Mechanism in Southern Thailand. *Int. J. Drug Policy* **2021**, *95*, 103197. [CrossRef]
26. Goh, Y.S.; Karunakaran, T.; Murugaiyah, V.; Santhanam, R.; Abu Bakar, M.H.; Ramanathan, S. Accelerated Solvent Extractions (ASE) of *Mitragyna speciosa Korth.* (Kratom) Leaves: Evaluation of Its Cytotoxicity and Antinociceptive Activity. *Molecules* **2021**, *26*, 3704. [CrossRef] [PubMed]
27. Eastlack, S.C.; Cornett, E.M.; Kaye, A.D. Kratom-Pharmacology, Clinical Implications, and Outlook: A Comprehensive Review. *Pain Ther.* **2020**, *9*, 55–69. [CrossRef]
28. Hassan, Z.; Muzaimi, M.; Navaratnam, V.; Yusoff, N.H.M.; Suhaimi, W.F.; Vadivelu, R.; Vicknasingam, K.B.; Amato, D.; Horsten, S.V.; Ismail, N.I.W.; et al. From Kratom to Mitragynine and Its Derivatives: Physiological and Behavioural Effects Related to Use, Abuse, and Addiction. *Neurosci. Biobehav. Rev.* **2013**, *37*, 138–151. [CrossRef]
29. Parthasarathy, S.; Ramanathan, S.; Murugaiyah, V.; Hamdan, M.R.; Mohd Said, M.I.; Lai, C.-S.; Mansor, S.M. A Simple HPLC–DAD Method for The Detection and Quantification of Psychotropic Mitragynine in *Mitragyna speciosa (ketum)* and Its Products for the Application in Forensic Investigation. *Forensic. Sci. Int.* **2013**, *226*, 183–187. [CrossRef]
30. Leong Bin Abdullah, M.F.I.; Tan, K.L.; Isa, S.M.; Yusoff, N.S.; Yeou Chear, N.J.; Singh, D. Lipid Profile of Regular Kratom (*Mitragyna speciosa* Korth.) Users in the Community Setting. *PLoS ONE* **2020**, *15*, e0234639. [CrossRef]
31. Purintrapiban, J.; Keawpradub, N.; Kansenalak, S.; Chittrakarn, S.; Janchawee, B.; Sawangjaroen, K. Study on Glucose Transport in Muscle Cells by Extracts from *Mitragyna speciosa* (*Korth*) and Mitragynine. *Nat. Prod. Res.* **2011**, *25*, 1379–1387. [CrossRef] [PubMed]
32. Anwar, R. In vitro Effect of Mitragynine on Activity of Drug Metabolizing Enzymes, N-demethylase and Glutathuione S-transferase in Streptozotocin-Induced Diabetic Rats. *Pharmacologyonline* **2012**, *1*, 68–75.
33. Lee, S.Y.; Mediani, A.; Nur Ashikin, A.H.; Azliana, A.B.S.; Abas, F. Antioxidant and α-Glucosidase Inhibitory Activities of the Leaf and Stem of Selected Traditional Medicinal Plants. *Int. Food. Res. J.* **2014**, *21*, 379–386.
34. Chen, L.; Fei, S.; Olatunji, O.J. LC/ESI/TOF-MS Characterization, Anxiolytic and Antidepressant-like Effects of *Mitragyna speciosa* Korth Extract in Diabetic Rats. *Molecules* **2022**, *27*, 2208. [CrossRef] [PubMed]
35. La-up, A.; Saengow, U.; Aramrattana, A. High Serum High-Density Lipoprotein and Low Serum Triglycerides in Kratom Users: A Study of Kratom Users in Thailand. *Heliyon* **2021**, *7*, e06931. [CrossRef]
36. Vernon, L.S.; Orthofer, R.; Lamuela-Raventós, R.M. Analysis of total phenols and other oxidation substrates and antioxidants by means of folin-ciocalteu reagent. *Methods Enzymol.* **1999**, *299*, 152–178. [CrossRef]
37. Shraim, A.M.; Ahmed, T.A.; Rahman, M.M.; Hijji, Y.M. Determination of total flavonoid content by aluminum chloride assay: A critical evaluation. *LWT* **2021**, *150*, 111932. [CrossRef]
38. John, B.; Sulaiman, C.T.; George, S.; Reddy, V.R.K. Spectrophotometric estimation of total alkaloids in selected justicia species. *Int. J. Pharm. Pharm. Sci.* **2014**, *6*, 647–648.
39. Inthongkaew, P.; Chatsumpun, N.; Supasuteekul, C.; Kitisripanyam, T.; Putalun, W.; Likhitwitayawuid, K.; Sritularak, B. α-Glucosidase and Pancreatic Lipase Inhibitory Activities and Glucose Uptake Stimulatory Effect of Phenolic Compounds from *Dendrobium formosum*. *Rev. Bras. Farmacogn.* **2017**, *27*, 480–487. [CrossRef]
40. Prevete, E.; Kuypers, K.P.C.; Theunissen, E.L.; Corazza, C.; Bersani, G.; Ramaekers, J.G. A systematic review of (pre)clinical studies on the therapeutic potential and safety profile of kratom in humans. *Hum. Psychopharmacol. Clin. Exp.* **2021**, *37*, e2805. [CrossRef]
41. Zhang, Q.-W.; Lin, L.-G.; Ye, W.-C. Techniques for Extraction and Isolation of Natural Products: A Comprehensive Review. *Chin. Med.* **2018**, *13*, 20. [CrossRef] [PubMed]
42. Kruegel, A.C.; Gassaway, M.M.; Kapoor, A.; Varadi, A.; Majumar, S.; Filizola, M.; Javitch, A.J.; Sames, D. Synthetic and Receptor Signaling Explorations of the Mitragyna Alkaloids: Mitragynine as an Atypical Molecular Framework for Opioid Receptor Modulators. *J. Am. Chem. Soc.* **2016**, *138*, 6754–6764. [CrossRef] [PubMed]

43. Wen, H.; Stewart, A.J.; Tao, Y.; Shao, Y.; Cui, Y.; Yue, H.; Pei, J.; Liu, Z.; Mei, L.; Yu, R.; et al. Erythritol Attenuates Postprandial Blood Glucose by Inhibiting α-Glucosidase. *J. Agric. Food. Chem.* **2018**, *66*, 1401–1407. [CrossRef] [PubMed]
44. Mateș, L.; Popa, D.S.; Rusu, M.E.; Fizeșan, I.; Leucuța, D. Walnut Intake Interventions Targeting Biomarkers of Metabolic Syndrome and Inflammation in Middle-Aged and Older Adults: A Systematic Review and Meta-Analysis of Randomized Controlled Trials. *Antioxidants* **2022**, *11*, 1412. [CrossRef]
45. Veeramachaneni, G.K.; Raj, K.K.; Chalasani, L.M.; Annamraju, S.K.; Js, B.; Talluri, V.R. Shape Based Virtual Screening and Molecular Docking Towards Designing Novel Pancreatic Lipase Inhibitors. *Bioinformation* **2015**, *11*, 535–542. [CrossRef]
46. Nguyen, P.T.V.; Huynh, H.A.; Truong, D.V.; Tran, T.-D.; Thi Vo, C.-V. Exploring Aurone Derivatives as Potential Human Pancreatic Lipase Inhibitors through Molecular Docking and Molecular Dynamics Simulations. *Molecules.* **2020**, *25*, 4657. [CrossRef]
47. Swogger, M.T.; Smith, K.E.; Garcia-Romeu, A.; Grundmann, O.; Veltri, C.A.; Henningfield, J.E.; Busch, L.Y. Understanding Kratom Use: A Guide for Healthcare Providers. *Front. Pharmacol.* **2022**, *13*, 801855. [CrossRef]

nutrients

MDPI

Article

Rosmarinic Acid Attenuates Rotenone-Induced Neurotoxicity in SH-SY5Y Parkinson's Disease Cell Model through Abl Inhibition

Xiao Han [1,†], Bing Han [1,†], Yue Zhao [2], Gang Li [1], Tian Wang [3], Jie He [1], Wenxiao Du [4], Xiaolin Cao [4], Jing Gan [1], Zhenhua Wang [1,*] and Wei Zheng [1,*]

1 Center for Mitochondria and Healthy Aging, College of Life Sciences, Yantai University, Yantai 264005, China
2 Shandong Hongli Medical Animal Experimental Research Co., Jinan 250000, China
3 School of Pharmacy, Yantai University, Yantai 264005, China
4 College of Life Science, Yantai University, Yantai 264005, China
* Correspondence: skywzh@ytu.edu.cn (Z.W.); weizheng@ytu.edu.cn (W.Z.)
† These authors contributed equally to this work.

Abstract: Rosmarinic acid (RA) is a natural polyphenolic compound with antioxidative property. With the present study, we aimed to evaluate the neuroprotective role of RA on Parkinson's disease using rotenone induced SH-SY5Y cell model of Parkinson's disease, the underlying mechanism of action of RA was also investigated. Cell viability, cell morphology, apoptosis, signaling protein phosphorylation and expression, cellular reactive oxygen species (ROS) production, ATP content, and mitochondrial membrane potential were tested in SH-SY5Y cells. RA showed a neuroprotective effect in a rotenone-induced SH-SY5Y cell model of Parkinson's disease with dose-dependent manner, it reduced cell apoptosis and restored normal cell morphology. RA not only decreased levels of α-synuclein and Tau phosphorylation but also elevated the contents of AMPK phosphorylation, Akt phosphorylation, and PGC-1α. RA restored the reduced mitochondrial membrane potential and ATP content as well as inhibited rotenone-induced ROS overproduction. Further findings demonstrated that the neuroprotective role of RA was partially due to the inhibition of Abl tyrosine kinase. RA treatment suppressed the hyperphosphorylation of Abl Y412 and CrkII Y221 induced by rotenone. Nilotinib, a specific inhibitor of Abl, elicited a similar neuroprotective effect as that of RA. The present study indicates that RA has a property of neuroprotection against rotenone, and the neuroprotective effect is partially attributed to the inhibition of Abl.

Keywords: rosmarinic acid; Parkinson's disease; mitochondrial function; Abl tyrosine kinase; rotenone

Citation: Han, X.; Han, B.; Zhao, Y.; Li, G.; Wang, T.; He, J.; Du, W.; Cao, X.; Gan, J.; Wang, Z.; et al. Rosmarinic Acid Attenuates Rotenone-Induced Neurotoxicity in SH-SY5Y Parkinson's Disease Cell Model through Abl Inhibition. *Nutrients* **2022**, *14*, 3508. https://doi.org/10.3390/nu14173508

Academic Editors: Haixia Yang and Jianjun Deng

Received: 31 July 2022
Accepted: 22 August 2022
Published: 26 August 2022

1. Introduction

Parkinson's disease (PD) is the second most prevalent neurodegenerative disease. There are currently more than five million PD patients worldwide. Up to date, no curing treatment is available [1]. The main characteristics of PD are the degeneration of dopaminergic neurons in the substantia nigra pars compacta and the lack of dopamine in the striatum [2], which result in the clinical features including tremor, bradykinesia, and muscle stiffness [3]. A few molecular mechanisms have been identified to be responsible for the neuronal cell death in PD, including mitochondrial dysfunction, increased iron content, and oxidative stress. Previous studies showed an increase of oxidative stress in the cerebrospinal fluid of PD patients and suggested that excess formation of free radicals was involved in the progression of PD [4]. Therefore, several research studies were carried out to investigate the roles of antioxidative agents on treating PD animal or cell models [5,6].

Rosmarinic acid (RA) is a natural polyphenolic compound that exists in several kinds of plants including Lamiaceae and Boraginaceae families [7] (Figure 1). It has antiviral, antibacterial, anti-inflammatory, and antioxidative properties [8]. A PD cell model study with SH-SY5Y showed that RA activated antioxidant enzyme heme oxygenase-1 to suppress reactive oxygen species (ROS) production and cell death induced by H_2O_2 exposure [9]. RA also reduced the oxidative stress and neurotoxic effects induced by 6-hydroxydopamine (6-OHDA) and reversed the mitochondrial membrane potential reduction in 6-OHDA challenged cells [10]. Another study further proved the neuroprotective effect of RA by re-establishing the mitochondrial complex I function and recovering the dopamine content level and cell viability in 1-methyl-4-phenylpyridinium (MPP+) exposed MES23.5 dopaminergic cells [11]. In PD animal model, RA was demonstrated to protect against 1-methyl-4-phenyl-1,2,3,6-tetrahydropyridine (MPTP) induced neurotoxicity in zebrafish embryos [6]. These studies indicated RA being a potential compound for PD treatment.

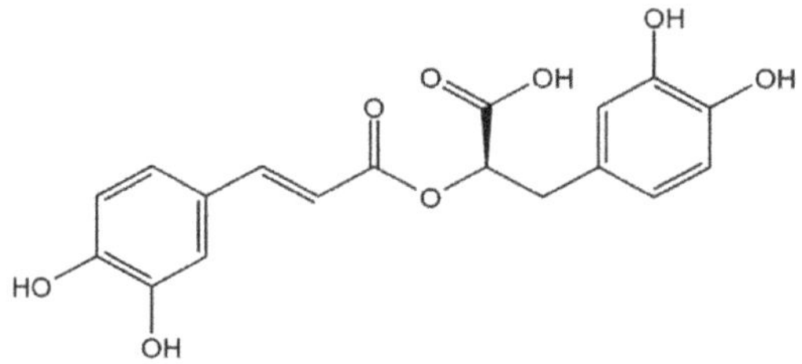

Figure 1. Chemical structure of rosmarinic acid (RA).

Rotenone is a natural compound present in plants including Tephrosia and Lonchocarpus [12]. Since the rotenone challenged animal models display two typical characteristics of PD, Lewy bodies in the substantia nigra neurons and degeneration of these cells [13], rotenone exposed animal or cell model gives one of the best experimental models for research on neuroprotective therapies for PD [14–16]. In neuronal cells, rotenone has an inhibitory effect on mitochondrial complex I and mediates α-synuclein aggregation, it causes mitochondrial dysfunction and overproduction of ROS, which ultimately induces neuronal cell death [17–19]. Up to date, the molecular mechanism of action of RA protecting neuronal cells has not been fully elucidated, and the neuroprotective role of RA on rotenone-exposed cells has not been evaluated.

Abelson tyrosine kinase (Abl) is a nonreceptor tyrosine kinase. Upon being activated by oxidative stress, Abl phosphorylates α-synuclein and promotes its accumulation, aggregation, and plays an essential role in neuron degeneration [20]. Abl inhibition rescued mitochondrial function reduction and dopaminergic neuronal degeneration in several mouse PD models [20]. A potent Abl inhibitor, nilotinib, was first developed to treat chronic myeloid leukemia [21]. Later study showed that nilotinib protected dopaminergic neurons in MPTP-induced preclinical mice model of PD [22].

The aim of the current study was to explore the neuroprotective effects of RA on rotenone challenged SH-SY5Y cell model of PD. Furthermore, the role of Abl in the neuroprotective mechanism of RA against rotenone was also assessed.

2. Materials and Methods

2.1. Cell Culture

Human neuroblastoma SH-SY5Y cells were purchased from Shanghai Cell Bank (Shanghai, China). The cells were cultured in DMEM media (Gibco, Waltham, MA, USA) with 10% (*v*/*v*) fetal bovine serum (FBS) (TIANHANG, Zhejiang, China), 100 U/mL penicillin, and 100 mg/mL streptomycin (Gibco, Waltham, MA, USA). Cells were maintained at 37 °C in a saturated humidity atmosphere containing 95% air and 5% CO_2. Cells grew to around 70% confluences before they were used for experiments.

2.2. Cell Viability Assay and Morphological Observation

Cell viability was evaluated using 3-(4,5-dimethylthiazol-2-yl)-2,5-diphenyltetrazolium bromide (MTT) assay. In previous research, RA was pre-administered to animals or cells for different lengths of time before they were challenged with the disease-modeling reagents. RA pretreatment for 2 h protects against MPTP-induced neurotoxicity in zebrafish embryos dopaminergic neurons [6]. Pretreatment for 30 min with RA protects SK-N-SH cells against MPTP induced toxicity [8]. Carnosic acid is a natural antioxidant compound derived from rosemary from which rosmarinic acid was also derived [23]. 12 h pretreatment with carnosic acid has neuroprotective effect on SH-SY5Y cells against 6-OHDA induced neurotoxicity [23]. Based on these data, a medium time length, 3 h, was chosen as the RA pretreatment time length before SH-SY5Y cells were exposed to rotenone. SH-SY5Y cells were exposed to 0, 0.1, 1, 10, 50, 100, and 200 μM rotenone for 24 h at 37 °C for the rotenone toxicity evaluation. SH-SY5Y cells were pretreated with rosmarinic acid (RA, Sigma-Aldrich, St. Louis, MO, USA) before they were then either exposed to 50 μM rotenone, or 0, 1, 10, or 100 μM RA together with rotenone, or 30 nM nilotinib with rotenone, or RA alone, or nilotinib alone for 24 h at 37 °C. Then, 10 μL of MTT (5 mg/mL) was added to each well in a 96-well plate and incubated for 4 h. The insoluble blue formazan was then solubilized with 100 μL/well dimethyl sulfoxide (DMSO), and optical density (OD) values of the mixture were measured at 595 nm with a SpectraMax Paradigm Multi-Mode Microplate Reader (Molecular Devices, Sacramento, CA, USA). Each sample in MTT assay was measured at least in triplicate. For cell morphology analysis, the cells were observed and photographed using a Leica Microsystems microscope (Leica Microsystems CMS GmbH, Wetzlar, Germany) after exposed to 50 μM rotenone and 100 μM RA.

2.3. Apoptosis Assay

SH-SY5Y cells were seeded in 6-well plates and then exposed to 50 μM rotenone and 100 μM RA for 24 h before they were harvested by trypsin solution to produce a single cell suspension. The cells were pelleted by centrifugation and washed twice with PBS to remove trypsin. The cell suspension was then prepared and stained with a YF488-Annexin V and PI Cell Apoptosis Assay Kit (Biorigin, Beijing, China) according to the instruction. The stained cells were analyzed using a ACEA Novocyte flow cytometer (ACEA Biosciences, San Diego, CA, USA) in combination with Novo Express software.

2.4. Western Blot Assay

For Western blot experiments, SH-SY5Y cells were treated as indicated, in the presence or absence of 50 μM rotenone, 100 μM RA, and 30 nM of Nilotinib. Cells were treated for 90 min for the analysis of phosphorylation levels and for 24 h for the analysis of protein levels. Cells were then rinsed with PBS before they were lysed with RIPA buffer (150 mmol/L NaCl, 50 mmol/L Tris–HCl (pH 7.4), 100 mg/mL Phenylmethyl Sulphonyl Fluoride (PMSF), 1% NP-40, 0.1% sodium dodecyl sulfate (SDS)) for 15 min. The cells were scraped and collected before they were centrifuged at 13,000× g for 15 min at 4 °C. The supernatants were collected for the following analysis. A bicinchoninic acid (BCA) Protein Assay Kit was then used to measure the protein concentration in the collected cell lysates. The cell lysates were mixed with loading buffer and boiled for 5 min. 30 μg protein from each sample was used for SDS polyacrylamide gel electrophoresis (SDS-PAGE) analysis (Miniprotean; Bio-Rad). Proteins separated on the gel were then transferred onto Immunobilon-P polyvinylidene difluoride (PVDF) membranes (Millipore, Bedford, MA, USA). 5% BSA (Sigma-Aldrich, St. Louis, MO, USA) in TBS Tris-buffered saline (68 mmol/L NaCl, 10 mmol/L Tris-base, pH 7.5) containing 0.1% Tween 20 was used for 1 h membrane blocking at room temperature. Recommended concentration of primary antibodies were incubated together with the blocked membrane at 4 °C overnight. Membranes were washed and incubated with peroxidase conjugated antibodies for 1 h at room temperature. After the washing step, bound antibodies were visualized using a 5200 Multi Luminescent image analyzer (Tanon Science & Technology Co., Ltd., Shanghai, China) and quantified with Image pro plus 6.0 software. The antibodies used in this study were against: Tau (Abcam, Cambridge, UK), Phospho-Tau Ser404 (Abcam, Cambridge, UK), AMPK (Cell Signaling Technology, Danvers, MA, USA), Phospho-AMPK Thr172 (Cell Signaling Technology, Danvers, MA, USA), Akt (Cell Signaling Technology, Danvers, MA, USA), Phospho-Akt Ser473 (Cell Signaling Technology, Danvers, MA, USA), c-Abl (Cell Signaling Technology, Danvers, MA, USA), Phospho-c-Abl Tyr412 (Cell Signaling Technology, Danvers, MA, USA), Phospho-CrkII Tyr221 (Cell Signaling Technology, Danvers, MA, USA), α-synuclein (Cell Signaling Technology, Danvers, MA, USA), PGC-1α (Cell Signaling Technology, Danvers, MA, USA), β-actin (Cell Signaling Technology, Danvers, MA, USA). The secondary antibody used was horseradish peroxidase (HRP) Goat anti-Rabbit immunoglobulin G (IgG).

2.5. Measurement of Mitochondrial Membrane Potential

Fluorescent probe JC-1 (Thermo Fisher Scientific, Waltham, MA, USA) was used to evaluate the mitochondrial membrane potential. After being exposed to 50 μM rotenone, 100 μM RA or 30 nM nilotinib for 12 h, cells were incubated with JC-1 staining solution (5 μg/mL) for 30 min at 37 °C. A single cell suspension was then obtained by cell trypsinization. The cells were collected by centrifugation before they were rinsed with PBS. The cell-bound JC-1 fluorescent signal was analyzed using a ACEA Novocyte flow cytometer (ACEA Biosciences, San Diego, CA, USA) in combination with Novo Express software. The mitochondrial membrane potential of SH-SY5Y cells was calculated as the ratio of red fluorescence to green fluorescence.

Cells were cultured on coverslips at a density of 1×10^5 cells/mL in 24 well plates for 24 h. After the indicated treatments, the cells were stained with JC-1 for 30 min at 37 °C. Fluorescence microscope (Leica Microsystems CMS GmbH, Wetzlar, Germany) was used to capture fluorescence images.

2.6. ROS Assay

After being exposed to 50 μM rotenone, 100 μM RA, or 30 nM nilotinib for 1 h, the cells were co-incubated with DMEM containing 10 μM of DCFH-DA for 30 min at 37 °C. A single cell suspension was then obtained using cell trypsinization. The cells were collected using centrifugation before they were rinsed with PBS. The cell-bound DCFH-DA fluorescent signal was analyzed using a ACEA Novocyte flow cytometer (ACEA Biosciences, San Diego, CA, USA) in combination with Novo Express software.

Cells were cultured on coverslips at a density of 1×10^5 cells/mL in 24 well plates for 24 h. After the indicated treatments, cells were coincubated with DMEM containing 10 μM of DCFH-DA for 30 min at 37 °C. A fluorescence microscope (Leica Microsystems CMS GmbH, Wetzlar, Germany) was used to capture fluorescence images.

2.7. ATP Assay

ATP content in SH-SY5Y cells was measured using a luciferase-based luminescence enhanced ATP assay kit (Beyotime, Shanghai, China). Cells were washed with ice-cold PBS before they were lysed with 100 μL ice-cold ATP releasing buffer. The cell lysates obtained were centrifuged at 12,000× g for 5 min at 4 °C. The supernatants were incubated with the ATP testing working solution provided by the kit. ATP content in cell lysates was determined using a SpectraMax Paradigm Multi-Mode Microplate Reader (Molecular Devices, Sacramento, CA, USA).

2.8. Statistical Analyses

All data were expressed as the Mean ± SEM. Analysis was performed using GraphPad Prism software 8.0. The data were analyzed using one-way ANOVA followed by Tukey's post hoc test. Statistical significance was defined as $p < 0.05$.

3. Results

3.1. *RA Restored Cell Viability and Morphology in Rotenone Challenged SH-SY5Y Cells*

To investigate the role of RA in protecting neurons against the toxic effect of rotenone, SH-SY5Y cell viability and morphology were tested in the absence and presence of rotenone and RA. The results showed that 1–200 μM rotenone significantly reduced cell viability (Figure 2C). Cotreatment with rotenone and increasing concentration of RA led to gradual recovery of cell viability (Figure 2D). RA at a concentration of 100 μM significantly rescued SH-SY5Y cells from the toxic effect of rotenone (Figure 2A,D).

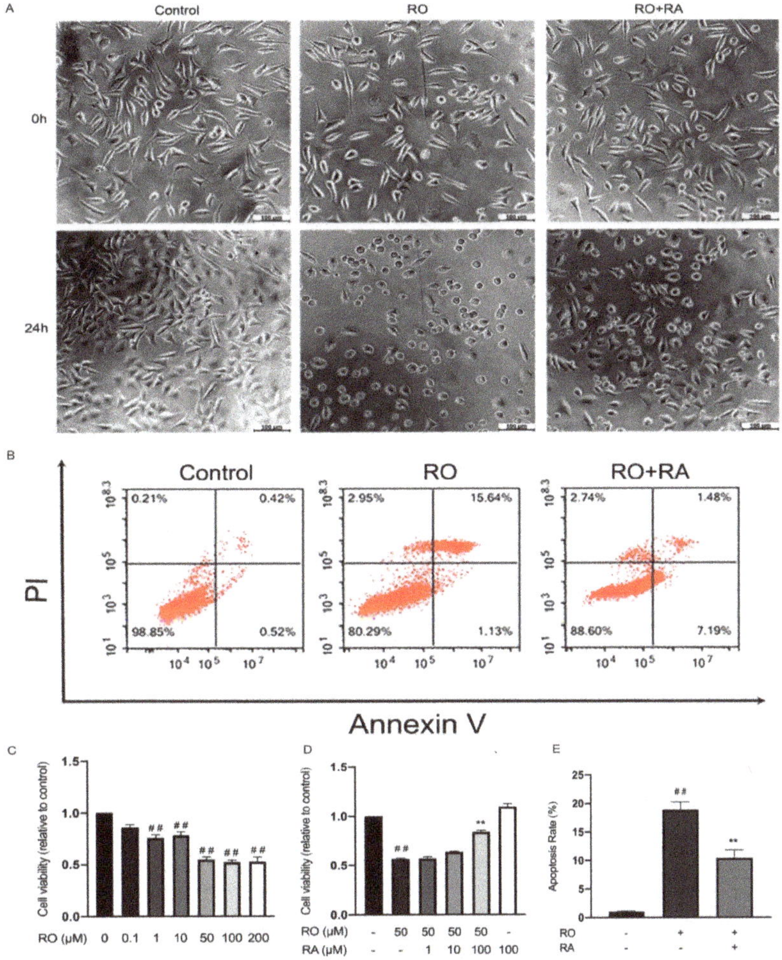

Figure 2. Rosmarinic acid protected SH-SY5Y cells from the toxicity effect of rotenone (RO). (**A**) Cell morphology microscopic photograph of SH-SY5Y cells in the absence or presence of 50 μM rotenone and 100 μM RA. Scale bar = 100 μm. (**B**) Cell apoptosis evaluated with flow cytometry in the absence or presence of rotenone and RA. (**C**) Bar graph quantification analysis of SH-SY5Y cell viability after being exposed to the indicated concentrations of rotenone evaluated with MTT assay. (**D**) Bar graph quantification analysis of SH-SY5Y cell viability in the absence and presence of indicated concentration of rotenone and RA evaluated with MTT assay. (**E**) Bar graph quantification analysis of SH-SY5Y cell apoptosis rate in the absence or presence of rotenone and RA evaluated with flow cytometry. All data were expressed as the Mean ± SEM of at least three independent experiments. ## $p < 0.01$ compared with the control group, ** $p < 0.01$ compared with the rotenone group.

3.2. RA Reduced Cell Apoptosis Rate in Rotenone Challenged SH-SY5Y Cells

Apoptosis analysis showed that 24 h rotenone exposure led to increased cell apoptosis rate, while RA treatment significantly reduced cell apoptosis induced by rotenone (Figure 2B,E).

3.3. RA Suppressed Tau Phosphorylation and α-Synuclein Expression Elevated by Rotenone

Phosphorylation of Tau at Ser404 and accumulation of α-synuclein are considered to be the indicators of PD and induce neurodegeneration [24]. To analyze the molecular mechanism of RA's neuroprotective effect, Ser404 phosphorylation of Tau and α-synuclein protein level in SH-SY5Y cells were examined. Rotenone exposure increased Tau phosphorylation and α-synuclein level. RA significantly suppressed the elevated Tau phosphorylation and α-synuclein level induced by rotenone (Figure 3A).

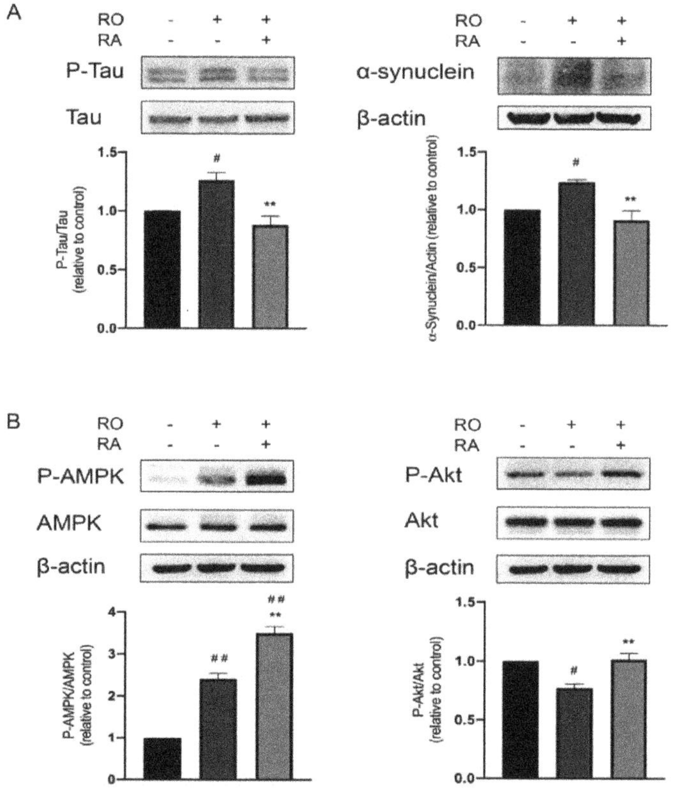

Figure 3. RA treatment modulated the protein level of α-synuclein, the phosphorylation level of Tau Ser404, AMPK Thr172, and Akt Ser473. (**A**) Representative Western blot photographs of the phosphorylation level of Tau, protein levels of α-synuclein, Tau, and β-actin, and the corresponding bar graph quantification analyses of P-Tau relative to Tau protein level and α-synuclein relative to β-actin protein level. (**B**) Representative Western blot photographs of P-AMPK, P-Akt, AMPK, Akt, and β-actin levels, and the corresponding bar graph quantification analyses of P-AMPK relative to AMPK and P-Akt relative to Akt. All data were expressed as the mean ± SEM of three independent experiments. # $p < 0.05$, ## $p < 0.01$ compared with the control group, ** $p < 0.01$ compared with the rotenone group.

3.4. RA Rescued Rotenone Induced P-Akt Reduction and Further Promoted the Phosphorylation of AMPK Thr172

AMPK kinase plays important roles in neuronal protection in PD, Thr172 is located in the activation loop of AMPK, and its phosphorylation is required for AMPK activation [25]. Akt signaling is critical for cell survival in PD [26]. Results in Figure 3B showed that rotenone exposure caused upregulation of AMPK Thr172 phosphorylation, while treatment of RA further elevated AMPK phosphorylation. Rotenone exposure also downregulated Akt phosphorylation, and RA recovered Akt phosphorylation.

3.5. RA and Nilotinib Reduced Abl Y412 and CrkII Y221 Phosphorylation Elevated by Rotenone and Restored SH-SY5Y Cell Viability under Rotenone Treatment

Tyrosine kinase Abl plays an essential role in neurodegeneration. CrkII is a well-known substrate of Abl and can be phosphorylated at Tyr221 by Abl [27]. To investigate if Abl kinase is involved in RA-mediated neuroprotective effect against rotenone, cellular phosphorylation of Abl Y412 and CrkII Y221 were evaluated in the presence and absence of rotenone, RA and Abl specific inhibitor, nilotinib. Rotenone exposure induced increases in Abl Y412 and CrkII Y221 phosphorylation, while treatment with nilotinib or RA significantly reduced Abl and CrkII phosphorylation (Figure 4A). As shown in Figure 4B, rotenone reduced the SH-SY5Y cell viability, and RA or nilotinib treatments increased cell viability level in the rotenone-exposed cells.

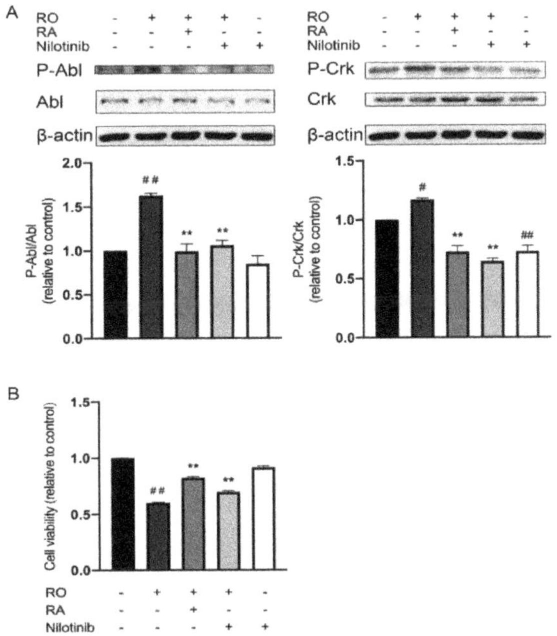

Figure 4. RA and nilotinib reduced Abl Tyr412 and CrkII Tyr221 phosphorylation induced by rotenone and restored SH-SY5Y cell viability under rotenone treatment. (**A**) Representative Western blot photographs of the phosphorylation level of Abl Tyr412, CrkII Tyr221, protein levels of Abl, CrkII and β-actin, and the corresponding bar graphs of quantification analyses of P-Abl relative to Abl protein and P-CrkII relative to Crk protein. (**B**) Bar graph quantification analysis of SH-SY5Y cell viability in the absence and presence of 50 μM rotenone, 100 μM RA, and 30 nM nilotinib evaluated with MTT assay. All data were expressed as the mean ± SEM of three independent experiments. # $p < 0.05$, ## $p < 0.01$ compared with the control group, ** $p < 0.01$ compared with the rotenone group.

3.6. Abl Inhibition and RA Increased PGC-1α Expression, Akt, and AMPK Phosphorylation under Rotenone Exposure

Results in Figure 5A showed that rotenone exposure downregulated Akt phosphorylation, PGC-1α expression, and upregulated AMPK phosphorylation. RA and Abl inhibition increased Akt phosphorylation and PGC-1α expression level and further increased AMPK phosphorylation.

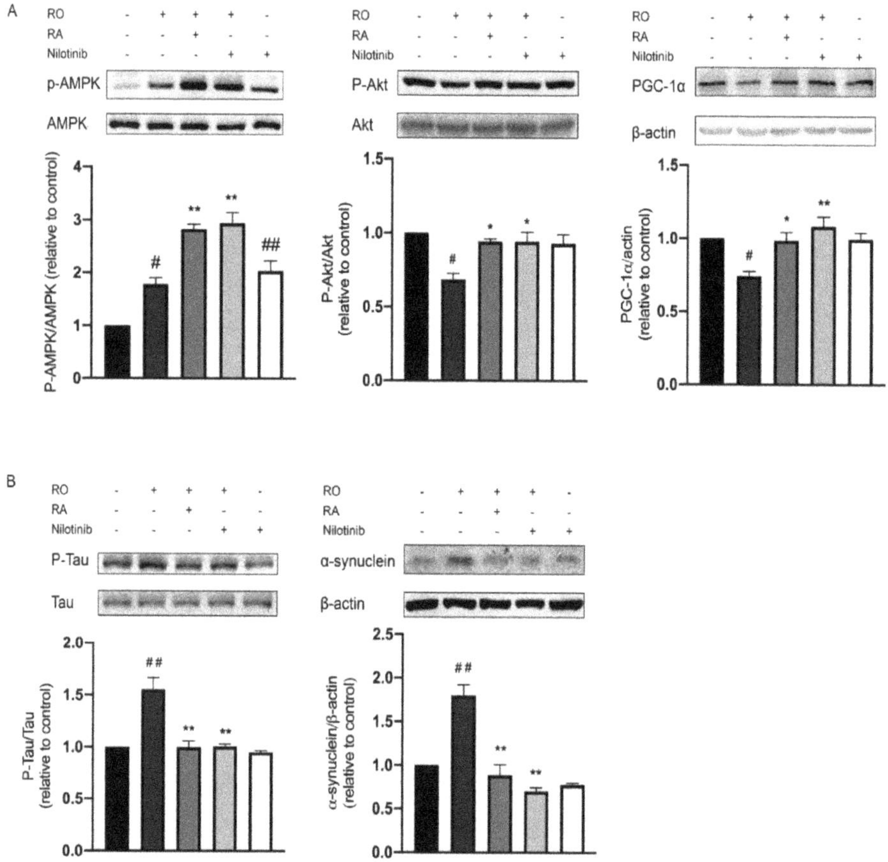

Figure 5. Abl inhibition modulated the protein level of α-synuclein, PGC-1α, and the phosphorylation level of Tau, AMPK, and Akt. (**A**) Representative Western blot photographs of the phosphorylation level of AMPK Thr172, Akt Ser473, and protein level of PGC-1α, AMPK, Akt, and β-actin, and the corresponding bar graphs quantification analyses of P-AMPK relative to AMPK, P-Akt relative to Akt and PGC-1α relative to β-actin. (**B**) Representative Western blot photographs of the phosphorylation level of Tau Ser404, the protein level of α-synuclein, Tau, and β-actin, and the corresponding bar graphs quantification analyses of P-Tau relative to Tau and α-synuclein relative to β-actin. All data were expressed as the mean ± SEM of three independent experiments. # $p < 0.05$, ## $p < 0.01$ compared with the control group, * $p < 0.05$, ** $p < 0.01$ compared with the rotenone group.

3.7. Abl Inhibition Reduced Tau Phosphorylation and α-Synuclein Expression Level Elevated by Rotenone

To investigate the mechanism of the neuroprotective effect of RA, Tau Ser404 protein phosphorylation, and α-synuclein level in SH-SY5Y cells were examined. Like RA, nilotinib significantly reduced the elevation of Tau phosphorylation and α-synuclein expression induced by rotenone exposure, while nilotinib alone did not lead to any significant changes in cells in the absence of rotenone (Figure 5B).

3.8. Abl Inhibition and RA Suppressed Rotenone-Induced ROS Overproduction

ROS production was measured in the cells. ROS level was markedly increased upon rotenone exposure. RA or nilotinib significantly reduced the ROS production induced by rotenone. Treatment with nilotinib alone did not lead to any significant changes of ROS production when compared with that of the control group (Figure 6A–C).

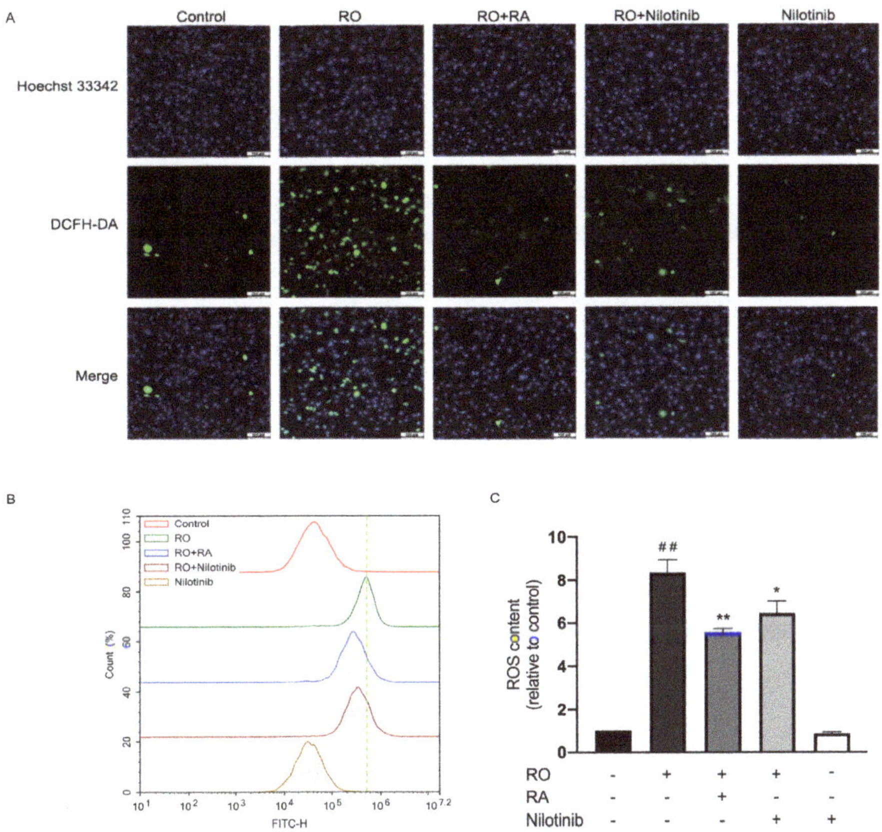

Figure 6. RA and Abl inhibition suppressed rotenone induced ROS production. (**A**) SH-SY5Y cells were exposed to rotenone, RA, and nilotinib for 1 h; then, cells were incubated with 10 μM DCFH-DA for 30 min, images were acquired using fluorescence microscope. Scale bar = 100 μm. (**B**) After treatments indicated, DCFH2-DA fluorescence intensity was evaluated using flow cytometry. (**C**) Bar graph quantification analysis of ROS level evaluated with flow cytometry. All data were expressed as the mean ± SEM of three independent experiments. ## $p < 0.01$ compared with the control group, * $p < 0.05$, ** $p < 0.01$ compared with the rotenone group.

3.9. Abl Inhibition and RA Restored Mitochondrial Membrane Potential and ATP Content in Rotenone Exposed SH-SY5Y Cells

Mitochondrial membrane potential in SH-SY5Y cells was measured. After rotenone exposure, mitochondrial membrane potential was reduced. RA or nilotinib treatment significantly restored the mitochondrial membrane potential in rotenone exposed cells (Figure 7A–C). Treatment with nilotinib alone did not lead to any significant changes of mitochondrial membrane potential when compared with control. Figure 7D showed the cellular ATP content measured. Rotenone reduced ATP content of the cells. RA or nilotinib treatment significantly restored the reduced ATP content. Nilotinib alone did not lead to any significant changes of ATP amount when compared with control.

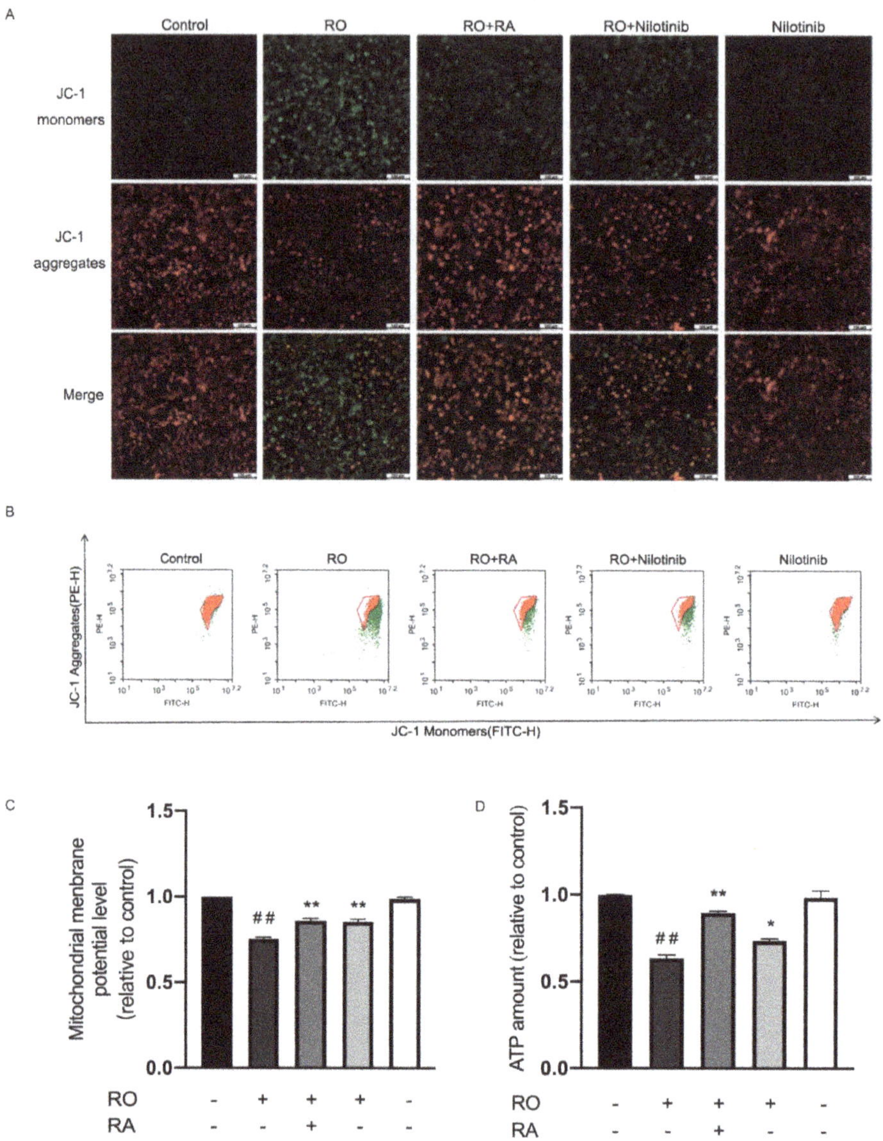

Figure 7. Abl inhibition and RA restored mitochondrial membrane potential and ATP content in rotenone exposed SH-SY5Y cells. (**A**) Cells were exposed to rotenone, RA, and nilotinib for 12 h, the cells were then stained with JC-1 for 30 min, images were acquired using fluorescence microscope. Scale bar = 100 µm. (**B**) After treatments, JC-1 fluorescent signal was detected using flow cytometry (Green: JC-1 monomers, Red: JC-1 aggregates). (**C**) Bar graph quantification analysis of mitochondrial membrane potential evaluated with flow cytometry. (**D**) Bar graph quantification analysis of cellular ATP level measured with luminescence microplate reader. All data were expressed as the mean ± SEM of three independent experiments. ## $p < 0.01$ compared with the control group, * $p < 0.05$, ** $p < 0.01$ compared with the rotenone group.

4. Discussion

In the present study, we evaluated the neuroprotective role of RA in the rotenone challenged SH-SY5Y model and the corresponding molecular mechanism of actions of RA. We demonstrated that RA protected SH-SY5Y cells against rotenone by restoring mitochondrial membrane potential, cellular ATP content and suppressing ROS overproduction. The mechanism of action of RA involves the regulation of Akt, AMPK, Tau phosphorylation, and α-synuclein as well as PGC-1α protein expression. RA exerted this neuroprotective effect partially by attenuating Abl kinase activity.

Rotenone exposure induces cell apoptosis by inhibiting mitochondrial respiratory chain complex I and increasing ROS production in HL60 and HT1080 cell lines [18]. Superoxide dismutase and antioxidants inhibit rotenone-induced apoptosis [18]. In the cell apoptosis analysis of our study, apoptosis rate was calculated as the percentage of Annexin V staining positive cells. The result showed that 50 μM rotenone markedly increased the apoptosis rate of SH-SY5Y cells. RA, as an antioxidant, also significantly reduced rotenone-induced cell apoptosis. Microscopic observation showed that SH-SY5Y cell morphology changed obviously after rotenone exposure. Rotenone exposure made cells become round, RA restored the normal shape of the cells. Cell viability was reduced by rotenone, and 100 μM RA significantly rescued cell viability in the presence of rotenone. These findings indicated a neuroprotective function of RA against rotenone. In the substantia nigra pars compacta of brain, metabolism of dopamine leads to the generation of reactive oxygen species (ROS) including hydrogen peroxide and hydroxyl radicals. This makes this dopamine rich areas particularly vulnerable to oxidative stress [28]. Thus, oxidative stress plays an important role in mediating neuronal cell death in PD [29]. Oxidative stress disrupts the activity of mitochondrial complex I. Loss of mitochondrial complex I activity in the dopaminergic neurons is considered to be a hallmark of PD and is closely related to mitochondrial function [30]. Mitochondrial function can be reflected by the changes in mitochondrial transmembrane potential, the collapse of which further induces ROS overproduction [31]. The disruption of complex I activity and mitochondrial transmembrane potential lead to the reduction of ATP production. In the current study, we demonstrated that rotenone induced the overproduction of ROS, which is effectively attenuated by RA. Rotenone downregulated mitochondrial membrane potential and ATP level, while RA treatment leads to the partial restoration of them. These findings suggested that RA reversed the mitochondrial function reduction induced by rotenone in SH-SY5Y cells.

One of the characteristics of PD is the abnormal accumulation and aggregation of α-synuclein in the form of Lewy bodies [32]. α-synuclein induces neuronal toxicity and death through causing mitochondrial dysfunction, lysosomal impairment, and membrane disturbance [33]. Pathological aggregation and phosphorylation of Tau protein at Ser404 are also involved in many neurodegenerative diseases including PD [34]. Studies have shown that PD patients with cognitive impairment have elevated level of phosphorylated Tau in Lewy bodies along with α-synuclein [35]. In the current study, rotenone caused overexpression of α-synuclein and hyperphosphorylation of Tau, and RA markedly reduced α-synuclein expression and Tau phosphorylation. These results implied that RA protected neuronal cells potentially through suppression α-synuclein accumulation, aggregation, and reduction of Tau protein phosphorylation.

Disruption of mitochondrial biogenesis has been shown to be associated with PD development. The peroxisome proliferator-activated receptor gamma coactivator 1 (PGC-1α) is considered as a master regulator of mitochondrial biogenesis and antioxidant response [36]. PGC-1α function is down-regulated in PD [37]. PGC-1α knock out mice developed higher vulnerability to the neurodegenerative effects of MPTP [38]. Serine/threonine kinase AMPK is activated by falling energy levels and plays an important role in restoring cellular energy balance [25]. Activating AMPK has multiple effects in PD model including changing cellular metabolism, promoting autophagy, enhancing mitochondrial quality control, and increasing antioxidant capacity [25]. AMPK also promotes mitochondrial biogenesis and reduces ROS through activation of PGC-1α [39]. Several studies showed that AMPK activation facilitated the clearance of α-synuclein and promoted neuronal survival [40]. Akt signaling is central to cell survival and is impaired in PD [26]. Selective loss of dopaminergic neurons was accompanied by a decrease of Akt phosphorylation at Ser473 in the PD brain [41]. In the current study, PGC-1α protein and Akt phosphorylation levels were evidently reduced in the presence of rotenone, RA led to restorations of PGC-1α expression and Akt phosphorylation. Rotenone induced an elevation of AMPK Thr172 phosphorylation, while RA treatment further promoted AMPK Thr172 phosphorylation. The initial upregulation of AMPK phosphorylation induced by rotenone could be due to the cellular self-adapting mechanism. Under stress conditions, cells would initially activate certain pro-survival signaling pathways including AMPK to attempt to overcome the encountered stress condition [42]. These observations highlighted that RA restored the protein expression and phosphorylation of the prosurvival, neuroprotective, and mitochondrial function-promoting proteins in rotenone induced PD cell model.

Multiple animal models of PD suggested that activation of Abl played an important role in the initiation and progression of neurodegeneration [1]. To investigate if the suppression of Abl is involved in the neuroprotective effect of RA against rotenone, we analyzed the tyrosine phosphorylation level of Abl Tyr412, which is located at the kinase activation loop of Abl. The phosphorylation of this site is required for Abl kinase activation [43]. In the present study, rotenone markedly induced Abl Tyr412 phosphorylation. RA treatment suppressed this hyperphosphorylation induced by rotenone, which mimicked the effect of Abl specific inhibitor, nilotinib. Nilotinib treatment also partially restored SH-SY5Y cell viability reduced by rotenone. To further assess the activity of Abl, the phosphorylation state of an Abl substrate, CrkII, was analyzed at its Abl phosphorylation site Tyr221 [44]. Rotenone exposure increased CrkII Tyr221 phosphorylation, RA treatment suppressed this hyperphosphorylation induced by rotenone. Like RA, nilotinib also significantly restored cell viability in rotenone exposed SH-SY5Y cells. These results suggested that RA exerted the protective effect on SH-SY5Y cells partially through the suppression of Abl activation. Moreover, Abl inhibition with nilotinib also had similar effects to that of RA on the regulation of the phosphorylation state and protein expression level of several key signaling proteins involved in PD and mitochondrial function regulation. Nilotinib reduced the protein level of α-synuclein, phosphorylation level of Tau Ser404, and promoted phosphorylation of Akt Ser473, AMPK Thr172, and expression of PGC-1α. Both nilotinib and RA treatment effectively suppressed rotenone induced ROS overproduction and increased mitochondrial membrane potential and cellular ATP level reduced by rotenone. These observations indicated that at least part of the neuroprotective function of RA is due to its inhibitory effect on Abl. Nilotinib alone reduced CrkII Tyr221 phosphorylation and did not lead to any significant changes in Abl Tyr412 phosphorylation. This suggests that the basal activity of Abl is low, and it is slightly further inhibited by nilotinib. However, since this activity reduction is small, it did not further modify the ROS, mitochondrial membrane potential, ATP level and the phosphorylation of other PD-related signaling proteins.

As a mitochondrial complex I inhibitor, rotenone causes mitochondrial dysfunction and overproduction of ROS [18]. ROS activates Abl protein, which causes accumulation of α-synuclein, further disrupts mitochondrial function and eventually leads to neuronal cell death [4]. In addition to α-synuclein, studies have shown that Abl also phosphorylates and inactivates an E3 ubiquitin ligase, Parkin [45]. Since Parkin interacts with Pink1 and plays an important role in the clearance of dysfunctional mitochondria, cell death mediating protein PARP-1 and other unwanted intracellular proteins, the inactivation of Parkin leads to the loss of neuronal cells [16]. Abl can phosphorylate PKC [46]. PKC activation leads to the phosphorylation of AMPK Ser487, which results in the reduction of AMPK activity [47]. Thus, inhibition of Abl could possibly increase AMPK activity through this pathway and promote cell survival. This explains the reason for which nilotinib treatment alone induced AMPK Thr172 phosphorylation in our experiment. Moreover, Abl activation also leads to dopaminergic neuron loss by phosphorylating and activating p38α MAPK [48]. In the current study, we speculate that RA inhibits Abl protein signaling by reducing the intracellular ROS level. However, further study is needed to investigate other potential pathways through which RA leads to the inhibition of Abl.

5. Conclusions

In conclusion, this is the first study indicating that natural compound RA attenuated the neurotoxicity induced by rotenone in SH-SY5Y Parkinson's disease cell model through inhibiting Abl and ameliorating the mitochondrial dysfunction. This study also demonstrated that downregulation of α-synuclein protein, Tau phosphorylation, and increases of signaling proteins Akt Ser473, AMPK Thr172 phosphorylation and PGC-1α level were involved in the neuroprotective effect of RA. This finding contributes to a new understanding with regard to the molecular mechanisms underlying the neuroprotective effect of RA in PD and will promote the further use of RA as a natural neuroprotective reagent.

Author Contributions: Conceptualization, W.Z. and X.H.; methodology, X.H., Z.W. and W.Z.; validation, W.Z., X.H., B.H., Z.W., T.W. and J.G.; investigation, X.H., W.Z., B.H. and Y.Z.; resources, W.Z., Z.W., B.H., G.L. and J.G.; data curation, X.H., B.H. and Y.Z.; writing—original draft preparation, W.Z. and X.H.; writing—review and editing, Z.W., T.W., B.H., J.G., X.H., G.L., W.D., X.C. and J.H.; visualization, X.H., Y.Z. and J.H.; supervision, W.Z. and Z.W.; project administration, W.Z.; funding acquisition, W.Z. and Z.W. All authors have read and agreed to the published version of the manuscript.

Funding: This work was supported by the Natural Science Foundation of Shandong Province (no. ZR2020MH377) and Yantai University Research Grant (Grant numbers 2219015), China.

Institutional Review Board Statement: Not applicable.

Informed Consent Statement: Not applicable.

Data Availability Statement: Not applicable.

Conflicts of Interest: The authors declare no conflict of interest.

References

1. Werner, M.H.; Olanow, C.W. Parkinson's disease modification through Abl kinase inhibition: An opportunity. *Mov. Disord.* **2022**, *37*, 6–15. [CrossRef] [PubMed]
2. Thomas, B.; Beal, M.F. Parkinson's disease. *Hum. Mol. Genet.* **2007**, *16*, R183–R194. [CrossRef] [PubMed]
3. Titova, N.; Qamar, M.A.; Chaudhuri, K.R. The nonmotor features of Parkinson's disease. *Int. Rev. Neurobiol.* **2017**, *132*, 33–54. [PubMed]
4. Jenner, P.; Olanow, C.W. Oxidative stress and the pathogenesis of Parkinson's disease. *Neurology* **1996**, *47* (Suppl. 3), S161–S170. [CrossRef]
5. Chen, X.-C.; Zhou, Y.-C.; Chen, Y.; Zhu, Y.-G.; Fang, F.; Chen, L.-M. Ginsenoside Rg1 reduces MPTP-induced substantia nigra neuron loss by suppressing oxidative stress1. *Acta Pharmacol. Sin.* **2005**, *26*, 56–62. [CrossRef]
6. Zhao, Y.; Han, Y.; Wang, Z.; Chen, T.; Qian, H.; He, J.; Li, J.; Han, B.; Wang, T. Rosmarinic acid protects against 1-methyl-4-phenyl-1,2,3,6-tetrahydropyridine-induced dopaminergic neurotoxicity in zebrafish embryos. *Toxicol. Vitr.* **2020**, *65*, 104823. [CrossRef]
7. Babaei, M.; Borja Zamfir, G.M.; Chen, X.; Christensen, H.B.; Kristensen, M.; Nielsen, J.; Borodina, I. Metabolic engineering of Saccharomyces cerevisiae for Rosmarinic Acid Production. *ACS Synth. Biol.* **2020**, *9*, 1978–1988. [CrossRef]
8. Qu, L.; Xu, H.; Jia, W.; Jiang, H.; Xie, J. Rosmarinic acid protects against MPTP-induced toxicity and inhibits iron-induced α-synuclein aggregation. *Neuropharmacology* **2019**, *144*, 291–300. [CrossRef]
9. Lee, H.J.; Cho, H.-S.; Park, E.; Kim, S.; Lee, S.-Y.; Kim, C.-S.; Kim, D.K.; Kim, S.-J.; Chun, H.S. Rosmarinic acid protects human dopaminergic neuronal cells against hydrogen peroxide-induced apoptosis. *Toxicology* **2008**, *250*, 109–115. [CrossRef]
10. Ren, P.; Jiang, H.; Li, R.; Wang, J.; Song, N.; Xu, H.-M.; Xie, J.-X. Rosmarinic acid inhibits 6-OHDA-induced neurotoxicity by anti-oxidation in MES23.5 cells. *J. Mol. Neurosci.* **2009**, *39*, 220–225. [CrossRef]
11. Du, T.; Li, L.; Song, N.; Xie, J.; Jiang, H. Rosmarinic acid antagonized 1-methyl-4-phenylpyridinium (MPP+)-induced neurotoxicity in MES23.5 dopaminergic cells. *Int. J. Toxicol.* **2010**, *29*, 625–633. [CrossRef] [PubMed]
12. Heinz, S.; Freyberger, A.; Lawrenz, B.; Schladt, L.; Schmuck, G.; Ellinger-Ziegelbauer, H. Mechanistic investigations of the mitochondrial complex I inhibitor rotenone in the context of pharmacological and safety evaluation. *Sci. Rep.* **2017**, *7*, 45465. [CrossRef] [PubMed]
13. Alam, M.; Schmidt, W. Rotenone destroys dopaminergic neurons and induces parkinsonian symptoms in rats. *Behav. Brain Res.* **2002**, *136*, 317–324. [CrossRef]

14. Betarbet, R.; Sherer, T.B.; MacKenzie, G.; Garcia-Osuna, M.; Panov, A.V.; Greenamyre, J.T. Chronic systemic pesticide exposure reproduces features of Parkinson's disease. *Nat. Neurosci.* **2000**, *3*, 1301–1306. [CrossRef] [PubMed]
15. Johnson, M.E.; Stringer, A.; Bobrovskaya, L. Rotenone induces gastrointestinal pathology and microbiota alterations in a rat model of Parkinson's disease. *NeuroToxicology* **2018**, *65*, 174–185. [CrossRef] [PubMed]
16. Dauer, W.; Przedborski, S. Parkinson's disease: Mechanisms and models. *Neuron* **2003**, *39*, 889–909. [CrossRef]
17. Betarbet, R.; Canet-Aviles, R.M.; Sherer, T.B.; Mastroberardino, P.G.; McLendon, C.; Kim, J.H.; Lund, S.; Na, H.-M.; Taylor, G.; Bence, N.F.; et al. Intersecting pathways to neurodegeneration in Parkinson's disease: Effects of the pesticide rotenone on DJ-1, alpha-synuclein, and the ubiquitin-proteasome system. *Neurobiol. Dis.* **2006**, *22*, 404–420. [CrossRef]
18. Li, N.; Ragheb, K.; Lawler, G.; Sturgis, J.; Rajwa, B.; Melendez, J.A.; Robinson, J.P. Mitochondrial complex I inhibitor rotenone induces apoptosis through enhancing mitochondrial reactive oxygen species production. *J. Biol. Chem.* **2003**, *278*, 8516–8525. [CrossRef]
19. Watabe, M.; Nakaki, T. Rotenone induces apoptosis via activation of bad in human dopaminergic SH-SY5Y cells. *J. Pharmacol. Exp. Ther.* **2004**, *311*, 948–953. [CrossRef]
20. Brahmachari, S.; Ge, P.; Lee, S.H.; Kim, D.; Karuppagounder, S.S.; Kumar, M.; Mao, X.; Shin, J.H.; Lee, Y.; Pletnikova, O.; et al. Activation of tyrosine kinase c-Abl contributes to α-synuclein–induced neurodegeneration. *J. Clin. Investig.* **2016**, *126*, 2970–2988. [CrossRef]
21. Plosker, G.L.; Robinson, D.M. Nilotinib. *Drugs* **2008**, *68*, 449–459. [CrossRef] [PubMed]
22. Karuppagounder, S.S.; Brahmachari, S.; Lee, Y.; Dawson, V.L.; Dawson, T.M.; Ko, H.S. The c-Abl inhibitor, nilotinib, protects dopaminergic neurons in a preclinical animal model of Parkinson's disease. *Sci. Rep.* **2014**, *4*, 4874. [CrossRef] [PubMed]
23. Chen, J.-H.; Ou, H.-P.; Lin, C.-Y.; Lin, F.-J.; Wu, C.-R.; Chang, S.-W.; Tsai, C.-W. Carnosic Acid Prevents 6-Hydroxydopamine-Induced Cell Death in SH-SY5Y Cells via Mediation of Glutathione Synthesis. *Chem. Res. Toxicol.* **2012**, *25*, 1893–1901. [CrossRef]
24. Hu, S.; Hu, M.; Liu, J.; Zhang, B.; Zhang, Z.; Zhou, F.H.; Dong, J. Phosphorylation of Tau and alpha-Synuclein induced neurodegeneration in MPTP mouse model of Parkinson's Disease. *Neuropsychiatr. Dis. Treat.* **2020**, *16*, 651–663. [CrossRef] [PubMed]
25. Curry, D.W.; Stutz, B.; Andrews, Z.B.; Elsworth, J.D. Targeting AMPK signaling as a neuroprotective strategy in Parkinson's Disease. *J. Parkinsons Dis.* **2018**, *8*, 161–181. [CrossRef] [PubMed]
26. Furlong, R.M.; Lindsay, A.; Anderson, K.E.; Hawkins, P.T.; Sullivan, A.M.; O'Neill, C. The Parkinson's disease gene PINK1 activates Akt via PINK1 kinase-dependent regulation of the phospholipid PI(3,4,5)P3. *J. Cell Sci.* **2019**, *132*, 20.
27. Feller, S.; Knudsen, B.; Hanafusa, H. c-Abl kinase regulates the protein binding activity of c-Crk. *EMBO J.* **1994**, *13*, 2341–2351. [CrossRef]
28. Lotharius, J.; Brundin, P. Pathogenesis of Parkinson's disease: Dopamine, vesicles and alpha-synuclein. *Nat. Rev. Neurosci.* **2002**, *3*, 932–942. [CrossRef]
29. Shibata, N.; Kobayashi, M. The role for oxidative stress in neurodegenerative diseases. *Brain Nerve* **2008**, *60*, 157–170.
30. Surmeier, D.J.; Obeso, J.A.; Halliday, G.M. Selective neuronal vulnerability in Parkinson disease. *Nat. Rev. Neurosci.* **2017**, *18*, 101–113. [CrossRef]
31. Chan, P.H. Mitochondria and neuronal death/survival signaling pathways in cerebral ischemia. *Neurochem. Res.* **2004**, *29*, 1943–1949. [CrossRef] [PubMed]
32. Xu, L.; Pu, J. Alpha-Synuclein in Parkinson's Disease: From pathogenetic dysfunction to potential clinical application. *Park. Dis.* **2016**, *2016*, 1–10. [CrossRef] [PubMed]
33. Gómez-Benito, M.; Granado, N.; García-Sanz, P.; Michel, A.; Dumoulin, M.; Moratalla, R. Modeling Parkinson's disease with the alpha-synuclein protein. *Front. Pharmacol.* **2020**, *11*, 356. [CrossRef] [PubMed]
34. Zhang, X.; Gao, F.; Wang, D.; Li, C.; Fu, Y.; He, W.; Zhang, J. Tau pathology in Parkinson's Disease. *Front Neurol.* **2018**, *9*, 809. [CrossRef]
35. Pan, L.; Meng, L.; He, M.; Zhang, Z. Tau in the Pathophysiology of Parkinson's Disease. *J. Mol. Neurosci.* **2021**, *71*, 2179–2191. [CrossRef]
36. Piccinin, E.; Sardanelli, A.; Seibel, P.; Moschetta, A.; Cocco, T.; Villani, G. PGC-1s in the Spotlight with Parkinson's Disease. *Int. J. Mol. Sci.* **2021**, *22*, 3487. [CrossRef]
37. Corona, J.C.; Duchen, M.R. PPARgamma and PGC-1alpha as therapeutic targets in Parkinson's. *Neurochem. Res.* **2015**, *40*, 308–316. [CrossRef]
38. St-Pierre, J.; Drori, S.; Uldry, M.; Silvaggi, J.M.; Rhee, J.; Jager, S.; Handschin, C.; Zheng, K.; Lin, J.; Yang, W.; et al. Suppression of reactive oxygen species and neurodegeneration by the PGC-1 transcriptional coactivators. *Cell* **2006**, *127*, 397–408. [CrossRef]
39. Scarpulla, R.C. Metabolic control of mitochondrial biogenesis through the PGC-1 family regulatory network. *Biochim. Biophys. Acta* **2011**, *1813*, 1269–1278. [CrossRef]
40. Moors, T.E.; Hoozemans, J.J.M.; Ingrassia, A.; Beccari, T.; Parnetti, L.; Chartier-Harlin, M.-C.; van de Berg, W.D.J. Therapeutic potential of autophagy-enhancing agents in Parkinson's disease. *Mol. Neurodegener.* **2017**, *12*, 11. [CrossRef]
41. Greene, L.A.; Levy, O.; Malagelada, C. Akt as a victim, villain and potential hero in Parkinson's disease pathophysiology and treatment. *Cell Mol. Neurobiol.* **2011**, *31*, 969–978. [CrossRef] [PubMed]
42. Hou, W.L.; Yin, J.; Alimujiang, M.; Yu, X.Y.; Ai, L.G.; Bao, Y.Q.; Liu, F.; Jia, W.-P. Inhibition of mitochondrial complex I improves glucose metabolism independently of AMPK activation. *J. Cell Mol. Med.* **2018**, *22*, 1316–1328. [CrossRef] [PubMed]

43. Brasher, B.B.; van Etten, R.A. c-Abl has high intrinsic tyrosine kinase activity that is stimulated by mutation of the Src homology 3 domain and by autophosphorylation at two distinct regulatory tyrosines. *J. Biol. Chem.* **2000**, *275*, 35631–35637. [CrossRef] [PubMed]
44. Amoui, M.; Miller, W.T. The substrate specificity of the catalytic domain of Abl plays an important role in directing phosphorylation of the adaptor protein Crk. *Cell Signal* **2000**, *12*, 637–643. [CrossRef]
45. Ko, H.S.; Lee, Y.; Shin, J.H.; Karuppagounder, S.S.; Gadad, B.S.; Koleske, A.J.; Pletnikiva, O.; Troncoso, J.C.; Dawson, V.L.; Dawson, T.M. Phosphorylation by the c-Abl protein tyrosine kinase inhibits parkin's ubiquitination and protective function. *Proc. Natl. Acad. Sci. USA* **2010**, *107*, 16691–16696. [CrossRef]
46. Yuan, Z.M.; Utsugisawa, T.; Ishiko, T.; Nakada, S.; Huang, Y.; Kharbanda, S.; Weichselbaum, R.; Kufe, D. Activation of protein kinase C delta by the c-Abl tyrosine kinase in response to ionizing radiation. *Oncogene* **1998**, *16*, 1643–1648. [CrossRef]
47. Heathcote, H.R.; Mancini, S.J.; Strembitska, A.; Jamal, K.; Reihill, J.A.; Palmer, T.M.; Gould, G.W.; Salt, I.P. Protein kinase C phosphorylates AMP-activated protein kinase α1 Ser487. *Biochem. J.* **2016**, *473*, 4681–4697. [CrossRef]
48. Wu, R.; Chen, H.; Ma, J.; He, Q.; Huang, Q.; Liu, Q.; Li, M.; Yuan, Z. c-Abl-p38alpha signaling plays an important role in MPTP-induced neuronal death. *Cell Death Differ.* **2016**, *23*, 542–552. [CrossRef]

nutrients

Article

Impact of Clarified Apple Juices with Different Processing Methods on Gut Microbiota and Metabolomics of Rats

Lei Xu [1,2], Shini Yang [1,2], Kewen Wang [1,2], Anjing Lu [3,*], Xue Wang [1] and Zhenzhen Xu [1,*]

1 Institute of Quality Standard & Testing Technology for Agro-Products, Chinese Academy of Agricultural Sciences, Key Laboratory of Agro-Food Safety and Quality, Ministry of Agriculture and Rural Affairs, Beijing 100081, China
2 College of Food Science and Nutritional Engineering, China Agricultural University, Beijing Key Laboratory for Food Nonthermal Processing, Key Lab of Fruit and Vegetable Processing, Ministry of Agriculture and Rural Affairs, Beijing 100083, China
3 China Center for Information Industry Development, Beijing 100048, China
* Correspondence: luanjing@ccidthinktank.com (A.L.); xuzhenzhen@caas.cn (Z.X.); Tel.: +86-1068205639 (A.L.); +86-1082106560 (Z.X.)

Abstract: The consumption of processed foods has increased compared to that of fresh foods in recent years, especially due to the coronavirus disease 2019 pandemic. Here, we evaluated the health effects of clarified apple juices (CAJs, devoid of pectin and additives) processed to different degrees, including not-from-concentrate (NFC) and from-concentrate (FC) CAJs. A 56-day experiment including a juice-switch after 28 days was designed. An integrated analysis of 16S rRNA sequencing and untargeted metabolomics of cecal content were performed. In addition, differences in the CAJs tested with respect to nutritional indices and composition of small-molecule compounds were analyzed. The NFC CAJ, which showed a higher phenolic content resulting from the lower processing degree, could improve microbiota diversity and influence its structure. It also reduced bile acid and bilirubin contents, as well as inhibited the microbial metabolism of tryptophan in the gut. However, we found that these effects diminished with time by performing experiment extension and undertaking juice-switching. Our study provides evidence regarding the health effects of processed foods that can potentially be applied to public health policy decision making. We believe that NFC juices with a lower processing degree could potentially be healthier than FC juice.

Keywords: clarified apple juices; food processing degree; gut microbiota; metabolomics; rats

Citation: Xu, L.; Yang, S.; Wang, K.; Lu, A.; Wang, X.; Xu, Z. Impact of Clarified Apple Juices with Different Processing Methods on Gut Microbiota and Metabolomics of Rats. *Nutrients* **2022**, *14*, 3488. https://doi.org/10.3390/nu14173488

Academic Editors: Haixia Yang, Jianjun Deng and Rosa Casas

Received: 30 July 2022
Accepted: 12 August 2022
Published: 25 August 2022

1. Introduction

According to the degree and purpose of processing, a novel food classification system called NOVA (meaning novel in Portuguese) was proposed in 2017, which divides processed foods into four categories: unprocessed or minimally processed foods, processed culinary ingredients, processed foods, and ultra-processed foods [1,2]. NOVA's public health advice is that excessive intake of processed foods should be avoided in order to improve diet nutrition [2]. Many studies have also demonstrated the relationship between ultra-processed foods and non-communicable diseases, including diabetes, obesity, cardiovascular disease, coronary heart disease, and cerebrovascular diseases [3–5]. In-depth scientific research is now crucial to fully understanding the connection between the degree of food processing and public health [6].

The consumption of fresh fruits has decreased, while that of fruit juices has increased in recent years, especially since the outbreak of coronavirus disease 2019 (COVID-19) [7,8]. Apple juice is the most popular juice worldwide due to its flavor and taste. It has been reported that the administration of phenolics and pectin [9], as well as cloudy apple juice, could significantly regulate the gut microbiota. The effects of clarified apple juice (CAJ)—which is characterized by high energy density and the presence of food additives, as well as a lack of dietary fiber—on the gut microbiota remains unknown.

The gut microbiota is a complex microbial ecosystem that is essential for human health via a reciprocal relationship, and its important role can be explained by the correlation between composition and health status. However, it can also induce diseases, such as obesity, diabetes, and cardiovascular disease [10,11]. Research has suggested that gut microbiota disorders could be a mechanism explaining the correlation between processed foods and metabolic syndromes [4,12].

In this study, CAJs with different processing degrees, not-from-concentrate and from-concentrate (NFC and FC, respectively, pectin and additives eliminated), were selected. First, the effect of CAJs was demonstrated. Then, the effects of CAJs with different processing degrees were evaluated separately. In addition, a juice-switch experiment was performed. This study could demonstrate the comprehensive influence of different CAJs, which could provide additional evidence for the health effect of foods with different processing degrees and instructions for consumption behavior and public health policy designation.

2. Materials and Methods

2.1. CAJ

CAJs, including the unsterilized NFC and concentrated apple juices, from four factories located in Shaanxi and Shandong provinces in 2019 and 2020 were collected. After sampling, CAJs were frozen immediately, transported by cold chain to the laboratory, and stored at −20 °C. NFC includes juices processed by high-pressure processing (HPP) sterilization and pasteurization. For HPP, unsterilized NFC was added to a polyethylene terephthalate (PET) plastic bottle and treated at 550 MPa for 5 min (CQC30L-600, Beijing Suyuan Zhongtian technology limited company, Beijing, China). For pasteurization, unsterilized NFC was pasteurized at 85 °C for 15 s using an FT74 UHT/HTST processing system (Armfield, Ringwood, UK) and then added to PET bottles. For FC, concentrated apple juice was restored to the total solid sugar content before concentration (±0.1 °Bx), pasteurized under 85 °C for 15 s, and added to PET bottles. The 2019 samples from factories in Shaanxi were used for difference analysis and biomarker discovery, whereas the 2020 samples from one of the factories in Shaanxi were used for animal intervention.

2.2. Animal Intervention

Fifty-four male Sprague Dawley rats aged 6–8 weeks provided by Charles River (Wilmington, MA, USA) were randomly and evenly divided into nine groups housed in 18 cages. The temperature of the animal facility was 20–23°C with a relative humidity of 45–60% and was under a 12 h dark–light cycle. The rats were free to drink and eat, and the main nutrients in the animal feed used in the experiment are provided in Supplemental Table S1. Food intake amount was calculated 3 times/week, with the total food weight loss of one cage divided by the number of rats in the cage. NFC (processed by HPP) and FC CAJs were used, and other detailed experimental designs are shown in Figure 1. The rats were gavaged 20 mL/kg body weight saline or FC or NFC 3 times per day. The gavage needle was slowly inserted into the mouth and then the esophagus. The juice was injected slowly to prevent returning to the mouth. Fasting glucose and plasma lipids were measured from blood from the eye socket after 12 h fasting and anesthesia by isoflurane.

2.3. 16 S rRNA Sequencing

Genomic DNA of the microbial community was extracted from the cecal content and analyzed using the Illumina MiSeq PE300 platform (Illumina, San Diego, CA, USA). Please refer to the online supplemental material for further details.

2.4. Untargeted Metabolomics

Metabolites extracted from the cecal content were analyzed using an AB SCIEX 6600 mass spectrometer coupled with ExionLC ultra-high-performance liquid chromatography

(AB SCIEX, Framingham, MA, USA). Please see the online supplemental material methods section for further details.

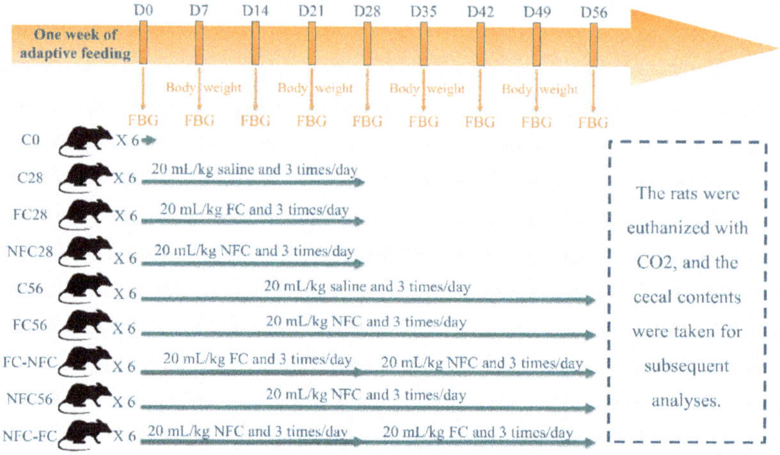

Figure 1. Schematic diagram of animal experimental scheme.

2.5. *CAJ Analyses*

The glucose, fructose, sucrose, total phenol, amino acids, pH, and nutritional ingredients of the CAJs were analyzed. Table 1 presents the results. Furthermore, the small-molecule compounds were analyzed using UHPLC-QTOF. Please see the online supplemental material for details on the methods.

Table 1. Basic indices of NFC and FC.

Index	FC	NFC
Total phenol (GAE mg/100 mL)	1073.32 ± 72.33 [b]	2169.03 ± 116.84 [a]
pH	3.57 ± 0.02 [b]	3.63 ± 0.02 [a]
Sugar content (°Bx)	10.83 ± 1.41 [a]	10.49 ± 0.11 [a]
Glucose (g/100 mL)	7.81 ± 0.92 [a]	6.13 ± 1.05 [b]
Fructose (g/100 mL)	4.24 ± 0.79 [a]	2.85 ± 0.74 [b]
Sucrose (g/100 mL)	0.16 ± 0.06 [a]	0.10 ± 0.02 [a]
Malic acid (mg/100 mL)	186.16 ± 2.86 [a]	187.10 ± 0.31 [a]
Tartaric acid (mg/100 mL)	21.33 ± 0.08 [a]	18.80 ± 0.15 [b]
Asparagine (mg/100 mL)	4.57 ± 0.39 [a]	10.82 ± 6.58 [a]
Alanine (mg/100 mL)	4.43 ± 0.39 [a]	10.73 ± 6.61 [a]
Serine (mg/100 mL)	0.69 ± 0.09 [a]	1.66 ± 1.08 [a]
Arginine (mg/100 mL)	0.14 ± 0.02 [a]	0.31 ± 0.19 [a]
Glutamine (mg/100 mL)	0.07 ± 0.01 [a]	0.15 ± 0.10 [a]

GAE: gallic acid equivalent; Data marked with different letters indicate significant difference of $p < 0.05$; $n = 3$.

2.6. *Data Processing, Statistical Analysis, and Visualization*

For 16S rRNA sequencing, the data were analyzed with online tools from Majorbio (http://www.majorbio.com/, accessed on 25 July 2022). For metabolomic analyses, the peak extraction, alignment, and correction were performed using MS-DIAL ver. 4.36 software. MetaboAnalyst ver. 4.0 and SIMCA 14.1 were used for data analysis. MS-FINDER ver. 3.50, MassHunter PCDL ver. B.07.00, and LibraryView™ ver. 1.1 were used for compound identification. Please see the online supplemental material methods section for further details.

3. Results

3.1. Continuous Intake of CAJ Has Limited Effect on Body Weight and No Effect on the Intake Amount, Gut Microbiota, and Blood Lipids

As shown in Figure 2A, the body weight of the CAJ group was higher than that of the C group throughout the experimental period, and it was significantly different after D30 ($p < 0.05$, except for D32). However, the administration of fruit juice had no significant influence on the food intake amount (Figure 2B, where significance was only observed on D9 and D14). These results suggest that the energy intake of the CAJ group was approximately 24 kcal/kg body weight more per day (Supplemental Table S2). For fasting blood glucose (Figure 2C), no significant difference was observed except on D7 ($p < 0.05$). We also analyzed four parameters of blood lipids; the triglyceride levels (Figure 2D) in the CAJ group were always higher than those of the C group, but were significantly different only on D7, D28, and D35 ($p < 0.05$). The levels of the other three proteins (cholesterol, high-density lipoprotein, and low-density lipoprotein) are shown in Supplemental Figure S1. Only the high-density lipoprotein content of the C group on D7 was significantly higher than that of the AJ group ($p < 0.05$).

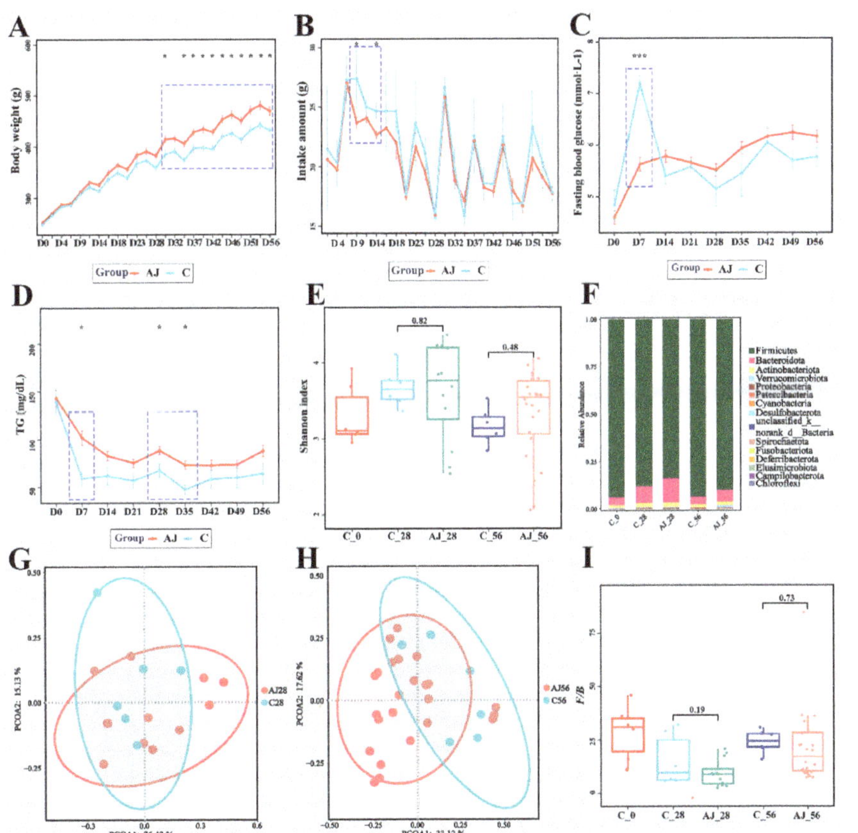

Figure 2. Effect of CAJ on rats: (**A**) Body weight. (**B**) Intake amount. (**C**) Fasting blood glucose. (**D**) Triglyceride content. (**E**) Shannon index. (**F**) Bar plot of gut microbiota on phylum level. (**G**) PCoA scores plot on OTU level of D28. (**H**) PCoA scores plot on OTU level of D56. (**I**) Box plot of F/B ratio. C, control group; AJ, apple juice groups, including NFC and FC groups. * indicates significant difference ($p < 0.05$) between CAJ and C groups. The "*" means $p < 0.05$, and "***" means $p < 0.001$.

According to the α-diversity results shown in Figure 2E, no significant difference was observed between the groups at the same time point. Moreover, no obvious separation between the CAJ and C groups was observed in principal coordinate analysis (PCoA) score plots of β-diversity at the operational taxonomic unit (OTU) level on D28 and D56 (Figure 2G,H). In the gut microbiota at the phylum level, the dominant phyla were *Firmicutes* and *Bacteroidota*, whose relative abundance was higher than 80.8% in all samples (Figure 2F). According to numerous studies on gut microbiota and obesity, changes in the relative abundance of these two dominant bacteria, that is, an increase in *Firmicutes* and a decrease in *Bacteroidota*, can lead to an improvement in the host's ability to obtain energy from foods, thus affecting the host's energy balance and body weight [13]. This improvement may not immediately lead to obesity; however, a slight change in energy balance can cause significant changes in body weight over a long period [14,15]. Therefore, the ratio of *Firmicutes* to *Bacteroidota* could be a good biomarker for obesity [16]. As shown in Figure 2I, the ratio of the two dominant bacteria, *Firmicutes*/*Bacteroidota* (F/B), was not significantly different between groups at the same time point. Prolonged and continuous intake of CAJ may incrementally improve body weight, as revealed by the significantly higher body weight of the CAJ group. However, as the fasting blood glucose and F/B ratio were not significantly affected, and their energy intake was considerably higher, we conclude that the continuous intake of CAJ presents no health risk in rats.

3.2. *CAJ with Lower Processing Degree Could Improve Microbiota Diversity and Inhibit the Metabolism of Bile Acids, Bilirubin, and Tryptophan in the Gut*

First, we focused on the first 28 days of administration. As shown in Figure 3A, the Shannon index of the FC group was not significantly different ($p > 0.05$) from that of the C group. In contrast, the NFC group with a lower processing degree was significantly different ($p < 0.05$) from the other two groups. The results indicated that NFC significantly improved the diversity of the gut microbiota in rats after 28 days of administration. Obesity is often accompanied by a decrease in the diversity of the gut microbiota [17]. In view of the β-diversity at the OTU level shown in Figure 3B, the NFC group was distinctly separated from the FC and C groups, which indicated that NFC could affect the gut microbiota structure, while FC had no such effect.

To further understand this discrepancy, we analyzed the abundance at multiple levels. More specifically, 15 phyla were detected at the phylum level (Figure 3C). *Firmicutes* abundance in the NFC group was significantly lower than that in the FC and C groups (Supplemental Figure S2A). In contrast, *Bacteroidota* abundance (shown in Figure 3D) in the NFC group was significantly higher than that in the other two groups ($p < 0.05$). As shown in Figure 3E, the F/B ratio was significantly lower in the NFC group than in the FC group. These results indicate that NFC could affect the dominant bacteria in the gut, thus reducing the risk of obesity. *Spirochaetota* abundance in the NFC group was significantly higher than in the FC and C groups ($p < 0.05$, Supplemental Figure S2B), and there was no significant difference between the FC and C groups. *Spirochaetota*, whose relative abundance was less than 0.04% in all samples, was detected in many environmental samples, with an unnoted proportion of 54.8% [18]. Furthermore, it is rich in carbohydrate hydrolases and is mainly involved in cellulose degradation [19]. In addition, its abundance in the gut of Hadza people from Tanzania was significantly higher than that of people from developed coastal areas [20]. In order to find biomarkers among NFC and FC groups, LDA effect size (LEfSe) analysis was used. As shown in Figure 3F, the dominant bacteria in the NFC group were primarily composed of *p_Bacteroidota*, *p_Actinobacteriota*, and *p_Spirochaetota*, while the dominant bacteria in the FC group were mainly located in *c_Clostridia* and *o_Lachnospirales*.

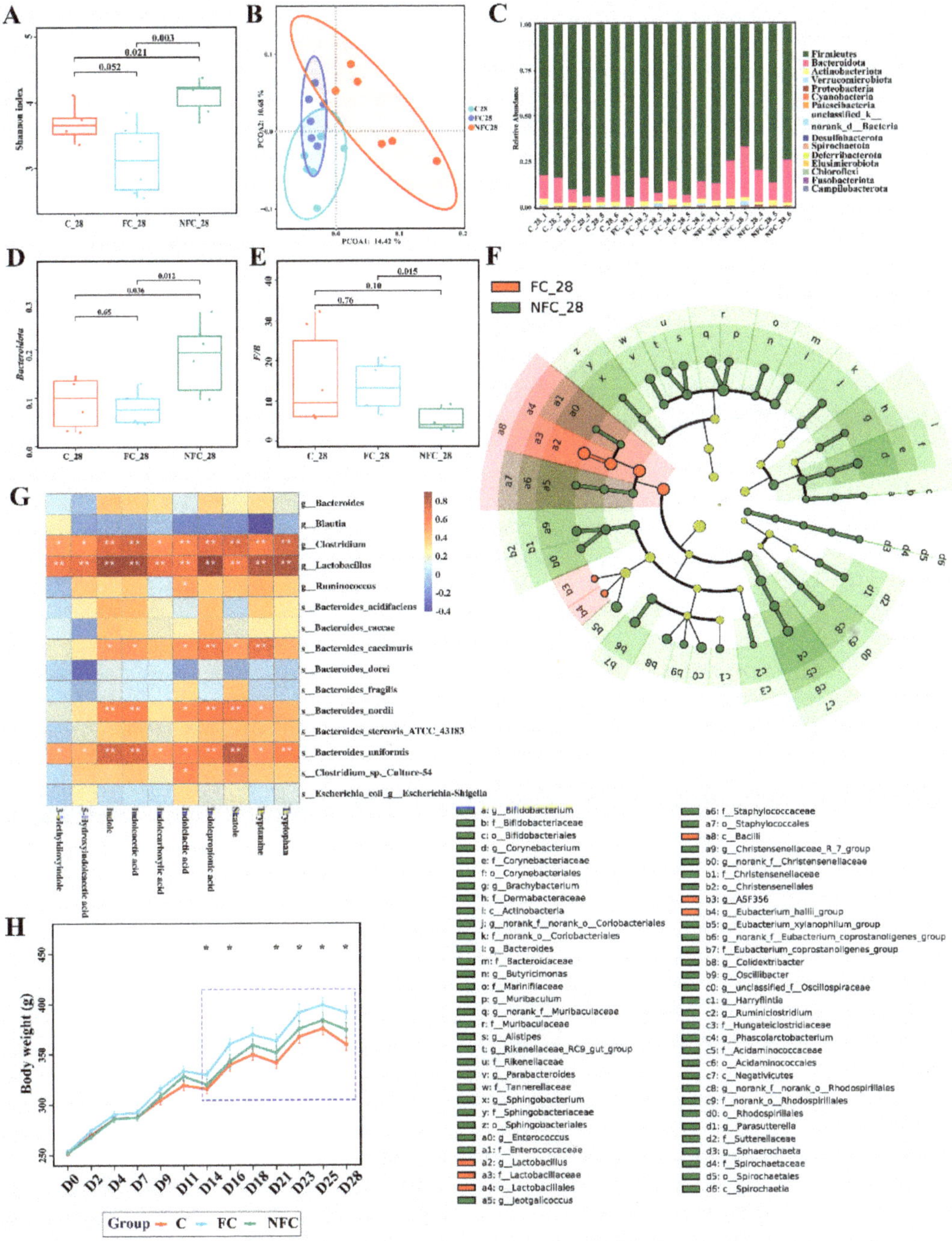

Figure 3. Effect of NFC and FC CAJs on rats: (**A**) Shannon index. (**B**) PCoA scores plot on OTU level. (**C**) Bar plot of gut microbiota on phylum level. (**D**) Box plot of *Bacteroidota* abundance. (**E**) Box plot of F/B ratio. (**F**) Cladogram from LEfSe analysis. (**G**) Correlation heatmap of microbiota and metabolites (The "*" means $p < 0.05$, and "**" means $p < 0.01$). (**H**) Body weight; * indicates significant difference ($p < 0.05$) between FC and C groups.

The effect of NFC on microbiota could also cause metabolite changes in the gut. The results of differential metabolites analysis between the NFC and FC groups, as well as their related metabolites, are shown in Table 2. The abundance of bile acids in the FC group (including cholic acid, deoxycholic acid, chenodeoxycholic acid, taurocholic acid, 7-ketodeoxycholic acid, and 3-oxo-4,6-choladienoic acid) was higher than that in the NFC group and comparable to that in the C group. Although there was no significant difference, the same trend of lithocholic acid content was observed.

Table 2. Annotation results of differential and their related metabolites from cecal content.

Class or Name	Formula	C (%)	FC (%)	NFC (%)
Bile acids				
Cholic acid	$C_{24}H_{40}O_5$	43.32 [a]	47.92 [a]	8.77 [b]
Deoxycholic acid	$C_{24}H_{40}O_4$	48.24 [a]	38.52 [a]	13.24 [b]
Chenodeoxycholic acid	$C_{24}H_{40}O_4$	41.40 [a]	39.77 [a]	18.83 [b]
Taurocholic acid	$C_{26}H_{45}NO_7S$	48.19 [a]	37.35 [a]	14.45 [b]
Lithocholic acid	$C_{24}H_{40}O_3$	35.38 [a]	35.79 [a]	28.83 [a]
7-Ketodeoxycholic acid	$C_{24}H_{38}O_5$	28.33 [a,b]	49.52 [a]	22.14 [b]
3-Oxo-4,6-choladienoic acid	$C_{24}H_{34}O_3$	34.36 [a,b]	43.91 [a]	21.73 [b]
Bilirubinoids				
Mesobilirubinogen	$C_{33}H_{44}N_4O_6$	22.30 [b]	60.23 [a]	17.47 [b]
D-Urobilinogen	$C_{33}H_{42}N_4O_6$	21.14 [a]	57.88 [a]	20.97 [b]
(−)-Stercobilin	$C_{33}H_{46}N_4O_6$	20.85 [b]	49.25 [a]	29.90 [b]
Tryptophan and its metabolites				
Tryptophan	$C_{11}H_{12}N_2O_2$	35.46 [b]	43.08 [a]	21.46 [c]
Indole	C_8H_7N	29.30 [b]	39.01 [a]	31.69 [a,b]
Tryptamine	$C_{10}H_{12}N_2$	27.75 [a]	29.21 [a]	43.04 [a]
Indolecarboxylic acid	$C_9H_7NO_2$	13.77 [b]	72.01 [a]	14.22 [b]
Indolelactic acid	$C_{10}H_9NO_2$	34.41 [a]	45.44 [a]	20.15 [b]
Indolepropionic acid	$C_{11}H_{11}NO_2$	29.88 [b]	42.13 [a]	27.99 [b]
Indoleacrylic acid	$C_{11}H_9NO_2$	30.13 [b]	38.90 [a]	30.97 [a,b]
Skatole	C_9H_9N	27.00 [a]	39.77 [a]	33.23 [a]
3-Methyldioxyindole	$C_9H_9NO_2$	13.77 [b]	72.01 [a]	14.22 [b]
5-Hydroxyindoleacetic acid	$C_{10}H_9NO_3$	49.36 [a]	28.60 [a,b]	22.05 [b]
Kynurenic acid	$C_{10}H_7NO_3$	18.87 [b]	63.20 [a]	17.94 [b]

Last three columns show the relative percent contents between the three groups; Different letters indicate significant difference of $p < 0.05$; $n = 6$.

As shown in Table 2, the bilirubin content in the FC group was higher than that in the NFC group, whereas the levels in the C group were comparable to those in the NFC group. Bilirubin is an important component of the heme catabolic pathway, which can be reduced to urobilinoids and/or urobilinogens by the gut microbiota. Urobilinogens are then either deposited into the feces as bile pigments or reabsorbed into the hepatic portal circulation. Urobilinogens are taken up by the kidneys, oxidized to urobilin, and excreted in the urine [21]. Bilirubin reductase may be derived from *Clostridium ramosum*, *Clostridium perfringens*, *Clostridium difficile*, or *Bacteroides fragilis*. Bilirubin can reduce hepatic fat accumulation, and increased levels of unconjugated bilirubin in the plasma have been suggested as treatments for obesity and type 2 diabetes mellitus [21].

As shown in Table 2, tryptophan levels in the FC group were significantly higher than that in the NFC group. In addition, its related metabolites, indole and its derivatives, also showed higher abundance in the FC group. The indolecarboxylic acid, indolelactic acid, indolepropionic acid, and 3-methyldioxyindole content in the FC group were significantly higher than those in the NFC group. Tryptophan is an essential amino acid that must be supplemented in the diet [22]. Tryptophan accounted for 0.26% of the rat diet (Supplemental Table S1), and tryptophan levels in CAJs were low and not significantly different ($p > 0.075$, Supplemental Figure S6).

Figure 3G shows a correlation heatmap between the abundance of metabolites and bacteria related to tryptophan metabolism [23]. *Clostridium*, *Lactobacillus*, and *Bacteroides_uniformis* were positively correlated with all tryptophan-related metabolites, and *Lactobacillus* abundance in the NFC group was significantly lower than that in the FC group. *Clostridium* and *Bacteroides_uniformis* also showed decreasing trends. In addition, *Bacteroides_nordii* and *Bacteroides_caecimuris* were positively correlated with six metabolites, whereas *Clostridium_*sp.*_cultural-54* was positively correlated with indolelactic acid and skatole, and *Ruminococcus* was positively correlated with only indolelactic acid. These results indicate that differences between NFC and FC juices affected tryptophan metabolism in the gut. *Clostridium*, *Lactobacillus*, and *Ruminococcus* can convert tryptophan to tryptamine via tryptophan decarboxylase [23].

All rats displayed a continual increase in body weight during the experimental period, and there was no significant difference between the NFC and C groups at the same time point (Figure 3H). However, the body weight of the FC group was significantly higher than that of the C group after D14, except on D18 ($p < 0.05$). The weight of the FC group was the highest where the upward trend was the most evident.

3.3. Latter Intervention of NFC Did Not Show the Same Effect

To observe the long-term effects of NFC and FC, we extended the experimental period to 56 days. As shown in Figure 4A, body weight was not significantly different between the NFC and FC groups during the entire prolonged period. At D35, D39, and D46, body weight of the NFC group was significantly higher than that in the C group, while in almost all the days except D32, the weight of the FC group was significantly higher than that of the C group ($p < 0.05$). For the Shannon index (Figure 4B), there was no significant difference between the three groups. As for *Firmicutes*, *Bacteroidota*, *Spirochaetota*, and *F/B* shown in Supplemental Figure S7, no significant difference was observed between the NFC and FC groups. For β-diversity, there was no distinct separation between the NFC and FC groups in the PCoA scores plot (Figure 4C). Based on the above results, as the experimental period was prolonged, the effect of NFC compared to FC was diminished.

NFC could improve microbiota diversity and inhibit several intestinal metabolic processes in the former intervention, but we also observed that these beneficial effects would diminish over a prolonged period. We therefore designed the juice-switch groups. The rats that were treated with NFC were administered FC, and vice versa. The body weights of the juice-switch groups were similar to those before the juice-switch, as shown in Figure 4D. The NFC and NFC-FC groups had higher body weights compared to that of the C group, whereas weights in the FC and FC-NFC groups were much higher. The FC-NFC group was always significantly higher than the C group ($p < 0.05$). The Shannon index between the five groups showed no significant differences (Supplemental Figure S8). As for hierarchical clustering analysis based on differential metabolites (Figure 4E,F), the clustering of groups on D56 (including juice-switch groups) was not as obvious as groups on D28 (Figure 4E).

3.4. The Beneficial Effect of NFC Could Come from Polyphenol Compounds

In the present study, we analyzed the basic indices and nutritional ingredients of CAJ used in animal experiments. As shown in Table 1, the total phenol content of NFC juice was more than two times higher than that of FC samples, which indicated the mass loss of phenols during the concentration process. However, the nutritional ingredients, such as energy, remained the same (Supplemental Table S2).

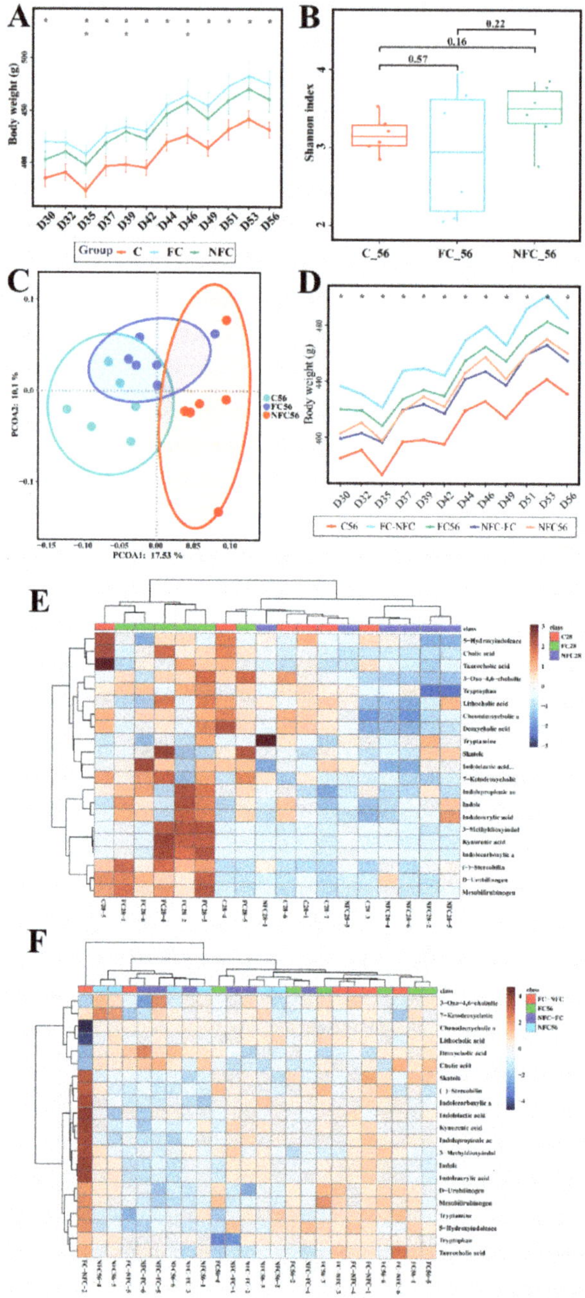

Figure 4. The effect of NFC and FC CAJ in rats in the prolonged and juice-switch period: (**A**) Body weight in prolonged period; * ($p < 0.05$) and ** ($p < 0.01$) in the first/second indicates significant difference between FC56/NFC56 and C56 groups. (**B**) Shannon index. (**C**) PCoA scores plot on OTU level. (**D**) Body weight in juice-switch period; * indicates the significant difference ($p < 0.05$) between FC-NFC and C56 groups. (**E**,**F**) Heatmaps from hierarchical clustering analysis for differential metabolites on D28 and D56.

Furthermore, we also analyzed the small-molecule compounds (including all the samples collected, not only the samples used for animal experiments) and determined the differences between them. It was observed from the PCA score plots in Supplemental Figure S3 that CAJs with different processing methods were distinctly separated, indicating a significant difference between them. Simultaneously, we observed that samples from the same processing method were clustered into multiple clusters based on differences in origin. However, the differences resulting from origin were no more than those from the processing method. The corresponding heatmaps from the hierarchical clustering analysis also indicated similar results (Supplemental Figure S3). Student's *t*-test, fold change, and orthogonal projection to latent structures discriminant analysis (OPLS-DA) were conducted to identify the latent constant differential compounds. A compound could be defined as characteristic based on the following parameters: $p < 0.05$, $\geq$2-fold change, and variable importance for projection (VIP, from OPLS-DA) values > 1. Therefore, compounds were annotated and identified using chemical standards. The results for 18 phenolic compounds, whose contents in NFC were significantly higher than those in FC, are shown in Table 3. The beneficial effects of NFC could be attributed to phenolic compounds.

Table 3. Polyphenol compounds of NFC and FC.

Name	Fold Change (NFC/FC)	*p* Value	VIP
Chlorogenic acid	2.21	2.96×10^{-7}	1.19
5-Methoxysalicylic acid	46.46	5.28×10^{-17}	2.16
p-Coumaric acid	18.84	3.82×10^{-12}	2.19
Caffeic acid	3.32	4.94×10^{-6}	1.44
Ferulic acid	31.44	5.56×10^{-14}	1.92
Phloretin	10.21	7.02×10^{-7}	1.53
(+)-Catechin	16.76	1.71×10^{-16}	2.24
(−)-Epicatechin	50.53	1.04×10^{-16}	2.40
(−)-gallocatechin	2.92	7.55×10^{-19}	1.85
Phlorizin	69.98	2.21×10^{-16}	2.20
Isoquercitrin	59.83	3.96×10^{-11}	2.09
Rutin	27.15	4.01×10^{-12}	2.22
Naringenin	8.52	1.84×10^{-9}	1.42
Eriodictyol	30.07	3.76×10^{-12}	1.70
(+−)-Taxifolin	8.05	6.21×10^{-13}	1.64
Quercetin-3-O-galactoside/Hyperoside	50.38	2.19×10^{-12}	2.07
Procyanidin B1	102.51	2.84×10^{-16}	2.51
p-Coumaraldehyde	67.65	2.98×10^{-19}	2.90

VIP: variable importance for projection.

4. Discussion

This study was conducted to evaluate the effects of CAJs (thus eliminating pectin), including NFC and FC, on the gut microbiota of rats. This research suggests that substances in fruit juices other than pectin could have a significant effect. FC, which was treated using enzymatic hydrolysis for pectin, thermal concentration, and restoration, could be classified as an ultra-processed food based on the NOVA classification system [1]. Studies have shown that ultra-processed foods reduce the diversity of the gut microbiota and disrupt microbial functions, further affecting the health status of the host [4,12]. Meanwhile, thermal processing of food leads to the destruction of heat-sensitive vitamins and phytochemicals or generation of harmful substances [12], which will further affect the community characteristics of the gut microbiota and reduce its diversity [24].

In our study, although with limited health effects, CAJ intake over a long-term period significantly increased the body weight of rats (shown in Figure 2A). However, there was no distinct difference in gut microbiota diversity between the CAJ and control groups at the same time point (shown in Figure 2E). Furthermore, body weight, microbiota, and metabolomics of cecal content were affected by the administration of CAJ from different processing degrees in the first 28 days. NFC could affect the microbiome structure, significantly increase α-diversity and *Bacteroidetes* abundance, and reduce *Firmicutes* abundance and the F/B ratio (shown in Figure 3A,D,E). In β-diversity analysis shown in Figure 3B, a distinct separation was observed between the NFC and FC groups. Meanwhile, NFC reduced the levels of bile acids, tryptophan, bilirubin, and their related metabolites in the gut (shown in Table 2). In addition, we analyzed and compared the basic indices and small-molecule compounds of NFC and FC, showing that the multi-polyphenol content (18 phenolic compounds in Table 3) and total phenolic content (in Table 1) in the NFC group were both two times higher than those in the FC group.

As reported, only 5–10% of phenolic compounds ingested into the body through diet can be directly absorbed in the small intestine, while the vast majority (90–95%) arrive in the gut and play a role through decomposition and metabolism by microorganisms [25]. Meanwhile, apple phenolic extract can inhibit inflammatory pathway activation, protect intestinal mucosa integrity, restore the disorder of bile acid metabolism, and improve the diversity of gut microbiota. Among the polyphenols observed as differential ones between NFC and FC, most can change gut microbiota abundance and regulate bacterial structure and inflammation, obesity, and energy metabolism, thus improving health status. In mice with colitis, phloretin can reduce *Firmicutes* abundance and improve *Bacteroidota* abundance to achieve bacterial community rebalancing [26]. Quercetin can increase bacterial diversity, reduce the F/B ratio, and restore bacterial imbalance caused by dextran sodium sulfate [27]. It was reported that catechin can improve *Bacteroidetes* abundance in obese rats [28]. Chlorogenic acid can reduce plasma lipids, reverse obesity and metabolic disorders induced by high-carbohydrate and high-fat diets, as well as improve microbiota diversity [29,30]. Moreover, it can prevent type 2 diabetes by affecting glucose absorption and carbohydrate metabolism [31]. Phlorizin can significantly reduce energy intake, body weight gain, fasting blood glucose, triglyceride and total cholesterol levels and improve fecal microbial diversity [32]. Caffeic acid can significantly improve obesity induced by a high-fat diet, promote lipid metabolism, reduce body weight and fat accumulation, improve lipid structure, increase energy consumption, restore gut microbiota imbalance, and increase the abundance of anti-obesity-related and butyrate-producing bacteria [33].

Bile acids are synthesized from cholesterol by liver cells, stored in the gallbladder, and then released into the gut, which can promote the absorption of dietary fat and vitamins, as well as regulate glucose metabolism, lipid metabolism, energy homeostasis, etc. [34–36]. The composition of bile acids is closely related to obesity and is affected by gut microbiota [36]. As such, disturbance of the gut microbiota can alter the composition of bile acids, which may further alter important bile acid signaling pathways and affect host metabolism [36]. Unhealthy dietary habits (such as a long-term high-fat diet) increase the level of bile acids, aggravate the proliferation of stem cells, disrupt homeostasis in the gut, and may even lead to cancer [37]. The results shown in Table 2 indicated that NFC apple juice can reduce bile acid content and inhibit lipid absorption. Studies have shown that disorders of bile acid metabolism and the gut microbiota can increase intestinal permeability and aggravate the intestinal inflammatory response [38]. In addition, apple polyphenol extract can improve the intestinal inflammatory response by alleviating the disorder of bile acids and gut microbiota [38].

Tryptophan can be absorbed in the small intestine through food protein digestion and can enter blood circulation [39], while unabsorbed tryptophan travels to the large intestine and is broken down by bacteria to produce various indole derivatives, including indole, tryptamine, indoleethanol, indoleacetic acid, indolepropionic acid, indoleacrylic acid, indolealdehyde, and skatole [22]. Some bacteria in *Clostridium* and *Bacteroides* also metabolize tryptophan through tryptophanase. Indoleacetic acid can be generated by *Bacteroides* through indole-acetamide generated by tryptophan monooxygenase. *Lactobacillus* can also produce indole-3-lactic acid from aromatic amino acid aminotransferase and indolelactic acid dehydrogenase [23]. In our study, tryptophan and its related metabolite contents were found to be significantly correlated with tryptophan-metabolizing bacteria (shown in Figure 3G) [23].

Furthermore, there is a large amount of pectin in apple and cloudy apple juice, and some studies have found that these effects on gut microbiota could mainly be attributed to pectin [9]. The pectin contents of these two CAJs used were no more than 0.04 g/100 g because of the enzymatic hydrolysis process (Supplemental Table S2), which was far lower than the limit (5 g/100 g) at which it could have an obvious influence on gut microbiota [40]. Furthermore, the administration of pectin is usually accompanied with an increase in short-chain fatty acids; however, no such significant difference was observed in our results (Supplemental Figure S3).

During the prolonged experiment period, the effect diminished rather than accumulated (shown in Figure 4). This was not only indicated by the continuous administration of the same CAJ, but also by the juice-switch experiment. This may be because the intestinal sensitivity to phenolic compounds decreased as the rats aged. In addition, this effect could be counteracted by the sugar in CAJs. Sugar consumption is linked to an increase in obesity and non-communicable diseases; obesity is not the cause but rather a marker of metabolic dysfunction [41].

The results presented herein demonstrate the comprehensive effect of CAJ with different processing degrees. The evidence suggests that CAJ (with pectin eliminated) with a lower processing degree could lessen excessive increases in body weight by regulating the gut microbiota due to its higher multi-polyphenol contents. In a prolonged experiment, the combined effect of phenolics and pectin in cloudy apple juices may still be effective, while phenolics alone in CAJ could not counteract the effect of sugar. These results could be conducive to consumer behavior and public health policy design. Since the consumption of fruit juices has increased during the COVID-19 pandemic [7,8], more convenient and easy-to-transport fruit juices with a lower processing degree (with fewer additives and less processing) should be recommended.

5. Conclusions

In conclusion, NFC with higher phenolic content can significantly improve gut microbiota diversity and influence its structure. Simultaneously, it can reduce bile acids and bilirubin, as well as inhibit the microbial metabolism of tryptophan in the gut. However, these effects diminished with an extension of the experimental period. Furthermore, the juice-switch experiment confirmed that. The health outcomes and metabolomic differences between NFC and FC mainly originate from the phenolic differences caused by different processing methods and degrees. We believe that NFC juices that are processed with a lesser degree, although naturally containing a certain amount of sugar, could have more health benefits than we originally believed. In particular, with little attention paid to the total energy intake, NFC juices could be a viable option for consumption as natural polyphenol-rich foods.

Supplementary Materials: The following supporting information can be downloaded at: https://www.mdpi.com/article/10.3390/nu14173488/s1. Figure S1: The cholesterol, high-density lipoprotein and low-density lipoprotein contents; Figure S2: The abundance of *Firmicutes* and *Spirochaetota* on D28; Figure S3: The PCA analysis of apple juices in positive and negative ion modes; Figure S4: The HCA analysis of apple juices in positive and negative ion modes; Figure S5: The OPLS-DA analysis of apple juices; Figure S6: The relative tryptophan contents in NFC and FC clarified apple juices; Figure S7: The abundance of *Firmicutes, Bacteroidota, Spirochaetota* and *F/B* on D56; Figure S8: Shannon index of five groups on D56; Table S1: The main nutrients of animal feed used in this experiment; Table S2: Nutrition ingredients of NFC and FC clarified apple juices; Table S3: The relative SCFA contents of D28 [42–57].

Author Contributions: Conceptualization, L.X. and Z.X.; Data curation, L.X., S.Y. and X.W.; Funding acquisition, Z.X.; Methodology, L.X.; Supervision, Z.X.; Visualization, L.X., S.Y. and Z.X.; Writing—original draft, L.X.; Writing—review and editing, S.Y., K.W., A.L. and Z.X. All authors have read and agreed to the published version of the manuscript.

Funding: This work was funded by the National Natural Science Foundation of China (32072234) and the International Postdoctoral Exchange Fellowship Program (PC2021089).

Institutional Review Board Statement: All experimental procedures were conducted in compliance with the institutional guidelines for the care and use of laboratory animals in China and were approved by the Institutional Animal Care and Use Committee of Beijing Vital River Laboratory Animal Technology Co., Ltd. under permit number ID P2020072. Animal welfare and experimental protocols were in accordance with guidelines for the care and use of laboratory animals.

Informed Consent Statement: Not applicable.

Data Availability Statement: The sequencing data of the 16S rRNA gene from this study are available in the Sequence Read Archive (SRA) under the project number PRJNA793573. Metabolomics data from mass spectrometry are available via MetaboLights with the identifier MTBLS4054.

Acknowledgments: The authors would like to thank Yinfei Ma from Jinan Fruit Research Institute, All China Federation of Supply and Marketing Cooperatives for contributions to CAJ sampling.

Conflicts of Interest: The authors declare no conflict of interest.

References

1. Monteiro, C.A.; Levy, R.B.; Claro, R.M.; Castro, I.R.R.D.; Cannon, G. A new classification of foods based on the extent and purpose of their processing. *Cad. Saúde Pública* **2010**, *26*, 2039–2049. [CrossRef]
2. Monteiro, C.A.; Cannon, G.; Moubarac, J.-C.; Levy, R.B.; Louzada, M.L.C.; Jaime, P.C. The UN Decade of Nutrition, the NOVA food classification and the trouble with ultra-processing. *Public Health Nutr.* **2018**, *21*, 5–17. [CrossRef]
3. Lawrence, M.A.; Baker, P.I. Ultra-processed food and adverse health outcomes. *BMJ-Br. Med. J.* **2019**, *365*, l2289. [CrossRef]
4. Wu, Q.; Chen, T.; El-Nezami, H.; Savidge, T.C. Food ingredients in human health: Ecological and metabolic perspectives implicating gut microbiota function. *Trends Food Sci. Technol.* **2020**, *100*, 103–117. [CrossRef]
5. Hur, J.; Otegbeye, E.; Joh, H.-K.; Nimptsch, K.; Ng, K.; Ogino, S.; Meyerhardt, J.A.; Chan, A.T.; Willett, W.C.; Wu, K.; et al. Sugar-sweetened beverage intake in adulthood and adolescence and risk of early-onset colorectal cancer among women. *Gut* **2021**, *70*, 2330–2336. [CrossRef]
6. Fardet, A. Minimally processed foods are more satiating and less hyperglycemic than ultra-processed foods: A preliminary study with 98 ready-to-eat foods. *Food Funct.* **2016**, *7*, 2338–2346. [CrossRef]
7. Marty, L.; de Lauzon-Guillain, B.; Labesse, M.; Nicklaus, S. Food choice motives and the nutritional quality of diet during the COVID-19 lockdown in France. *Appetite* **2021**, *157*, 105005. [CrossRef]
8. Bracale, R.; Vaccaro, C.M. Changes in food choice following restrictive measures due to COVID-19. *Nutr. Metab. Cardiovasc. Dis.* **2020**, *30*, 1423–1426. [CrossRef]
9. Licht, T.R.; Hansen, M.; Bergstrom, A.; Poulsen, M.; Krath, B.N.; Markowski, J.; Dragsted, L.O.; Wilcks, A. Effects of apples and specific apple components on the cecal environment of conventional rats: Role of apple pectin. *BMC Microbiol.* **2010**, *10*, 13. [CrossRef]
10. Tomas-Barberan, F.A.; Espin, J.C. Effect of Food Structure and Processing on (Poly)phenol-Gut Microbiota Interactions and the Effects on Human Health. *Annu. Rev. Food Sci. Technol.* **2019**, *10*, 221–238. [CrossRef]
11. Yang, Q.; Liang, Q.; Balakrishnan, B.; Belobrajdic, D.P.; Feng, Q.J.; Zhang, W. Role of Dietary Nutrients in the Modulation of Gut Microbiota: A Narrative Review. *Nutrients* **2020**, *12*, 381. [CrossRef]
12. Miclotte, L.; Van de Wiele, T. Food processing, gut microbiota and the globesity problem. *Crit. Rev. Food Sci. Nutr.* **2020**, *60*, 1769–1782. [CrossRef]
13. Grigor'eva, I.N. Gallstone Disease, Obesity and the Firmicutes/Bacteroidetes Ratio as a Possible Biomarker of Gut Dysbiosis. *J. Pers. Med.* **2020**, *11*, 13. [CrossRef]
14. Turnbaugh, P.J.; Ley, R.E.; Mahowald, M.A.; Magrini, V.; Mardis, E.R.; Gordon, J.I. An obesity-associated gut microbiome with increased capacity for energy harvest. *Nature* **2006**, *444*, 1027–1031. [CrossRef]
15. Flegal, K.M.; Troiano, R.P. Changes in the distribution of body mass index of adults and children in the US population. *Int. J. Obes.* **2000**, *24*, 807–818. [CrossRef]
16. de Souza, A.Z.; Zambom, A.Z.; Abboud, K.Y.; Reis, S.K.; Tannihao, F.; Guadagnini, D.; Saad, M.J.; Prada, P.O. Oral supplementation with L-glutamine alters gut microbiota of obese and overweight adults: A pilot study. *Nutrition* **2015**, *31*, 884–889. [CrossRef]
17. Fu, J.; Wang, Y.; Tan, S.; Wang, J. Effects of Banana Resistant Starch on the Biochemical Indexes and Intestinal Flora of Obese Rats Induced by a High-Fat Diet and Their Correlation Analysis. *Front. Bioeng. Biotechnol.* **2021**, *9*, 575724. [CrossRef]
18. Lobb, B.; Tremblay, B.J.; Moreno-Hagelsieb, G.; Doxey, A.C. An assessment of genome annotation coverage across the bacterial tree of life. *Microb. Genom.* **2020**, *6*, e000341. [CrossRef]
19. Zhou, S.; Luo, R.; Gong, G.; Wang, Y.; Gesang, Z.; Wang, K.; Xu, Z.; Suolang, S. Characterization of Metagenome-Assembled Genomes and Carbohydrate-Degrading Genes in the Gut Microbiota of Tibetan Pig. *Front. Microbiol.* **2020**, *11*, 595066. [CrossRef]

20. Zhao, L.; Zhang, Q.; Ma, W.; Tian, F.; Shen, H.; Zhou, M. A combination of quercetin and resveratrol reduces obesity in high-fat diet-fed rats by modulation of gut microbiota. *Food Funct.* **2017**, *8*, 4644–4656. [CrossRef]
21. Hamoud, A.R.; Weaver, L.; Stec, D.E.; Hinds, T.D., Jr. Bilirubin in the Liver-Gut Signaling Axis. *Trends Endocrinol. Metab.* **2018**, *29*, 140–150. [CrossRef]
22. Roager, H.M.; Licht, T.R. Microbial tryptophan catabolites in health and disease. *Nat. Commun.* **2018**, *9*, 3294. [CrossRef]
23. Gao, K.; Mu, C.L.; Farzi, A.; Zhu, W.Y. Tryptophan Metabolism: A Link between the Gut Microbiota and Brain. *Adv. Nutr.* **2020**, *11*, 709–723. [CrossRef]
24. Zhang, Z.; Li, D. Thermal processing of food reduces gut microbiota diversity of the host and triggers adaptation of the microbiota: Evidence from two vertebrates. *Microbiome* **2018**, *6*, 99. [CrossRef]
25. Kawabata, K.; Yoshioka, Y.; Terao, J. Role of Intestinal Microbiota in the Bioavailability and Physiological Functions of Dietary Polyphenols. *Molecules* **2019**, *24*, 370. [CrossRef]
26. Wu, M.; Li, P.; An, Y.; Ren, J.; Yan, D.; Cui, J.; Li, D.; Li, M.; Wang, M.; Zhong, G. Phloretin ameliorates dextran sulfate sodium-induced ulcerative colitis in mice by regulating the gut microbiota. *Pharmacol. Res.* **2019**, *150*, 104489. [CrossRef]
27. Hong, Z.; Piao, M. Effect of Quercetin Monoglycosides on Oxidative Stress and Gut Microbiota Diversity in Mice with Dextran Sodium Sulphate-Induced Colitis. *Biomed Res. Int.* **2018**, *2018*, 8343052. [CrossRef]
28. Luo, J.; Han, L.; Liu, L.; Gao, L.; Xue, B.; Wang, Y.; Ou, S.; Miller, M.; Peng, X. Catechin supplemented in a FOS diet induces weight loss by altering cecal microbiota and gene expression of colonic epithelial cells. *Food Funct.* **2018**, *9*, 2962–2969. [CrossRef]
29. Bhandarkar, N.S.; Brown, L.; Panchal, S.K. Chlorogenic acid attenuates high-carbohydrate, high-fat diet-induced cardiovascular, liver, and metabolic changes in rats. *Nutr. Res.* **2019**, *62*, 78–88. [CrossRef]
30. Wang, Z.; Lam, K.L.; Hu, J.; Ge, S.; Zhou, A.; Zheng, B.; Zeng, S.; Lin, S. Chlorogenic acid alleviates obesity and modulates gut microbiota in high-fat-fed mice. *Food Sci. Nutr.* **2019**, *7*, 579–588. [CrossRef]
31. Williamson, G. Protection against developing type 2 diabetes by coffee consumption: Assessment of the role of chlorogenic acid and metabolites on glycaemic responses. *Food Funct.* **2020**, *11*, 4826–4833. [CrossRef] [PubMed]
32. Zhang, X.Y.; Chen, J.; Yi, K.; Peng, L.; Xie, J.; Gou, X.; Peng, T.; Tang, L. Phlorizin ameliorates obesity-associated endotoxemia and insulin resistance in high-fat diet-fed mice by targeting the gut microbiota and intestinal barrier integrity. *Gut Microbes* **2020**, *12*, 1842990. [CrossRef] [PubMed]
33. Xu, J.; Ge, J.; He, X.; Sheng, Y.; Zheng, S.; Zhang, C.; Xu, W.; Huang, K. Caffeic acid reduces body weight by regulating gut microbiota in diet-induced-obese mice. *J. Funct. Foods* **2020**, *74*, 104061. [CrossRef]
34. Jia, W.; Xie, G.; Jia, W. Bile acid-microbiota crosstalk in gastrointestinal inflammation and carcinogenesis. *Nat. Rev. Gastroenterol. Hepatol.* **2018**, *15*, 111–128. [CrossRef]
35. Molinaro, A.; Wahlstrom, A.; Marschall, H.U. Role of Bile Acids in Metabolic Control. *Trends Endocrinol. Metab.* **2018**, *29*, 31–41. [CrossRef]
36. Wei, M.; Huang, F.; Zhao, L.; Zhang, Y.; Yang, W.; Wang, S.; Li, M.; Han, X.; Ge, K.; Qu, C.; et al. A dysregulated bile acid-gut microbiota axis contributes to obesity susceptibility. *EBioMedicine* **2020**, *55*, 102766. [CrossRef]
37. Sorrentino, G.; Perino, A.; Yildiz, E.; El Alam, G.; Bou Sleiman, M.; Gioiello, A.; Pellicciari, R.; Schoonjans, K. Bile Acids Signal via TGR5 to Activate Intestinal Stem Cells and Epithelial Regeneration. *Gastroenterology* **2020**, *159*, 956–968.e8. [CrossRef]
38. Liu, F.; Wang, X.; Li, D.; Cui, Y.; Li, X. Apple polyphenols extract alleviated dextran sulfate sodium-induced ulcerative colitis in C57BL/6 male mice by restoring bile acid metabolism disorder and gut microbiota dysbiosis. *Phytother. Res.* **2021**, *35*, 1468–1485. [CrossRef]
39. Lin, L.; Huang, Z.; Gao, Y.; Yan, X.; Xing, J.; Hang, W. LC-MS based serum metabonomic analysis for renal cell carcinoma diagnosis, staging, and biomarker discovery. *J. Proteome Res.* **2011**, *10*, 1396–1405. [CrossRef]
40. Aprikian, O.; Duclos, V.; Guyot, S.; Besson, C.; Manach, C.; Bernalier, A.; Morand, C.; Rémésy, C.; Demigné, C. Apple Pectin and a Polyphenol-Rich Apple Concentrate Are More Effective Together than Separately on Cecal Fermentations and Plasma Lipids in Rats. *J. Nutr.* **2003**, *133*, 1860–1865. [CrossRef]
41. Lustig, R.H.; Schmidt, L.A.; Brindis, C.D. The toxic truth about sugar. *Nature* **2012**, *482*, 27–29. [CrossRef] [PubMed]
42. Chen, S.; Zhou, Y.; Chen, Y.; Gu, J. fastp: An ultra-fast all-in-one FASTQ preprocessor. *Bioinformatics* **2018**, *34*, i884–i890. [CrossRef] [PubMed]
43. Magoc, T.; Salzberg, S.L. FLASH: Fast length adjustment of short reads to improve genome assemblies. *Bioinformatics* **2011**, *27*, 2957–2963. [CrossRef] [PubMed]
44. Edgar, R.C. UPARSE: Highly accurate OTU sequences from microbial amplicon reads. *Nat. Methods* **2013**, *10*, 996–998. [CrossRef]
45. Stackebrandt, E.; Goebel, B.M. Taxonomic note: A place for DNA-DNA reassociation and 16S rRNA sequence analysis in the present species definition in bacteriology. *Int. J. Syst. Bacteriol* **1994**, *44*, 846–849. [CrossRef]
46. Wang, Q.; Garrity, G.M.; Tiedje, J.M.; Cole, J.R. Naive Bayesian classifier for rapid assignment of rRNA sequences into the new bacterial taxonomy. *Appl. Environ. Microbiol.* **2007**, *73*, 5261–5267. [CrossRef]
47. Tian, Y.; Zhang, L.; Wang, Y.; Tang, H. Age-related topographical metabolic signatures for the rat gastrointestinal contents. *J. Proteome Res.* **2012**, *11*, 1397–1411. [CrossRef]
48. Xie, G.; Zhong, W.; Zheng, X.; Li, Q.; Qiu, Y.; Li, H.; Chen, H.; Zhou, Z.; Jia, W. Chronic ethanol consumption alters mammalian gastrointestinal content metabolites. *J. Proteome Res.* **2013**, *12*, 3297–3306. [CrossRef]

49. Xu, L.; Xu, Z.; Strashnov, I.; Liao, X. Use of information dependent acquisition mass spectra and sequential window acquisition of all theoretical fragment-ion mass spectra for fruit juices metabolomics and authentication. *Metabolomics* **2020**, *16*, 81. [CrossRef]
50. Zhang, Y.; Bilbao, A.; Bruderer, T.; Luban, J.; Strambio-De-Castillia, C.; Lisacek, F.; Hopfgartner, G.; Varesio, E. The Use of Variable Q1 Isolation Windows Improves Selectivity in LC-SWATH-MS Acquisition. *J. Proteome Res.* **2015**, *14*, 4359–4371. [CrossRef]
51. Sabina, B.; Ivana, B.; Verheij, E.R.; Raymond, R.; Sunil, K.; Macdonald, I.A.; Ben, V.O.; Smilde, A.K. Large-scale human metabolomics studies: A strategy for data (pre-) processing and validation. *Anal. Chem.* **2006**, *78*, 567–574.
52. Wang, B.; Wang, X.; Bei, J.; Xu, L.; Zhang, X.; Xu, Z. Development and Validation of an Analytical Method for the Quantification of Arabinose, Galactose, Glucose, Sucrose, Fructose, and Maltose in Fruits, Vegetables, and Their Products. *Food Anal. Methods* **2021**, *14*, 1227–1238. [CrossRef]
53. Wei-de, S.; Li-jun, D.; Yao-hua, G.; Chuan-jie, Z.; Xiang-ying, Y.; Ming-qin, S. Simultaneous determination of twenty amino acids in fruit juice by ion chromatography with integrated pulsed amperometric detection. *China Food Addit.* **2019**, *30*, 132–138.
54. Lai, Z.; Tsugawa, H.; Wohlgemuth, G.; Mehta, S.; Mueller, M.; Zheng, Y.; Ogiwara, A.; Meissen, J.; Showalter, M.; Takeuchi, K.; et al. Identifying metabolites by integrating metabolome databases with mass spectrometry cheminformatics. *Nat. Methods* **2018**, *15*, 53–56. [CrossRef] [PubMed]
55. Tsugawa, H.; Kind, T.; Nakabayashi, R.; Yukihira, D.; Tanaka, W.; Cajka, T.; Saito, K.; Fiehn, O.; Arita, M. Hydrogen Rearrangement Rules: Computational MS/MS Fragmentation and Structure Elucidation Using MS-FINDER Software. *Anal. Chem.* **2016**, *88*, 7946–7958. [CrossRef]
56. Wohlgemuth, G.; Mehta, S.S.; Mejia, R.F.; Neumann, S.; Pedrosa, D.; Pluskal, T.; Schymanski, E.L.; Willighagen, E.L.; Wilson, M.; Wishart, D.S.; et al. SPLASH, a hashed identifier for mass spectra. *Nat. Biotechnol.* **2016**, *34*, 1099–1101. [CrossRef]
57. Tautenhahn, R.; Cho, K.; Uritboonthai, W.; Zhu, Z.; Patti, G.J.; Siuzdak, G. An accelerated workflow for untargeted metabolomics using the METLIN database. *Nat. Biotechnol.* **2012**, *30*, 826–828. [CrossRef]

nutrients

MDPI

Article

Valorization of Grape Pomace and Berries as a New and Sustainable Dietary Supplement: Development, Characterization, and Antioxidant Activity Testing

Adina Frum [1], Carmen Maximiliana Dobrea [1,*], Luca Liviu Rus [1,*], Lidia-Ioana Virchea [1], Claudiu Morgovan [1], Adriana Aurelia Chis [1], Anca Maria Arseniu [1], Anca Butuca [1], Felicia Gabriela Gligor [1], Laura Gratiela Vicas [2], Ovidiu Tita [3] and Cecilia Georgescu [3]

1 Faculty of Medicine, "Lucian Blaga" University of Sibiu, 550169 Sibiu, Romania; adina.frum@ulbsibiu.ro (A.F.); lidia.virchea@ulbsibiu.ro (L.-I.V.); claudiu.morgovan@ulbsibiu.ro (C.M.); adriana.chis@ulbsibiu.ro (A.A.C.); anca.arseniu@ulbsibiu.ro (A.M.A.); anca.butuca@ulbsibiu.ro (A.B.); felicia.gligor@ulbsibiu.ro (F.G.G.)

2 Faculty of Medicine and Pharmacy, University of Oradea, 410028 Oradea, Romania; laura.vicas@gmail.com

3 Faculty of Agriculture Science, Food Industry and Environmental Protection, "Lucian Blaga" University of Sibiu, 550012 Sibiu, Romania; ovidiu.tita@ulbsibiu.ro (O.T.); cecilia.georgescu@ulbsibiu.ro (C.G.)

* Correspondence: carmen.dobrea@ulbsibiu.ro (C.M.D.); liviu.rus@ulbsibiu.ro (L.L.R.)

Citation: Frum, A.; Dobrea, C.M.; Rus, L.L.; Virchea, L.-I.; Morgovan, C.; Chis, A.A.; Arseniu, A.M.; Butuca, A.; Gligor, F.G.; Vicas, L.G.; et al. Valorization of Grape Pomace and Berries as a New and Sustainable Dietary Supplement: Development, Characterization, and Antioxidant Activity Testing. *Nutrients* **2022**, *14*, 3065. https://doi.org/10.3390/nu14153065

Academic Editors: Haixia Yang and Jianjun Deng

Received: 30 June 2022
Accepted: 25 July 2022
Published: 26 July 2022

Abstract: Grape pomace and berries represent natural sources of phytochemicals that can increase the quality of life of consumers by contributing to the prevention of chronic diseases; thus, the development of a dietary supplement was necessary. The raw material (r.m.) used for the development of the dietary supplement consisted of dried and powdered bilberries (*Vaccinium myrtillus* L.), red currants (*Ribes rubrum* L.), and red fermented pomaces (*Vitis vinifera* L.) from Feteasca Neagra and Cabernet Sauvignon cultivars. The particle size distribution, powder flow, total phenolic content (TPC), HPLC-DAD phenolic profile assessment, and radical scavenging assay (RSA) were employed for the analysis of the raw material. After encapsulation, the average mass and uniformity of mass, the disintegration, and the uniformity of content for the obtained capsules were performed to obtain a high-quality dietary supplement. All the assays performed complied to the compendial requirements and the TPC was determined at 9.07 ± 0.25 mg gallic acid equivalents/g r.m. and RSA at 48.32 ± 0.74%. The highest quantities of phenolic compounds determined were 333.7 ± 0.50 μg/g r.m. for chlorogenic acid, followed by rutin, ferulic acid, and (+)-catechin with 198.9 ± 1.60 μg/g r.m., 179.8 ± 0.90 μg/g r.m. and 118.7 ± 0.75 μg/g r.m., respectively. The results of this study can be used for the manufacturing and assessing of pilot scale-up capsule batches and thinking of quality assurance, we recommend that the industrial batch extracts should be standardized in polyphenols, and the manufacturing process should be validated.

Keywords: antioxidants; bilberry; by-products; food supplement; HPLC; phenolic compounds; polyphenols; pomace; red currant

1. Introduction

Free radicals are unstable compounds that destroy biomolecules in the body, being responsible for the occurrence of oxidative stress. They have been associated with the development of cardiovascular disease, cancer, or other conditions [1,2]. By fighting free radicals, antioxidants protect the body from their harmful effects and can contribute to the prevention of such diseases [3,4].

For example, the consumption of natural polyphenols reduces the risk of myocardial infarctions and coronary heart disease due to their cardioprotective effect [5,6]. Flavonoid consumption prevents advanced stages of prostate cancer [7,8].

Metabolic syndrome includes a number of factors, such as chronic inflammation, hypertension, insulin resistance, high values of triglycerides, and low HDL (high-density lipoprotein) cholesterol levels that can lead to type 2 diabetes and cardiovascular disease. Atherosclerosis, a cause of cardiovascular disease, is associated with oxidative stress (OS) [6,9,10]. Diet plays an important role in the prevention of metabolic syndrome. The consumption of fruits and vegetables decreases the risk of metabolic syndrome and cardiovascular disease [6,11].

Berries contain phenolic compounds such as polyphenols, flavonoids, anthocyanins, flavonols, flavanols, stilbenes, and tannins [7,12,13]. Bilberries (*Vaccinium myrtillus* L. fruits) are considered to be valuable berries, as they capture a large amount of solar energy that they turn into compounds of importance for human health [3,14]. They have a nutritional value of approximately 399 kcal/ 100 g and contain mainly carbohydrates (94.5%), proteins (3%), lipids (1%), ash (1.5%), and also present bioactive compounds such as phenolic compounds, carotenoids, and vitamins. Bilberries are rich in anthocyanins, which are phenolic compounds responsible for fruit color [6,15]. The American Herbal Products Association recommends the administration of 160–480 mg of powdered extract divided into several doses per day or a daily dose of 20–60 g of dried bilberries [16,17]. Bilberry polyphenols have a role in controlling blood sugar in patients with type 2 diabetes, improving vision, inhibiting lipid peroxidation, and preventing the oxidative destruction of DNA [3].

Currants have an important energy value. They contain 9.2% carbohydrates, 4.2% fiber, 1.7% lipids, and 1.3% protein in the dry raw material. Red currants (*Ribes rubrum* L. fruits) are a rich source of vitamin C, polyphenols, and anthocyanins, which possess antioxidant properties [18,19]. Due to the rich content of polyphenols, currants (*Ribes* sp. L. fruits) are used in the management of diseases such as hypertension and other cardiovascular diseases, inflammation, osteoporosis, and cancer. They are important elements in the diet, due to their beneficial effects for the human body [7,20].

Both bilberries and currants are consumed fresh or processed in the form of juices, syrups, jams, but also in the form of extracts in food supplements or beverages [3,7,21,22]. The consumption of fresh berries brings a higher intake of polyphenols because some of them are lost through processing. However, processing offers the advantage of consuming bioactive compounds from berries throughout the year [3,23].

Over 67 million tons of grapes are produced annually, with grapes being the most cultivated fruit in the world [24,25]. Numerous losses occur as a result of food processing. Fortunately, the resulting waste from food processing is rich in bioactive compounds [26,27]. Many countries are investing in the extraction of bioactive compounds from by-products. The residue obtained when processing the grapes in order to make wine is called grape pomace and it represents approximately 20–25% of the mass of grapes subjected to processing [26,28,29]. It consists of peels, seeds, and stem fragments [26,30]. It also contains yeast results from the fermentation process [28]. In the past, pomace was considered a product without economic value, because it had no use [28], but now there is a tendency to obtain sustainable wines by the valorification of pomace [26]. Grape pomace is used to extract tartaric acid and obtain ethanol. The use of grape pomace as a fertilizer is disadvantageous because the polyphenolic compounds in its composition inhibit seed germination. Grape pomace has also been added to animal feed, but it has been observed that polymeric polyphenols inhibit cellulolytic and proteolytic enzymes and slow down digestion [24,31]. Grape pomace contains phenolic compounds (phenolic acids, anthocyanins, resveratrol, and procyanidins) [6,32] with important health benefits such as antioxidant, anti-inflammatory, antiproliferative and antimicrobial properties [26,28]. Grape pomace is used in the pharmaceutical, food, and cosmetics industries [26,33]. An interesting application of grape pomace is the production of ecocyanin, an anthocyanin used by the food industry as a food coloring [28].

An adequate diet could provide all the necessary nutrients for a healthy life, but because of lifestyles or other reasons people need to supplement their diet with the necessary nutrients by using dietary supplements [34]. They should be properly consumed when needed, because an inappropriate intake of dietary supplements can lead to adverse events [35,36].

The aim of this study is to develop a dietary supplement based on berries and a by-product of the vinification process that possesses antioxidant properties and can be used to supplement a faulty diet and to increase the quality of life by contributing to the prevention of chronic diseases.

2. Materials and Methods

2.1. Chemicals

The standards of gallic acid (purity > 99%), ferulic acid (purity > 99%), syringic acid (purity > 95%), cinnamic acid (purity > 99%), caffeic acid (purity > 99%), (+)-catechin (purity > 98%), resveratrol (purity > 99%), chlorogenic acid (purity > 95%), quercetin (purity > 95%), rutin (purity > 94%), and methanol suitable for HPLC analysis (purity ≥ 99.9%) were purchased from Sigma-Aldrich (St. Louis, MO, USA). DPPH (2,2-Diphenyl-1-picrylhydrazyl) and Folin–Ciocalteu reagent, were purchased from Sigma-Aldrich (St. Louis, MO, USA) too and sodium acetate, glacial acetic acid, hydrochloric acid, and ethanol, all analytical grades, were purchased from the Chemical Company (Iasi, Romania). In all analyses, ultrapure water was used (conductivity 0.05 μS/cm).

2.2. Sample Preparation

The pulverulent raw material (r.m) used for the development of the dietary supplement consisted of bilberries (*Vaccinium myrtillus* L.), red currant (*Ribes rubrum* L.), and red fermented pomaces (*Vitis vinifera* L.) from Feteasca Neagra (FN) and Cabernet Sauvignon (CS) cultivars, added in equal parts (1:1:1). The berries were harvested from unpopulated areas of Sibiu County, Romania, and the pomaces were obtained from grapes harvested from a vineyard from Alba County, Romania. The berries and the pomace were dried in a Memmert UNB100 oven with continuous airflow at 40 °C and 20 °C, respectively, until they reached a constant mass and then they were ground on a domestic mill. The powder was calibrated using a 1.0 mm sieve. The mixing of the obtained powders was made by energic shaking for 10 min. The obtained powder was stored in amber glass containers at room temperature, away from sunlight until it was analyzed and encapsulated.

2.3. Raw Material Analysis

2.3.1. Particle Size Distribution

Each sieve of the Retsch AS200 analytical sieve shaker was weighed and then the apparatus was assembled by putting the sieves in decreasing order of their sizes (710 μm, 224 μm, 125 μm, 90 μm, and 63 μm) and the collecting pan on the bottom. Exactly 100 g of raw material was weighed and put uniformly into the top sieve. After placing the cap and securing the assembled sieves the apparatus was set at 1.50 mm amplitude for 15 min. Each sieve was weighed, and the results were expressed as percentages (%).

Regarding the particle size distribution assessed by the sieve test, the powders are classified as coarse if through a 710 μm sieve passes a minimum 95% of the analyzed powder and through a 90 μm sieve passes a maximum 40% of it, moderately fine if through a 224 μm sieve passes a minimum 95% of the analyzed powder and through a 125 μm sieve passes a maximum 40% of it, fine if through a 125 μm sieve passes a minimum 95% of the analyzed powder and through a 90 μm sieve passes a maximum 40% of it, and very fine if through a 90 μm sieve passes a minimum 95% of the analyzed powder and through a 63 μm sieve passes a maximum 40% of it [37].

2.3.2. Powder Flow

A total of 100 g of raw material was weighed and introduced into a 250 mL graduated cylinder. The unsettled apparent volume was registered (V_0) and the cylinder was put on the Erweka SVM 102 tapped density tester. The apparatus was set at 10 beats, then 500 beats and then 1000 beats, until the registered volume remains the same. Then, the final tapped volume (V_f) was registered. The Hausner ratio and the compressibility index was calculated by using the following formulas [37]:

$$\text{Hausner Ratio} = \frac{V_0}{V_f}$$

$$\text{Compressibility Index} = 100 \cdot \frac{V_0 - V_f}{V_f}$$

The interpretation of the results was made by using the European Pharmacopoeia's instructions, as follows: an excellent flow was characterized by a compressibility index between 1 and 10% and a Hausner ratio between 1.00 and 1.11, a good one by a compressibility index of 11–15% and a Hausner ratio of 1.12–1.18, a fair one by a compressibility index of 16–20% and a Hausner ratio of 1.19–1.25, a passable one by a compressibility index of 21–25% and a Hausner ratio of 1.26–1.34 and poor, very poor and very, very poor ones by compressibility indexes of 26–31, 32–37 and lower than 38, respectively and Hausner ratios of 1.35–1.45, 1.46–1.59 and below 1.60, respectively [37].

2.3.3. Phytochemical Extraction

A total of 500 mg of r.m. was weighed and added into a flask together with 10 mL of solvent (methanol: water: 0.12M hydrochloric acid = 70:29:1 (*V*/*V*/*V*)). The mixture was covered and put in an ultrasound bath at 40 °C for 30 min. After the time elapsed, the flask was cooled, and the mixture was filtered and brought to 10 mL into a volumetric flask with the same solvent. To determine the antioxidant activity of the r.m., the same extraction method was used by using methanol as the solvent [38].

2.3.4. Phytochemical Analysis

Total Polyphenolic Content (TPC) Assay

A total of 0.4 mL sample solution, 1 mL of Folin–Ciocalteu reagent, 15 mL of water, and 2 mL of a 290 g/L Na_2CO_3 solution were added into a test tube, shaken for 10 min, and kept at 40 °C for 20 min in a water bath. After cooling, the extinction was recorded at λ = 760 nm by using a Thermo Scientific Evolution 300 Spectrophotometer [38,39]. The calibration curve was linear for the range of 0.9–4.5 µg gallic acid/mL, the equation was y = 0.258x + 0.022 (R^2 = 0.998) where y = extinction at λ = 760 nm and x = concentration expressed as µg gallic acid/ mL, and the results were expressed as mg gallic acid equivalents (GAE)/g r.m. and the analysis was performed in triplicate.

HPLC-DAD Phenolic Profile Assessment

The HPLC-DAD method for the qualitative and quantitative assessment of the ten phenolic compounds analyzed was obtained by using methods already applied on plants and food supplements [38,40,41]. The analysis was carried out on an Agilent technologies 1200 series HPLC system equipped with a degasser, a quaternary pump, a diode array detector, a thermostatted column compartment, and an autosampler. The column that was used was Zorbax Eclipse Plus C18 (250 mm × 4.6 mm i.d. × 5 µm), which was kept at a temperature of 25 °C. The elution was performed by using three mobile phases in gradient (mobile phase A = purified water, mobile phase B = methanol, and mobile phase C = purified water: acetic acid = 96:4 *V*/*V*). The gradient program was used with 15% B and 85% C at 0 min, 75% A and 25% B at 15 min, 15% A and 85% B at 20 min, 40% A and 60% B at 40 min, 5% A and 95% B at 45 min, 5% A and 95% B at 55 min, 85% A and 15% B at 60 min, and 85% A and 15% B at 70 min. The flow rate program was 0.5 mL/min for the first 15 min and from minute 15 to minute 70, 0.8 mL/min and the injection volume was 5 µL. The detection was performed at 280 nm for gallic acid, (+)-catechin, syringic acid, and cinnamic acid, 303 nm for resveratrol, 330 nm for chlorogenic acid, caffeic acid, and ferulic acid, and 360 nm for rutin and quercetin. The results were expressed as µg phenolic compound/g r.m. and the analysis was performed in triplicate.

Antioxidant Activity Assay

The antioxidant activity assay was performed by using the 2,2-diphenyl-1-picrylhydrazyl free radical scavenging assay (RSA). A stock solution of 25 μg/mL DPPH in methanol was prepared and kept at a low temperature and in the dark for 2 h before use. A total of 970 μL of DPPH stock solution was added into 30 μL of sample solution. The absorbance was recorded at 517 nm, using a Thermo Scientific Evolution 300 Spectrophotometer. The results were expressed as percentages and the analysis was performed in triplicate [42,43]. The calibration curve was linear for the range of DPPH concentrations of 0.25–250 μg/mL. The equation based on the calibration curve was y = 0.0411x + 0.0349 (R^2 = 0.999), where y = extinction at λ = 593 nm and x = concentration expressed as μg DPPH/mL. The DPPH radical scavenging activity was determined by using the following formula:

$$\text{RSA (\%)} = \frac{C_0 - C_1}{C_0} \cdot 100$$

where: RSA = DPPH radical scavenging activity (%), C_0 = concentration of the DPPH stock solution (μg/mL), and C_1 = DPPH concentration in the sample (μg/mL).

2.4. *Encapsulation*

A total of 35 g of r.m. was filled into hard capsules size 00 and then the capsules were closed by locking the caps by using a pharmaceutical grade manual capsule filler.

2.5. *Analysis of Capsules*

2.5.1. Average Mass and Uniformity of Mass

Twenty full capsules were individually weighed and then they were opened without losing any part of the shell and the content was removed as completely as possible. The shell was weighed and the difference between the full capsule and the shell represents the weight of the capsule's content. The average mass was calculated by using the following formula:

$$\text{Average mass (mg/capsule)} = \frac{\sum_{i=1}^{n} M_i}{n}$$

where: M_i = the individual weighing of each capsule and n = the number of weight capsules [37].

In order for the capsules to comply to the requirements of the European Pharmacopoeia, the 10th edition, the average mass of the full capsules has to be 820.0 mg/capsule ± 7.5% (758.5–881.5 mg/capsule), the average mass of the content of the capsules has to be 700.0 mg/capsule± 7.5% (647.5–752.5 mg/capsule), and at least 18 capsules must have a deviation of at most ± 7.5% from the average mass and no more than 2 capsules can have a deviation of at most ± 15% from the average mass.

2.5.2. Disintegration

Six capsules were placed each in one of the 6 tubes of the basket of an Erweka ZT 72 Disintegratin Tester and a sensor disk was placed on top of each capsule. The determination was employed by using water as the immersion fluid at 37 ± 2 °C. The full disintegration of the capsules was observed. The test is considered finished after all the analyzed capsules are disintegrated. For the capsules to comply to the requirements of the European Pharmacopoeia, the 10th edition, the disintegration time must be less than 30 min for all the capsules. If 1 or 2 capsules are not completely disintegrated, the test will be repeated for 12 capsules. The test is compliant if 16 of 18 capsules are disintegrated in less than 30 min [37].

2.5.3. Uniformity of Content

To determine the uniformity of the content of phytochemicals within the obtained capsules, the same extraction and analysis as for the r.m. were performed as presented in Sections 2.3.3 and 2.3.4. The powder to be analyzed was the content of 20 capsules that was mixed properly, and the results were expressed as content of phytochemical per capsule [37].

3. Results

3.1. Raw Material analysis

3.1.1. Particle Size Distribution

From the analysis of the results obtained, it can be observed that the analyzed r.m. is a coarse powder, because 95.36% of the added r.m. passed through the 710 μm sieve and 0.37% through the 224 μm sieve (Table 1).

Table 1. R.m. particle size distribution.

Sieve (μm)	Weight (g)	Weight (%)
710	4.25	4.27
224	94.93	95.36
125	0.37	0.37
90	0.00	0.00
63	0.00	0.00

3.1.2. Powder Flow

The powder flow was good according to the obtained results (Table 2) that were interpreted by using Section 2.3.2.

Table 2. R.m powder flow assessment.

Number of Beats	Volume (mL)	Compressibility Index	Hausner Ratio
0	208	14.71%	1.17
10	184		
500	174		
1000	174		

This parameter shows the flowability of the r.m. and assesses the possibility of this powder to be encapsulated without any difficulties even if no excipients that improve the flowability were to be added.

3.1.3. Phytochemical Analysis

The analyzed r.m. presented 9.07 ± 0.25 mg GAE/g r.m. TPC and 48.32 ± 0.74% RSA. The analyzed phenolic compounds can be observed in Figure 1. The results were expressed as mean ± standard deviation (SD) for three determinations (n = 3).

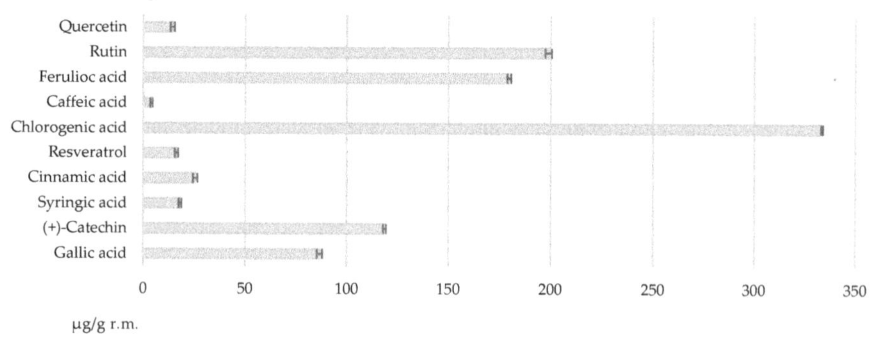

Figure 1. Quantification of phenolic compounds assessed for the r.m.

The highest quantity of phenolic compound determined was 333.7 ± 0.50 μg/g r.m. for chlorogenic acid. Values between 100 and 200 μg/g r.m. had rutin, ferulic acid, and (+)-catechin with 198.9 ± 1.60 μg/g r.m., 179.8 ± 0.90 μg/g r.m., and 118.7 ± 0.75 μg/g r.m., respectively. The values obtained for gallic acid, cinnamic acid, resveratrol, syringic acid, quercetin, and caffeic acid were beneath 100 μg/g r.m. (Figure 1).

The percentages calculated for the phenolic compounds analyzed were: 33.53% for chlorogenic acid, 19.98% for rutin, 18.06% for ferulic acid, 11.93 for (+)-catechin, 8.72% for gallic acid, 2.53% for cinnamic acid, 1.62% for resveratrol, 1.80% for syringic acid, 1.43% for quercetin, and 0.40% for caffeic acid.

3.2. *Analysis of Capsules*

3.2.1. Average Mass and Uniformity of Mass

Given that all the analyzed capsules fell into the ±7.5% deviation regarding the average mass and uniformity of mass (Table 3), it can be concluded that the analyzed capsules comply to the requirements of the European Pharmacopoeia.

Table 3. Average mass and uniformity of mass assessment.

Capsule No.	Full Capsule (mg)	Content (mg)
1	820	703
2	827	707
3	817	701
4	815	700
5	823	702
6	820	701
7	835	713
8	827	712
9	822	702
10	827	707
11	831	713
12	817	698
13	826	707
14	830	709
15	815	700
16	826	708
17	819	699
18	824	707
19	815	696
20	823	703
Average	822.95	704.4

3.2.2. Disintegration

The disintegration of the six capsules analyzed was performed in under 3 min (Table 4), thus the requirements of the European Pharmacopoeia regarding this analysis were met.

Table 4. The time elapsed for the disintegration of capsules.

Basket No.	1	2	3	4	5	6
Time (Minutes:Seconds)	2:10	2:03	1:20	2:13	2:35	1:59

3.2.3. Uniformity of Content

This test is performed to assess the uniformity of phytochemicals within the obtained capsules. The capsules analyzed presented 6.60 ± 0.44 mg GAE/capsule TPC. The highest quantity of phytochemical obtained was for chlorogenic acid, which had 218.33 ± 1.12 μg/capsule, followed by rutin that presented an amount of 121.03 ± 1.22 μg/capsule and then by ferulic acid with a quantity of 107.75 ± 0.79 μg/capsule. The rest of the analyzed phenolic compounds presented values that were below 100 μg/capsule (Figure 2). The results were expressed as mean ± SD for three determinations (n = 3).

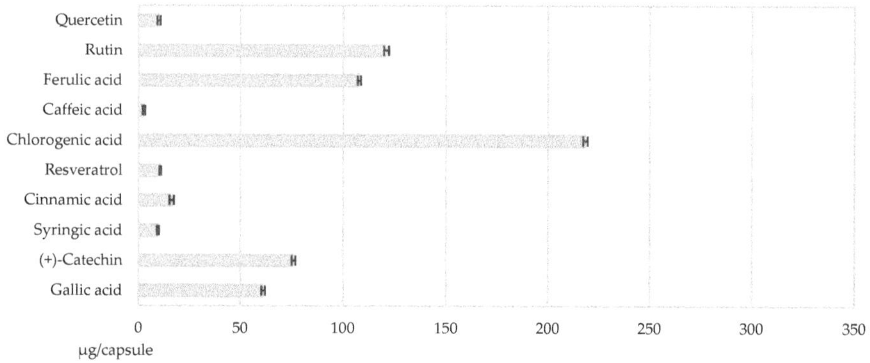

Figure 2. The uniformity of content in phenolic compounds of the capsules analyzed.

The percentages of the phenolic compounds analyzed per capsule were as follows: 34.51% chlorogenic acid, 19.13% rutin, 17.03% ferulic acid, 12.04% (+)-catechin, 9.66% gallic acid, 2.54% cinnamic acid, 1.65% resveratrol, 1.50% syringic acid, 1.55% quercetin, and 0.40% caffeic acid.

4. Discussion

Grape pomace is represented by peels, seeds, and stem fragments of the grapes used for industrial processes, mainly the vinification process [28,44,45]. It can be considered a burden for the environment, and it causes financial losses if it is not valorized [46,47]. Iqbal et al. (2021) and Antonic et al. (2020) state that grape pomace can be considered a source of dietary fiber, unsaturated fatty acids, mono- and polysaccharides, and a significant source of polyphenols that can be used as functional foods or ingredients that add value to food products [46,47], thus the valorization of this by-product can be beneficial to humanity and the environment as well. Ferreira et al. (2022) states that grape pomace can be used in the cosmetic industry as skin hydration products, in the food industry as functional foods or additives, in agriculture as fertilizers or animal feed, and in the pharmaceutical industry as dietary supplements or drug administration systems [48].

The tests performed on the raw material and the obtained capsules are mandatory for the development of any proper dietary supplement. These tests provide information regarding the efficiency of the proposed vegetal material as a dietary supplement [49,50].

Studies conducted before for the assessment of phenolic compounds in bilberries, red currants, and fermented red pomaces from the CS and FN cultivar reveal higher quantities than 400 μg (+)-catechin/g vegetal product dry weight (d.w.) for FN and CS pomaces and red currants. Rutin had the greatest amount in bilberries (greater than 350 μg/g d.w.) followed by CS pomace with a concentration greater than 100 μg/g d.w. [38,51]. Bilberries had amounts greater than 900 μg/g d.w. for chlorogenic acid and greater than 350 μg/g d.w. for ferulic acid and rutin. Quantities greater than 300 μg/g d.w. of syringic acid were determined for red currants and greater than 150 μg/g d.w. of gallic acid for bilberries and red currants. Caffeic acid and chlorogenic acid were not detected in both the analyzed pomaces, and quercetin, caffeic acid, and resveratrol in red currants [38,51]. The mixture of these vegetal products provides a wider range of phenolic compounds in higher quantities than each compound would have alone, thus providing a better intake of phenolic compounds to consumers. In most of the cases, antioxidants from multiple sources have proven superior to an individual compound [52].

The pathogenesis of chronic disease has often been related with OS [53–55]. Several molecules of major interest such as DNA, proteins, and lipids can be damaged by reactive oxygen species (ROS) and reactive nitrogen species (RNS) [56,57]. The mechanism of action of antioxidants has been extensively studied, and their beneficial effect against DNA methylation [58], lipid peroxidation [59], and protein damage [60] has been reported. Polyphenols and other antioxidants have successfully influenced the health status of patients diagnosed with a variety of chronic conditions, such as cardiovascular disease [61,62], cancer [63,64], endometriosis [65], osteoarthritis [66,67], Parkinson's [68], and Alzheimer's disease [69,70]. In most of the cases, antioxidants from multiple sources have proven superior to an individual compound [52,71].

In vitro, grape pomace displays high antioxidant potential [72], lowers PGE2 levels, and downregulates COX-2 gene expression [73]. Furthermore, grape pomace polyphenols have proven their efficacy in vivo, by delivering antiglycation agents and preventing the formation of toxic glycation end-products [74].

Even though polyphenols present antioxidant activity, with numerous benefits in preserving human health, the pro-oxidant effect of these compounds must be recognized [75–77]. When consumed in low or moderate concentrations, pro-oxidants can be considered as weapons for the defense system of the organism and if consumed in high doses, they have toxic potential by causing OS. On the other hand, the ability of pro-oxidants to produce cell apoptosis can be exploited in cancer therapy [76–78]. Thus, polyphenol-rich dietary supplements could offer health benefits to consumers when administered in appropriate doses [79].

During the technological development of our dietary supplement, we followed the high quality standard guidance of good manufacturing practices for pharmaceuticals, because two thirds of small molecule-active pharmaceutical ingredients are of natural origin [80]. The quality of the finished product is a sum of all the factors contributing to its manufacturing. Through proper documentation, traceability ensures high quality levels for each stage of the manufacturing process, starting with the raw materials and continuing with the manufacturing process and equipment, intermediary or bulk products, and analytical methods and equipment [81].

The pharmaco-technical methods used for the evaluation of the pharmaceutical form and raw materials were in accordance with the European Pharmacopoeia. This compendium standardizes a mechanical method for particle size distribution estimation by analytical sieving and recommends it for powders where the majority of the particles are larger than approximately 75 μm [37]. Modern methods for assessing the particle size of fine powder use particle size analyzers or light scattering spectrophotometers [82].

The dimensions of particles play a crucial role in the quality of the finished product on multiple levels. First, they affect the production process by determining flow properties [83]. Second, they impact disintegration and dissolution, influencing both bioavailability and pharmacodynamics [84]. The decision to continue the study without further reduction of raw material to nano-size particles was based on a series of factors: (i) the avoidance of a supplementary process stage; (ii) the negative impact fines have on rheological properties during manufacturing [83]; (iii) the cost efficiency of the formulation by avoiding unnecessary excipients for enhancing the flowability.

The results of this study can be rapidly employed to manufacturing and assessing pilot scale-up capsule batches. Having in mind quality assurance, it is recommended that the industrial batch extracts should be standardized in polyphenols, and the manufacturing process should be validated [85]. Such a dietary supplement would combine several benefits: a high acceptance by the population, a precise dosage and a stability of oral solid dosage, a production requiring regular manufacturing equipment, and compendial quality control methods. A good adherence to polyphenol-rich supplements ensures the bioavailability of these active compounds and constitutes the first stage towards the onset of their pharmacodynamic potential [86].

5. Conclusions

The valorization of pomace, a by-product of the vinification process, is of utmost importance for the preservation of the environment and for financial purposes. The obtained dietary supplement combines bilberries, red currants, and fermented red pomaces obtained from two cultivars, it complies to quality requirements, and it provides a wide range of phenolic compounds, thus providing an increase in the quality of life of the consumers by contributing to the prevention of chronic diseases.

Although polyphenol-rich dietary supplements could possess beneficial properties for human health, tests regarding their bioavailability, in vivo studies, and clinical trials that provide evidence of their therapeutic potential and toxicity remain to be assessed.

Author Contributions: Conceptualization, A.F., C.G. and F.G.G.; methodology, A.F., C.M.D., L.L.R. and C.M.; software, C.M.D., L.-I.V. and C.M.; validation, A.A.C., A.M.A., A.B. and L.G.V.; formal analysis, A.F., C.M.D. and L.L.R.; investigation, A.F.; resources, A.F., F.G.G. and C.G.; data curation, C.M.D., L.L.R., C.M. and A.M.A.; writing—original draft preparation, A.F., C.M.D. and L.-I.V.; writing—review and editing, A.F., C.M.D., L.-I.V., C.G., F.G.G., C.M. and L.G.V.; visualization, A.F., A.A.C., A.M.A. and A.B.; supervision, C.G. and O.T.; project administration, A.F.; funding acquisition, A.F. All authors have read and agreed to the published version of the manuscript.

Funding: Project financed by Lucian Blaga University of Sibiu & Hasso Plattner Foundation research grants LBUS-IRG-2019-05/No. 4105, 31 July 2019.

Institutional Review Board Statement: Not applicable.

Informed Consent Statement: Not applicable.

Data Availability Statement: Not applicable.

Conflicts of Interest: The authors declare no conflict of interest.

Abbreviations

CS	Cabernet Sauvignon
d.w.	dry weight
DPPH	2,2-Diphenyl-1-picrylhydrazyl
FN	Feteasca Neagra
GAE	gallic acid equivalents
OS	oxidative stress
r.m.	raw material
RNS	reactive nitrogen species
ROS	reactive oxygen species
RSA	radical scavenging assay
SD	standard deviation
TPC	total phenolic content
V_0	apparent volume
V_f	final volume

References

1. Phaniendra, A.; Jestadi, D.B.; Periyasamy, L. Free Radicals: Properties, Sources, Targets, and Their Implication in Various Diseases. *Indian J. Clin. Biochem.* **2015**, *30*, 11–26. [CrossRef]
2. Sharifi-Rad, M.; Anil Kumar, N.V.; Zucca, P.; Varoni, E.M.; Dini, L.; Panzarini, E.; Rajkovic, J.; Tsouh Fokou, P.V.; Azzini, E.; Peluso, I.; et al. Lifestyle, Oxidative Stress, and Antioxidants: Back and Forth in the Pathophysiology of Chronic Diseases. *Front. Physiol.* **2020**, *11*, 694. [CrossRef]
3. Ciulca, S.; Roma, G.; Alexa, E.; Radulov, I.; Cocan, I.; Madosa, E.; Ciulca, A. Variation of Polyphenol Content and Antioxidant Activity in Some Bilberry (*Vaccinium myrtillus* L.) Populations from Romania. *Agronomy* **2021**, *11*, 2557. [CrossRef]
4. Lobo, V.; Patil, A.; Phatak, A.; Chandra, N. Free Radicals, Antioxidants and Functional Foods: Impact on Human Health. *Pharmacogn. Rev.* **2010**, *4*, 118–126. [CrossRef] [PubMed]
5. Kaur, C.; Kapoor, H.C. Antioxidant Activity of Some Fruits in Indian Diet. *Acta Hortic.* **2005**, *696*, 563–565. [CrossRef]
6. Chan, S.W.; Tomlinson, B. Effects of Bilberry Supplementation on Metabolic and Cardiovascular Disease Risk. *Molecules* **2020**, *25*, 1653. [CrossRef]

7. Laczkó-zöld, E.; Komlósi, R.; Ülkei, T.; Fogarasi, E.; Croitoru, M.; Fülöp, I.; Domokos, E.; Ştefănescu, R.; Varga, E. Extractability of Polyphenols from Black Currant, Red Currant and Gooseberry and Their Antioxidant Activity. *Acta Biol. Hung.* **2018**, *69*, 156–169. [CrossRef] [PubMed]
8. Geybels, M.S.; Verhage, B.A.J.; Arts, I.C.W.; Van Schooten, F.J.; Alexandra Goldbohm, R.; Van Den Brandt, P.A. Dietary Flavonoid Intake, Black Tea Consumption, and Risk of Overall and Advanced Stage Prostate Cancer. *Am. J. Epidemiol.* **2013**, *177*, 1388–1398. [CrossRef]
9. Ghibu, S.; Craciun, C.E.; Rusu, R.; Morgovan, C.; Mogosan, C.; Rochette, L.; Gal, A.F.; Dronca, M. Impact of Alpha-Lipoic Acid Chronic Discontinuous Treatment in Cardiometabolic Disorders and Oxidative Stress Induced by Fructose Intake in Rats. *Antioxidants* **2019**, *8*, 636. [CrossRef]
10. Ghibu, S.; Decea, N.; Morgovan, C.; Mogosan, C. An Experimental Model to Induce Metabolic Syndrome in Rats. The Fructose-Enriched Diet. *Farmacia* **2013**, *61*, 420–426.
11. Aune, D.; Giovannucci, E.; Boffetta, P.; Fadnes, L.T.; Keum, N.N.; Norat, T.; Greenwood, D.C.; Riboli, E.; Vatten, L.J.; Tonstad, S. Fruit and Vegetable Intake and the Risk of Cardiovascular Disease, Total Cancer and All-Cause Mortality-A Systematic Review and Dose-Response Meta-Analysis of Prospective Studies. *Int. J. Epidemiol.* **2017**, *46*, 1029–1056. [CrossRef] [PubMed]
12. Pap, N.; Fidelis, M.; Azevedo, L.; do Carmo, M.A.V.; Wang, D.; Mocan, A.; Pereira, E.P.R.; Xavier-Santos, D.; Sant'Ana, A.S.; Yang, B.; et al. Berry Polyphenols and Human Health: Evidence of Antioxidant, Anti-Inflammatory, Microbiota Modulation, and Cell-Protecting Effects. *Curr. Opin. Food Sci.* **2021**, *42*, 167–186. [CrossRef]
13. Golovinskaia, O.; Wang, C.K. Review of Functional and Pharmacological Activities of Berries. *Molecules* **2021**, *26*, 3904. [CrossRef] [PubMed]
14. Jaakola, L.; Määttä-Riihinen, K.; Kärenlampi, S.; Hohtola, A. Activation of Flavonoid Biosynthesis by Solar Radiation in Bilberry (*Vaccinium myrtillus* L.) Leaves. *Planta* **2004**, *218*, 721–728. [CrossRef] [PubMed]
15. Pires, T.C.S.P.; Caleja, C.; Santos-Buelga, C.; Barros, L.; Ferreira, I.C.F.R. *Vaccinium myrtillus* L. Fruits as a Novel Source of Phenolic Compounds with Health Benefits and Industrial Applications—A Review. *Curr. Pharm. Des.* **2020**, *26*, 1917–1928. [CrossRef]
16. Graff, A.; Petrone, C.; Associate, R.; Ba, D.S.; Upton, R.; Pizzorno, J.; Myrtillus, V. American Herbal Pharmacopoeia^{TM} and Herapeutic Ompendium. *Bilberry* Fruit. 2001. Available online: http://naturamedica.lt/wp-content/uploads/2015/07/AHP-Bilberry.pdf (accessed on 29 June 2022).
17. Chu, W.K.; Cheung, S.C.M.; Lau, R.A.W. Bilberry (*Vaccinium myrtillus* L.). *Herb. Med. Biomol. Clin. Asp.* **2011**, *20115386*, 55–71. [CrossRef]
18. Šumić, Z.; Vakula, A.; Tepić, A.; Čakarević, J.; Vitas, J.; Pavlić, B. Modeling and Optimization of Red Currants Vacuum Drying Process by Response Surface Methodology (RSM). *Food Chem.* **2016**, *203*, 465–475. [CrossRef] [PubMed]
19. Nour, V.; Trandafir, I.; Ionica, M.E. Ascorbic Acid, Anthocyanins, Organic Acids and Mineral Content of Some Black and Red Currant Cultivars. *Fruits* **2011**, *66*, 353–362. [CrossRef]
20. Zdunić, G.; Šavikin, K.; Pljevljakušić, D.; Djordjević, B. *Black (Ribes nigrum L.) and Red Currant (Ribes rubrum L.) Cultivars*; Elsevier Inc.: Amsterdam, The Netherlands, 2015; ISBN 9780124081178.
21. Vesela Shopska, A.; Georgieva, R.; Nedyalkov, P.; Shopska, V.; Kaneva, M. Effect of Blueberries Addition during Beer Maturation on Yeast Metabolism. *Food Sci. Appl. Biotechnol.* **2021**, *4*, 105–110. [CrossRef]
22. Nedyalkov, P.; Bakardzhiyski, I.; Dinkova, R.; Shopska, V.; Kaneva, M. Influence of the Time of Bilberry (*Vaccinium myrtillus* L.) Addition on the Phenolic and Protein Profile of Beer. *Acta Sci. Pol. Technol. Aliment.* **2022**, *21*, 5–15. [CrossRef] [PubMed]
23. Olas, B. Berry Phenolic Antioxidants—Implications for Human Health? *Front. Pharmacol.* **2018**, *9*, 78. [CrossRef] [PubMed]
24. Fontana, A.R.; Antoniolli, A.; Bottini, R. Grape Pomace as a Sustainable Source of Bioactive Compounds: Extraction, Characterization, and Biotechnological Applications of Phenolics. *J. Agric. Food Chem.* **2013**, *61*, 8987–9003. [CrossRef] [PubMed]
25. Grassi, F.; De Lorenzis, G. Back to the Origins: Background and Perspectives of Grapevine Domestication. *Int. J. Mol. Sci.* **2021**, *22*, 4518. [CrossRef] [PubMed]
26. Sirohi, R.; Tarafdar, A.; Singh, S.; Negi, T.; Gaur, V.K.; Gnansounou, E.; Bharathiraja, B. Green Processing and Biotechnological Potential of Grape Pomace: Current Trends and Opportunities for Sustainable Biorefinery. *Bioresour. Technol.* **2020**, *314*, 123771. [CrossRef] [PubMed]
27. Ferri, M.; Bin, S.; Vallini, V.; Fava, F.; Michelini, E.; Roda, A.; Minnucci, G.; Bucchi, G.; Tassoni, A. Recovery of Polyphenols from Red Grape Pomace and Assessment of Their Antioxidant and Anti-Cholesterol Activities. *N. Biotechnol.* **2016**, *33*, 338–344. [CrossRef] [PubMed]
28. García-Lomillo, J.; González-SanJosé, M.L. Applications of Wine Pomace in the Food Industry: Approaches and Functions. *Compr. Rev. Food Sci. Food Saf.* **2017**, *16*, 3–22. [CrossRef]
29. Theagarajan, R.; Malur Narayanaswamy, L.; Dutta, S.; Moses, J.A.; Chinnaswamy, A. Valorisation of Grape Pomace (Cv. Muscat) for Development of Functional Cookies. *Int. J. Food Sci. Technol.* **2019**, *54*, 1299–1305. [CrossRef]
30. Chowdhary, P.; Gupta, A.; Gnansounou, E.; Pandey, A.; Chaturvedi, P. Current Trends and Possibilities for Exploitation of Grape Pomace as a Potential Source for Value Addition. *Environ. Pollut.* **2021**, *278*, 116796. [CrossRef]
31. Ping, L.; Pizzi, A.; Guo, Z.D.; Brosse, N. Condensed Tannins Extraction from Grape Pomace: Characterization and Utilization as Wood Adhesives for Wood Particleboard. *Ind. Crops Prod.* **2011**, *34*, 907–914. [CrossRef]

32. Averilla, J.N.; Oh, J.; Kim, H.J.; Kim, J.S.; Kim, J.S. Potential Health Benefits of Phenolic Compounds in Grape Processing By-Products. *Food Sci. Biotechnol.* **2019**, *28*, 1607–1615. [CrossRef]
33. Venkitasamy, C.; Zhao, L.; Zhang, R.; Pan, Z. *Grapes*; Elsevier Inc.: Amsterdam, The Netherlands, 2019; ISBN 9780128141397.
34. Directive 2002/46/EC of the European Parliament and of the Council of 10 June 2002 on the Approximation of the Laws of the Member States Relating to Food Supplements. *Off. J. Eur. Communities* **2002**, *183*, 51–57.
35. Morgovan, C.; Ghibu, S.; Juncan, A.M.; Rus, L.L.; Butucă, A.; Vonica, L.; Muntean, A.; Moş, L.; Gligor, F.; Olah, N.K. Nutrivigilance: A New Activity in the Field of Dietary Supplements. *Farmacia* **2019**, *67*, 537–544. [CrossRef]
36. Chiba, T.; Sato, Y.; Kobayashi, E.; Ide, K.; Yamada, H.; Umegaki, K. Behaviors of Consumers, Physicians and Pharmacists in Response to Adverse Events Associated with Dietary Supplement Use. *Nutr. J.* **2017**, *16*, 18. [CrossRef] [PubMed]
37. Council of Europe. European Pharmacopoeia 10th Edition | EDQM—European Directorate for the Quality of Medicines. Available online: https://www.edqm.eu/en/european-pharmacopoeia-ph-eur-10th-edition- (accessed on 29 June 2022).
38. Frum, A.; Georgescu, C.; Gligor, F.G.; Stegarus, D.I.; Dobrea, C.M.; Tita, O.; Blaga, L.; Sciences, A.; Industry, F.; Street, I.R.; et al. Identification and Quantification of Phenolic Compounds from Red Grape Pomace. *Sci. Study Res. Chem. Chem. Eng. Biotechnol. Food Ind.* **2018**, *19*, 45–52.
39. Georgescu, C.; Bratu, I. Tamas M Studiul Unor Polifenoli Din Rhododendron Kotschyi. *Rev. Chim.* **2005**, *56*, 779–780.
40. Gligor, F.G.; Frum, A.; Vicaș, L.G.; Totan, M.; Roman-Filip, C.; Dobrea, C.M. Determination of a Mixture of *Plantago lanceolata* L. and *Salvia officinalis* L. by High-Performance Liquid Chromatography with Ultraviolet Detection (HPLC-UV). *Anal. Lett.* **2020**, *53*, 1391–1406. [CrossRef]
41. Craciun, V.I.; Gligor, F.G.; Juncan, A.M.; Chis, A.A.; Rus, L.L. A New, Rapid and Efficient HPLC Method to Assay Resveratrol in Food Supplements. *Rev. Chim.* **2019**, *70*, 3202–3205. [CrossRef]
42. Popovici, C.; Saykova, I.; Tylkowski, B. Evaluation de l'activité Antioxydant Des Composés Phénoliques Par La Réactivité Avec Le Radical Libre DPPH. *Rev. Génie Ind.* **2009**, *4*, 25–39.
43. Tița, O.; Constantinescu, M.A.; Tița, M.A.; Georgescu, C. Use of Yoghurt Enhanced with Volatile Plant Oils Encapsulated in Sodium Alginate to Increase the Human Body's Immunity in the Present Fight Against Stress. *Int. J. Environ. Res. Public Health* **2020**, *17*, 7588. [CrossRef]
44. Bordiga, M.; Travaglia, F.; Locatelli, M.; Arlorio, M.; Coïsson, J.D. Spent Grape Pomace as a Still Potential By-Product. *Int. J. Food Sci. Technol.* **2015**, *50*, 2022–2031. [CrossRef]
45. Spinei, M.; Oroian, M. The Potential of Grape Pomace Varieties as a Dietary Source of Pectic Substances. *Foods* **2021**, *10*, 867. [CrossRef] [PubMed]
46. Iqbal, A.; Schulz, P.; Rizvi, S.S.H. Valorization of Bioactive Compounds in Fruit Pomace from Agro-Fruit Industries: Present Insights and Future Challenges. *Food Biosci.* **2021**, *44*, 101384. [CrossRef]
47. Antonić, B.; Jančíková, S.; Dordević, D.; Tremlová, B. Grape Pomace Valorization: A Systematic Review and Meta-Analysis. *Foods* **2020**, *9*, 1627. [CrossRef] [PubMed]
48. Ferreira, S.M.; Santos, L.; Serratosa, M.P.; Angeles, M.; Santos, V.; Morikawa, T.; Ferreira, S.M.; Santos, L. A Potential Valorization Strategy of Wine Industry By-Products and Their Application in Cosmetics—Case Study: Grape Pomace and Grapeseed. *Molecules* **2022**, *27*, 969. [CrossRef]
49. Gligor, F.G.; Dobrea, C.M.; Georgescu, C.; Vonica Gligor, L.A.; Frum, A.; Totan, M. Silymarin Food Supplements—Oral Solid Dosage Forms. *Sci. Study Res. Chem. Chem. Eng. Biotechnol. Food Ind.* **2016**, *17*, 349–354.
50. Mititelu, M.; Moroșan, E.; Nicoară, A.C.; Secăreanu, A.A.; Musuc, A.M.; Atkinson, I.; Cusu, J.P.; Nițulescu, G.M.; Ozon, E.A.; Sarbu, I.; et al. Development of Immediate Release Tablets Containing Calcium Lactate Synthetized from Black Sea Mussel Shells. *Mar. Drugs* **2022**, *20*, 45. [CrossRef]
51. Frum, A.; Lengyel, E.; Georgescu, C.; Gligor, F.; Tita, O. Analysis of Phenolic Compounds Extracted from Three Types of Berries. In Proceedings of the 6th BIOATLAS Conference, Brașov, Romania, 27–28 May 2016; Volume 12, pp. 5–9.
52. Costa, C.; Tsatsakis, A.; Mamoulakis, C.; Teodoro, M.; Briguglio, G.; Caruso, E.; Tsoukalas, D.; Margina, D.; Dardiotis, E.; Kouretas, D.; et al. Current Evidence on the Effect of Dietary Polyphenols Intake on Chronic Diseases. *Food Chem. Toxicol.* **2017**, *110*, 286–299. [CrossRef]
53. Liguori, I.; Russo, G.; Curcio, F.; Bulli, G.; Aran, L.; Della-Morte, D.; Gargiulo, G.; Testa, G.; Cacciatore, F.; Bonaduce, D.; et al. Oxidative Stress, Aging, and Diseases. *Clin. Interv. Aging* **2018**, *13*, 757–772. [CrossRef]
54. García-Sánchez, A.; Miranda-Díaz, A.G.; Cardona-Muñoz, E.G. The Role of Oxidative Stress in Physiopathology and Pharmacological Treatment with Pro- and Antioxidant Properties in Chronic Diseases. *Oxid. Med. Cell. Longev.* **2020**, *2020*, 2082145. [CrossRef]
55. Lourenço, S.C.; Moldão-Martins, M.; Alves, V.D. Antioxidants of Natural Plant Origins: From Sources to Food Industry Applications. *Molecules* **2019**, *24*, 4132. [CrossRef]
56. Daenen, K.; Andries, A.; Mekahli, D.; Van Schepdael, A.; Jouret, F.; Bammens, B. Oxidative Stress in Chronic Kidney Disease. *Pediatr. Nephrol.* **2019**, *34*, 975–991. [CrossRef] [PubMed]
57. Di Meo, S.; Reed, T.T.; Venditti, P.; Victor, V.M. Role of ROS and RNS Sources in Physiological and Pathological Conditions. *Oxid. Med. Cell. Longev.* **2016**, *2016*, 1245049. [CrossRef] [PubMed]
58. Beetch, M.; Harandi-Zadeh, S.; Shen, K.; Lubecka, K.; Kitts, D.D.; O'Hagan, H.M.; Stefanska, B. Dietary Antioxidants Remodel DNA Methylation Patterns in Chronic Disease. *Br. J. Pharmacol.* **2020**, *177*, 1382–1408. [CrossRef] [PubMed]

59. Cicero, A.F.G.; Colletti, A. Polyphenols Effect on Circulating Lipids and Lipoproteins: From Biochemistry to Clinical Evidence. *Curr. Pharm. Des.* **2018**, *24*, 178–190. [CrossRef] [PubMed]
60. Ali, S.S.; Ahsan, H.; Zia, M.K.; Siddiqui, T.; Khan, F.H. Understanding Oxidants and Antioxidants: Classical Team with New Players. *J. Food Biochem.* **2020**, *44*, e13145. [CrossRef]
61. Leri, M.; Scuto, M.; Ontario, M.L.; Calabrese, V.; Calabrese, E.J.; Bucciantini, M.; Stefani, M. Healthy Effects of Plant Polyphenols: Molecular Mechanisms. *Int. J. Mol. Sci.* **2020**, *21*, 1250. [CrossRef] [PubMed]
62. Quiñones, M.; Miguel, M.; Aleixandre, A. Beneficial Effects of Polyphenols on Cardiovascular Disease. *Pharmacol. Res.* **2013**, *68*, 125–131. [CrossRef]
63. Forman, H.J.; Zhang, H. Targeting Oxidative Stress in Disease: Promise and Limitations of Antioxidant Therapy. *Nat. Rev. Drug Discov.* **2021**, *20*, 689–709. [CrossRef]
64. Briguglio, G.; Costa, C.; Pollicino, M.; Giambò, F.; Catania, S.; Fenga, C. Polyphenols in Cancer Prevention: New Insights (Review). *Int. J. Funct. Nutr.* **2020**, *1*, 1–11. [CrossRef]
65. Meresman, G.F.; Götte, M.; Laschke, M.W. Plants as Source of New Therapies for Endometriosis: A Review of Preclinical and Clinical Studies. *Hum. Reprod. Update* **2021**, *27*, 367–392. [CrossRef]
66. Ansari, M.Y.; Ahmad, N.; Haqqi, T.M. Oxidative Stress and Inflammation in Osteoarthritis Pathogenesis: Role of Polyphenols. *Biomed. Pharmacother.* **2020**, *129*, 110452. [CrossRef] [PubMed]
67. Valsamidou, E.; Gioxari, A.; Amerikanou, C.; Zoumpoulakis, P.; Skarpas, G.; Kaliora, A.C. Dietary Interventions with Polyphenols in Osteoarthritis: A Systematic Review Directed from the Preclinical Data to Randomized Clinical Studies. *Nutrients* **2021**, *13*, 1420. [CrossRef] [PubMed]
68. Singh, A.; Tripathi, P.; Yadawa, A.K.; Singh, S. Promising Polyphenols in Parkinson's Disease Therapeutics. *Neurochem. Res.* **2020**, *45*, 1731–1745. [CrossRef] [PubMed]
69. Szczechowiak, K.; Diniz, B.S.; Leszek, J. Diet and Alzheimer's Dementia—Nutritional Approach to Modulate Inflammation. *Pharmacol. Biochem. Behav.* **2019**, *184*, 172743. [CrossRef] [PubMed]
70. Sathya, S.; Devi, K.P. The Use of Polyphenols for the Treatment of Alzheimer's Disease. In *Role of the Mediterranean Diet in the Brain and Neurodegerative Diseases*; Farooqui, T., Farooqui, A.A., Eds.; Academic Press: Cambridge, MA, USA, 2018; pp. 239–252. [CrossRef]
71. Qin, X.; Lu, Y.; Peng, Z.; Fan, S.; Yao, Y. Systematic Chemical Analysis Approach Reveals Superior Antioxidant Capacity via the Synergistic Effect of Flavonoid Compounds in Red Vegetative Tissues. *Front. Chem.* **2018**, *6*, 9. [CrossRef] [PubMed]
72. Mollica, A.; Scioli, G.; Valle, A.D.; Cichelli, A.; Novellino, E.; Bauer, M.; Kamysz, W.; Llorent-Martínez, E.J.; Córdova, M.L.F.; Castillo-López, R.; et al. Phenolic Analysis and In Vitro Biological Activity of Red Wine, Pomace and Grape Seeds Oil Derived from Vitis Vinifera l. Cv. Montepulciano d'abruzzo. *Antioxidants* **2021**, *10*, 1704. [CrossRef] [PubMed]
73. Chiavaroli, A.; Balaha, M.; Acquaviva, A.; Ferrante, C.; Cataldi, A.; Menghini, L.; Rapino, M.; Orlando, G.; Brunetti, L.; Leone, S.; et al. Phenolic Characterization and Neuroprotective Properties of Grape Pomace Extracts. *Molecules* **2021**, *26*, 6216. [CrossRef] [PubMed]
74. Sri Harsha, P.S.C.; Lavelli, V. Use of Grape Pomace Phenolics to Counteract Endogenous and Exogenous Formation of Advanced Glycation End-Products. *Nutrients* **2019**, *11*, 1917. [CrossRef]
75. León-González, A.J.; Auger, C.; Schini-Kerth, V.B. Pro-Oxidant Activity of Polyphenols and Its Implication on Cancer Chemoprevention and Chemotherapy. *Biochem. Pharmacol.* **2015**, *98*, 371–380. [CrossRef] [PubMed]
76. Seck, I.; Hosu, A.; Cimpoiu, C.; Ndoye, S.F.; Ba, L.A.; Sall, C.; Seck, M. Phytochemicals Content, Screening and Antioxidant/pro-Oxidant Activities of Carapa Procera (Barks) (Meliaceae). *S. Afr. J. Bot.* **2021**, *137*, 369–376. [CrossRef]
77. Dlamini, L.M.; Tata, C.M.; Djuidje, M.C.F.; Ikhile, M.I.; Nikolova, G.D.; Karamalakova, Y.D.; Gadjeva, V.G.; Zheleva, A.M.; Njobeh, P.B.; Ndinteh, D.T. Antioxidant and Prooxidant Effects of Piptadeniastrum Africanum as the Possible Rationale behind Its Broad Scale Application in African Ethnomedicine. *J. Ethnopharmacol.* **2019**, *231*, 429–437. [CrossRef] [PubMed]
78. Fernando, W.; Rupasinghe, H.P.V.; Hoskin, D.W. Dietary Phytochemicals with Anti-Oxidant and pro-Oxidant Activities: A Double-Edged Sword in Relation to Adjuvant Chemotherapy and Radiotherapy? *Cancer Lett.* **2019**, *452*, 168–177. [CrossRef] [PubMed]
79. Martin, K.R.; Appel, C.L. Polyphenols as Dietary Supplements: A Double-Edged Sword. *Nutr. Diet. Suppl.* **2010**, *2*, 1–12. [CrossRef]
80. Canning, A.D.; Death, R.G.; Waltham, N.J. Pharmaceutical Companies Should Pay for Raiding Nature's Medicine Cabinet. *Lancet* **2021**, *398*, 840–841. [CrossRef]
81. European Medicines Agency. Good Manufacturing Practice. 2018. Available online: https://www.ema.europa.eu/en/human-regulatory/research-development/compliance/good-manufacturing-practice (accessed on 21 June 2022).
82. Adnan, M.; Azad, M.O.K.; Ju, H.S.; Son, J.M.; Park, C.H.; Shin, M.H.; Alle, M.; Cho, D.H. Development of Biopolymer-Mediated Nanocomposites Using Hot-Melt Extrusion to Enhance the Bio-Accessibility and Antioxidant Capacity of Kenaf Seed Flour. *Appl. Nanosci.* **2020**, *10*, 1305–1317. [CrossRef]
83. Tang, X.; Zakhvatayeva, A.; Zhang, L.; Wu, Z.F.; Sun, P.; Wu, C.Y. Flow Behaviour of Pharmaceutical Powders during Rotary Die Filling with a Paddle Feeder. *Int. J. Pharm.* **2020**, *585*, 119547. [CrossRef] [PubMed]
84. Ngeacharernkul, P.; Stamatis, S.D.; Kirsch, L.E. Particle Size Distribution Equivalency as Novel Predictors for Bioequivalence. *AAPS PharmSciTech* **2018**, *19*, 2787–2800. [CrossRef]

85. European Medicines Agency. Guideline on Process Validation for Finished Products-Information and Data to be Provided in Regulatory Submissions. 2016. Available online: https://www.ema.europa.eu/en/documents/scientific-guideline/guideline-process-validation-finished-products-information-data-be-provided-regulatory-submissions_en.pdf (accessed on 21 June 2022).
86. Di Lorenzo, C.; Colombo, F.; Biella, S.; Stockley, C.; Restani, P. Polyphenols and Human Health: The Role of Bioavailability. *Nutrients* **2021**, *13*, 273. [CrossRef] [PubMed]

nutrients

MDPI

Article

Ginsenoside Rh4 Suppresses Metastasis of Esophageal Cancer and Expression of c-Myc via Targeting the Wnt/β-Catenin Signaling Pathway

Jun Chen [1,2,3], Zhiguang Duan [1,2,3], Yannan Liu [1,2,3], Rongzhan Fu [1,2,3] and Chenhui Zhu [1,2,3,*]

1 Shaanxi Key Laboratory of Degradable Biomedical Materials, School of Chemical Engineering, Northwest University, 229 North Taibai Road, Xi'an 710069, China; cehnjun151@163.com (J.C.); duanzhiguang@nwu.edu.cn (Z.D.); liuyannan@nwu.edu.cn (Y.L.); furongzhan@nwu.edu.cn (R.F.)

2 Shaanxi R&D Center of Biomaterials and Fermentation Engineering, School of Chemical Engineering, Northwest University, 229 North Taibai Road, Xi'an 710069, China

3 Biotech & Biomed Research Institute, Northwest University, 229 North Taibai Road, Xi'an 710069, China

* Correspondence: zch2005@nwu.edu.cn; Tel./Fax: +86-29-8830-5118

Abstract: The metastasis of esophageal squamous cell carcinoma (ESCC) is a leading cause of death worldwide, however, it has a poor prognosis. Ginsenoside Rh4 is a rare saponin that has been shown to have potential antitumor effectiveness in ESCC. However, the utility of Rh4 in ESCC metastasis and its undiscovered mode of action has not yet been explored. In this study, we found that Rh4 could inhibit ESCC metastasis by regulating the Wnt/β-catenin signaling pathway and the level of c-Myc, which is an important transcription factor in cancer. In in vitro experiments, Rh4 could inhibit the migration and invasion of ESCC cells without affecting cell viability. In in vivo experiments, Rh4 restrained ESCC metastasis to the lymph nodes and lungs via the suppression of epithelial-mesenchymal transition (EMT). The Wnt agonist HLY78 promoted EMT and migration of ESCC cells, whereas treatment of Rh4 can attenuate the promotion effect of HLY78. The siRNA knocking out c-Myc can also significantly reduce the expression of EMT-related marker proteins. This study illustrates a new concept for further research on the mechanism of Rh4 in ESCC.

Keywords: metastasis; ginsenoside Rh4; ESCC; Wnt/β-catenin; c-Myc

Citation: Chen, J.; Duan, Z.; Liu, Y.; Fu, R.; Zhu, C. Ginsenoside Rh4 Suppresses Metastasis of Esophageal Cancer and Expression of c-Myc via Targeting the Wnt/β-Catenin Signaling Pathway. *Nutrients* **2022**, *14*, 3042. https://doi.org/10.3390/nu14153042

Academic Editor: Khalid A. El Sayed

Received: 6 June 2022
Accepted: 17 July 2022
Published: 25 July 2022

1. Introduction

Esophageal cancer, as one of the most invasive cancers, has the seventh greatest prevalence worldwide with more than 540,000 cases and ranks sixth in the world for mortality rates [1]. Esophageal squamous cell carcinoma (ESCC) is the most common type of esophageal cancer, comprising more than 90% of cases, and it has a higher incidence in Asia [2]. Due to the lack of effective markers in esophageal cancer screening, most patients are already in stages II or III when esophageal cancer is identified [3]. A large number of patients with primary esophageal cancer have been found to have lymph node metastasis at the first diagnosis [4]. Tumor metastasis endangers the life of patients even after effective treatment [5,6]. Therefore, finding molecular markers and therapeutic intervention targets involved in esophageal cancer metastasis is crucial for guiding treatment and improving the survival rate of esophageal cancer patients.

Recently, research has discerned that the Wnt/β-catenin signal transduction pathway is abnormally activated in cancer stem cells in breast cancer, liver cancer, colon cancer, and other tumor tissues [7]. The Wnt/β-catenin signaling pathway is involved in many cellular activities, consisting of apoptosis, differentiation, senescence, invasion, migration, and epithelial to mesenchymal transition (EMT) [8–10]. When the Wnt signal is activated, the increased content of β-catenin enters the cell nucleus to deactivate other proteins, such as vimentin, metal matrix protein MMP-2, etc. [11]. Increasing evidence shows that the

Wnt/β-catenin signaling pathway is the main oncogenic pathway of ESCC [12]. Wnt signaling induces EZH2 binding to β-catenin, regulating of the epitranscriptome, including c-Myc [13]. c-Myc is a gene that is highly correlated with cancer, it is involved in tumor initiation, growth, and progression. Moreover, c-Myc contributes to angiogenesis, invasion, and migration [14,15].

EMT is a crucial factor affecting the development and progression of ESCC [16]. Its main characteristic is its ability to decrease the utterances of cell adhesion molecules (such as E-cadherin) and there is also a conversion in the cytoskeleton, closely linked to the expression of vimentin, and the morphological characteristics of mesenchymal cells [17]. Through EMT, converted epithelial cells acquire higher motility, along with MMP2 and MMP9, which are involved in degrading the extracellular matrix in this process [18].

Ginseng is a rare perennial herb and is cyclopedically used in traditional Chinese medicine. Ginsenosides play a crucial part in ginseng, which have antitumor [19], anti-inflammatory [20,21], anti-obesity [22], antidiabetic [23], neuroprotection [24] and immune enhancement effects [25]. Ginsenoside Rh4 is a kind of tetracyclic triterpene saponins and it is composed of triterpene aglycones and glycosides [26]. Previous studies have shown that ginsenoside Rh4 inhibits glycolysis by targeting AKT, which is involved in the suppression of the cell cycle, thereby is effectual in restraining the proliferation of esophageal cancer [27]. We have also previously reported that ginsenoside CK can curb the metastasis of liver cancer [28]. However, whether Rh4 is inhibitory to the metastasis of ESCC remains unknown.

In the study, we evaluated the ability of ginsenoside Rh4 at non-cytotoxic concentrations to inhibit migration, invasion, and EMT of ESCC in vitro and in vivo. Our findings suggested that Rh4 restrains the migration and invasion of ESCC as well as its metastasis to the lymph nodes and lungs through the Wnt/β-catenin pathway-regulated EMT process. Our research shows that Rh4 might be a potential drug for ESCC metastasis measurement.

2. Materials and Methods

2.1. Cell Culture

The ESCC cell lines KYSE30, KYSE150, and KYSE410 were provided by the Shanghai Institute of Cell Biology. The above cell lines were cultured in a mixed RPMI-1640 medium, containing 10% FBS and 1% penicillin and streptomycin in an incubator at 37 °C with 5% CO_2.

2.2. Antibodies and Reagents

Ginsenoside Rh4 (Figure 1A) (purity $\geq$ 99%) was provided by the Northwestern University Institute of Biomedical Sciences (Xi'an, China). Dimethyl sulfoxide (DMSO) was obtained from Aladdin Biotechnology (Shanghai, China) and methylthiazolyldiphenyl tetra-zolium bromide (MTT) and BCA protein assay reagent kits and phosphatase inhibitor cocktails were obtained from Solarbio Science & Technology Co., Ltd. (Beijing, China). Trypsin, penicillin and streptomycin, crystal violet, and protease inhibitor cocktails were also purchased from Solarbio. Rabbit antibodies against E-cadherin, snail, MMP2, MMP9, Wnt, β-catenin, and c-Myc and mouse antibodies against N-cadherin and Vimentin were obtained from Proteintech Group, Inc (Chicago, MI, USA). Anti-phospho-β-catenin rabbit antibodies, primary antibodies against β-actin, goat anti-mouse IgG, and goat anti-rabbit IgG were purchased from Cell Signaling Technology (Boston, MA, USA). The ratio of related antibodies has been shown (Table S1).

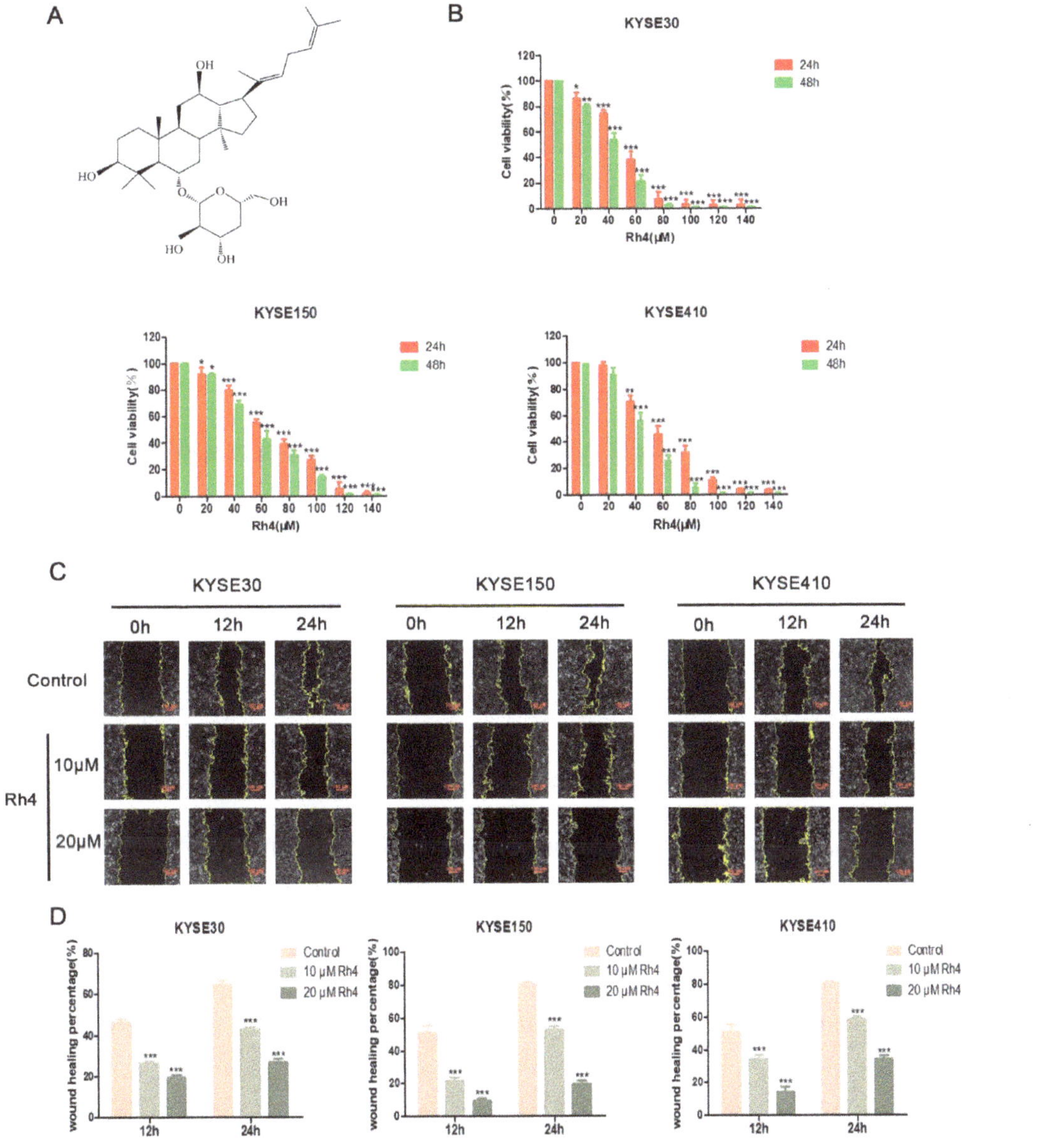

Figure 1. (**A**)The chemical structure of ginsenoside Rh4. (**B**) KYSE30, KYSE150, and KYSE410 were measured with designated content (0–140 μM) of Rh4 cell viability (**C**) Migration images of Rh4 cells with or without the addition of Rh4 cells at 0, 12, and 24 h. The yellow line is the migration edge. Scale bar = 100 μm (**D**) Percent wound healing, treatment group vs. control group. All data are presented as mean ± SD, * $p < 0.05$, ** $p < 0.01$, and *** $p < 0.001$.

2.3. Cell Viability Assays

Cell proliferation was quantified by MTT assays. KYSE30, KYSE150, and KYSE410 cells were seeded onto 1×10^4 cells per well in a 96-well plate. After 24 h, designated destines of Rh4 (0–140 μM) were also placed inside. After the cells were attached, they were measured with Rh4 (0–140 μM) for 24 or 48 h; then, we incubated them with a 50 μL MTT (5 mg/mL) solution for 2–4 h. MTT was then aspirated and replaced with 150 μL of

DMSO. The microplate reader (Power Wave XS2, Bio-Tek Instruments Inc., Burlington, VE, USA) was used to measure absorbance at 490 nm.

2.4. In Vitro Scratch Assay

We took the KYSE30, KYSE150, and KYSE410 cells in the logarithmic growth phase and inoculated them on a 6-well plate with 1×10^6 cells per well. They were cultivated in an incubator for 24 h. The liter pipette was scratched in the well plate in a three-horizontal and two-vertical manner and the suspended cells were washed with PBS and taken under a microscope for recording. Then, processed cells were measured with ginsenoside Rh4 serum-free medium at a designated content of 0, 10, and 20 μM for 24 h; we took pictures in the same position, with at least four pictures of each area. Cell migration efficiency was assessed by the wound healing rate.

The wound healing rate was calculated by Image J.

2.5. Transwell Assays

Migration and invasion assays were assayed using 24-well plates (Corning Incorporated, Corning, NY, USA). In migration experiments, KYSE30, KYSE150, and KYSE410 cells (2×105 cells per well) were seeded in the upper compartment of the Transwell chamber with a serum-free conditioned medium (Millipore Corporation, Billerica, MA, USA). A total of 20% FBS medium (600 μL) was added to the lower chamber. Cells were cultured in the presence or absence of Rh4 (10 μM and 20 μM) for 24 h. After incubation, the methanol-fixed filters were stained with crystal violet (0.1%). Then, we took pictures and counted in three randomly selected areas.

The method of the cell invasion assay differs from that of the cell migration assay in that the upper chamber needs to be precoated with 50 μL of Matrigel (Corning, NY, USA).

2.6. Quantitative Real-Time PCR (qRT-PCR)

Rh4-treated cells were lysed using TRIzol reagent. They were reverse transcribed with QuantScript RT kit (Tiangen, Beijing, China) and then reverse transcribed using PowerUp SYBR Green Master Mix (Applied Biosystems) on an ABI StepOnePlus Real-Time PCR System (Thermo Fisher Scientific, qRT-PCR was performed in the USA). The $2^{-\Delta\Delta}$Ct method was used to evaluate the relative expression.

The primer sequences used were as follows:

E-cadherin (5′-GACGCCATCAACACCGAGTT-3′; 5′-AAATTGCCAGGCTCAATGAC-3′), N-cadherin (5′-GGTGGAGGAGAAGAAGACCAG-3′; 5′-GGCATCAGGCTCCACAGT-3′), Vimentin (5′-GACGCCATCAACACCGAGTT-3′; 5′-CTTTGTCGTTGGTTAGCTGGT-3′), snail (5′-CTTCCAGCAGCCCTACGAC-3′; 5′-CGGTGGGGTTGAGGATCT-3′), Wnt (5′-CAGAGCCACGAGTTTGGAGTT-3′; 5′-GATTGGGTTTGGGTTGGAGGT-3′), β-catenin (5′-GGGATTGGCTTTAGGCCTGT-3′; 5′-GAAATTGCCGTAGCGGGTTC-3′), c-Myc (5′-TGCATGATCAAATGCAACCT-3′; 5′-TCTTTTATGCCCAAAGTCCAA-3′) β-actin (5′-TTGTTACAGGAAGTCCCTTGCC-3′; 5′-ATGCTATCACCTCCCCTGTGTG-3′).

2.7. Western Blotting

After washing the cells with PBS buffer, we added RIPA buffer to lyse the cell samples and lymph nodes on ice for 20 min. The supernatant protein concentration was then determined using the BCA protein assay kit (Thermo Scientific, Fremont, CA, USA) by centrifugation at 4 °C and 12,000 rpm for 20 min. Proteins were transferred to polyvinylidene fluoride (PVDF) membranes using SDS-PAGE. Membranes were blocked in TBST containing 5% nonfat milk for 2 h and incubated overnight after adding the primary antibody. Then, visualization was performed using the ECL system (PerkinElmer, Waltham, MA, USA).

2.8. Plasmids and siRNA Transfection

According to the human c-myc gene sequence searched in the GenBank database, the corresponding c-myc-siRNA sense: (5′-AGGAAGUCAUCGUGGCCAATT-3′ and antisense: 5′-UUUCUCACUCUCAUACACCTT-3′) primer sequences were devised and commissioned by GenePharma (Shanghai, China). The KYSE30, KYSE150, and KYSE410 cells were diluted in the logarithmic growth phase to 5×10^4/mL, and they were cultured in an antibiotic-free medium. After 48 h of transfection, the cells were exposed to Rh4 in further experiments.

2.9. Animal Experiments

Our experiments obtained Northwestern University Animal Ethics Committee approval (NWU-AWC-20210301M). Male BALB/c nude mice (about 20 days, n = 8) were obtained from Gem Pharmatech (Jiangsu, China) and fostered under specific pathogen-free (SPF) conditions. After one week of acclimation, the mice were inoculated with 1×10^6 KYSE30 cells to build a footpad transfer model [29]. After the inoculation, the nude mice were left for about 5 days, and then they were placed into different groups: (a) control group (saline, intraperitoneally, i.p.); (b) low-dose Rh4 group (30 mg/kg/day, i.p.); (c) high-dose Rh4 group (60 mg/kg/day, i.p.); (d) capecitabine group (200 mg/kg/day, gavage). In addition, unvaccinated nude mice were differentiated into two groups: (e) normal group (saline, i.p.); (f) normal + Rh4 group (40 mg/kg/day, i.p.). Changes in the body weight of mice were recorded and the body weight was measured every 3 days. After our experiment, the nude mice were sacrificed, and the lymph nodes in the popliteal fossa and main organs were collected for H&E staining.

2.10. Hemogram Assay and Measurement of Biochemical Parameters

The peripheral blood we used was obtained from the periocular vein of nude mice and gathered in EDTA-containing specific containers. After blood collection, hematological parameters in each sample, including white blood cell (WBC) count, lymphocyte (LYM) count, and granulocyte (GRAN) count, were measured using an automated hematology analyzer (HC2200, Merrill Lynch, Suzhou, China).

Commercial kits (Shanghai Enzyme Biotechnology Co., Ltd., Shanghai, China) were used to analyze parameters including ALT and AST (liver function) and urea, uric acid, and creatinine (renal function).

2.11. Immunofluorescence Assay (IF)

Antigen retrieval in paraffin sections of lymph nodes and lung tissues was performed after specific processing (deparaffinization and dehydration). Fluorescent antibodies were incubated with the second antibodies for 2 h. Fluorescence intensity analysis was visualized and analyzed using a confocal microscope (Tokyo, Japan).

2.12. Histopathology and Immunohistochemistry

The lymph nodes and major organs that we needed were isolated from the mice during the course of the disease to form paraffin-embedded sections that were used for hematoxylin and eosin (H&E) staining. For the immunohistochemistry assay, lungs and lymph nodes were stained with E-cadherin, N-cadherin, Vimentin, Snail, c-Myc, Wnt, β-catenin, and p-β-catenin. The images observed were captured with a light microscope (Nikon, Japan).

2.13. Statistical Analysis

Experimental data represent mean ± standard deviation (SD). The experimental data were analyzed with SPSS and GraphPad Prism software using a one-way analysis of variance (ANOVA). A value of $p < 0.05$ was considered statistically significant for all experiments.

3. Results

3.1. Ginsenoside Rh4 Suppresses the Migration and Invasion of ESCC Cells

To examine the effect of ginsenoside Rh4 on the motility of ESCC without affecting cell viability, we conducted an MTT test in vitro of ginsenoside Rh4 on KYSE30, KYSE150, and KYSE410. Three cell lines were measured with unequal contents of Rh4 (0–140 μM) for 24 or 48 h and subjected to MTT. These results demonstrated that the IC50 values of KYSE30, KYSE150, and KYSE410 cells that were treated with Rh4 for 24 h are 52.28 μM, 64.88 μM, and 54.72 μM (Figure 1B). We chose to incubate with ginsenoside Rh4 (10 and 20 μM) to explore its effect on cell motility because the cell viability at those concentrations exceeded 80%.

The migration and invasion ability of ESCC were assessed using in vitro scratch and Transwell assays. In the in vitro scratch experiment, ginsenoside Rh4 (10 and 20 μM) significantly delayed wound closure ($p < 0.01$) at 12 and 24 h compared to untreated cells in all cell lines tested (Figure 1C). Through in vitro cell movement analysis, we confirmed that ginsenoside Rh4 restrained the migration and invasion of KYSE30, KYSE150, and KYSE410 cells. The quantity of invading cells in the tested cells decreased with increasing Rh4 concentration (Figure 2A). In contrast to the corresponding cells in the control group, the quantity of penetrating cells in the low-dose Rh4 group was reduced by at least 30%, and the quantity of penetrating cells in the high-dose Rh4 group was reduced by more than 50% (Figure 2B). Furthermore, the MMP2 and MMP9 in the cells were also probed. In contrast to the control group, ginsenoside Rh4 can significantly down-regulate the expression of these two proteins (Figure 2C). This demonstrates that ginsenoside Rh4 can potentially inhibit the invasion ability of ESCC.

3.2. Ginsenoside Rh4 Inhibits the Metastasis of ESCC by Regulating EMT

As previously reported, EMT is involved in the metastasis of multiple cancers [30]. Therefore, to study if Rh4 inhibits the metastasis of ESCC by regulating EMT, we detected the expression of KYSE30, KYSE150, and KYSE410 related proteins including E-cadherin, N-cadherin, vimentin, and snail (with or without Rh4) (Figure 2C). As shown in Figure 2D, treatment with ginsenoside Rh4 decreased the content of N-cadherin, vimentin, and snail, and it improved E-cadherin. To explore the effect of Rh4 on the gene levels of EMT in ESCC, qRT-PCR was used to examine the level of related mRNA in KYSE30, KYSE150, and KYSE410 (with or without Rh4). The results demonstrated that Rh4 also increased mRNA expression. Rh4 had a significant impact reducing N-cadherin, vimentin, and snail, and improving E-cadherin' expression (Figure 3A). This is consistent with Western blotting experiments.

In order to investigate whether Rh4 also affects ESCC metastasis by curbing EMT in vivo, a mouse footpad lymphatic metastasis model was used to evaluate the role of Rh4. The immunohistochemical results of lymph node and lung tissue illustrated that Rh4 treatment could improve the content of E-cadherin and reduce the positive increase in N-cadherin, Vimentin, and snail (Figure 3B). In the results of immunofluorescence experiments on mouse lymph nodes and lung tissues, the fluorescence intensity of E-cadherin was also significantly heightened, whereas the fluorescence intensity of N-cadherin was significantly lowered, both of these results were in contrast to the control group (Figure 3C). In addition, the expression level of E-cadherin, N-cadherin, Vimentin, and snail in lymph nodes attained the same change trend results in in vitro experiments and were exanimated (Figure 4A,B). The above results demonstrated that Rh4 inhibits the metastasis of ESCC carcinoma by inhibiting EMT.

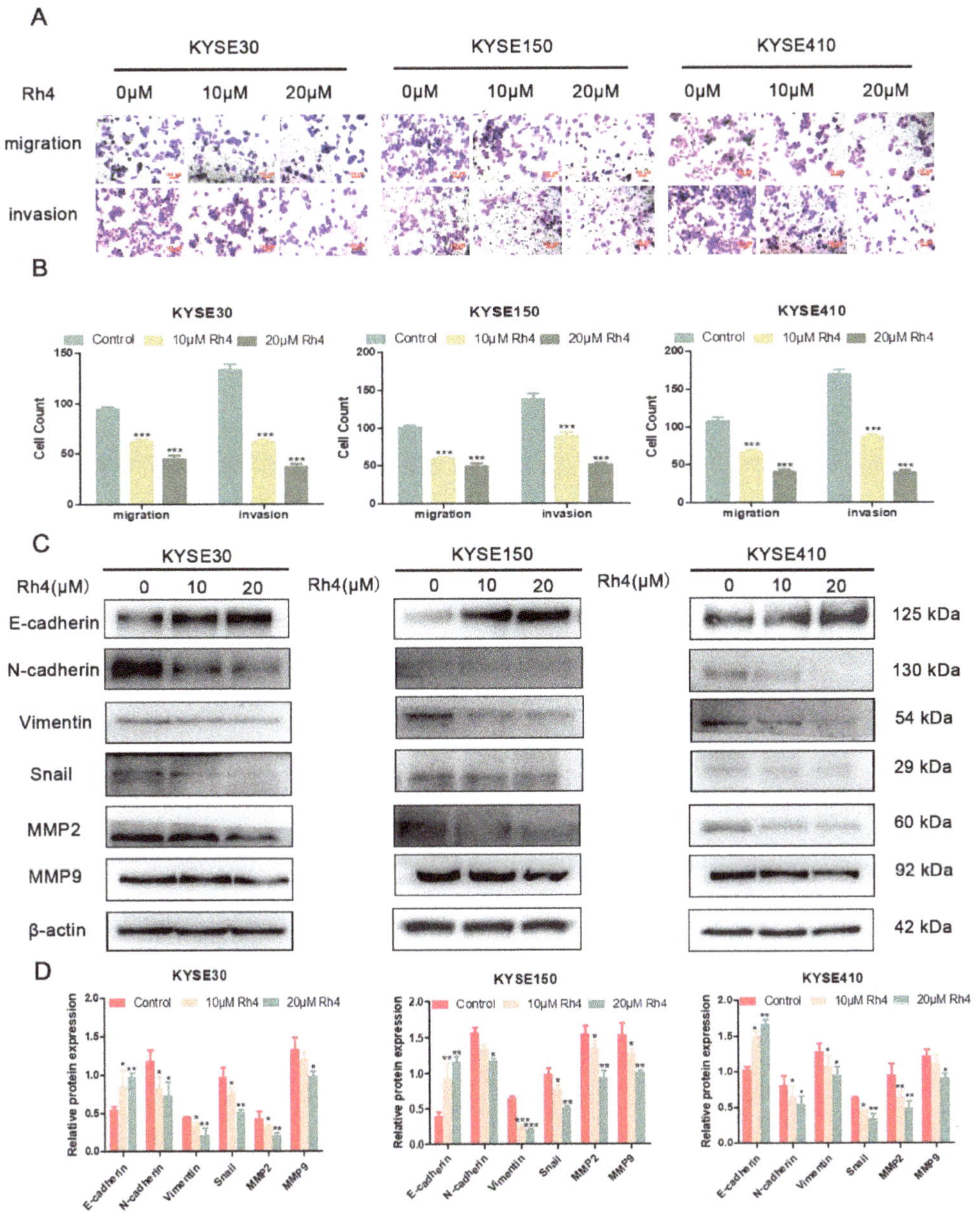

Figure 2. (**A**) Migration and invasion images of KYSE30, KYSE150, and KYSE410 cells. (**B**) Counting cells after crystal violet staining (**C**,**D**) Protein level of E-cadherin, N-cadherin, Vimentin, snail, MMP2, and MMP9 in KYSE30, KYSE150, and KYSE410 after Rh4 treatment (10 and 20 μM). All data are presented as mean ± SD, * $p < 0.05$, ** $p < 0.01$, and *** $p < 0.001$.

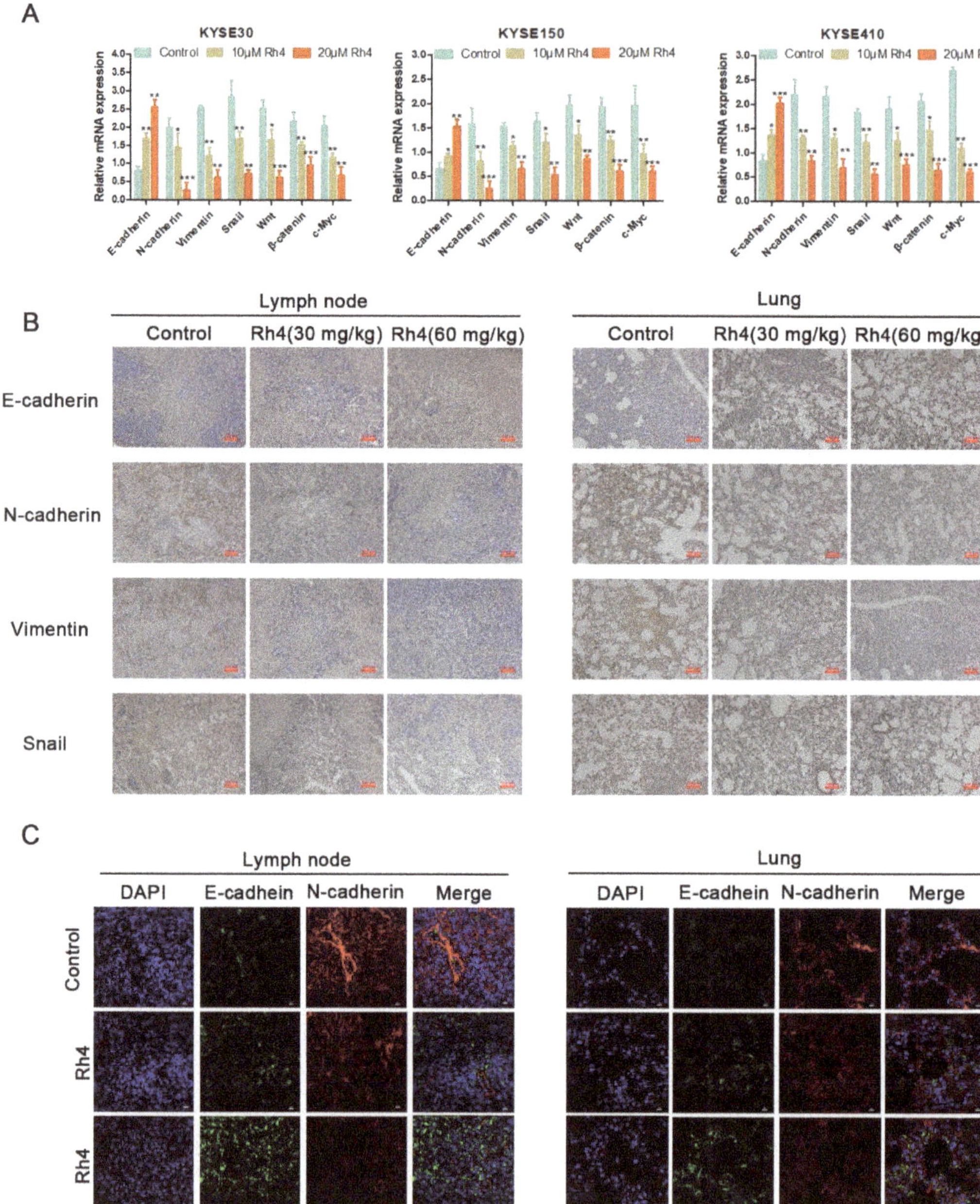

Figure 3. (**A**) Transcript of E-cadherin, N-cadherin, Vimentin, snail, wnt, β-catenin, and c-Myc after Rh4 treatment (10 and 20 μM). (**B**) Immunohistochemical images of E-cadherin, N-cadherin, Vimentin, and snail. In lymph nodes and lungs. Scale bar = 100 μm. (**C**) Immunofluorescence analysis showed that Rh4 promoted E-cadherin in lymph nodes and lungs and inhibited the expression of N-cadherin. Scale bar = 10 μm. All data are presented as mean ± SD, * $p < 0.05$, ** $p < 0.01$, *** and $p < 0.001$.

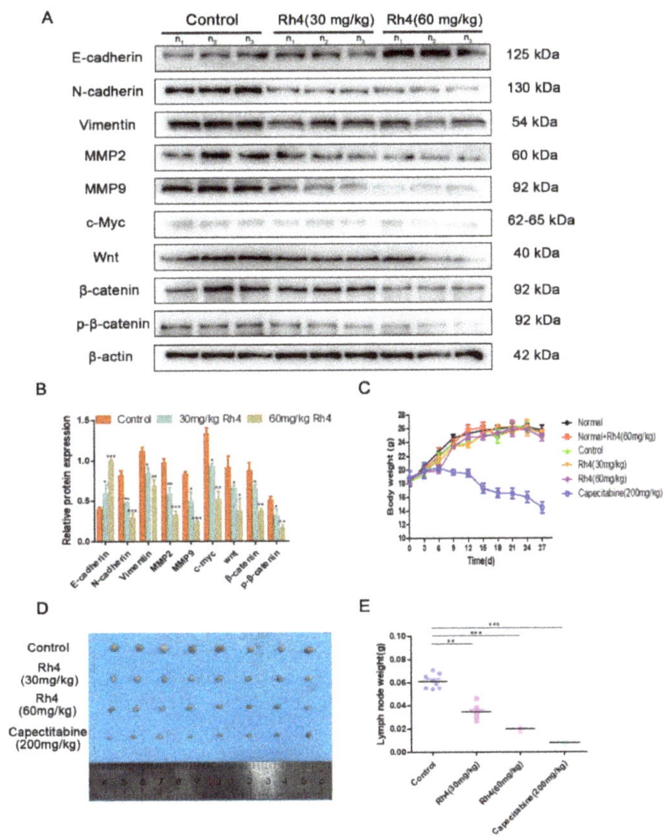

Figure 4. (**A**,**B**) Protein level of E-cadherin, N-cadherin, Vimentin, MMP2, MMP9, Wnt, β-catenin, and c-Myc in the lymph node. (**C**) Body weight change curve of six groups of mice. (**D**) Representative image of KYSE30 popliteal lymph nodes of measured groups. (**E**) Popliteal lymph nodes were measured after sacrifice. All data are presented as mean ± SD, * $p < 0.05$, ** $p < 0.01$, and *** $p < 0.001$.

3.3. Ginsenoside Rh4 Inhibits ESCC Metastasis In Vivo and Has Low Toxicity

Next, we studied the effect of ginsenoside Rh4 in the transfer of ESCC in vivo. We injected KYSE30 cells into the left foot pad of male BALB/c mice. While determining lymph-node-associated protein expression, we discerned that MMP2 and MMP9 in the Rh4 measurement group were also notably reduced compared to the control group (Figure 4A). Moreover, no notable diversification in the weight change of mice in the Rh4 treatment group was found in contrast to the normal group, but there was a conspicuous decrease in the capecitabine group (Figure 4C). The lymph nodes were collected and weighed after the experiment was completed. The growth of lymph nodes in the Rh4 group and capecitabine group was restrained. In contrast to the control group, the size and weight of lymph nodes in the 30 mg/kg Rh4 group and 60 mg/kg Rh4 group decreased (Figure 4D,E). Moreover, the number of lung nodules in these two groups was also significantly less than that in the control group (Figure S1). These results indicate that the addition of Rh4 can reduce the metastasis of ESCC cells in the proximal popliteal lymph nodes and distal lung tissue.

Therefore, to explore the safety of Rh4 in foot pad transfer model mice, we explored the peripheral blood parameters of six groups of mice (WBC, LYM, and GRAN). There were no significant differences in the hematology index of WBC, LYM, and GRAN (Figure 5A). Next, we studied the effect of Rh4 on organ activity. Compared with the normal group, the renal function indexes (urea, uric acid, and creatinine) and liver function indexes (ALT, AST) of mice in the Rh4 treatment group did not change significantly (Figure 5B,C). However, the related indicators in the capecitabine group were conspicuously increased, which suggested that capecitabine has obvious toxicity. H&E staining illustrated that in the normal group, the physiological appearance of important organs in the normal + Rh4 (60 mg/kg) and Rh4 measurement groups showed no obvious pathological changes (Figure 5D). However, in contrast to the control group, liver slices of the capecitabine group showed inflammatory infiltration, and the kidney slices also appeared to be disordered (Figure 5D). Additionally, in the organ index, the spleen index of the capecitabine group also showed an apparent decrease (Table S2). In summary, the results indicated that Rh4 treatment was not significantly toxic in vivo.

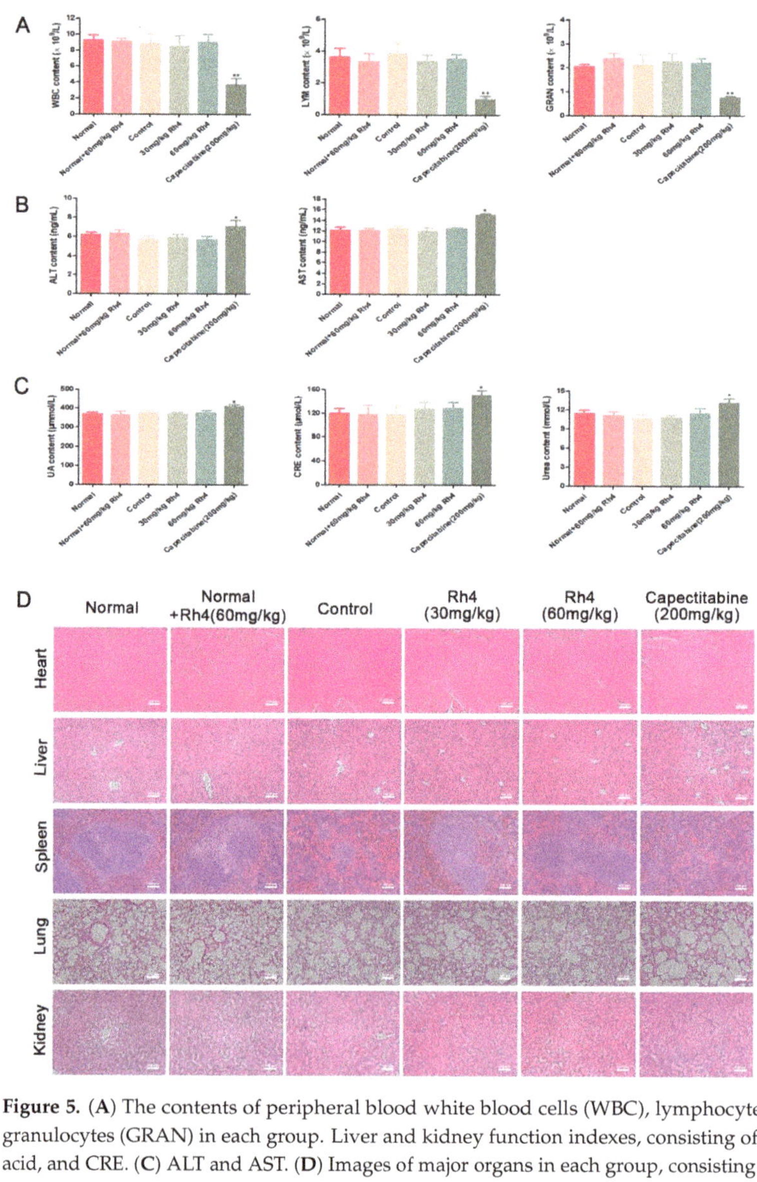

Figure 5. (**A**) The contents of peripheral blood white blood cells (WBC), lymphocytes (LYM), and granulocytes (GRAN) in each group. Liver and kidney function indexes, consisting of (**B**) urea, uric acid, and CRE. (**C**) ALT and AST. (**D**) Images of major organs in each group, consisting of heart, liver, spleen, lung, and kidney. Scale bar = 100 μm. All data are presented as mean ± SD, * $p < 0.05$, and ** $p < 0.01$.

3.4. Rh4 Inhibits ESCC Metastasis through Wnt/β-Catenin Signaling Pathway

To investigate if Rh4 inhibits ESCC metastasis through the Wnt/β-catenin pathway, we tested the protein levels of the Wnt/β-catenin pathway. In contrast to the control group, KYSE30, KYSE150, and KYSE410 cells demonstrated different densities of Rh4 reduced Wnt, β-catenin, and p-β-catenin levels (Figure 6A), similar results were also found in the Western blotting experiment of tumor tissues (Figure 6B). In addition, we also found from the immunohistochemical results of lymph node and lung tissue that Rh4 decreased the Wnt/β-catenin interrelated proteins, which was the same as the results of Western blotting experiment (Figure 6C). The immunofluorescence results for lymph node and lung tissue also suggested the Wnt and β-catenin in the Rh4 measurement group were noticeably reduced compared to the control group (Figure 6D).

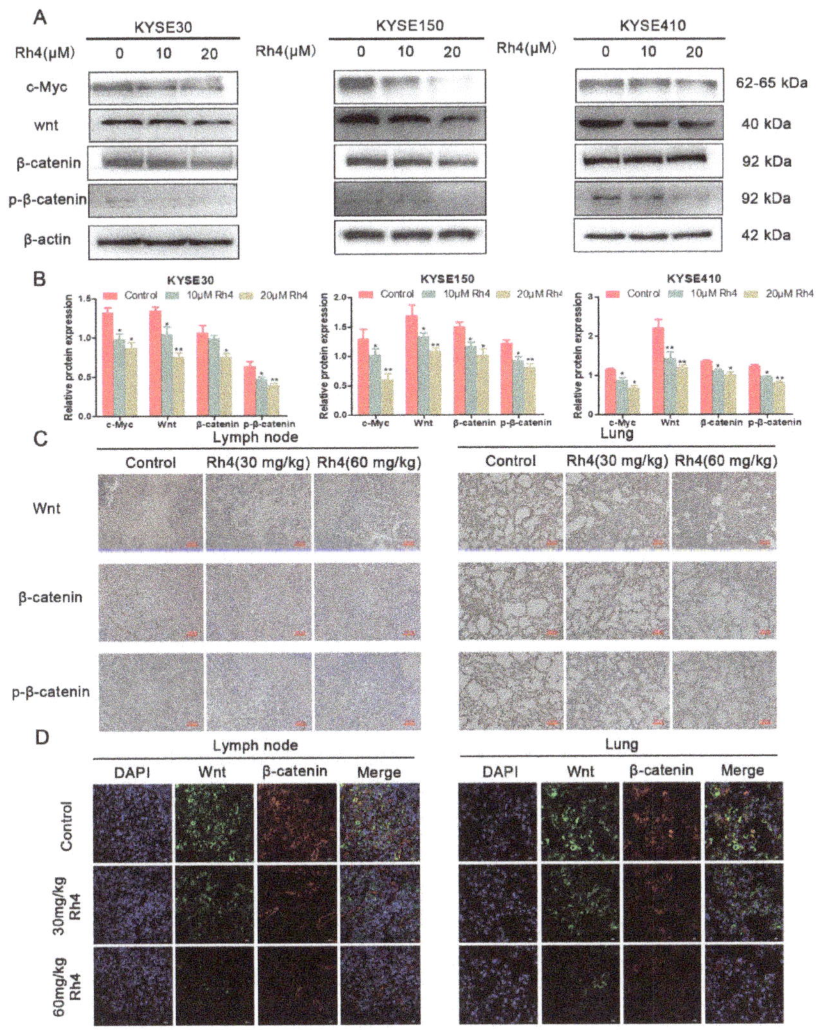

Figure 6. (**A**,**B**) Relative protein levels of Wnt, β-catenin, and p-β-catenin after KYSE30, KYSE150, and KYSE410 were treated by Rh4. (**C**) Immunohistochemical analysis showed that Rh4 attenuated Wnt, β-catenin, and p-β-catenin in lymph nodes and lungs. Scale bar = 100 μm. (**D**) Immunofluorescence analysis showed that Rh4 also attenuated the level of Wnt and β-catenin in lymph nodes and lungs. Scale bar = 10 μm. All data are presented as mean ± SD, * $p < 0.05$, and ** $p < 0.01$.

In order to further explore the utility of Rh4 in the Wnt/β-catenin signaling pathway, we introduced HLY78, which is an agonist of Wnt. The results we obtained from the scratch test showed that the addition of HLY78 could improve the migration ability of KYSE30, KYSE150, and KYSE410 cells. However, with the combined action of Rh4 and HLY78, the wound healing efficiency of the three types of cells was betwixt the control group and the HLY78 group (Figure 7A,B). In addition, we also examined the results in the Wnt/β-catenin pathway proteins after adding HLY78 (Figure 7C). The results showed that the levels of Wnt, β-catenin, and p-β-catenin interrelated proteins are consistent with the expression trend of the scratch experiment (Figure 7D), which indicated that Rh4 restrained the metastasis of ESCC by inhibiting the Wnt/β-catenin pathway.

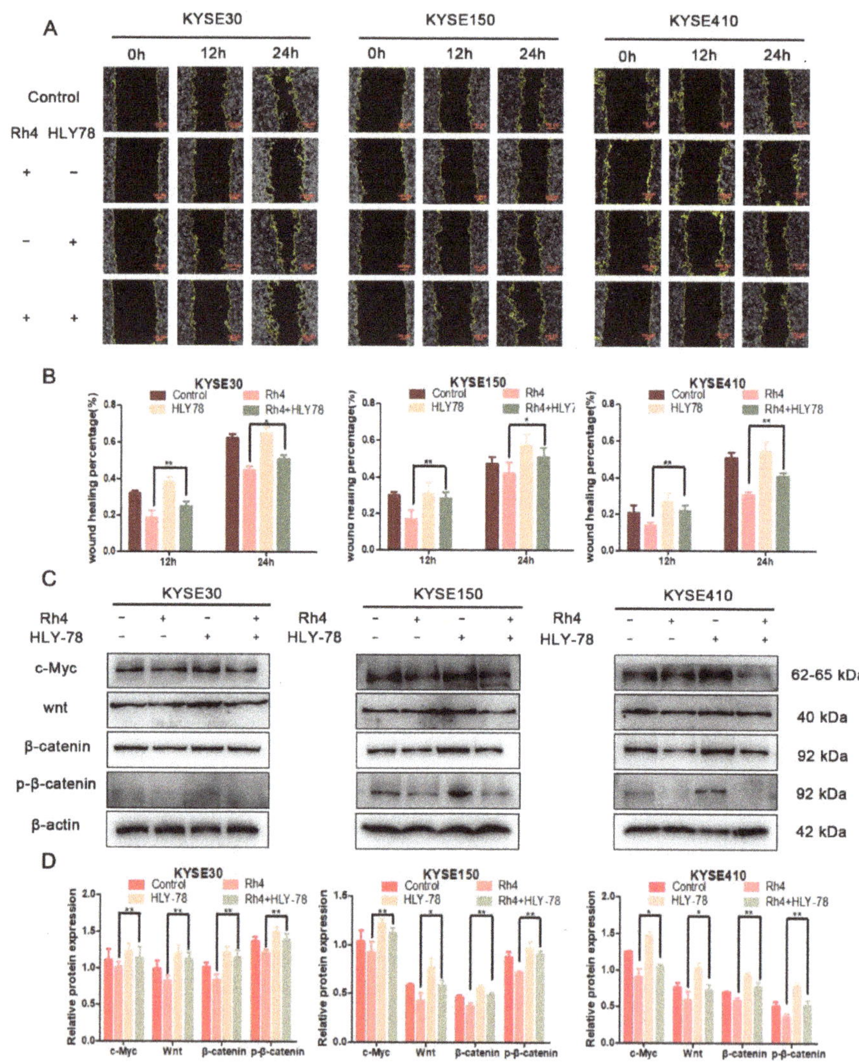

Figure 7. (**A**) Migration images with or without the addition of Rh4(HLY78) at 0, 12, and 24 h. The yellow line is the migration edge. Scale bar = 100 μm (**B**) Percent wound healing, treatment group vs. control group. (**C**) Protein levels after Rh4 and HLY-78 combined treatment of KYSE30, KYSE150, and KYSE410. (**D**) Bar graphs showing protein expression level. All data are presented as mean ± SD, * $p < 0.05$ and ** $p < 0.01$.

3.5. c-Myc Is Essential in the Inhibition of ESCC Metastasis by Rh4

It has been reported that the expression of c-myc can be adjusted by the Wnt/β-catenin signaling pathway. For our experiments, we confirmed that Rh4 could decrease c-Myc in KYSE30, KYSE150, and KYSE410 (Figure 6A). In order to investigate the utility of Rh4 on c-Myc in ESCC, we used c-Myc-siRNA to silence c-Myc, and then determine the protein level of EMT. These results demonstrated that c-Myc-siRNA significantly decreased c-Myc's expression. Silencing c-Myc reduced N-cadherin, Vimentin, and snail and improved the E-cadherin. Additionally, the combined effect of Rh4 and c-Myc-siRNA was more obvious than that of c-Myc (Figure 8A,B). The immunohistochemistry of the lymph nodes and lung tissues also illustrated the same results (Figure 8C). The results revealed that c-Myc is an essential transcription factor in the inhibition of ESCC metastasis by Rh4.

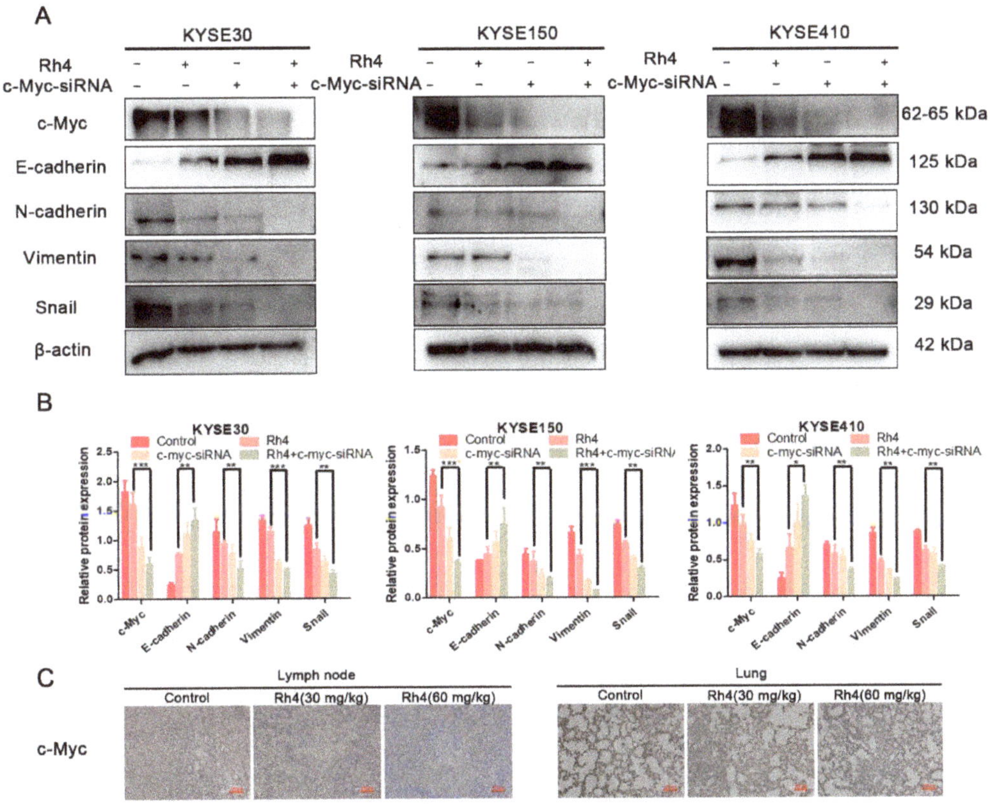

Figure 8. (**A**,**B**) Addition of c-Myc-siRNA enhanced Rh4 inhibition of E-cadherin, N-cadherin, Vimentin, and snail (**C**) Rh4 reduces c-Myc expression in lymph nodes and lungs in vivo. Scale bar = 100 μm. All data are presented as mean ± SD, * $p < 0.05$, ** $p < 0.01$, *** $p < 0.001$.

4. Discussion

Metastasis, which accompanies the diffusion of cancer cells from the initial part of the disease to distant sites [31], includes angiogenesis, cell adhesion, ECM degradation, migration, invasion, and transportation to additional areas [32]. ESCC is powerfully aggressive, and it causes high mortality [33]. A large number of patients with primary esophageal cancer have been found to have lymph node metastasis at the initial diagnosis [34]. Cancer has invaded the muscle layer in most patients, and at least half of all cancers have metastasized to other tissues and organs, most notably the lymph nodes and lungs [35]. It has been shown in recent studies that Ginsenoside Rk1 has an inhibitory effect on the expression of PD-L1 in lung adenocarcinoma [36]. Ginsenoside CK inhibits migration and invasion of human osteosarcoma cells and TGF-β-induced A549 cells via PI3K/mTOR/p70S6K1 and SIRT [37,38]. Our previous research has discerned that Rh4 has a significant effect on the suppression of the proliferation of esophageal cancer. In our research, we verified that Rh4 possesses an inhibitory function on ESCC metastasis. Scratch experiments showed that ginsenoside Rh4 had an obvious inhibitory effect on the migration of ESCC. When 20μM Rh4 was applied for 24 h, the wound healing rate of all cells was reduced by more than 60%. In the KYSE30 transplanted foot pad mouse model, we determined that 30 and 60 mg/kg Rh4 significantly curbed the growth of lymph nodes. Moreover, the ginsenoside group had a reduced expression of metastasis-related proteins in lymph nodes and lung tissues. Based on these results, Rh4 has a significant outcome for ESCC metastasis in vitro and in vivo.

Capecitabine is a first-line drug in the measurement of ESCC metastasis, but its many secondary actions, including liver toxicity and nephrotoxicity, limit the safe dose of capecitabine for patients [39]. If the adverse reaction is too strong, the drug may even be stopped [40]. Natural medicine may offer a new approach to the measurement of esophageal squamous cell carcinoma metastasis. Previous reports have demonstrated that capecitabine could cause liver damage [41–43], kidney damage, and even cardiotoxicity. The relevant serum biochemical indicators we obtained revealed organ damage, with liver and kidney activity (CRE, UA, and urea). H&E staining of various organs and tissues after capecitabine treatment also showed liver damage and spleen damage. Moreover, the measurement of blood parameters illustrated that capecitabine significantly reduced the number of immune cells (LYM, WBC, and GRAN), indicating capecitabine caused blood toxicity and affected the immune function of mice. Relevant indicators and H&E staining pictures of Rh4 group mice indicated that Rh4 did not influence the organ function of mice. Additionally, it had no obvious effect on the number of immune cells, which reveals Rh4 has no hematological toxicity. The results of the organ index showed that the spleen and lung indices of the mice were significantly reduced. The limitations of capecitabine on the immune responses and lung function in mice are noteworthy [44]. These indicators did not change significantly in the Rh4 group. Thus, these results suggest that Rh4 has minimal side effects in the treatment of ESCC metastasis in ESCC.

As previously reported, the Wnt/β-catenin signaling pathway is an oncogenic driving pathway in ESCC, which has been widely reported [45]. Wnt is a crucial part of cell axis patterning, including movement. β-catenin (β series protein) is an adhesion factor, and recent studies have found that it has the dual function of regulating and coordinating cell adhesion and gene transcription [46,47]. HLY78 is an agonist of the Wnt/β-catenin signaling pathway. HLY78 can enhance Wnt signaling by targeting the DIX domain of Axin in the channel and enhancing Axin-LRP6 cross-linking [48–50]. We pretreated with the Wnt agonist HLY78 to investigate the crucial utility of ginsenoside Rh4. The results of the scratch test discerned that after adding HLY78, the motility of ESCC cells was improved in contrast to the control group. Under the joint action of Rh4, the wound healing rate was between that of the Rh4 group and the HLY78 group. Western blotting illustrated that Wnt, β-catenin and p-β-catenin were consistent with the results of the scratch experiment. Previously, c-Myc expression was shown to be adjusted by the Wnt/β-catenin signaling pathway [51,52]. Moreover, after adding c-Myc-si-RNA, the expression of several EMT-related proteins showed a more obvious trend of change, compared with when only Rh4 was added. This indicates that Rh4 can restrain the metastasis of ESCC by inhibiting the Wnt/β-catenin signaling pathway.

5. Conclusions

To summarize, the research reveals for the first time that ginsenoside Rh4 restrains the metastasis of ESCC cells by adjusting the Wnt/β-catenin signaling pathway (Figure 9). In the mouse footpad lymph node metastasis model, Rh4 shows obvious antitumor metastasis activity with few side effects. Our study offers new perspectives on Rh4 as a latent anti-metastatic medicine.

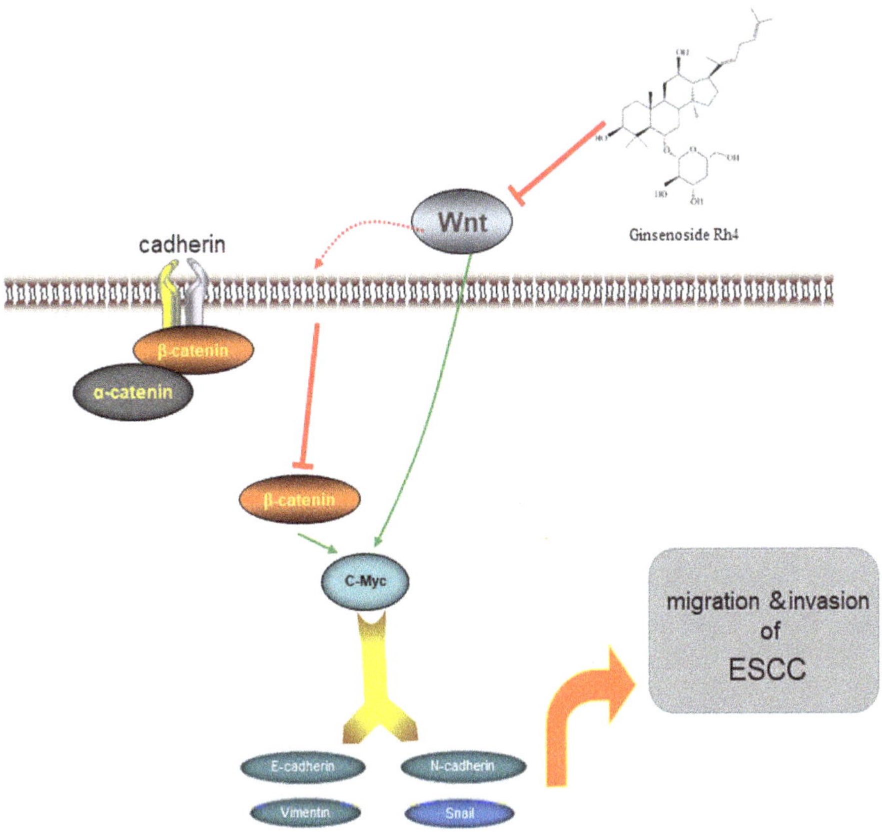

Figure 9. Proposed molecular mechanism of Rh4 anti-metastasis of ESCC.

Supplementary Materials: The following supporting information can be downloaded at: https://www.mdpi.com/article/10.3390/nu14153042/s1. Table S1: Information of antibodies; Table S2 Date of organ index; Figure S1:(A) KYSE30 cells were injected into the footpads of BALB/c mice, and lungs were harvested for staining. (B)Tumor nodule counts in lung tissue.

Author Contributions: J.C.: Conceptualization, Methodology, Validation, Formal Analysis, Investigation, Data Curation, Writing—Original Draft, Visualization, Project Administration, Writing—Review & Editing. Z.D.: Resources, Funding Acquisition. Y.L.: Resources, Funding Acquisition. R.F.: Resources, Funding acquisition. C.Z.: Resources, Funding Acquisition, Writing—Review & Editing, Project Administration, Supervision. All authors have read and agreed to the published version of the manuscript.

Funding: This study was financially supported by the National Key R&D Program of China (2021YFC2101500) and the National Natural Science Foundation of China (22178287, 22108224).

Institutional Review Board Statement: The animal experiments in this study operated in strict accordance with the "Chinese Animal Ethics Standards" and "Chinese Animal Ethics Guidelines", which were approved by the Northwestern University Animal Ethics Committee (NWU-AWC-20210301M).

Data Availability Statement: The original contributions presented in the study are included in the article and supplementary materials. Further inquiries can be directed to the corresponding author(s).

Acknowledgments: This work was financially supported by the Shaanxi Key Laboratory of the Degradable Biomedical Materials Program (2014SZS07-Z01).

Conflicts of Interest: The authors declare no conflict of interest.

References

1. Bray, F.; Ferlay, J.; Soerjomataram, I.; Siegel, R.L.; Torre, L.A.; Jemal, A. Global cancer statistics 2018: GLOBOCAN estimates of incidence and mortality worldwide for 36 cancers in 185 countries. *CA A Cancer J. Clin.* **2018**, *68*, 394–424. [CrossRef]
2. Siegel, R.; Miller, K.D.; Jemal, A. Cancer statistics 2016. *CA A Cancer J. Clin.* **2015**, *66*, 7–30. [CrossRef]
3. Ohashi, S.; Miyamoto, S.i.; Kikuchi, O.; Goto, T.; Amanuma, Y.; Muto, M. Recent Advances From Basic and Clinical Studies of Esophageal Squamous Cell Carcinoma. *Gastroenterology* **2015**, *149*, 1700–1715. [CrossRef]
4. Lin, D.-C.; Wang, M.-R.; Koeffler, H.P. Genomic and Epigenomic Aberrations in Esophageal Squamous Cell Carcinoma and Implications for Patients. *Gastroenterology* **2018**, *154*, 374–389. [CrossRef]
5. Kai, F.; Drain, A.P.; Weaver, V.M. The Extracellular Matrix Modulates the Metastatic Journey. *Dev. Cell* **2019**, *49*, 332–346. [CrossRef]
6. Orang, A.V.; Petersen, J.; McKinnon, R.A.; Michael, M.Z. Micromanaging aerobic respiration and glycolysis in cancer cells. *Mol. Metab.* **2019**, *23*, 98–126. [CrossRef]
7. Liu, H.-L.; Liu, D.; Ding, G.-R.; Liao, P.-F.; Zhang, J.-W. Hypoxia-inducible factor-1α and Wnt/β-catenin signaling pathways promote the invasion of hypoxic gastric cancer cells. *Mol. Med. Rep.* **2015**, *12*, 3365–3373. [CrossRef]
8. Yang, S.; Li, X.; Shen, W.; Hu, H.; Li, C.; Han, G. MicroRNA-140 Represses Esophageal Cancer Progression via Targeting ZEB2 to Regulate Wnt/β-Catenin Pathway. *J. Surg. Res.* **2021**, *257*, 267–277. [CrossRef]
9. Wang, W.; He, S.; Zhang, R.; Peng, J.; Guo, D.; Zhang, J.; Xiang, B.; Li, L. ALDH1A1 maintains the cancer stem-like cells properties of esophageal squamous cell carcinoma by activating the AKT signal pathway and interacting with β-catenin. *Biomed. Pharmacother.* **2020**, *125*, 109940. [CrossRef]
10. Ren, Y.; Guo, T.; Xu, J.; Liu, Y.; Huang, J. The novel target of esophageal squamous cell carcinoma: lncRNA GASL1 regulates cell migration, invasion and cell cycle stagnation by inactivating the Wnt3a/β-catenin signaling. *Pathol. Res. Pract.* **2021**, *217*, 153289. [CrossRef]
11. Fayard, B.; Bianchi, F.; Dey, J.; Moreno, E.; Djaffer, S.; Hynes, N.E.; Monard, D. The serine protease inhibitor protease nexin-1 controls mammary cancer metastasis through LRP-1-mediated MMP-9 expression. *Cancer Res.* **2009**, *69*, 5690–5698. [CrossRef] [PubMed]
12. Chen, D.; Zhang, H.; Jing, C.; He, X.; Yang, B.; Cai, J.; Zhou, Y.; Song, X.; Li, L.; Hao, X. Efficient synthesis of new phenanthridine Wnt/β-catenin signaling pathway agonists. *Eur. J. Med. Chem.* **2018**, *157*, 1491–1499. [CrossRef] [PubMed]
13. Yang, X.; Shao, F.; Guo, D.; Wang, W.; Wang, J.; Zhu, R.; Gao, Y.; He, J.; Lu, Z. WNT/β-catenin-suppressed FTO expression increases m(6)A of c-Myc mRNA to promote tumor cell glycolysis and tumorigenesis. *Cell Death Dis.* **2021**, *12*, 462. [CrossRef] [PubMed]
14. Lee, H.; Cha, J.; Kim, S.; Park, J.; Song, K.; Kim, P.; Kim, M.-Y. c-MYC Drives Breast Cancer Metastasis to the Brain, but Promotes Synthetic Lethality with TRAIL. *Mol. Cancer Res.* **2018**, *17*, 544–554. [CrossRef] [PubMed]
15. Meskyte, E.M.; Keskas, S.; Ciribilli, Y. MYC as a Multifaceted Regulator of Tumor Microenvironment Leading to Metastasis. *Int. J. Mol. Sci.* **2020**, *21*, 7710. [CrossRef]
16. Hseu, Y.C.; Lin, Y.C.; Rajendran, P.; Thigarajan, V.; Mathew, D.C.; Lin, K.Y.; Way, T.D.; Liao, J.W.; Yang, H.L. Antrodia salmonea suppresses invasion and metastasis in triple-negative breast cancer cells by reversing EMT through the NF-κB and Wnt/β-catenin signaling pathway. *Food Chem. Toxicol. Int. J. Publ. Br. Ind. Biol. Res. Assoc.* **2019**, *124*, 219–230. [CrossRef]
17. Wang, Q.; Cheng, Y.; Wang, Y.; Fan, Y.; Li, C.; Zhang, Y.; Wang, Y.; Dong, Q.; Ma, Y.; Teng, Y.-E.; et al. Tamoxifen reverses epithelial–mesenchymal transition by demethylating miR-200c in triple-negative breast cancer cells. *BMC Cancer* **2017**, *17*, 492. [CrossRef]
18. Theveneau, E.; Mayor, R. Cadherins in collective cell migration of mesenchymal cells. *Curr. Opin. Cell Biol.* **2012**, *24*, 677–684. [CrossRef]
19. Li, B.; Zhao, J.; Wang, C.Z.; Searle, J.; He, T.C.; Yuan, C.S.; Du, W. Ginsenoside Rh2 induces apoptosis and paraptosis-like cell death in colorectal cancer cells through activation of p53. *Cancer Lett.* **2011**, *301*, 185–192. [CrossRef]
20. Yayeh, T.; Jung, K.H.; Jeong, H.Y.; Park, J.H.; Song, Y.B.; Kwak, Y.S.; Kang, H.S.; Cho, J.Y.; Oh, J.W.; Kim, S.K.; et al. Korean Red Ginseng Saponin Fraction Downregulates Proinflammatory Mediators in LPS Stimulated RAW264.7 Cells and Protects Mice against Endotoxic Shock. *J. Ginseng Res.* **2012**, *36*, 263–269. [CrossRef]
21. Yi, Y.S. Roles of ginsenosides in inflammasome activation. *J. Ginseng Res.* **2019**, *43*, 172–178. [CrossRef] [PubMed]
22. Fan, X.; Zhang, C.; Niu, S.; Fan, B.; Gu, D.; Jiang, K.; Li, R.; Li, S. Ginsenoside Rg1 attenuates hepatic insulin resistance induced by high-fat and high-sugar by inhibiting inflammation. *Eur. J. Pharmacol.* **2019**, *854*, 247–255. [CrossRef] [PubMed]
23. Kang, K.S.; Ham, J.; Kim, Y.J.; Park, J.H.; Cho, E.J.; Yamabe, N. Heat-processed Panax ginseng and diabetic renal damage: Active components and action mechanism. *J. Ginseng Res.* **2013**, *37*, 379–388. [CrossRef] [PubMed]

24. Ying, Y.; Zhang, Y.L.; Ma, C.J.; Li, M.Q.; Tang, C.Y.; Yang, Y.F.; Zeng, J.H.; Huang, X.Y.; Yi, J.; Wang, X.M.; et al. Neuroprotective Effects of Ginsenoside Rg1 against Hyperphosphorylated Tau-Induced Diabetic Retinal Neurodegeneration via Activation of IRS-1/Akt/GSK3β Signaling. *J. Agric. Food Chem.* **2019**, *67*, 8348–8360. [CrossRef] [PubMed]
25. Kim, M.Y.; Cho, J.Y. 20S-dihydroprotopanaxadiol, a ginsenoside derivative, boosts innate immune responses of monocytes and macrophages. *J. Ginseng Res.* **2013**, *37*, 293–299. [CrossRef]
26. Coon, J.T.; Ernst, E. Panax ginseng: A systematic review of adverse effects and drug interactions. *Drug Saf.* **2002**, *25*, 323–344. [CrossRef]
27. Deng, X.; Zhao, J.; Qu, L.; Duan, Z.; Fu, R.; Zhu, C.; Fan, D. Ginsenoside Rh4 suppresses aerobic glycolysis and the expression of PD-L1 via targeting AKT in esophageal cancer. *Biochem. Pharmacol.* **2020**, *178*, 114038. [CrossRef]
28. Zhang, J.; Ma, X.; Fan, D. Ginsenoside CK Inhibits Hypoxia-Induced Epithelial-Mesenchymal Transformation through the HIF-1α/NF-κB Feedback Pathway in Hepatocellular Carcinoma. *Foods* **2021**, *10*, 1195. [CrossRef]
29. Chen, C.; He, W.; Huang, J.; Wang, B.; Li, H.; Cai, Q.; Su, F.; Bi, J.; Liu, H.; Zhang, B.; et al. LNMAT1 promotes lymphatic metastasis of bladder cancer via CCL2 dependent macrophage recruitment. *Nat. Commun.* **2018**, *9*, 3826. [CrossRef]
30. Lo, H.C.; Zhang, X.H. EMT in Metastasis: Finding the Right Balance. *Dev. Cell* **2018**, *45*, 663–665. [CrossRef]
31. Gao, Y.; Bado, I.; Wang, H.; Zhang, W.; Rosen, J.; Zhang, X. Metastasis Organotropism: Redefining the Congenial Soil. *Dev. Cell* **2019**, *49*, 375–391. [CrossRef] [PubMed]
32. Chen, W.; Zheng, R.; Baade, P.D.; Zhang, S.; Zeng, H.; Bray, F.; Jemal, A.; Yu, X.Q.; He, J. Cancer statistics in China, 2015. *CA Cancer J. Clin.* **2016**, *66*, 115–132. [CrossRef] [PubMed]
33. Strilic, B.; Offermanns, S. Intravascular Survival and Extravasation of Tumor Cells. *Cancer Cell* **2017**, *32*, 282–293. [CrossRef] [PubMed]
34. Huang, T.X.; Fu, L. The immune landscape of esophageal cancer. *Cancer Commun.* **2019**, *39*, 79. [CrossRef]
35. Huang, F.L.; Yu, S.J. Esophageal cancer: Risk factors, genetic association, and treatment. *Asian J. Surg.* **2018**, *41*, 210–215. [CrossRef]
36. Hu, M.; Yang, J.; Qu, L.; Deng, X.; Duan, Z.; Fu, R.; Liang, L.; Fan, D. Ginsenoside Rk1 induces apoptosis and downregulates the expression of PD-L1 by targeting the NF-κB pathway in lung adenocarcinoma. *Food Funct.* **2020**, *11*, 456–471. [CrossRef]
37. Chen, K.; Jiao, J.; Xue, J.; Chen, T.; Hou, Y.; Jiang, Y.; Qian, L.; Wang, Y.; Ma, Z.; Liang, Z.; et al. Ginsenoside CK induces apoptosis and suppresses proliferation and invasion of human osteosarcoma cells through the PI3K/mTOR/p70S6K1 pathway. *Oncol. Rep.* **2020**, *43*, 886–896. [CrossRef]
38. Sun, M.; Zhuang, X.; Lv, G.; Lin, Z.; Huang, X.; Zhao, J.; Lin, H.; Wang, Y. Ginsenoside CK Inhibits TGF-β-Induced Epithelial-Mesenchymal Transition in A549 Cell via SIRT1. *BioMed Res. Int.* **2021**, *2021*, 9140191. [CrossRef]
39. Cunningham, D.; Starling, N.; Rao, S.; Iveson, T.; Nicolson, M.; Coxon, F.; Middleton, G.; Daniel, F.; Oates, J.; Norman, A.R. Capecitabine and oxaliplatin for advanced esophagogastric cancer. *New Engl. J. Med.* **2008**, *358*, 36–46. [CrossRef]
40. Hecht, J.R.; Bang, Y.J.; Qin, S.K.; Chung, H.C.; Xu, J.M.; Park, J.O.; Jeziorski, K.; Shparyk, Y.; Hoff, P.M.; Sobrero, A.; et al. Lapatinib in Combination With Capecitabine Plus Oxaliplatin in Human Epidermal Growth Factor Receptor 2-Positive Advanced or Metastatic Gastric, Esophageal, or Gastroesophageal Adenocarcinoma: TRIO-013/LOGiC–A Randomized Phase III Trial. *J. Clin. Oncol. Off. J. Am. Soc. Clin. Oncol.* **2016**, *34*, 443–451. [CrossRef]
41. Lam, S.W.; Guchelaar, H.J.; Boven, E. The role of pharmacogenetics in capecitabine efficacy and toxicity. *Cancer Treat. Rev.* **2016**, *50*, 9–22. [CrossRef] [PubMed]
42. Primrose, J.N.; Fox, R.P.; Palmer, D.H.; Malik, H.Z.; Prasad, R.; Mirza, D.; Anthony, A.; Corrie, P.; Falk, S.; Finch-Jones, M.; et al. Capecitabine compared with observation in resected biliary tract cancer (BILCAP): A randomised, controlled, multicentre, phase 3 study. *Lancet Oncol.* **2019**, *20*, 663–673. [CrossRef]
43. Neoptolemos, J.P.; Palmer, D.H.; Ghaneh, P.; Psarelli, E.E.; Valle, J.W.; Halloran, C.M.; Faluyi, O.; O'Reilly, D.A.; Cunningham, D.; Wadsley, J.; et al. Comparison of adjuvant gemcitabine and capecitabine with gemcitabine monotherapy in patients with resected pancreatic cancer (ESPAC-4): A multicentre, open-label, randomised, phase 3 trial. *Lancet* **2017**, *389*, 1011–1024. [CrossRef]
44. Vaflard, P.; Ederhy, S.; Torregrosa, C.; André, T.; Cohen, R.; Lopez-Trabada, D. [Fluoropyrimidines cardiac toxicity: 5-fluorouracil, capecitabine, compound S-1 and trifluridine/tipiracil]. *Bull. Cancer* **2018**, *105*, 707–719. [CrossRef] [PubMed]
45. Wang, Z.; Hu, Q.; Chen, H.; Shi, L.; He, M.; Liu, H.; Li, T.; Lü, M.; Deng, M.; Luo, G. Inhibition of Growth of Esophageal Cancer by Alantolactone via Wnt/β- Catenin Signaling. *Anti-Cancer Agents Med. Chem.* **2021**, *21*, 2525–2535. [CrossRef]
46. Wei, W.; Liu, H.; Yuan, J.; Yao, Y. Targeting Wnt/β-catenin by anthelmintic drug niclosamide overcomes paclitaxel resistance in esophageal cancer. *Fundam. Clin. Pharmacol.* **2021**, *35*, 165–173. [CrossRef]
47. Wang, Q.; Lv, Q.; Bian, H.; Yang, L.; Guo, K.L.; Ye, S.S.; Dong, X.F.; Tao, L.L. A novel tumor suppressor SPINK5 targets Wnt/β-catenin signaling pathway in esophageal cancer. *Cancer Med.* **2019**, *8*, 2360–2371. [CrossRef]
48. Luo, X.; Li, L.; Xu, W.; Cheng, Y.; Xie, Z. HLY78 Attenuates Neuronal Apoptosis via the LRP6/GSK3β/β-Catenin Signaling Pathway After Subarachnoid Hemorrhage in Rats. *Neurosci. Bull.* **2020**, *36*, 1171–1181. [CrossRef]
49. Luo, X.; Li, L.; Zheng, W.; Gu, L.; Zhang, X.; Li, Y.; Xie, Z.; Cheng, Y. HLY78 protects blood-brain barrier integrity through Wnt/β-catenin signaling pathway following subarachnoid hemorrhage in rats. *Brain Res. Bull.* **2020**, *162*, 107–114. [CrossRef]
50. Si, J.; Zhang, J.; Gan, L.; Guo, M.; Xie, Y.; Di, C.; Sun, C.; Wang, F.; Yan, J.; Zhang, H. The effects of the Wnt/β-catenin signaling pathway on apoptosis in HeLa cells induced by carbon ion irradiation. *Oncol. Rep.* **2020**, *44*, 303–312. [CrossRef]

51. Bisso, A.; Filipuzzi, M.; Gamarra Figueroa, G.P.; Brumana, G.; Biagioni, F.; Doni, M.; Ceccotti, G.; Tanaskovic, N.; Morelli, M.J.; Pendino, V.; et al. Cooperation Between MYC and β-Catenin in Liver Tumorigenesis Requires Yap/Taz. *Hepatology* **2020**, *72*, 1430–1443. [CrossRef] [PubMed]
52. Vallée, A.; Guillevin, R.; Vallée, J.N. Vasculogenesis and angiogenesis initiation under normoxic conditions through Wnt/β-catenin pathway in gliomas. *Rev. Neurosci.* **2018**, *29*, 71–91. [CrossRef] [PubMed]

nutrients MDPI

Article

Role of Phytoestrogen-Rich Bioactive Substances (*Linum usitatissimum* L., *Glycine max* L., *Trifolium pratense* L.) in Cardiovascular Disease Prevention in Postmenopausal Women: A Systematic Review and Meta-Analysis

Agata Błaszczuk [1], Agnieszka Barańska [2,*], Wiesław Kanadys [3], Maria Malm [2], Monika Elżbieta Jach [4], Urszula Religioni [5], Rafał Wróbel [6], Jolanta Herda [7] and Małgorzata Polz-Dacewicz [1]

1 Department of Virology with SARS Laboratory, Medical University of Lublin, 20-093 Lublin, Poland; agata.blaszczuk@umlub.pl (A.B.); malgorzata.polz.dacewicz@umlub.pl (M.P.-D.)
2 Department of Medical Informatics and Statistics with e-Health Lab, Medical University of Lublin, 20-090 Lublin, Poland; mariamalm@umlub.pl
3 Specialistic Medical Center Czechow, 20-848 Lublin, Poland; wieslaw.kanadys@wp.pl
4 Department of Molecular Biology, Faculty of Science and Health, John Paul II Catholic University of Lublin, 20-708 Lublin, Poland; monijach@kul.lublin.pl
5 School of Public Health, Centre of Postgraduate Medical Education of Warsaw, 01-826 Warsaw, Poland; urszula.religioni@gmail.com
6 Department of Developmental Dentistry, Medical University of Lublin, 20-081 Lublin, Poland; rafal.wrobel@umlub.pl
7 Department of Public Health, Medical University of Lublin, 20-090 Lublin, Poland; jolantaherda@umlub.pl
* Correspondence: agnieszkabaranska@umlub.pl

Citation: Błaszczuk, A.; Barańska, A.; Kanadys, W.; Malm, M.; Jach, M.E.; Religioni, U.; Wróbel, R.; Herda, J.; Polz-Dacewicz, M. Role of Phytoestrogen-Rich Bioactive Substances (*Linum usitatissimum* L., *Glycine max* L., *Trifolium pratense* L.) in Cardiovascular Disease Prevention in Postmenopausal Women: A Systematic Review and Meta-Analysis. *Nutrients* **2022**, *14*, 2467. https://doi.org/10.3390/nu14122467

Academic Editor: Winston Craig

Received: 22 April 2022
Accepted: 12 June 2022
Published: 14 June 2022

Publisher's Note: MDPI stays neutral with regard to jurisdictional claims in published maps and institutional affiliations.

Abstract: The aim of this report was to determine the impact of flaxseed, soy and red clover, and their bioactive substances on the lipid profile in postmenopausal women in cardiovascular diseases prevention. We used the following databases: MEDLINE (PubMed), EMBASE and the Cochrane Library. Meta-analysis indicates that the intake of flaxseed by postmenopausal women is associated with a statistically significant reduction in total cholesterol (TC) levels (weighted-mean difference (WMD) = −0.26; 95% confidence interval (95% CI): −0.38 to −0.13; $p = 0.0001$), low-density lipoprotein cholesterol (LDL-C) levels (WMD = −0.19; 95% CI: −0.30 to −0.08; $p = 0.0006$), and high-density lipoprotein cholesterol (HDL-C) levels (WMD = −0.06; 95% CI: −0.11 to −0.01; $p = 0.0150$). The effect of soy protein on the lipid profile showed a significant decrease in TC levels: WMD = −0.15; 95% CI: −0.25–0.05; $p = 0.0048$, LDL-C levels: WMD = −0.15; 95% CI: −0.25–0.05; $p = 0.0067$, as well as a significant increase in HDL-C levels: WMD = 0.05; 95% CI: 0.02–0.08; $p = 0.0034$. Changes in the lipid profile showed a significant reduction in TC levels after the use of red clover (WMD = −0.11; 95% CI: −0.18–−0.04; $p = 0.0017$) and a significant increase in HDL-C levels (WMD = 0.04; 95% CI: 0.01 to 0.07; $p = 0.0165$). This meta-analysis provides evidence that consuming flaxseed, soy and red clover can have a beneficial effect on lipids in postmenopausal women and suggest a favorable effect in preventing cardiovascular diseases.

Keywords: flaxseed; soy; red clover; lipid profile; meta-analysis; cardiovascular disease; botanical supplements; postmenopausal woman

1. Introduction

Cardiovascular disease (CVD) is collection of disorders affecting the vasculature of the heart, brain and peripheral tissues, and remains the leading cause of death globally [1,2]. The most common cause of CV is atherosclerosis, which is initiated by an inflammatory reaction of the vascular endothelium [3]. The origins of these endothelial lesions are still not fully explained, but involved factors include: chronic elevations in blood pressure [4]; prolonged hyperglycemia and the resulting formation of advanced glycation end-products [5];

elevated lipoproteins, particularly molecules that have undergone oxidized modification [6]; and oxidative stress and inflammation [7]. With aging, a number of changes occur in the metabolism, known as the 'metabolic syndrome' [8]. Among others, these include the accumulation of fat mass in the abdominal compartment, transition to a more atherogenic lipid profile, hyperinsulinemia, insulin resistance and glucose intolerance [9,10]. The consequence of these changes is an enhanced risk of coronary heart disease, stroke and other atherosclerotic vascular diseases, including peripheral arterial disease, atherosclerotic aortic disease and carotid artery disease [11].

A bioactive effect on lipid metabolism involving lowering the level of total cholesterol (TC), low-density lipoprotein cholesterol (LDL-C) and triglycerides (TG), has been demonstrated during studies of some plant dietary items, such as: almonds [12], artichokes [13], barberry [14], curcumin [15], ginger [16], psyllium [17], sesame [18], cacao [19] and walnuts [20].

Women are at a higher risk of developing CVDs after menopause due to estrogen deficiency and dysregulated lipid metabolism [21]. Loss of ovarian endocrine function as a result of chronic hypoestrogenism is the main physiological symptom associated with menopause. The daily production of estrogen in postmenopausal women is 0.045 mg, compared with 0.35 mg during the reproductive period, which is reflected in serum estrogen concentrations of 10–20 μg/mL and 40–400 μg/mL, respectively [22]. Observed menopause-induced estrogen deficiency leads to various metabolic disorders including lipid metabolism. TC, LDL-C, and TG levels increase during the menopause and during the postmenopausal period. In turn, high-density lipoprotein cholesterol (HDL-C) levels, after an initial rise during the menopausal transition, gradually decline during late menopause [23–25] (of note, there were also studies showing no difference in HDL-C levels between premenopausal and postmenopausal women [26]). Dyslipidemia is one of the most important risk factors for CVD, which can be corrected and prevented. Botanical supplements as flaxseed, soybean and red clover are rich sources of bioactive compounds affecting lipid metabolism [27].

The benefits of consuming whole fractions of flaxseed (*Linum usitatissimum* L.) such as its protein, oil and mucilage, are related to the presence of specific bioactive substances. The flaxseed content of protein ranges from 10 to 31%, including higher amounts of arginine, aspartic and glutamic acids than other amino acids. Flaxseed also consists of 40% fat; and 25–28% fiber, of which 25% is in soluble form. Moreover, approximately 38–45% of flaxseed mass contains oil and 55–68% is meal. Flaxseed is a rich source of bioactive ingredients such as α-linolenic acid (ALA) and linoleic acid. Additionally, it contains phytochemicals such as lignan complex: secoisolariciresinol diglucoside (SDG), cinnamic acid glucoside and hydroxymethyl glutaric acid [28,29]. Flaxseed oil and active compounds, especially SDG and its metabolites, suppresses the inflammatory tissue damage caused by oxidative stress [30]. SDG may also directly lower serum cholesterol by modulating the enzymes 7α-hydroxylase and acyl-coenzyme A:cholesterol acyltransferases, both of which are involved in cholesterol metabolism [31]. The supplied ALA reduces the production of arachidonic acid (AA) and consequently, by decreasing proinflammatory eicosanoid, leads to a reduction in the inflammation process [32].

The soybean (*Glycine max* L.) is a significant source of protein (~36–40%), lipids (~20%) and dietary fiber (~9%) (based on the dry weight of mature raw seeds), and phytochemicals such as isoflavones, phytosterols and lecithins, which may act collectively or through independent mechanisms. The two major protein peptides, β-conglycinin (βCG) and glycinin, comprise 80–90% of the total protein in soybean, and affect lipid metabolism [33,34]. Additionally, soybeans are rich sources of essential fatty acids. Polyunsaturated (primarily linoleic acid, alpha-linolenic acid), monounsaturated (oleic acid) and saturated (primarily palmitic acid) fatty acids comprise approximately 63%, 23%, and 14%, respectively, of the total fat content of soybeans, and have an impact on the level of lipids [35]. The other major bioactive compounds in soybeans are isoflavones, which are associated with soy proteins. Isoflavones occur in large values in soybean as glycoside, such as genistin, daidzin and glycitin, or their aglycone forms, genistein, diadzein and glycitein [36]. Soy isoflavones, with structural similarities to the endogenous 17β-estradiol, reveal their biological effects via activating estrogen receptors (ER) with a higher affinity to ER-β, in comparison to ER-α. Although the affinity for the estrogen receptor by soy isoflavones is 100–1000 times less than that of natural estrogen, more than a thousand-fold greater isoflavone concentrations can appear in the plasma than those of endogenous estrogen [37]. Isoflavones, by binding to ERs, lead to gene activation and beneficial effects on lipid metabolism [38].

A number of other mechanisms regulating lipid metabolism without the mediation of the estrogen receptor have been recorded—including the increased expression of 3-hydroxy−3-methylglutaryl-CoA reductase (HMGCR), which leads to decreased cholesterol and TG levels; the enhanced expression of peroxisome proliferator-activated receptor (PPAR) and the activation of AMP-activated protein kinase (AMPK), which results in increased expression of genes involved in lipoprotein metabolism; the decreased expression of sterol regulatory-element binding protein-lc (SREBP-1) and increased expression of SREBP-2, which suppresses cholesterol synthesis and absorption in the liver; the inhibition of the expression and activity of the sterol regulatory element binding protein-1c (SREBP-1c) and carbohydrate response element binding protein-1 (ChREBP), which are proteins that enhance the expression of lipogenic genes and key enzymes involved in de novo lipogenesis; the promotion of the HDL-C metabolism and of the uptake, utilization and catabolism of fatty acids; and the modulation of the effects on several enzymes important in lipid transformation, such as lipoprotein lipase (LPL), hepatic lipase (HL) (also called hepatic triglyceride lipase (HTGL)), and 7alpha-hydroxylase [39–44].

Red clover (*Trifolium pratense* L.) contains a certain amount of protein and fat that is irrelevant from the point of view of human nutrition. It is also rich in bioactive substance-sused in medicine. Red clover isoflavones show a different mechanism of action on lipid metabolism than that of soy isoflavones, which is due to the different composition of the contained isoflavones. Grains of red clover contain higher concentrations of formononetin and biochanin A and lower concentrations of daidzein and genistein than soy [45]. This composition suggests that an equal production status may be less relevant [46]. Isoflavones with structural similarities to endogenous 17-β-estradiol reveal their biological effects via activating estrogen receptors (ER) with a higher affinity to ER-β, in comparison to ER-α, which mediates the cholesterol metabolism [47,48]. In addition, a number of non-hormonal effects have been reported in its isoflavones, including tyrosine kinase inhibition, antioxidant activity, and effects on ion transport [49]. Red clover extract and the isoflavones genistein and biochanin A can also regulate lipid metabolism without the mediation of estrogen receptors, as well as increase the expression of PPAR alpha and activate AMPK, which results in the enhanced activity of genes involved in lipoprotein metabolism [50].

The purpose of this study was to determine the impact of flaxseed, soy and red clover and their bioactive substanceson the lipid profile in postmenopausal women in cardiovascular prevention.

2. Materials and Methods

2.1. Search Strategy and Study Selection

This systematic review and meta-analysis was designed in accordance with The Preferred Reporting Items for Systematic Reviews and Meta-analysis (PRISMA) statement [51] to identify randomized controlled trials (RCTs) assessing the effects of flaxseed, soy protein, soy isoflavones and red clover isoflavones on the level of serum lipids.

The electronic databases MEDLINE (PubMed), Embase, and the Cochrane Library were searched for the identification of randomized controlled trials until December 2018. The following search terms were used for all databases in various combinations: ("flax" OR "flaxseed" OR "linseed" OR "Linum usitatissimum" OR "soybean" OR "Glycine max" OR "soy proteins" OR "soy isoflavones" OR "red clover" OR "Trifolium pratense") AND ("lipid profile" OR "lipids" OR "total cholesterol" OR "HDL cholesterol" OR "LDL cholesterol" OR "triglycerides") AND ("menopause" OR "postmenopause").

The search was limited to papers published in English and was conducted up to December 2018. References to selected research and review articles related to the topic of the work were also searched in order to identify additional studies.

The initial selection included the analysis of the titles and/or abstracts of all citations. After an independent and double analysis of the full texts of selected works, a decision was then made to include or exclude them. In turn, works were qualified for meta-analysis and collection of data on the clinical and methodological characteristics of the described clinical trials and for statistical evaluation.

Randomized controlled trials (RCTs) were considered eligible for inclusion if they met all of the following criteria: parallel-group design, or crossover design that contained data for the first period; a comparison with a placebo or with a no-intervention group; a follow-up period was at least 3 months; post-menopausal women as participants; appropriate interventions using flaxseed, soy or red clover and the presentation of sufficient information on plasma-lipid levels at baseline and after supplementation, or the net change values in both study arms. The exclusion criteria were as follows: men or premenopausal women as participants, no control group in the study, lack of sufficient information, and a study duration of less than 12 weeks. The results were reported as graphics or percent changes, and as duplicated reports.

2.2. Data Extraction

The data were extracted by the lead author and subsequently reviewed by co-authors for accuracy. Eligible studies were reviewed and the following data were abstracted: first author's name; year of publication; study location (country); follow-up period of the study; study design; number of participants in the intervention and control group; health characteristics of the population (age, menopausal status, body mass index); daily amount of flaxseed, soy protein, soy isoflavones and red clover isoflavones taken in the active arm; and data on baseline and follow-up TC, LDL-C, HDL-C and TG plasma levels.

2.3. Quality Assessment and Bias Risk of the Trials

The Jadad Scale is an Oxford system for assessing the quality of a clinical trial, designed to determine the minimum level of studies included in a systematic review/meta-analysis. The test may receive values from 0 (low quality) to 5 points (highest quality) [52]. This meta-analysis included studies that had a relatively high Jadad score. To explain the possible presence of bias publications, Begg's rank correlation test (Kendall Tau) and Egger's weighted regression test were applied [53,54].

2.4. Statistical Analysis and Meta-Analysis

The meta-analysis included all intervention groups from multi-arm studies. Moreover, to avoid the duplication of data from the same people in surveys covering multiple time points, only one such point was taken into account.

The data in each study were presented as numbers of subjects (N) and the mean $\pm$ standard deviations (SD). When the standard error of the mean (SEM) was employed, the conversion to SD was made according to the formula: SD = SEM $\times \sqrt{N}$. If a 95% confidence interval (95% CI) was applied, SD conversion was: SD = sqrt (N) $\times$ (upper bound–lower bound)/(2u) (equal to 3.96). When the results from the studies were presented in mg/dL, they were converted into mmol/L using standard conversion factors (the value in mg/dL was multiplied by 0.02586 for TC, LDL-C and HDL-C, and by 0.01143 for TG).

The outcome measures were the differences in the mean (MD) of components of the lipid profile between baseline and the end values for both the intervention and control groups. The missing SDs of MD were imputed using the formula: SD = sqrt ((SD "initial")2 + (SD "final")2 − (SD "initial" $\times$ SD "final") $\times$ 2R), where R is the correlation coefficient; we took an R value = 0.40 [55,56]. The outcome measures were the differences in the mean (net change in mmol/L) of elements of the lipid profile between the baseline and the end values for both the intervention and control groups.

Summary outcomes measures were presented as the mean differences between the intervention and control groups. A random-effects model was used to calculate the weighted-mean difference (WMD) and 95% confidence interval (CI) for each comparison, and the combined overall effect ($p < 0.05$ was considered statistically significant) according to DerSimonian and Laird [57]. Cochrane Q and I^2 statistics were used to assess the heterogeneity. The I^2 test determined whether the variance across studies was correct and not a result of a sampling error. The percentage of total variation indicated the degree of heterogeneity; I^2 values of ≤25% were considered low; >25% as moderate; and ≥75% as high heterogeneity [58]. STATISTICA Medical Software v. 11.0 StatSoft, Krakow, Poland was used for all statistical analyses.

3. Results

In total, a number of citations potentially related to the topic of work based on the key words—red clover = 3107; soy = 8074; and flaxseed = 4828—were identified. Building upon the title and/or abstract, exclusions were 3069 for red clover; 7991 for soy; and 4784 for flaxseed due to a lack of connection with the topic of this work. Consequently, 165 potentially relevant clinical trials qualified for further detailed qualitative analysis in the full-text assessment: red clover = 38; soy = 83; and flaxseed = 44. Among these, 130 studies were also discarded due to the failure to meet all inclusion criteria. As a result, 42 randomized controlled trials for meta-analysis. Detailed information about the literature search and study selection and identification can be found in Figure 1.

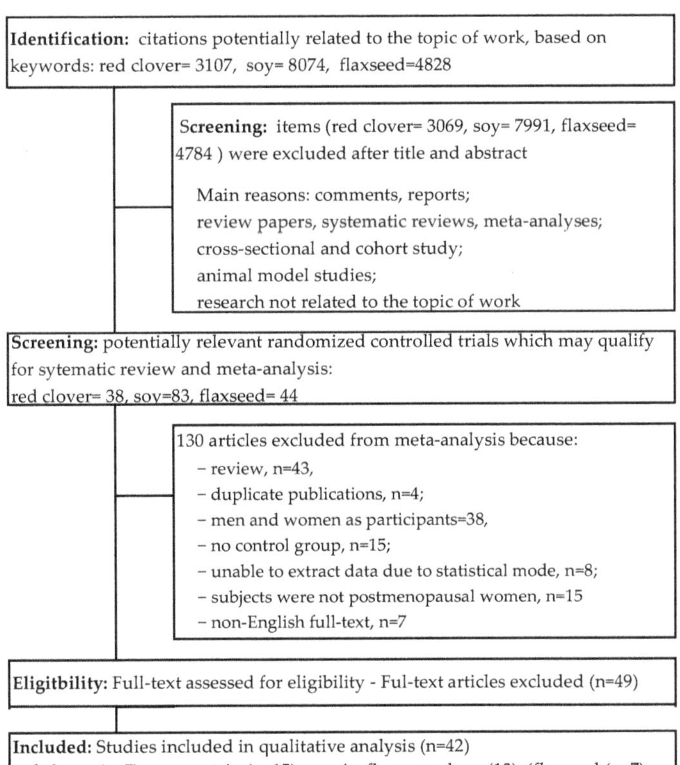

Figure 1. Flowchart of the selection procedure for studies included in the current review and meta-analysis.

3.1. Characteristics of Included Trials

The characteristics of selected randomized controlled studies assessing the influence of flaxseed, soy protein, soy isoflavones, and red clover on lipid profile in postmenopausal women are presented in Table 1. The meta-analysis included 42 studies published in English from 1998 to 2018 [59–100].

3.2. Associations between Flaxseed and Plasma Lipid Profiles

Changes in lipid profile after the use of flaxseed were analyzed on the basis of seven studies [59–65]. The results of the meta-analysis are presented in Figure 2. Compared to the control group, the use of flaxseed resulted in a statistically significant reduction in TC levels (WMD = −0.26; 95% CI: −0.38–−0.13; p = 0.0001), LDL-C levels (WMD = −0.19; 95% CI: −0.30–−0.08; p = 0.0006) and HDL-C levels (WMD = −0.06; 95% CI: −0.11–−0.01; p = 0.0150) and a slight, not statistically significant reduction in TG levels: WMD = −0.03; 95% CI: −0.12–0.07; p = 0.5452. The heterogeneity analysis performed for TC, LDL-C, HDL-C and TG did not show that the differences between the effects obtained in different studies were statistically significant. The Begg and Egger asymmetry tests showed no publication bias for TC (p-value 0.6523 and 0.3091, respectively), LDL-C (p-value 0.6523 and 0.1786, respectively), HDL-C (p-value 0.1765 and 0.1578, respectively) or TG (p-value 0.4527 and 0.9335, respectively).

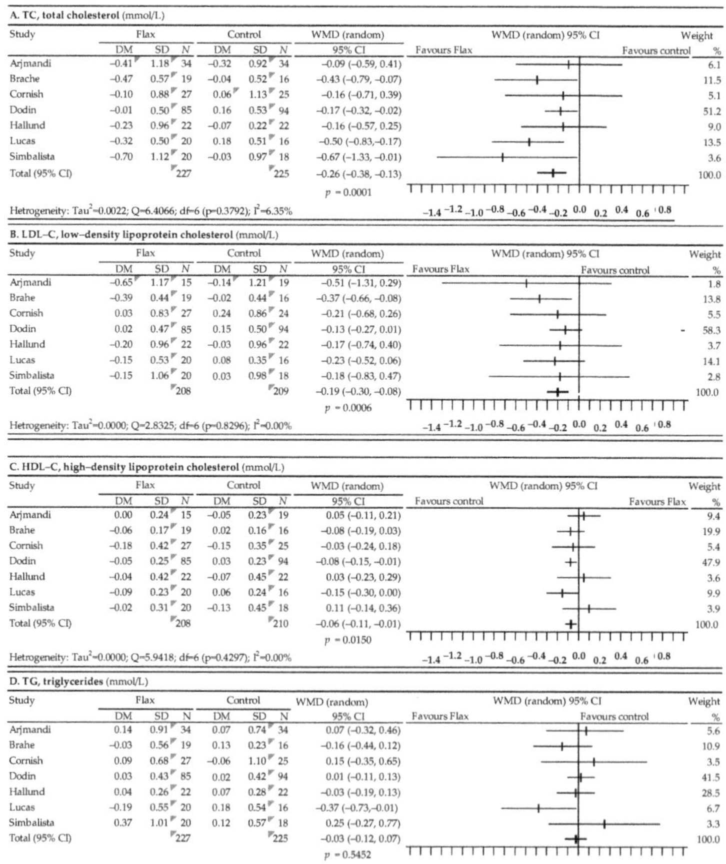

Figure 2. Forest plot representing the associations between flaxseed and lipid profiles. Data are presented as weighted mean difference with 95% CI.

Table 1. Characteristics of selected randomized controlled studies assessing the influence of flaxseed, soy protein, soy isoflavones, and red clover on lipid profile in postmenopausal women.

First Author [Ref.] Data Location	Study Design Trial Duration	Study Population Age (Mean ± SD) y, ysm, BMI, Health Condition	Intervention (Daily Dose)	GroUp Studied	Number Sample	Baseline Lipids Values Total-C mmol/L	LDL-C mmol/L	HDL-C mmol/L	TAG mmol/L	Jadad Score
			A. Flaxseed (*Linum usitatissimum* L.)							
Arjmandi [59] 1998 United States	Cross-over 6-week active phase 2-week washout.	Age 56.3 ± 6.5, ysm N/A, BMI 29.2 ± 7.4, obesity, hypercholesterolemia	WFX 38 g, ALA 8.5 g vs. placebo: sunflower seed (slice of bread or muffin)	FG CG	15 19	5.95 ± 1.44 5.92 ± 1.36	4.12 ± 1.39 4.06 ± 1.34	0.93 ± 0.23 1.08 ± 0.23	1.28 ± 0.92 1.27 ± 0.70	4
Lucas [64] 2002 United States	Parallel group 3-month follow-up	Age 54 ± 8, ysm N/A, BMI 29.1 ± 7.1 obesity	WFX 40 g vs. placebo, wheat-based 40 g	FG CG	20 16	5.76 ± 1.12 5.95 ± 1.12	3.21 ± 1.12 3.52 ± 1.12	1.89 ± 0.42 1.61 ± 0.40	1.48 ± 0.71 1.56 ± 0.76	4
Dodin [62] 2005 Canada	Parallel group 1-year follow-up	Age 54.0 ± 4.0, ysm 4.7 ± 5.2, BMI 25.5 ± 4.5 healthy	WFX 40 g, ALA 9.1 g vs. control, wheat germ (slice of bread or drinks)	FG CG	85 94	5.67 ± 0.75 5.78 ± 0.71	3.43 ± 0.69 3.50 ± 0.64	1.72 ± 0.33 1.74 ± 0.39	1.12 ± 0.45 1.16 ± 0.57	5
Hallund [63] 2006 Denmark	Cross-over 6-week active phase 6-week washout	Age 61 ± 7, ysm >24 mo, BMI 25.5 ± 4.5 healthy	Lignan complex, SDG 500 mg vs. control (in form muffins, 50 g)	FG CG	22 22	6.05 ± 1.03 6.03 ± 0.98	3.80 ± 1.03 3.79 ± 0.98	1.81 ± 0.42 1.82 ± 0.52	0.96 ± 0.28 0.93 ± 0.33	4
Cornish [61] 2009 Canada	Parallel group 6-month follow-up	Age 59.7 ± 5.3, ysm N/A, BMI 27.1 ± 5.3 healthy	Lignan complex, SGD 500 mg vs. placebo	FG CG	27 25	5.87 ± 0.88 6.14 ± 1.05	3.60 ± 0.88 3.77 ± 0.80	1.74 ± 0.42 1.54 ± 0.40	1.19 ± 0.68 1.77 ± 1.10	4
Simbalista [65] 2010 Brazil	Parallel group 3-month follow-up	Age 52.0 ± 2.9, ysm 3.8 ± 2.3, BMI 26 ± 3.6, healthy	GFX: WFX 25 g, SDG 46 mg, vs placebo: wheat bran (in form of slice bread)	FG CG	20 18	6.03 ± 0.87 5.18 ± 0.93	3.83 ± 0.89 2.87 ± 0.93	1.61 ± 0.31 1.86 ± 0.42	1.49 ± 0.80 1.00 ± 0.54	5
Brache [60] 2015 Denmark	Parallel group 6-week follow-up	Age 60.6 ± 6.4 y, ysm ≥1 y, BMI 35.2 ± 4.5, obesity	10 g flaxseed mucilage vs. placebo: maltodextrin (in form buns)	FG CG	19 16	6.39 ± 0.89 5.76 ± 0.69	4.11 ± 0.84 3.44 ± 0.74	1.40 ± 0.22 1.56 ± 0.42	1.51 ± 0.77 1.07 ± 0.32	3

Table 1. *Cont.*

First Author [Ref.] Data Location	Study Design Trial Duration	Study Population Age (Mean ± SD) y, ysm, BMI, Health Condition	Intervention (Daily Dose)	GroUp Studied	Number Sample	Baseline Lipids Values Total-C mmol/L	LDL-C mmol/L	HDL-C mmol/L	TAG mmol/L	Jadad Score
B. Soybean (*Glycine max* (L.) Merr.)										
B. 1. Soy protein without and with isoflavones										
Baum [68] 1998 United States	Parallel groups 2-week run-in/ 12-week follow-up	Age 60.8 ± 8.6 y, ysm N/A, BMI 27.8 ± 5.3, hypercholesterolemia	a. SP 40 g: a. IAE 90 mg; b. SP 40 g; IAE 56 mg vs. control, CP + MP 40 g	SG 90 SG 56 CG	21 23 22	6.47 ± 0.88 6.57 ± 0.85 6.26 ± 0.67	N/A N/A 4.9 ± 0.8	1.38 ± 0.32 1.34 ± 0.28 1.38 ± 0.31	1.74 ± 0.75 1.89 ± 1.02 1.75 ± 1.11	3
Vigna [80] 2000 Italy	Parallel groups 12-week follow-up	Age 53.4 ± 3.3, ysm 2.4 y, BMI 25.9 ± 3.5, healthy	SP 40 g, IF 76 mg vs. control, CP 40 g	SG CG	40 37	6.37 ± 1.01 6.55 ± 0.93	4.13 ± 0.87 4.33 ± 0.87	1.57 ± 0.36 1.61 ± 0.38	1.47 ± 0.90 1.32 ± 0.77	4
Gardner [72] 2001 United States	Parallel groups 4-week run-in/ 12-week follow-up	Age 59.9 ± 6.6, ysm N/A, BMI 26.3 ± 4.6, hypercholesterolemia	a. SP 42 g b. SP 42 g (52 mg Gen, 25 mg Dai, 4 mg Gly) vs. control, MP 42 g.	SG SG CG	33 31 30	5.9 ± 0.7 5.9 ± 0.6 6.1 ± 0.6	3.9 ± 0.6 3.9 ± 0.6 4.0 ± 0.5	1.4 ± 0.3 1.5 ± 0.3 1.5 ± 0.4	1.3 ± 0.5 1.3 ± 0.8 1.3 ± 0.7	4
Han [73] 2002 Brazil	Parallel groups 4-month follow-up	Age 48.5 ± 7.6, ysm 1.9 ± 1.6 y, BMI 24.3 ± 3.2, healthy	SP 50.3 mg, IAE 23.3 mg Gen, 3.8 mg Gly, 6.2 mg Dai) vs. placebo	SG CG	40 40	5.83 ± 0.88 5.86 ± 1.26	3.45 ± 0.87 3.45 ± 1.32	1.04 ± 0.23 1.03 ± 0.21	2.31 ± 1.66 1.99 ± 1.66	5
Dalais [71] 2003 Australia	Parallel groups 3-month follow-up	Age 60 ± 6.2, ysm N/A, BMI 25.3 ± 4.6, healthy	SP 40 g, IC 118 mg (69 mg Agl) vs. control, CP 40 g	SG CG	38 40	6.12 ± 0.92 5.92 ± 0.88	4.00 ± 0.86 3.69 ± 0.88	1.63 ± 0.49 1.72 ± 0.51	1.09 ± 0.68 1.01 ± 0.57	5
Steinberg [78] 2003 United States	Cross-over 6-week active phase 4-week washout	Age 5.49 ± 5.29, ysm N/A, BMI 24.6 ± 3.2, healthy	a. SP 25 g b. SP 25 g, IAE 107 mg (55 mg Gen, 47 mg Dai, 5 mg Gly) vs. control, MP 25 g	SG a SG b CG	24 24 24	4.91 ± 0.49 4.91 ± 0.49 4.91 ± 0.49	2.89 ± 0.49 2.89 ± 0.49 2.89 ± 0.49	1.55 ± 0.49 1.55 ± 0.49 1.55 ± 0.49	1.03 ± 0.49 1.03 ± 0.49 1.03 ± 0.49	4
Cuevas [70] 2003 Chile	Cross-over 8-week active phase 4-week washout	Age 59 y, ysm 10 y, BMI 29.3 ± 3.43, obesity, hypercholesterolemia	SP 40 g, IAE 80 mg (60% Gen, 30% Dai, 10% Gly) vs. control, caseinate 40 g	SG CG	18 18	7.90 ± 0.74 7.90 ± 0.74	5.04 ± 0.66 5.04 ± 0.66	1.39 ± 0.27 1.39 ± 0.27	2.18 ± 0.83 2.18 ± 0.83	4
Kreijkamp- Kaspers [75] 2004 Netherlands	Parallel groups 12-month follow-up	Age 66.6 ± 4.7, ysm 17.9 ± 6.9 y, BMI 26.1 ± 3.8, healthy	SP 25.6 g, IAE 99 mg (52 mg Gen, 6 mg Gly, 41 mg Dai) vs. control, MP 25,6 mg	SG CG	88 87	6.21 ± 0.73 6.11 ± 0.95	4.16 ± 0.99 4.12 ± 0.88	1.55 ± 0.41 1.53 ± 0.34	1.36 ± 0.72 1.25 ± 0.59	4
Teede [79] 2005 Australia	Parallel groups 3-day run-in/ 3-month follow-up	Age 59.5 ± 4.5, ysm N/A, BMI 25.9 ± 5.4, healthy	SP 40 g, IC 118 mg (54 mg Gen, 3.6 mg Gly, 26 mg Dai) vs. control, CP 40 g	SG CG	19 21	6.2 ± 1.30 5.8 ± 0.92	4.0 ± 0.87 3.6 ± 0,92	1.6 ± 0.43 1.6 ± 0.46	1.0 ± 0.48 1.0 ± 0.63	3

Table 1. *Cont.*

First Author [Ref.] Data Location	Study Design Trial Duration	Study Population Age (Mean ± SD) y, ysm, BMI, Health Condition	Intervention (Daily Dose)	GroUp Studied	Number Sample	Baseline Lipids Values Total-C mmol/L	LDL-C mmol/L	HDL-C mmol/L	TAG mmol/L	Jadad Score
Allen [66] 2007 United States	Parallel groups 4-week run-in/ 12-week follow-up	Age 56.8 ± 5.6, ysm 9.4 ± 8.3 y, BMI 27.9 ± 4.7, hypercholesterolemia	SP 20 g, IC 160 mg (~96 mg Agl) vs. control, MP 20 g	SG CG	93 98	5.80 ± 0.68 5.71 ± 0.64	3.67 ± 0.68 3.60 ± 0.57	1.56 ± 0,37 1.52 ± 0.31	1.25 ± 0.51 1.28 ± 0.60	5
Maesta [77] 2007 Brazil	Parallel group 16-week follow-up	Age 61.3 ± 5,2, ysm 10.7 ± 4.9 y, BMI 27.2 ± 5.3 healthy	SP 25 g, IAE 50 mg (32 mg Gen, 15 mg Dai, 3 mg Gly) vs. placebo, maltodextrine	SG CG	10 11	5.95 ± 0.71 5,76 ± 0.98	3.71 ± 0.72 3.56 ± 0.70	1.62 ± 0.34 1.32 ± 0.25	1.36 ± 0.52 1.95 ± 0.71	5
Basaria [67] 2009 United States	Parallel groups 12-week follow-up	Age 55.7 ± 1.3, ysm 5.7 ± 0.9, BMI 26.1 ± 0.8, healthy	SP 20 g, IC 160 mg (IAE: 64 mg Gen, 63 mg Dai, 34 mg Gly) vs. control, MP 20 g	SG CG	38 46	5.48 ± 0.14 5.69 ± 0.85	3.15 ± 0.75 3.21 ± 0.74	1.88 ± 0.46 2.02 0.46	1.03 ± 0.58 0.99 ± 0.46	4
Campbell [69] 2010 United States	Parallel groups 12-month follow-up	Age 54.7 ± 5.5, ysm 5.5 ± 5.0, BMI 27.9 ± 5.9, hypercholesterolemia	SP 25 g, 60 mg IF vs. control, CP 25 g	SG CG	35 27	5.97 ± 0,93 6.15 ± 0.91	3.88 ± 0.90 3.95 ± 0.87	1.47 ± 0.38 1.50 ± 0.36	1.34 ± 0.70 1.48 ± 0.67	4
Jassi [74] 2010 India	Parallel groups 12-week follow-up	Age 51.1 ± 8.6, ysm 2.3 ± 1.2, BMI 23.4 ± 2.7, healthy	SP 30 g, IF 60 mg vs. control, CP 30 g	SG CG	25 25	4.96 ± 0.36 4.69 ± 0.71	3.09 ± 0.37 2.83 ± 0.76	1.06 ± 0.15 1.06 ± 0.16	1.76 ± 0.28 1.76 ± 0.17	4
Liu [76] 2012 Hong Kong SAR	Parallel groups 2-week run-in/ 3-month follow-up	Age 56.3 ± 4.3, ysm 5.9 ± 5.4, BMI 24.4 ± 3.6, prediabetes	SP 15 g, IAE 100 mg (59 mg Gen,4 mg Gly, 35 mg Dai) vs. control, MP 15 g	SG CG	60 60	5.83 ± 0.94 5.63 ± 0.93	3.94 ± 0.67 3.81 ± 0.88	1.66 ± 0.31 1.65 ± 0.30	1.35 ± 1.19 1.30 ± 0.70	5
			B.2. Soy isoflavones preparations							
Dewell [85] 2002 USA	Parallel groups 2-month follow-up	Age 69.5 ± 4.2 y, ysm N/A, BMI 25.0 ± 4,2, moderate hypercholesterolemia	IC 150 mg (90 mg Agl: 45 mg Gen, 55% Dai and Gly) vs. placebo	SG CG	20 16	6.8 ± 0.9 6.3 ± 2.0	N/A N/A	1.2 ± 0.5 1.2 ± 0.4	0.8 ± 0.5 1.3 ± 0.8	4
Colacurci [93] 2005 Italy	Parallel groups 6-month follow-up	Age 55.1 ± 38 y, ysm 4.9 ± 0.6, BMI 25.9 ± 1.8, healthy	IAE 60 mg (30 mg Gen, 30 mg Dai) vs. placebo	SG CG	29 28	NR NR	3.7 ± 0.3 3.6 ± 0.4	1.06 ± 0.5 1.05 ± 0.5	1.5 ± 0.6 1.6 ± 0.8	4
Garrido [87] 2006 Chile	Parallel groups 12-week follow-up	Age 55.5 ± 4.0 y, ysm N/A, BMI 26.9 ± 2.3, healthy	IAE ~100 mg (46.8 mg Gen, 48.2 mg Dai) vs. placebo	SG CG	15 14	5.5 ± 1.0 4.8 ± 0.5	3.4 ± 0.4 2.9 ± 03	1.4 ± 0.3 1.8 ± 0.6	1.3 ± 0.2 1.4 ± 0.2	3

Table 1. *Cont.*

First Author [Ref.] Data Location	Study Design Trial Duration	Study Population Age (Mean ± SD) y, ysm, BMI, Health Condition	Intervention (Daily Dose)	GroUp Studied	Number Sample	Baseline Lipids Values Total-C mmol/L	LDL-C mmol/L	HDL-C mmol/L	TAG mmol/L	Jadad Score
Wu [92] 2006 Japan	Parallel group 6-month follow-up	Age 54.4 ± 2.9 y, ysm N/A, BMI 21.1 ± 2.4, healthy	IC 75 mg (47 mg Agl: 38.3 mg Dai, 8.6 mg, 1 mg Gly) vs. placebo	SG CG	25 29	5.90 ± 0.76 5.88 ± 0.86	3.52 ± 0.72 3.59 ± 0.76	1.92 ± 0.47 1.85 ± 0.38	0.95 ± 0.43 1.16 ± 0.53	3
Nahas [90] 2007 Brazil	Parallel groups 4-week run-in 4-month follow-up	Age 55.7 ± 6.8, ysm 6.9 ± 4.5, BMI 29.1 ± 5.0, obesity	IC 100 mg (50% Gen, 35% Dai), vs. placebo	SG CG	38 36	5.56 ± 0.92 5.37 ± 0.97	3.47 ± 0.82 3.26 ± 0.82	1.29 ± 0.27 1.35 ± 0.34	1.73 ± 0.74 1.67 ± 0.89	3
Ho [88] 2007 China	Parallel groups 6-month follow up	Age 54.2 ± 3.1, ysm 4,1 ± 2.4, BMI 24.1 ± 3.6, healthy	a. IAE 80 mg, b. IAE 40 mg (46.4% Dai, 38.8 Gly, 14.7% Gen) vs. placebo	SG 80 SG 40 CG	67 68 68	5.86 ± 0.83 5.83 ± 0.84 5.93 ± 0.89	3.19 ± 0.74 3.23 ± 0.68 3.25 ± 0.73	1.89 ± 0.41 1.80 ± 0.39 1.86 ± 0.42	1.13 ± 0.56 1.32 ± 0.93 1.29 ± 0.96	4
Aubertin-Leheudre [81] 2008 Canada	Parallel groups 6-month follow-up	Age 57.4 ± 5.4 y, ysm 8.6 ± 7.5, BMI 32.0 ± 12.5, obesity	IAE 70 mg (44 mg Dai, 16 mg Gly, 10 mg Gen) vs. placebo	SG CG	21 18	5.41 ± 0.88 5.33 ± 0.83	3.17 ± 0.81 3.17 ± 0.78	1.55 ± 0.49 1.45 ± 0.37	1.51 ± 0.69 1.52 ± 0.69	4
Özturk Turhan [91] 2009 Turkey	Parallel groups 6-month follow-up	Age 51.5 ± 5.1; ysm 3.6 ± 1.7, BMI 27.1 ± 3.1	IAE 40 mg (29.8 mg Gen, 7.8 mg Dai, 2.4 mg Gly) vs. placebo	SG CG	45 45	6.82 ± 0.96 6.30 ± 0.76	4.25 ± 0.73 4.01 ± 0.65	1.06 ± 0.15 1.06 ± 0.16	1.76 ± 0.28 1.76 ± 0.17	4
Choquette [84] 2011 Canada	Parallel groups 6-month follow-up	Age 58.5 ± 5.5 y, ysm 9.0 ± 7.0, BMI 30.1 ± 2.7, obesity	IAE 70 mg (44 mg Dai, 16 mg Gly, 10 mg Gen) vs. placebo	SG CG	23 22	5.40 ± 0.80 5.58 ± 0.86	3.34 ± 0.75 3.34 ± 0.81	1.49 ± 0.34 1.37 ± 0.32	1.47 ± 0.67 1.44 ± 0.73	5
Kim [89] 2013 Republic of Korea	Parallel groups 12-week follow-up	Age 53.6 ± 3.4 y, ysm 3.6 ± 2,4, BMI 23.3 ± 2.5, healthy	IC 70 mg (Glyc: 38 mg glycitin 20 mg daidzin, 12 mg genistin) vs. placebo	SG CG	42 43	5.13 ± 0.85 5.48 ± 1.03	2.97 ± 0.70 3.25 ± 0.92	1.48 ± 0.36 1.52 ± 0.37	1.26 ± 0.72 1.27 ± 0.66	4
Chilibec [83] 2013 Canada	Parallel groups 24-month follow-up	Age 56.6 ± 68 y, yms N/A, BMI 27.1 ± 4.1, healthy	IC 165 mg (150 mg Agl: Gen, Da and Gly in ratio of 1:1:0.5) vs. placebo	SG CG	72 73	5.87 ± 0.96 5.76 ± 0.91	3.68 ± 0.91 3.59 ± 0.89	1.58 ± 0.41 1.52 ± 0.44	1.41 ± 1.03 1.43 ± 0.79	4
Engelbert [86] 2016 Germany	Parallel groups 12-week follow-up	Age 59.5 ± 6.03 y, yms ≥ 1 y, BMI 25.2 ± 3.8, healthy	IAE 117.4 mg (49.7% Gen, 41.4% Dai, 9.0% Gly) vs. placebo, maltodextrin	SG CG	85 85	5.88 ± 0.89 5.80 ± 0.91	3.78 ± 0.89 3.67 ± 0.85	1.95 ± 0.44 1.99 ± 0.45	1.04 ± 0.39 1.04 ± 0.38	4
Barrasa [82] 2018 Chile	Parallel groups 1-week run-in 3-month follow-up	Age 64.7 ± 4.6 y, ysm N/A, BMI 27.6 ± 0.9, healthy	IAE 100 mg (52 mg Gen, 40 mg Dai, 8 mg Gly) vs. placebo	SG CG	20 15	5.13 ± 0.68 4.87 ± 0.62	3.10 ± 0.94 2.97 ± 0.50	1.30 ± 0.43 1.18 ± 0.38	1.53 ± 0.39 1.54 ± 0.36	4

Table 1. *Cont.*

First Author [Ref.] Data Location	Study Design Trial Duration	Study Population Age (Mean ± SD) y, ysm, BMI, Health Condition	Intervention (Daily Dose)	GroUp Studied	Number Sample	Baseline Lipids Values Total-C mmol/L	LDL-C mmol/L	HDL-C mmol/L	TAG mmol/L	Jadad Score
			C. Red clover (*Trifolium pratense* L.)							
Hale [96] 2001 Australia	Parallel groups 3-month follow-up	Age 47.2 ± 2.4 y, yms N/A, BMI 26.7 ± 4.6, healthy	IAE 30 mg (big amount of Bio and small amount of For (no data)) vs. placebo	RCG CG	14 14	4.64 ± 0.78 4.19 ± 0.85	2.89 ± 0.61 2.49 ± 0.73	1.29 ± 0.24 1.34 ± 0.43	1.46 ± 0.67 1.61 ± 1.04	4
Atkinson [94] 2004 United Kingdom	Parallel groups 12-month follow-up	Age 52.2 ± 4.8 y, yms N/A, BMI 25.3 ± 3.7, healthy	IAE 40 mg (24.5 mg Bio, 8.0 mg For, 1 mg Gen, 1 mg Dai) vs. placebo	RCG CG	77 86	6.34 ± 1.19 6.08 ± 1.04	4.21 ± 0.94 3.88 ± 1.00	1.61 ± 0.41 1.66 ± 0.48	1.24 ± 0.71 1.19 ± 0.66	3
Schult [100] 2004 USA	Parallel groups 2-week run-in 12-week follow-up	Age 52.3 ± 3.1 y, yms 3.2 ± 4.5, BMI 26.1 ± 4.9, healthy	IAE 82 mg (49 mg Bio, 14 mg For, 8 mg Gen, 7 mg Dai). IAE 57 mg (44.6 mg For, 5.8 mg Bio, 0.8 mg Dai, 0.8 mg Gly) vs. placebo	RCG 82 RCG 57 CG	81 81 83	5.76 ± 0.92 5.77 ± 1.01 5.72 ± 0.83	3.77 ± 1.01 3.81 ± 1.14 3.72 ± 0.79	1.36 ± 0.37 1.34 ± 0.34 1.38 ± 0.40	1.32 ± 0.65 1.31 ± 0.77 1.22 ± 0.56	4
Hilgado [97] 2005 Ecuador	Cross-over 90-day active phase 7-day washout	Age 51.3 ± 3.5 y, yms ≥ 1 y, BMI 26.1 ± 3.9, healthy	IAE 80 mg (49 mg Bio, 16 mg For, 8 mg Gen, 7 mg Dai) vs. placebo	RCG CG	53 53	5.79 ± 0.97 5.79 ± 0.97	3.80 ± 0.77 3.80 ± 0.77	1.03 ± 0.30 1.03 ± 0.30	2.28 ± 0.89 2.28 ± 0.89	4
Clifton-Bligh [95] 2015 Australia	Parallel groups 1-month run-in 12-month follow-up	Age 54.4 ± 3.9 y, yms ≥ 1 y, BMI 24.8 ± 4.3, healthy	IAE 57 mg (44.6 mg For, 5.8 mg Bio, 1.9 mg Dai, 0.8 mg Gen, 0.8 Gly) vs. placebo	RCG CG	56 47	5.91 ± 1.05 5.80 ± 0.88	3.68 ± 0.94 3.43 ± 0.86	1.67 ± 0.35 1.82 ± 0.49	1.33 ± 0.60 1.11 ± 0.63	5
Lambert [98] 2017 Denmark	Parallel groups 12-week follow-up	Age 52.5 ± 3.5 y, yms N/A, BMI 25.7 ± 4.3 healthy	IEA 33.8 mg (19 mg For, 9 mg Bio, 2.2 mg Gen, 1.6 Dai) vs. placebo	RCG CG	30 29	5.38 ± 0.19 5.63 ± 0.10	3.36 ± 0.16 3.40 ± 0.17	1.76 ± 0.15 1.73 ± 0.10	1.20 ± 0.09 1.18 ± 0.10	6
Lambert [99] 2017 Denmark	Parallel groups 12-month follow-up	Age 61.8 ± 6.4 y, amenorrhea ≥12 months, BMI 25.6 ± 4.5, healthy	IEA 55.8 mg (31.4 mg For, 14.9 mg Bio, 6.9 mg Gen, 2.6 mg Dai) vs. placebo	RCG CG	38 40	5.54 ± 0.86 5.64 ± 1.01	3.28 ± 0.86 3.37 ± 0.89	1.81 ± 0.43 1.82 ± 0.51	1.16 ± 0.37 1.38 ± 0.63	5

Data are presented as mean ± standard deviation (SD). Abbreviations: Agl, aglycone; ALA, α-linolenic acid; Bio, biochanin; BMI, body mass index (kg/m2); CG, control group; CP, casein protein; Dai, daidzein; FG, flaxseed group; For, formononetin; FXO, flaxseed oil; Gen, genistein; GFX, ground flaxseed; Gly, glycitein; Glyc, glycoside; HDL-C, high-density lipoprotein cholesterol; IAE; IC, isoflavone conjugate containing aglycone and glycoside; IF, isoflavones (form and composition unknown); LDL-C, low-density lipoprotein cholesterol; MP, milk protein; N/A, not available, RCG, red clover group; ref., reference; SDG, secoisolariciresinol diglucoside; SG, soy group; SP, soy protein; TAG, triacylglycerols; Total-C, total cholesterol; WFX, whole flaxseed; y, year or years; ysm, years since sine menopause.

3.3. Associations between Soy Protein without and with Isoflavones and Lipid Profiles

Fifteen studies were used in the analysis of the effect of soy protein on the lipid profile [66–80], but the data from the study by Baum et al. did not allow for a comparison of the effect in the case of LDL-C levels [68]. The results of the meta-analysis are presented in Figure 3. Statistical analysis showed a significant decrease in TC levels: WMD = −0.15; 95% CI: −0.25–0.05; $p = 0.0048$, LDL-C levels: WMD = −0.15; 95% CI: −0.25–0.05; $p = 0.0067$, and a significant increase in HDL-C levels: WMD = 0.05; 95% CI: 0.02–0.08; $p = 0.0034$. There was also a slight reduction in TG levels, which, however, was statistically non-significant (WMD = −0.08; 95% CI: −0.19 to 0.03; $p = 0.1462$). The performed analysis of heterogeneity did not show statistically significant differences between the effects of the included studies for TC, LDL-C and HDL-C, but in the case of TG, the heterogeneity was high ($I^2 = 61.43\%$). Begg's test gave a statistically non-significant result for TC ($p = 0.2403$), as well as LDL-C ($p = 0.4421$), HDL-C ($p = 0.8196$) and TG ($p = 0.0945$), which indicated no publication bias. Moreover, Egger's test showed no publication bias for TC: $p = 0.6815$, LDL-C: $p = 0.5596$, HDL-C: $p = 0.6843$, and TG: $p = 0.8158$.

A. TC, total cholesterol (mmol/L)

Study	SP			Control			WMD (random)	Weight
	DM	SD	*N*	DM	SD	*N*	95% CI	%
Allen	−0.055	0.75	93	0.03	0.74	98	−0.09 (−0.30, 0.13)	11.9
Basaria	−0.078	0.99	38	−0.06	1.04	46	−0.02 (−0.45, 0.42)	4.6
Baum (96)	−0.34	0.98	21	−0.18	0.76	40	−0.16 (−0.64, 0.32)	3.9
Baum (52)	−0.39	0.96	23	−0.18	0.76	40	−0.21 (−0.67, 0.25)	4.3
Campbell	0.30	1.02	35	0.41	1.00	27	−0.11 (−0.62, 0.40)	3.6
Cuevas	−1.17	0.86	18	−1.09	0.79	18	−0.08 (−0.62, 0.46)	3.2
Dalais	−0.81	1.17	38	−0.51	0.89	40	−0.30 (−0.76, 0.16)	4.2
Gardner	0.00	0.89	33	−0.20	0.72	30	0.20 (−0.24, 0.64)	4.5
Gardner	−0.20	0.61	31	−0.20	0.72	30	0.00 (−0.34, 0.34)	6.9
Han	−0.688	1.01	40	0.005	1.41	40	−0.69 (−1.23, −0.16)	3.2
Jassi	−0.56	0.39	25	−0.025	0.75	25	−0.54 (−0.87, −0.20)	7.0
Kaspers	−0.032	1.25	88	−0.178	0.97	87	0.15 (−0.19, 0.48)	7.0
Liu	−0.16	0.99	60	−0.20	1.03	60	0.04 (−0.32, 0.40)	6.2
Maesta	−0.75	0.46	10	−0.19	0.49	11	−0.56 (−0.97, −0.15)	5.2
Steinberg	0.01	1.25	24	0.09	0.29	24	−0.08 (−0.59, 0.43)	3.5
Steinberg	−0.09	0.29	24	0.09	0.29	24	−0.18 (−0.34, −0.02)	14.8
Teede	−1.00	1.42	19	−0.50	1.01	21	−0.50 (−1.27, 0.27)	1.7
Vigina	−0.41	1.07	40	−0.41	0.96	37	0.00 (−0.45, 0.45)	4.3
Total (95% CI)			660			662	−0.15 (−0.25, −0.05)	100.0
							p = 0.0048	

Hetrogeneity: Tau^2=0.0122; Q=23.0273; df=17 (p=0.1484); I^2=26.17%

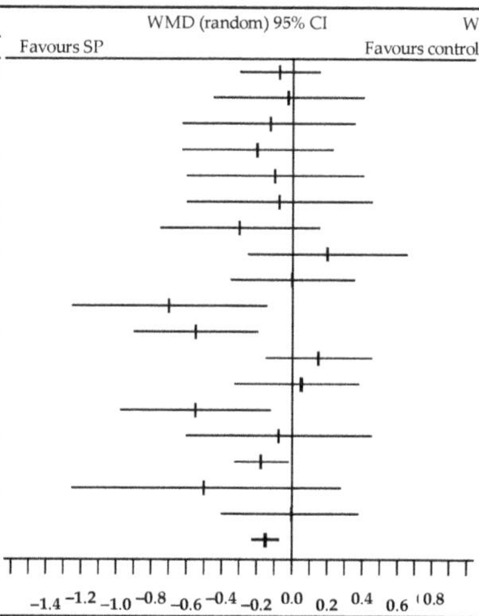

B. LDL–C, low–density lipoprotein cholesterol (mmol/L)

Study	SP			Control			WMD (random)	Weight
	DM	SD	*N*	DM	SD	*N*	95% CI	%
Allen	−0.122	0.68	93	0.006	0.71	98	−0.13 (−0.33, 0.07)	12.5
Basaria	−0.028	0.87	38	−0.034	0.83	46	0.01 (−0.36, 0.37)	4.7
Campbell	0.13	0.97	35	0.21	0.90	27	−0.08 (−0.55, 0.39)	3.0
Cuves	−0.90	0.77	18	−0.88	0.86	18	−0.02 (−0.55, 0.51)	2.4
Dalais	−0.59	0.89	38	−0.29	1.01	40	−0.30 (−0.72, 0.12)	3.7
Gardner	−0.10	0.78	33	−0.30	0.61	30	0.20 (−0.14, 0.54)	5.2
Gardner	−0.40	0.61	31	−0.30	0.61	30	−0.10 (−0.41, 0.21)	6.4
Han	−0.344	0.87	40	0.142	1.05	40	−0.49 (−0.91, −0.06)	3.6
Jassi	−0.592	0.43	25	−0.083	0.81	25	−0.51 (−0.87, −0.15)	4.9
Kaspers	−0.031	1.01	88	−0.171	0.89	87	0.14 (−0.14, 0.42)	7.3
Liu	−0.12	0.92	60	−0.13	0.87	60	0.01 (−0.31, 0.33)	5.9
Maesta	−0.62	0.50	10	−0.26	0.51	11	−0.36 (−0.79, 0.07)	3.5
Steinberg	−0.02	0.29	24	0.09	0.29	24	−0.11 (−0.27, 0.05)	15.7
Steinberg	−0.09	0.29	24	0.05	0.29	24	−0.14 (−0.30, 0.02)	15.7
Teede	−0.70	0.95	19	−0.30	1.01	21	−0.40 (−1.01, 0.21)	1.9
Vigina	−0.35	1.01	40	−0.32	0.88	37	−0.03 (−0.45, 0.39	3.6
Total (95% CI)			616			618	−0.12 (−0.20, −0.03)	100.0
							p = 0.0067	

Hetrogeneity: Tau^2=0.0049; Q=18.1738; df=15 (p=0.2536); I^2=17.46%

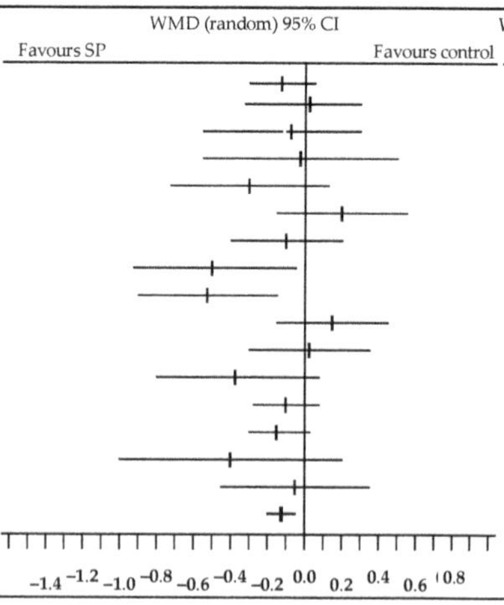

Figure 3. *Cont.*

C. HDL-C, high-density lipoprotein cholesterol (mmol/L)

Study	SP DM	SP SD	SP *N*	Control DM	Control SD	Control *N*	WMD (random) 95% CI	Weight %
Allen	0.068	0.42	93	0.036	0.35	98	0.03 (−0.08, 0.14)	7.8
Basaria	−0.103	0.53	38	0.008	0.51	46	−0.11 (−0.33, 0.11)	1.9
Baum (96)	0.04	0.35	21	−0.06	0.33	22	0.10 (−0.10, 0.30)	2.3
Baum (52)	0.08	0.35	23	−0.06	0.33	22	0.14 (−0.06, 0.34)	2.4
Campbell	0.09	0.42	35	0.12	0.40	27	−0.03 (−0.24, 0.18)	2.2
Cuevas	−0.02	0.21	18	−0.08	0.69	18	0.06 (−0.27, 0.39)	0.9
Dalais	−0.11	0.23	38	−0.25	0.54	40	0.14 (−0.04, 0.32)	2.8
Gardner	0.10	0.29	33	0.00	0.44	30	0.10 (−0.09, 0.28)	2.7
Gardner	0.10	0.33	31	0.00	0.44	30	0.10 (−0.10, 0.29)	2.5
Han	0.106	0.27	40	0.101	0.26	40	0.00 (−0.11, 0,12)	7.0
Jassi	0.192	0.20	25	0.041	0.19	25	0.15 (0.04, 0.26)	8.1
Kaspers	−0.01	0.44	88	−0.059	0.37	87	0.05 (−0.07, 0.17)	6.5
Liu	−0.02	0.16	60	−0.07	0.11	60	0.05 (0.00, 0.10)	39.1
Maesta	−0.05	0.20	10	−0.04	0.23	11	−0.01 (−0.19, 0.17)	2.8
Steinberg	0.00	0.29	24	0.06	0.29	24	−0.06 (−0.22, 0.11)	3.5
Steinberg	−0.06	0.29	24	0.06	0.29	24	−0.12 (−0.28, 0.04)	3.5
Tedde	−0.10	0.47	19	−0.20	0.50	21	0.10 (−0.20, 0.40)	1.0
Vigna	0.01	0.39	40	−0.03	0.41	37	0.04 (−0.14, 0.22)	3.0
Total (95% CI)			660			662	0.05 (0.02, 0.08)	100.0
							p = 0.0034	

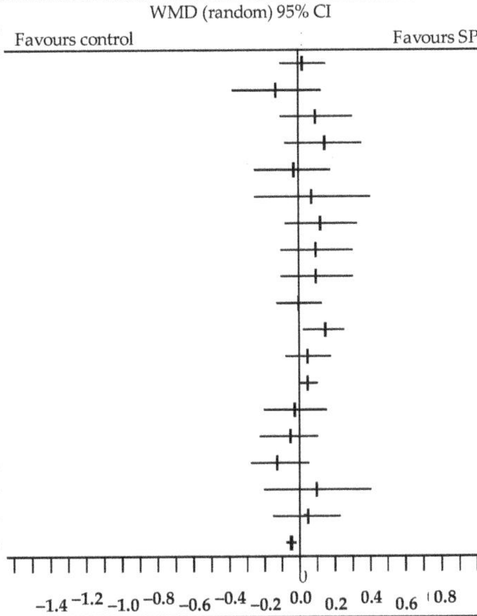

Hetrogeneity: Tau^2=0.0000; Q=15.3446; df=17 (p=0.5707); I^2=0.00%

D. TG, triglycerides (mmol/L)

Study	SP DM	SP SD	SP *N*	Control DM	Control SD	Control *N*	WMD (random) 95% CI	Weight %
Allen	−0.005	0.55	93	−0.023	0.59	98	0.02 (−0.14, 0.18)	8.8
Basaria	0.064	0.66	38	−0.039	0.51	46	0.10 (−0.15, 0.36)	6.8
Baum	0.00	0.95	21	0.01	1.32	22	−0.01 (−0.69, 0.68)	2.0
Baum	−0.16	1.68	23	0.01	1.32	22	−0.17 (−1.05, 0.71)	1.3
Campbell	0.19	0.77	35	0.22	0.75	27	−0.03 (−0.41, 0.35)	4.6
Cuevas	−0.63	0.88	18	−0.35	0.97	18	−0.28 (−0.89, 0.33)	2.4
Dalais	−0.22	0.47	38	0.05	0.50	40	−0.27 (−0.49, −0.05)	7.6
Gardner	0.00	0.79	33	0.10	0.96	30	−0.10 (−0.54, 0.34)	3.9
Gardner	0.00	0.82	31	0.10	0.96	30	−0.10 (−0.55, 0.35)	3.7
Han	0.101	0.26	40	0.116	1.54	40	−0.02 (−0.50, 0.47)	3.4
Jassi	−0.356	0.27	25	0.078	0.23	25	−0.43 (−0.57, −0.29)	9.3
Kaspers	0.021	0.73	88	0.121	0.64	87	−0.10 (−0.30, 0.10)	7.9
Liu	0.04	1.02	60	−0.02	0.62	60	0.06 (−0.24, 0.36)	5.8
Maesta	−0.18	0.37	10	−0.21	0.23	10	0.03 (−0.24, 0.30)	6.5
Steinberg	0.05	0.29	24	−0.05	0.29	24	0.10 (−0.06, 0.26)	8.7
Steinberg	0.01	0.29	24	−0.05	0.29	24	0.06 (−0.10, 0.22)	8.7
Teede	−0.40	0.60	19	0.10	0.85	21	−0.50, −0.95, −0.05)	3.7
Vigina	−0.16	0.86	40	−0.13	0.75	37	−0.03 (−0.39, 0.33)	4.9
Total (95% CI)			660			662	−0.08 (−0.19, 0.03)	100.0
							p = 0.1462	

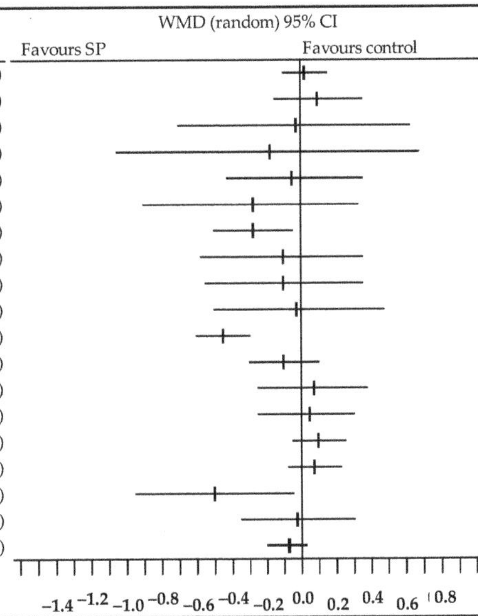

Hetrogeneity: Tau^2=0.0271; Q=44.0812; df=17 (p=0.0003); I^2=61.43%

Figure 3. Forest plot representing the associations between soy protein and lipid profiles. Data are presented as weighted mean difference with 95% CI.

3.4. *Associations between Soy Isoflavones Alone (Preparation) and Lipid Profiles*

A total of 13 studies were selected to analyze the effect of soy isoflavones on the lipid profile [81–92], among which the data from the Colacurici et al. [93] did not allow for the analysis of the effect of isoflavones on TC, while in the study by Dewell et al. [85], there were insufficient data on LDL-C. The results of the meta-analysis are shown in Figure 4. A slight, statistically insignificant decrease in TC levels was observed: WMD = −0.07; 95% CI: −0.18–0.05; p = 0.2428, as well as TG: WMD = −0.04; 95% CI: −0.13–0.05; p = 0.4200. On the other hand, no effect of the use of isoflavones on LDL-C levels was noticed: WMD = 0.00; 95% CI: −0.07–0.07; p = 0.9750 and HDL-C: WMD = 0.01; 95% CI: −0.03–0.05; p = 0.6449. The heterogeneity of the studies was not significant in the case of TC, LDL-C and HDL-C, but it turned out to be high in the case of TG (I^2 = 47.34%). The results for the asymmetry tests were not statistically significant for TC: Begg's test—p = 0.0672; Egger's test—p = 0.1619, LDL-C: Egger's test—p = 0.0872, HDL-C: Begg's test—p = 0.7016; Egger's test—p = 0.9451 and TG: Begg's test—p = 0.3520; Egger's test—p = 0.3281. However, Begg's test showed a statistically significant publication bias for LDL-C (p = 0.0281).

A. TC, total cholesterol (mmol/L)

Study	ISOF			Control			WMD (random)	Weight
	DM	SD	*N*	DM	SD	*N*	95% CI	%
Aubertin–Leheudre	−0.18	0.89	21	0.33	0.86	18	−0.51 (−1.06, 0.04)	3.9
Barrasa	−0.127	0.69	20	0.137	0.65	15	−0.26 (−0.71, 0.18)	5.6
Chilibeck	−0.21	0.78	74	−0.12	0.88	72	−0.09 (−0.36, 0.18)	11.7
Choquette	−0.08	0.88	23	0,06	0.90	22	−0.14 (−0.66, 0.38)	4.3
Dewell	−0.30	0.97	20	0.10	2.00	16	−0.40 (−1.47, 0.67)	1.1
Engelbert	0.21	0.97	85	0.11	1.02	85	0.10 (−0.20, 0.40)	10.2
Garrido	0.30	1.97	15	0.00	0.61	14	0.30 (−0.75, 1.35)	1.2
Ho	0.072	0.57	67	0.075	0.47	68	−0.00 (−0.18, 0.17)	18.5
Ho	0.145	0.50	68	0.075	0.47	68	0.07 (−0.09, 0.23)	19.7
Kim	−0.199	1.00	42	−0.243	1.06	43	0.04 (−0.39, 0.48)	5.7
Nahas	0.059	1.07	38	0.065	1.06	38	−0.01 (−0.48, 0.47)	4.9
Öztürk Turhan	−0.44	0.96	45	0.153	0.83	45	−0.59 (−0.96, −0.22)	7.5
Wu	0.129	0.94	33	0.098	0.88	33	0.03 (−0.41, 0.47)	5.7
Total (95% CI)			551			537	−0.07 (−0.18, 0.05)	100.0
							p = 0.2428	

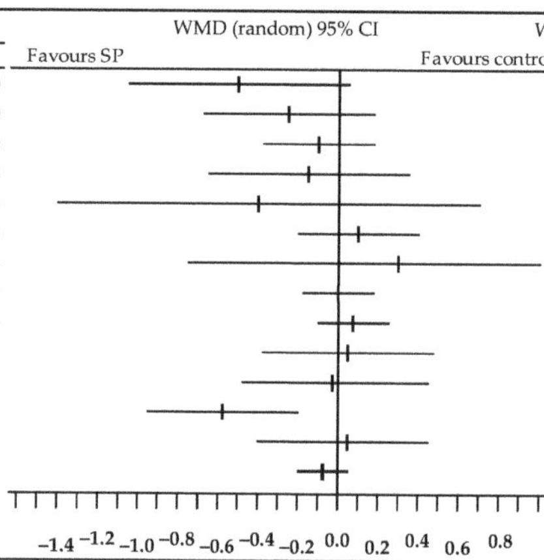

Hetrogeneity: Tau^2=0.0107; Q=16.4440; df=12 (p=0.1717); I^2=27.02%

B. LDL–C, low–density lipoprotein cholesterol (mmol/L)

Study	ISOF			Control			WMD (random)	Weight
	DM	SD	*N*	DM	SD	*N*	95% CI	%
Aubertin–Leheudre	0.00	0.83	21	0.31	0.79	18	−0.31 (−0.82, 0.20)	1.8
Barrasa	−0.181	0.95	20	0.129	0.55	15	−0.31 (−0.81, 0.19)	1.9
Chilibeck	−0.20	0.75	73	−0.09	0.75	73	−0.11 (−0.35, 0.13)	7.9
Choquette	0.01	0.83	23	0.11	0.81	22	−0.10 (−0.58, 0.38)	2.1
Colacurici	0.10	0.39	29	0.00	0.50	28	0.10 (−0.13, 0.33)	8.6
Engelbert	0.13	0.97	85	0.00	0.92	85	0.13 (−0.16, 0.41)	5.8
Garrido	0.30	0.39	15	0.20	0.39	14	0.10 (−0.18, 0.38)	5.8
Ho	−0.078	0.40	67	−0.096	0.36	68	0.02 (−0.11, 0.15)	26.3
Ho	−0.052	0.39	68	−0,096	0.36	68	0.04 (−0.08, 0.17)	27.1
Kim	−0.109	0.78	42	−0.087	0.95	43	−0.02 (−0.39, 0.35)	3.5
Nahas	0.039	0.93	38	0.034	0.99	38	0.01 (−0.43, 0.44)	2.6
Öztürk Turhan	−0.445	1.12	45	0.026	0.65	45	−0.47 (−0.85, −0.09)	3.3
Wu	−0.028	0.78	33	−0.068	0.80	33	0.04 (−0.34, 0.42)	3.3
Total (95% CI)			560			549	0.00 (−0.07, 0.07)	100.0
							p = 0.9750	

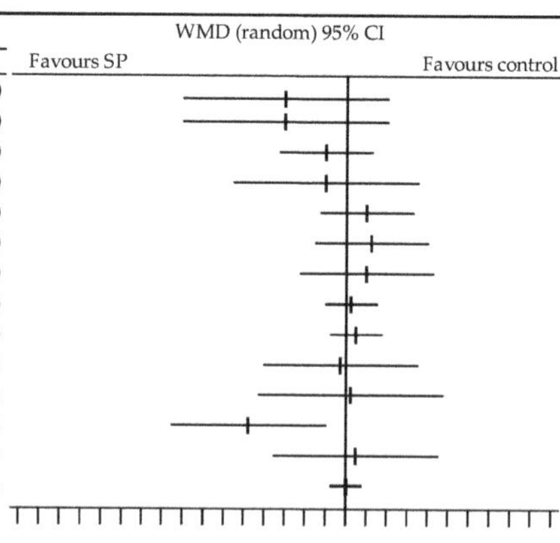

Hetrogeneity: Tau^2=0.0005; Q=12.3417; df=12 (p=0.4185); I^2=2.78%

Figure 4. *Cont.*

C. HDL-C, high-density lipoprotein cholesterol (mmol/L)

Study	ISOF DM	ISOF SD	ISOF N	Control DM	Control SD	Control N	WMD (random) 95% CI	Weight %
Aubertin-Leheudre	−0.14	0.46	21	0.10	0.39	18	−0.24 (−0.51, 0.03)	2.5
Barrasa	0.061	0.41	20	0.053	0.35	15	0.01 (−0.24, 0.26)	2.8
Chilibeck	−0.03	0.22	74	0.00	0.22	73	−0.03 (−0.10, 0.04)	16.8
Choquette	−0.04	0.35	23	−0.04	0.37	22	0.00 (−0.21, 0.21)	3.8
Colacurci	−0.10	0.61	29	0.00	0.39	28	−0.10 (−0.37, 0.16)	2.5
Dewell	−0.20	0.49	20	−0.20	0.44	16	0.00 (−0.30, 0.30)	1.9
Engelbert	0.01	0.48	85	0.05	0.53	85	−0.04 (−0.19, 0.11)	6.5
Garrido	0.40	0.39	15	−0.10	0.55	14	0.50 (0.15, 0.85)	1.5
Ho	−0.021	0.21	67	−0.026	0.15	68	0.01 (−0.06, 0.07)	18.8
Ho	0.013	0.17	68	−0.026	0.15	68	0.04 (−0.01, 0.09)	20.6
Kim	−0.008	0.41	42	0.018	0.42	43	−0.03 (−0.20, 0.15)	5.1
Nahas	0.057	0.27	38	−0.067	0.39	38	0.12 (−0.03, 0.27)	6.6
Öztürk Turhan	0.139	0.37	45	0.057	0.31	45	0.08 (−0.06, 0.22)	7.2
Wu	0.046	0.50	33	0.132	0.42	33	−0.09 (−0.31, 0.14)	3.4
Total (95% CI)			580			565	0.01 (−0.03, 0.05) p = 0.6449	100.0

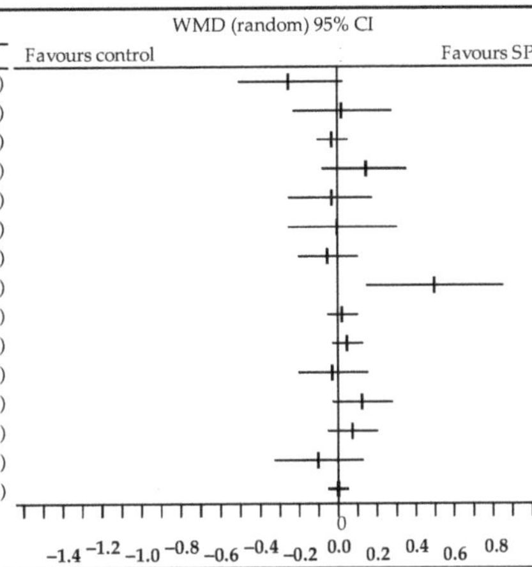

Hetrogeneity: Tau^2=0.0017; Q=18.5009; df=13 (p=0.1394); I^2=29.73%

D. TG, triglycerides (mmol/L)

Study	ISOF DM	ISOF SD	ISOF N	Control DM	Control SD	Control N	WMD (random) 95% CI	Weight %
Aubertin-Leheudre	−0.09	0.44	21	−0.18	0.42	18	0.09 (−0.18, 0.36)	7.1
Barrasa	−0.026	0.38	20	0.053	0.35	15	−0.08 (−0.32, 0.16)	8.0
Chilibeck	0.03	0.65	74	−0.06	0.63	73	0.09 (−0.12, 0.30)	9.4
Choquette	−0.13	0.69	23	0.00	0.90	22	−0.13 (−0.60, 0.34)	3.2
Colacurici	0.20	0.81	29	0.10	0.93	28	0.10 (−0.35, 0.55)	3.4
Dewell	0.40	0.89	20	0.00	0.82	16	0.40 (−0.16, 0.96)	2.4
Engelbert	0.06	0.46	85	−0.02	0.34	85	0.08 (−0.04, 0.20)	13.6
Garrido	0.1	0.61	15	0.00	0.93	14	0.10 (−0.48, 0.68)	2.2
Ho	0.042	0.45	67	0.069	0.43	68	−0.03 (−0.18, 0.12)	12.2
Ho	0.164	0.45	68	0.069	0.43	68	0.10 (−0.05, 0.24)	12.2
Kim	−0.183	0.76	42	0.093	0.94	43	−0.28 (−0.64, 0.09)	4.8
Nahas	−0.16	0.69	38	0.26	0.95	38	−0.42 (−0.79, −0.05)	4.6
Öztürk Turhan	−0.341	0.56	45	−0.007	0.73	45	−0.33 (−0.60, −0.07)	7.1
Wu	−0.17	0.34	33	0.04	0.47	33	−0.21 (−0.41, −0.01)	9.8
Total (95% CI)			580			565	−0.04 (−0.13, 0.05) p = 0.4200	100.0

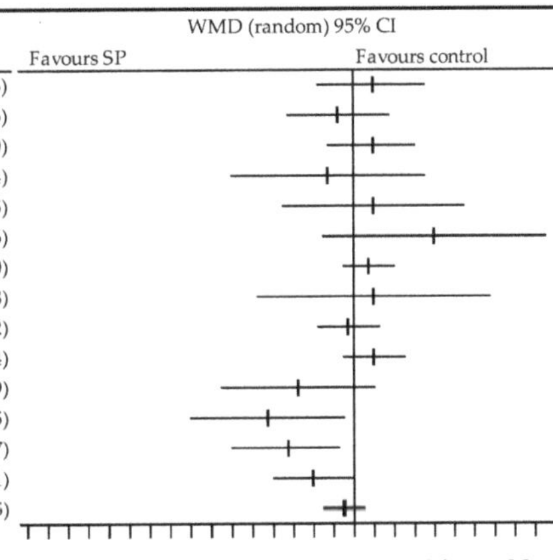

Hetrogeneity: Tau^2=0.0125; Q=24.6878; df=13 (p=0.0254); I^2=47.34%

Figure 4. Forest plot representing associations between isoflavones and lipid profiles. Data are presented as the weighted mean difference with 95% CI.

3.5. Associations between Red Clover and Lipid Profiles

The last analysis, presented in Figure 5, concerned the effect of red clover on the lipid profile, and included seven studies [94–100]. There was a significant reduction in TC levels after the use of red clover (WMD = −0.11; 95% CI: −0.18–−0.04; p = 0.0017) and a statistically significant increase in HDL-C levels (WMD = 0.04; 95% CI: 0.01 to 0.07; p = 0.0165). In the case of TC and HDL-C, no significant heterogeneity of the study effects was observed, and publication bias was not demonstrated. The p value of Begg's test was 0.4579 for TC and 0.6207 for HDL-C, while the p value of Egger's test was 0.3990 for TC and 0.5319 for HDL-C. In contrast, statistical analysis showed no significant changes in LDL-C levels after the use of red clover (WMD = −0.01; 95% CI: −0.13 to 0.10; p = 0.8230) and showed a slight decrease in TG levels, which was statistically insignificant (WMD = −0.05; 95% CI: −0.17–0.06; p = 0.3713). In the case of LDL-C and TG, the heterogeneity of the studies turned out to be high (I^2 = 49.57% and I^2 = 76.14%, respectively). The asymmetry tests showed no publication bias. The p value of Begg's test was 0.4527 for LDL-C and 0.4527 for TG, while the p value of Egger's test was 0.2560 for LDL-C and 0.6425 for TG.

A. TC, total cholesterol (mmol/L)

Study	RCE DM	RCE SD	RCE N	Control DM	Control SD	Control N	WMD (random) 95% CI	Weight %
Atkinson	0.00	1.52	77	0.16	1,44	86	−0.16 (−0.61, 0.30)	2.2
Clifton–Bligh	−0.32	0.83	56	−0.27	0.72	47	−0.05 (−0.35, 0.25)	5.1
Hale	−0.036	0.50	14	0.12	0.64	14	−0.16 (−0.58, 0.27)	2.6
Hildago	−0.25	0.85	53	−0.09	1.01	53	−0.16 (−0,52, 0.20)	3.7
Lambert	−0.26	0.22	30	−0.12	0.13	29	−0.14 (−0.23, −0.05)	54.5
Lambert	−0.11	1.12	38	−0.15	1.09	40	0.04 (−0.47, 0.55)	1.7
Shult (82 mg)	−0.132	0.537	81	−0.052	0.587	83	−0.08 (−0.25, 0.09)	15.5
Shult (57 mg)	−0.085	0.57	81	−0.052	0.587	83	−0.03 (−0.21, 0.14)	14.7
Total (95% CI)			439			435	−0.11 (−0.18, −0.04)	100.0
							p = 0.0017	

Hetrogeneity: Tau^2=0.0000; Q=1.8980; df=7 (p=0.9653); I^2=0.00%

B. LDL–C, low-density lipoprotein cholesterol (mmol/L)

Study	RCE DM	RCE SD	RCE N	Control DM	Control SD	Control N	WMD (random) 95% CI	Weight %
Atkinson	−0.2	2.31	77	−0.04	2.09	86	−0.16 (−0.84, 0.52)	2.6
Clifton–Bligh	−0.27	0.80	56	−0.14	0.79	47	−0.13 (−0.44, 0.18)	9.7
Hale	−0.26	0.59	14	−0.09	0.34	14	−0.17 (−0.53, 0.19)	7.9
Hildago	0.44	1.22	53	−0.18	0.93	53	0.62 (0.21, 1.03)	6.3
Lambert	−0.21	0.17	30	−0.11	0.17	29	−0.10 (−0.19, −0.01)	26.8
Lambert	−0.13	0.83	38	−0.30	0.95	40	0.17 (−0.23, 0.57)	6.7
Shult (82)	−0.174	0.406	81	−0.134	0.512	83	−0.04 (−0.18, 0.10)	21.4
Shult (57)	−0.122	0.605	81	−0.134	0.512	83	0.01 (−0.16, 0.18)	18.6
Total (95% CI)			439			435	−0.01 (−0.13, 0.10)	100.0
							p = 0.8230	

Hetrogeneity: Tau^2=0.0110; Q=13.8813; df=7 (p=0.0533); I^2=49.57%

C. HDL–C, high-density lipoprotein cholesterol (mmol/L)

Study	RCE DM	RCE SD	RCE N	Control DM	Control SD	Control N	WMD (random) 95% CI	Weight %
Atkinson	0.20	1.64	77	0.18	1.77	86	0.02 (−0.50, 0.54)	0.3
Clifton–Bligh	0.013	0.27	56	−0.04	0.23	47	0.05 (−0.04, 0.15)	9.7
Hale	0.10	0.34	14	0.05	0.25	14	0.05 (−0.17, 0.27)	1.8
Hildago	0.008	0.23	53	0.04	0.31	53	−0.03 (−0.14, 0.07)	8.3
Lambert	−0.02	0.17	30	−0.09	0.10	29	0.07 (0.00, 0.14)	17.9
Lambert	−0.14	0.31	38	−0.13	0.31	40	−0.01 (−0.15, 0.13)	4.8
Shult (82 mg)	0.035	0.156	81	0.003	0.196	83	0.03 (−0.02, 0.09)	30.7
Shult (57 mg)	0.046	0.185	81	0.003	0.196	83	0.04 (−0.02, 0.10)	26.5
Total (95% CI)			439			435	0.04 (0.01, 0.07)	100.0
							p = 0.0165	

Hetrogeneity: Tau^2=0.0000; Q=3.1698; df=7 (p=0.8689); I^2=0.00%

D. TG, triglycerides (mmol/L)

Study	RCE DM	RCE SD	RCE N	Control DM	Control SD	Control N	WMD (random) 95% CI	Weight %
Atkinson	0.05	1.52	77	0.03	1.44	86	0.02 (−0.44, 0.48)	5.1
Clifton–Bligh	−0.13	0.50	56	−0.05	0.61	47	−0.08 (−0.30, 0.14)	12.0
Hale	0.25	0.37	14	−0.014	0.36	14	0.26 (−0.01, 0.53)	9.8
Hildago	−0.21	0.88	53	0.49	1.08	53	−0.70 (−1.08, −0.32)	6.7
Lambert	−0.01	0.11	30	0.004	0.11	29	−0.01(−0.07, 0.04)	19.1
Lambert	0.10	0.20	38	−0.06	0.44	40	0.16 (0.01,0.31)	15.2
Shult (82)	−0.058	0.422	81	0.067	0.391	83	−0.13 (−0.25, 0.00)	16.4
Shult (57)	−0.098	0.508	81	0.067	0.391	83	−0.17 (−0.30, −0.03)	15.7
Total (95% CI)			474			482	−0.05 (−0.17, 0.06)	100.0
							p = 0.3713	

Hetrogeneity: Tau^2=0.0185; Q=29.3338; df=7 (p=0.0001); I^2=76.14%

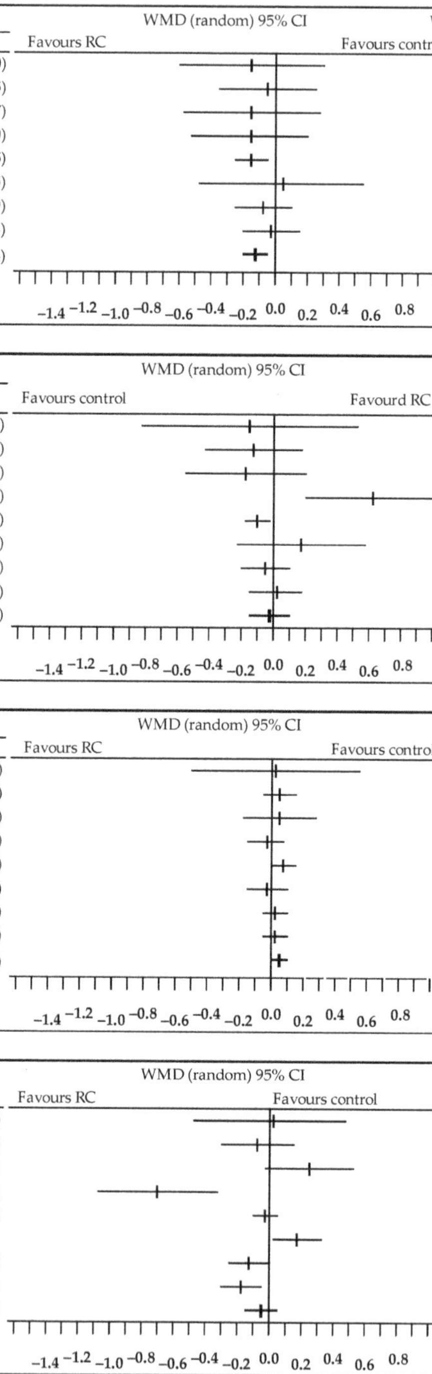

Figure 5. Forest plot representing associations between red clover and lipid profiles. Data are presented as weighted mean difference with 95% CI.

4. Discussion

The present meta-analysis indicates that the intake of flaxseed by postmenopausal women is associated with a statistically significant reduction in TC levels (WMD = −0.26; 95% CI: −0.38 to −0.13; p = 0.0001), LDL-C levels (WMD = −0.19; 95% CI: −0.30 to−0.08; p = 0.0006), HDL-C levels (WMD = −0.06; 95% CI: −0.11 to −0.01; p = 0.0150). These findings are consistent with previous published meta-analyses for the flaxseed effect. A meta-analysis by Hadi et al. incorporating 62 randomized trials involving dietary supplementation with flaxseed or flaxseed-derived products showed that flaxseed supplementation significantly reduced TC (WMD = −5.389 mg/dL; 95% CI: −9.483, −1.295, p = 0.010), TG (WMD = −9.422 mg/dL; 95% CI: −15.514, −3.330, p = 0.002), and LDL-C (WMD = −4.206 mg/dL; 95% CI: −7.260, −1.151, p = 0.007) concentrations. However, it had no effect on HDL-C (WMD = 0.047 mg/dL; 95% CI: −0.777, 0.872, p = 0.910) [101]. The meta-analysis of Yang et al. indicated that different flaxseed products showed different effects. Whole flaxseed supplementation significantly reduced TC (−11.85 mg/dL, 95% CI −20.12−−3.57, p = 0.005), LDL-C (− 10.51 mg/dL, 95% CI −14.96−−6.06, p < 0.001), TG (−19.77 mg/dL, 95% CI −33.61−−5.94, p = 0.005), TC/HDL-C (− 0.10, 95% CI −0.19−−0.003, p = 0.044), while lignans supplementation significantly reduced TC (− 17.86 mg/dL, p = 0.004), LDL-C (− 15.47 mg/dL, p < 0.001), and TC/HDL-C (− 0.45, p = 0.04). Flaxseed oil supplementation had no such lowering effect on lipid [102].

Our meta-analysis of the effect of soy protein on the lipid profile showed a significant decrease in TC levels: WMD = −0.15; 95% CI: −0.25–0.05; p = 0.0048, LDL-C levels: WMD = −0.15; 95% CI: −0.25–0.05; p = 0.0067, as well as a significant increase in HDL-C levels: WMD = 0.05; 95% from CI: 0.02 to 0.08; p = 0.0034. There was also a slight reduction in TG levels, which, however, was statistically non-significant (WMD = −0.08; 95% CI: from −0.19 to 0.03; p = 0.1462). The meta-analysis by Moradi et al. supports the hypercholesterolemic effect of soy lowering the serum TC levels. Soy consumption was associated with a significant decrease in TG: −5.04 mg/dL; 95% CI: −9.95, −0.13; p = 0.044), TC (MD: −3.02 mg/dL; 95% CI: −5.56, −0.47; p= 0.02), LDL-C (3.27 mg/dL; 95% CI: −6.01, −0.53; p = 0.019) and HDL-C (MD: −2.28 mg/dL; 95% CI: −4.27, −0.29; p = 0.025). The reductions in LDL-C, TG, and HDL-C were larger in subjects consuming isolated soy protein than taking-in isolated soy isoflavones [37]. The results of previous meta-analyses also revealed a significant decrease in serum TC, LDL-C, and TG concentrations after the consumption of soy protein containing isoflavones [103].

This meta-analysis showed a significant reduction in TC levels after the use of red clover (WMD = −0.11; 95% CI: from −0.18 to −0.04; p = 0.0017) and a significant increase in HDL-C levels (WMD = 0.04; 95% CI: from 0.01 to 0.07; p = 0.0165). However, the study demonstrated no significant changes in LDL-C levels (WMD = −0.01; 95% CI: from −0.13 to 0.10; p = 0.8230) and a slight statistically insignificant decrease in TG levels (WMD = −0.05; 95% CI: from −0.17 to 0.06; p = 0.3713) after the use of red clover. In their meta-analysis, Luis et al. verified that the consumption of red clover by perimenopausal and postmenopausal women results in a significant decrease in TC, LDL-C, and TG, together with a significant increase in HDL-C [104]. Furthermore, the meta-analysis by Kanadys et al. revealed changes in serum levels: TC, −0.29 (95 % CI: from −0.53 to −0.06) mmol/L, p = 0.0136; LDL-C, −0.13 (95 % CI: from −0.35 to 0.09) mmol/L, p = 0.2418; TG, −0.15 (95 % CI: from −0.32 to 0.01) mmol/L, p = 0.0592; and HDL-C, 0.14 (95 % CI: from −0.08 to 0.36) mmol/L, p = 0.2103—which suggest benefits from red clover consumption specific to correcting abnormal cholesterol levels [105].

Study Limitations

Despite the results obtained in this systematic review and its meta-analysis, some limitations were found. Because of the lack of standardization in some of the study designs, such as the ingredients and doses of isoflavones and the durations and outcomes of the trials, it currently remains difficult to draw overall conclusions for all aspects of isoflavone intake. These limitations warrant further investigation with regard to the use of isoflavone in women's health. Study limitations can be also be found due to individual differences in the bioavailability of individual components of preparations as these were prepared in a variety of ways that were suitable for each study. Moreover, limitations were posed by potential publication bias, which is revealed via the asymmetry of the funnel plot and the Egger's model. Publication bias suggests that some small studies with negative findings may have been missed or unpublished. Additionally, effects on vascular function have hardly been studied and more studies are needed to better establish what the effect of flaxseed, soy, red clover are on heart and vascular function.

5. Conclusions

This meta-analysis provides evidence that consuming flaxseed, soy, and red clover can have a beneficial effect on lipids in postmenopausal women. Their consumption could provide an important strategy to control dyslipidemia, and therefore, natural products can be an alternative to medicaments for preventing CVD, which has some clinical relevance in anti-atherosclerotic therapy. Our data also suggest that future well-designed studies with large sample sizes and adequate durations are needed to fully investigate the effectiveness of flaxseed, soy, and red clover.

Author Contributions: A.B. (Agata Błaszczuk): conceptualization, methodology, formal analysis, supervision, writing—original draft, review and editing; A.B. (Agnieszka Barańska): investigation, methodology, project administration, visualization, writing—review and editing; W.K.: conceptualization, investigation, formal analysis, methodology, visualization, writing—review and editing; M.M.: methodology and formal analysis. M.E.J.: resources; U.R. methodology and supervision; R.W.: investigation and project administration. J.H.: software and investigation. M.P.-D.: funding acquisition and supervision. All authors have read and agreed to the published version of the manuscript.

Funding: This research received no external funding.

Institutional Review Board Statement: Not applicable.

Informed Consent Statement: Not applicable.

Data Availability Statement: Not applicable.

Conflicts of Interest: The authors declare no conflict of interest.

References

1. Li, Z.; Lin, L.; Wu, H.; Yan, L.; Wang, H.; Yang, H.; Li, H. Global, Regional, and National Death, and Disability-Adjusted Life-Years (DALYs) for Cardiovascular Disease in 2017 and Trends and Risk Analysis From 1990 to 2017 Using the Global Burden of Disease Study and Implications for Prevention. *Front. Public Health* **2021**, *9*, 559751. [CrossRef] [PubMed]
2. McAloon, C.J.; Boylan, L.M.; Hamborg, T.; Stallard, N.; Osman, F.; Lim, P.B.; Hayat, S.A. The changing face of cardiovascular disease 2000−2012: An analysis of the world health organization global health estimates data. *Int. J. Cardiol.* **2016**, *224*, 256–264. [CrossRef]
3. Wautier, J.L.; Wautier, M.P. Endothelial cell participation in inflammatory reaction. *Int. J. Mol. Sci.* **2021**, *22*, 6341. [CrossRef]
4. Munakata, M. Clinical significance of stress-related increase in blood pressure: Current evidence in office and out-of-office settings. *Hypertens. Res.* **2018**, *41*, 553–569. [CrossRef] [PubMed]
5. Veiraiah, A. Hyperglycemia, lipoprotein glycation, and vascular disease. *Angiology* **2005**, *56*, 431–438. [CrossRef] [PubMed]
6. Pirro, M.; Bianconi, V.; Paciullo, F.; Mannarino, M.R.; Bagaglia, F.; Sahebkar, A. Lipoprotein(a) and inflammation: A dangerous duet leading to endothelial loss of integrity. *Pharmacol. Res.* **2017**, *119*, 178–187. [CrossRef] [PubMed]
7. Steven, S.; Frenis, K.; Oelze, M.; Kalinovic, S.; Kuntic, M.; Bayo Jimenez, M.T.; Vujacic-Mirski, K.; Helmstädter, J.; Kröller-Schön, S.; Münzel, T.; et al. Vascular inflammation and oxidative stress: Major triggers for cardiovascular disease. *Oxid. Med. Cell Longev.* **2019**, *2019*, 7092151. [CrossRef]
8. Papakonstantinou, E.; Lambadiari, V.; Dimitriadis, G.; Zampelas, A. Metabolic syndrome and cardiometabolic risk factors. *Curr. Vasc. Pharmacol.* **2013**, *11*, 858–879. [CrossRef]
9. Abraham, T.M.; Pedley, A.; Massaro, J.M.; Hoffmann, U.; Fox, C.S. Association between visceral and subcutaneous adipose depots and incident cardiovascular disease risk factors. *Circulation* **2015**, *132*, 1639–1647. [CrossRef] [PubMed]
10. Huang, Y.; Cai, X.; Mai, W.; Li, M.; Hu, Y. Association between prediabetes and risk of cardiovascular disease and all-cause mortality: Systematic review and meta-analysis. *BMJ* **2016**, *355*, i5953. [CrossRef]

11. Pearson, G.J.; Thanassoulis, G.; Anderson, T.J.; Barry, A.R.; Couture, P.; Dayan, N.; Francis, G.A.; Genest, J.; Grégoire, J.; Grover, S.A.; et al. Canadian Cardiovascular Society Guidelines for the Management of Dyslipidemia for the Prevention of Cardiovascular Disease in Adults. *Can. J. Cardiol.* **2021**, *37*, 1129–1150. [CrossRef] [PubMed]
12. Phung, O.J.; Makanji, S.S.; White, C.M.; Coleman, C.I. Almonds have a neutral effect on serum lipid profiles: A meta-analysis of randomized trials. *J. Am. Diet Assoc.* **2009**, *109*, 865–873. [CrossRef] [PubMed]
13. Shahinfar, H.; Bazshahi, E.; Amini, M.R.; Payandeh, N.; Pourreza, S.; Noruzi, Z.; Shab-Bidar, S. Effects of artichoke leaf extract supplementation or artichoke juice consumption on lipid profile: A systematic review and dose-response meta-analysis of randomized controlled trials. *Phytother. Res.* **2021**, *35*, 6607–6623. [CrossRef] [PubMed]
14. Hadi, A.; Arab, A.; Ghaedi, E.; Rafie, N.A.; Miraghajani, M.; Kafeshani, M. Barberry (*Berberis vulgaris* L.) is a safe approach for management of lipid parameters: A systematic review and meta—Analysis of randomized controlled trials. *Complement. Ther. Med.* **2019**, *43*, 117–124. [CrossRef] [PubMed]
15. Sahebkar, A. A systematic review and meta-analysis of randomized controlled trials investigating the effects of curcumin on blood lipid levels. *Clin. Nutr.* **2014**, *33*, 406–414. [CrossRef] [PubMed]
16. Pourmasoumi, M.; Hadi, A.; Rafie, N.; Najafgholizadeh, A.; Mohammadi, H.; Rouhani, M.H. The effect of ginger supplementation on lipid profile: A systematic review and meta-analysis of clinical trials. *Phytomedicine* **2018**, *43*, 28–36. [CrossRef] [PubMed]
17. Wei, Z.H.; Wang, H.; Chen, X.Y.; Wang, B.S.; Rong, Z.X.; Wang, B.S.; Su, B.H.; Chen, H.Z. Time- and dose-dependent effect of psyllium on serum lipids in mild-to-moderate hypercholesterolemia: A meta-analysis of controlled clinical trials. *Eur. J. Clin. Nutr.* **2009**, *63*, 821–827. [CrossRef] [PubMed]
18. Khalesi, S.; Paukste, E.; Nikbakht, E.; Khosravi-Boroujeni, H. Sesame fractions and lipid profiles: A systematic review and meta-analysis of controlled trials. *Br. J. Nutr.* **2016**, *115*, 764–773. [CrossRef] [PubMed]
19. Jia, L.; Liu, X.; Bai, Y.Y.; Li, S.H.; Sun, K.; He, C.; Hui, R. Short-term effect of cocoa product consumption on lipid profile: A meta-analysis of randomized controlled trials. *Am. J. Clin. Nutr.* **2010**, *92*, 218–225. [CrossRef] [PubMed]
20. Tuccinardi, D.; Farr, O.M.; Upadhyay, J.; Oussaada, S.M.; Klapa, M.I.; Candela, M.; Rampelli, S.; Lehoux, S.; Lázaro, I.; Sala-Vila, A.; et al. Mechanisms underlying the cardiometabolic protective effect of walnut consumption in obese people: A cross-over, randomized, double-blind, controlled inpatient physiology study. *Diabetes Obes. Metab.* **2019**, *21*, 2086–2095. [CrossRef] [PubMed]
21. da Silva, I.T.; de Almeida-Pititto, B.; Ferreira, S.R.G. Reassessing lipid metabolism and its potentialities in the prediction of cardiovascular risk. *Arch. Endocrinol. Metab.* **2015**, *59*, 171–180. [CrossRef] [PubMed]
22. Meldrum, D.R.; Davidson, B.J.; Tataryn, I.V.; Judd, J.L. Changes in circulating steroids with aging in postmenopausal women. *Obs. Gynecol.* **1981**, *57*, 624–628.
23. Derby, C.A.; Crawford, S.L.; Pasternak, R.C.; Sowers, M.; Sternfeld, B.; Matthews, K.A. Lipid changes during the menopause transition in relation to age and weight: The Study of Women's Health Across the Nation. *Am. J. Epidemiol.* **2009**, *119*, 1352–1361. [CrossRef] [PubMed]
24. Cho, E.J.; Min, Y.J.; Oh, M.S.; Kwon, J.E.; Kim, J.E.; Lee, W.S.; Lee, K.J.; Kim, S.W.; Kim, T.H.; Kim, M.A.; et al. Effects of the transition from premenopause to postmenopause on lipids and lipoproteins: Quantification and related parameters. *Korean J. Intern. Med.* **2011**, *26*, 47–53. [CrossRef] [PubMed]
25. Ambikairajah, A.; Walsh, E.; Cherbuin, N. Lipid profile differences during menopause: A review with meta-analysis. *Menopause* **2019**, *26*, 1327–1333. [CrossRef] [PubMed]
26. Anagnostis, P.; Stevenson, J.C.; Crook, D.; Johnston, D.G.; Godsland, I.F. Effects of menopause, gender and age on lipids and high-density lipoprotein cholesterol subfractions. *Maturitas* **2015**, *81*, 62–68. [CrossRef]
27. Ko, S.H.; Kim, H.S. Menopause-associated lipid metabolic disorders and foods beneficial for postmenopausal women. *Nutrients* **2020**, *12*, 202. [CrossRef]
28. Shim, Y.Y.; Gui, B.; Arnison, P.G.; Wang, Y. Flaxseed (*Linum usitatissimum* L.) bioactive compounds and peptide nomenclature: A review. *Trends Food Sci. Technol.* **2014**, *38*, 5–20. [CrossRef]
29. Campos, J.R.; Severino, P.; Ferreira, C.S.; Zielinska, A.; Santini, A.; Souto, S.B.; Souto, E.B. Linseed essential oil—Source of lipids as active ingredients for pharmaceuticals and nutraceuticals. *Curr. Med. Chem.* **2019**, *26*, 4537–4558. [CrossRef]
30. Kitts, D.D.; Yuan, Y.; Wijewickreme, A.N.; Thompson, L.U. Antioxidant activity of the flaxseed lignan secoisolariciresinol diglycoside and its mammalian lignan metabolites enterodiol and enterolactone. *Mol. Cell. Biochem.* **1999**, *202*, 91–100. [CrossRef]
31. Sanghvi, A.; Divven, W.; Seltman, H. Inhibition of rat liver cholesterol 7-alpha hydroxylase and Acylotransferaza acylo-CoA: Cholesterol activities by entrodiol and enterolactone. In *Proceedings of the Symposium on Drugs Affecting Lipid Metabolism*; Kritchevsky, D., Ed.; Plenum Press: New York, NY, USA, 1984; pp. 311–322.
32. Burdge, G. a-Linolenic acid metabolism in men and women: Nutritional and biological implications. *Curr. Opin. Clin. Nutr. Metab. Care* **2004**, *7*, 137–144. [CrossRef] [PubMed]
33. Agyei, A. Bioactive proteins and peptides from soybeans. *Recent Pat. Food Nutr. Agric.* **2015**, *7*, 100–107. [CrossRef]
34. Torres, N.; Torre-Villalvazo, I.; Tovar, A.R. Regulation of lipid metabolism by soy protein and its implication in diseases mediated by lipid disorders. *J. Nutr. Biochem.* **2006**, *17*, 365–373. [CrossRef] [PubMed]
35. Kim, I.S. Current perspectives on the beneficial effects of soybean isoflavones and their metabolites for humans. *Antioxidants* **2021**, *10*, 1064. [CrossRef] [PubMed]
36. Vitale, D.C.; Piazza, C.; Melilli, B.; Drago, F.; Salomone, S. Isoflavones: Estrogenic activity, biological effect and bioavailability. *Eur. J. Drug Metab. Pharmacokinet.* **2013**, *38*, 15–25. [CrossRef]

37. Moradi, M.; Daneshzad, E.; Azadbakht, L. The effects of isolated soy protein, isolated soy isoflavones and soy protein containing isoflavones on serum lipids in postmenopausal women: A systematic review and meta-analysis. *Crit. Rev. Food Sci. Nutr.* **2020**, *60*, 3414–3428. [CrossRef] [PubMed]
38. Butteiger, D.N.; Hibberd, A.A.; McGraw, N.J.; Napawan, N.; Hall-Porter, J.M.; Krul, E.S. Soy protein compared with milk protein in a western diet increases gut microbial diversity and reduces serum lipids in Golden Syrian Hamsters. *J. Nutr.* **2016**, *146*, 697–705. [CrossRef] [PubMed]
39. Dang, Z.C.; Audinot, V.; Papapoulos, S.E.; Boutin, J.A.; Löwik, C.W.G.M. Peroxisome proliferator-activated receptor gamma (PPARgamma) as a molecular target for the soy phytoestrogen genistein. *J. Biol. Chem.* **2003**, *278*, 962–967. [CrossRef] [PubMed]
40. Oliveira, L.P.M.; de Jesús, R.P.; Freire, T.O.; Oliveira, C.P.; Lyra, A.C.; Lyra, L.G.C. Possible molecular mechanisms soy-mediated in preventing and treating nonalcoholic fatty liver disease. *Nutr. Hosp.* **2012**, *27*, 991–998. [PubMed]
41. Dentin, R.; Girard, J.; Postic, C. Carbohydrate responsive element binding protein (ChREBP) and sterol 42. regulatory element binding protein-1c (SREBP-1c): Two key regulators of glucose metabolism and lipid synthesis in liver. *Biochimie* **2005**, *87*, 81–86. [CrossRef] [PubMed]
42. Demonty, I.; Lamarche, B.; Deshaies, Y.; Jacques, H. Role of soy isoflavones in the hypotriglyceridemic effect of soy protein in the rat. *J. Nutr. Biochem.* **2002**, *13*, 671–677. [CrossRef]
43. Xue, Z.; Zhang, Q.; Yu, W.; Wen, H.; Hou, X.; Li, D.; Kou, X. Potential lipid-lowering mechanisms of Biochanin A. *Agric. Food Chem.* **2017**, *65*, 3842–3850. [CrossRef]
44. Andres, S.; Hansen, U.; Niemann, B.; Palavinskas, R.; Lampen, A. Determination of the isoflavone composition and estrogenic activity of commercial dietary supplements based on soy or red clover. *Food Funct.* **2015**, *6*, 2017–2025. [CrossRef] [PubMed]
45. Lemežiene, N.; Padarauskas, A.; Butkuté, B.; Cesevičiené, J.; Taujenis, L.; Norkevičiené, E. The concentration of isoflavones in red clover (*Trifolium pratense* L.) at flowering stage. *Zemdirb. Agric.* **2015**, *102*, 443–448. [CrossRef]
46. Booth, N.L.; Overk, C.R.; Yao, P.; Burdette, J.E.; Nikolic, D.; Chen, S.N.; Bolton, J.L.; van Breemen, R.B.; Pauli, G.F.; Farnsworth, N.R. The chemical and biologic profile of a red clover (*Trifolium pratense* L.) phase II clinical extract. *J. Altern. Complement. Med.* **2006**, *12*, 113–133. [CrossRef] [PubMed]
47. Booth, N.L.; Overk, C.R.; Yao, P.; Totura, S.; Deng, Y.; Hedayat, A.S.; Bolton, J.L.; Pauli, G.F.; Farnsworth, N.R. Seasonal variation of red clover (*Trifolium pratense* L., Fabaceae) isoflavones and estrogenic activity. *J. Agric. Food Chem.* **2006**, *54*, 1277–1282. [CrossRef]
48. Akbaribazm, M.; Khazaei, F.; Naseri, L.; Pazhouhi, M.; Zamanian, M.; Khazaei, M. Pharmacological and therapeutic properties of the Red Clover (*Trifolium pratense* L.): An overview of the new findings. *J. Tradit. Chin. Med.* **2021**, *41*, 642–649.
49. Akiyama, T.; Ishida, J.; Nakagawa, S.; Ogawara, H.; Watanabe, S.; Itoh, N.; Shibuya, M.; Fukami, Y. Genistein, a specific inhibitor of tyrosine-specific protein kinases. *J. Biol. Chem.* **1987**, *262*, 5592–5595. [CrossRef]
50. Hobiger, S.; Jungbauer, A. Red clover extract: A source for substances that activate peroxisome proliferator-activated receptor alpha and ameliorate the cytokine secretion profile of lipopolysaccharide-stimulated macrophages. *Menopause* **2010**, *17*, 379–387.
51. Moher, D.; Shamseer, L.; Clarke, M.; Ghersi, D.; Liberati, A.; Mark Petticrew, M.; Shekelle, P.; Stewart, L.A. PRISMA-P Group. Preferred reporting items for systematic review and meta-analysis protocols (PRISMA-P) 2015 statement. *Syst. Rev.* **2015**, *4*, 1. [CrossRef]
52. Jadad, A.R.; Moore, R.A.; Carroll, D.; Jenkinson, C.; Reynolds, D.J.; Gavaghan, D.J.; McQuay, H.J. Assessing the quality of reports of randomized clinical trials: Is blinding necessary? *Control Clin. Trials.* **1996**, *17*, 100–107. [CrossRef]
53. Begg, C.B.; Mazumdar, M. Operating characteristics of a rank correlation test for publication bias. *Biometrics* **1994**, *50*, 1088–1101. [CrossRef] [PubMed]
54. Egger, M.; Smith, G.D.; Schneider, M.; Minder, C. Bias in meta-analysis detected by a simple, graphical test. *BMJ* **1997**, *315*, 629–634. [CrossRef]
55. Higgins, J.P.T.; Thomas, J.; Chandler, J.; Cumpston, M.; Li, T.; Page, M.J.; Welch, V.A. *Cochrane Handbook for Systematic Reviews of Interventions Version 6.0 (Updated July 2020)*; John Wiley: Hooboken, NJ, USA, 2019.
56. Follmann, D.; Elliott, P.; Suh, I.; Cutler, J. Variance imputation for overviews of clinical trials with continuous response. *J. Clin. Epidemiol.* **1992**, *45*, 769–773. [CrossRef]
57. DerSimonian, R.; Laird, N. Meta-analysis in clinical trials. *Control Clin. Trials.* **1986**, *7*, 177–188. [CrossRef]
58. Higgins, J.P.T.; Thompson, S.G.; Deeks, J.J.; Altman, D.G. Measuring inconsistency in meta-analyses. *BMJ* **2003**, *327*, 557–560. [CrossRef]
59. Arjmandi, B.H.; AKhan, D.A.; Juma, S.; Drum, M.L.; Venkatesh, S.; Sohn, E.; Wei, L.; Derman, R. Whole flaxseed consumption lowers serum LDL-cholesterol and lipoprotein(a) concentrations in postmenopausal women. *Nutr. Res.* **1998**, *18*, 1203–1204. [CrossRef]
60. Brahe, L.K.; Le Chatelier, E.; Prifti, E.; Pons, N.; Kennedy, S.; Blaedel, T.; Hakansson, J.; Dalsgaard, T.K.; Hansen, T.; Pedersen, O.; et al. Dietary modulation of the gut microbiota—A randomized controlled trial in obese postmenopausal women. *Br. J. Nutr.* **2015**, *114*, 406–417. [CrossRef] [PubMed]
61. Cornish, S.M.; Chilibeck, P.D.; Paus-Jennsen, L.; Biem, H.J.; Khozani, T.; Senanayake, V.; Vatanparast, H.; Little, J.P.; Whiting, S.J.; Pahwa, P. A randomized controlled trial of the effects of flaxseed lignan complex on metabolic syndrome composite score and bone mineral in older adults. *Appl. Physiol. Nutr. Metab.* **2009**, *34*, 89–98. [CrossRef] [PubMed]

62. Dodin, S.; Lemay, A.; Jacques, H.; Legare, F.; Forest, J.C.; Masse, B. The effects of flaxseed dietary supplement on lipid profile, bone mineral density, and symptoms in menopausal women: A randomized, double-blind, wheat germ placebo-controlled clinical trial. *J. Clin. Endocrinol. Metab.* **2005**, *90*, 1390–1397. [CrossRef]
63. Hallund, J.; Ravn-Haren, G.; Bügel, S.; Tholstrup, T.; Tetens, I. A lignan complex isolated from flaxseed does not affect plasma lipid concentrations or antioxidant capacity in healthy postmenopausal women. *J. Nutr.* **2006**, *136*, 112–116. [CrossRef] [PubMed]
64. Lucas, E.A.; Wild, R.D.; Hammond, L.J.; Khalil, D.A.; Juma, S.; Daggy, B.P.; Stoecker, B.J.; Arjmandi, B.H. Flaxseed improves lipid profile without altering biomarkers of bone metabolism in postmenopausal women. *J. Clin. Endocrinol. Metab.* **2002**, *87*, 1527–1532. [CrossRef]
65. Simbalista, R.L.; Sauerbronn, A.V.; Aldrighi, J.M.; Arêas, J.A.G. Consumption of a flaxseed-rich food is not more effective than a placebo in alleviating the climacteric symptoms of postmenopausal women. *J. Nutr.* **2010**, *140*, 293–297. [CrossRef] [PubMed]
66. Allen, J.K.; Becker, D.M.; Kwiterovich, P.O.; Lindenstruth, K.A.; Curtis, C. Effect of soy protein-containing isoflavones on lipoproteins in postmenopausal women. *Menopause* **2007**, *14*, 106–114. [CrossRef] [PubMed]
67. Basaria, S.; Wisniewski, A.; Dupree, K.; Bruno, T.; Song, M.Y.; Yao, F.; Ojumu, A.; John, M.; Dobs, A.S. Effect of high-dose isoflavones on cognition, quality of life, androgens, and lipoprotein in post-menopausal women. *J. Endocrinol. Investig.* **2009**, *32*, 150–155. [CrossRef]
68. Baum, J.A.; Teng, H.; Erdman, J.W., Jr.; Weigel, R.M.; Klein, B.P.; Persky, V.W.; Freels, S.; Surya, P.; Bakhit, R.M.; Ramos, E.; et al. Long-term intake of soy protein improves blood lipid profiles and increases mononuclear cell low-density-lipoprotein receptor messenger RNA in hypercholesterolemic, postmenopausal women. *Am. J. Clin. Nutr.* **1998**, *68*, 545–551. [CrossRef]
69. Campbell, S.C.; Khalil, D.A.; Payton, M.E.; Arjmandi, B.H. One-year soy protein supplementation does not improve lipid profile in postmenopausal women. *Menopause* **2010**, *17*, 587–593. [CrossRef]
70. Cuevas, A.M.; Irribarra, V.L.; Castillo, O.A.; Yañez, M.D.; Germain, A.M. Isolated soy protein improves endothelial function in postmenopausal hypercholesterolemic women. *Eur. J. Clin. Nutr.* **2003**, *57*, 889–894. [CrossRef]
71. Dalais, F.S.; Ebeling, P.R.; Kotsopoulos, D.; McGrath, B.P.; Teede, H.J. The effects of soy protein containing isoflavones on lipids and indices of bone resorption in postmenopausal women. *Clin. Endocrinol.* **2003**, *58*, 704–709. [CrossRef]
72. Gardner, C.D.; Newell, K.A.; Cherin, R.; Haskell, W.L. The effect of soy protein with or without isoflavones relative to milk protein on plasma lipids in hypercholesterolemic postmenopausal women. *Am. J. Clin. Nutr.* **2001**, *73*, 728–735. [CrossRef]
73. Han, K.K.; Soares, J.M., Jr.; Haidar, M.A.; de Lima, G.R.; Baracat, E.C. Benefits of soy isoflavone therapeutic regimen on menopausal symptoms. *Obstet. Gynecol.* **2002**, *99*, 389–394. [PubMed]
74. Jassi, H.K.; Jain, A.; Arora, S.; Chitra, R. Effect of soy proteins vs. soy isoflavones on lipid profile in postmenopausal women. *Indian J. Clin. Biochem.* **2010**, *25*, 201–207. [CrossRef] [PubMed]
75. Kreijkamp-Kaspers, S.; Kok, L.; Grobbee, D.E.; de Haan, E.H.; Aleman, A.; Lampe, J.W.; van der Schouw, Y.T. Effect of soy protein containing isoflavones on cognitive function, bone mineral density, and plasma lipids in postmenopausal women: A randomized controlled trial. *JAMA* **2004**, *292*, 65–74. [CrossRef] [PubMed]
76. Liu, Z.M.; Ho, S.C.; Chen, Y.M.; Ho, Y.P. The effects of isoflavones combined with soy protein on lipid profiles, C-reactive protein and cardiovascular risk among postmenopausal Chinese women. *Nutr. Metab. Cardiovasc. Dis.* **2012**, *22*, 712–719. [CrossRef] [PubMed]
77. Maesta, N.; Nahas, E.A.P.; Nahas-Neto, J.; Orsatti, F.L.; Fernandes, C.E.; Traiman, P.; Burini, R.C. Effects of soy protein and resistance exercise on body composition and blood lipids in postmenopausal women. *Maturitas* **2007**, *56*, 350–358. [CrossRef]
78. Steinberg, F.M.; Guthrie, N.L.; Villablanca, A.C.; Kumar, K.; Murray, M.J. Soy protein with isoflavones has favorable effects on endothelial function that are independent of lipid and antioxidant effects in healthy postmenopausal women. *Am. J. Clin. Nutr.* **2003**, *78*, 123–130. [CrossRef]
79. Teede, H.J.; Dalais, F.S.; Kotsopoulos, D.; McGrath, B.P.; Malan, E.; Gan, T.E.; Peverill, R.E. Dietary soy containing phytoestrogens does not activate the hemostatic system in postmenopausal women. *J. Clin. Endocrinol. Metab.* **2005**, *90*, 1936–1941. [CrossRef]
80. Vigna, G.B.; Pansini, F.; Bonaccorsi, G.; Albertazzi, P.; Donegà, P.; Zanotti, L.; De Aloysio, D.; Mollica, G.; Fellin, R. Plasma lipoproteins in soy-treated postmenopausal women: A double-blind, placebo-controlled trial. *Nutr. Metab. Cardiovasc. Dis.* **2000**, *10*, 315–322.
81. Aubertin-Leheudre, M.; Lord, C.; Khalil, A.; Dionne, I.J. Effect of 6 months of physical activity and isoflavone supplementation on clinical cardiovascular risk factors in obese postmenopausal women: A randomized, double-blind study. *Menopause* **2007**, *14*, 624–629. [CrossRef]
82. Barrasa, G.R.R.; Canete, N.G.; Boasi, L.E.V. Age of postmenopause women: Effect of soy isoflavone in lipoprotein and inflammation markers. *J. Menopausal. Med.* **2018**, *243*, 176–182. [CrossRef]
83. Chilibeck, P.D.; Vatanparast, H.; Pierson, R.; Case, A.; Olatunbosun, O.; Whiting, S.J.; Beck, T.J.; Pahwa, P.; Biem, H.J. Effect of exercise training combined with isoflavone supplementation on bone and lipids in postmenopausal women: A randomized clinical trial. *J. Bone Miner Res.* **2013**, *28*, 780–793. [CrossRef] [PubMed]
84. Choquette, S.; Riesco, É.; Cormier, É.; Dion, T.; Aubertin-Leheudre, M.; Dionne, I.J. Effects of soya isoflavones and exercise on body composition and clinical risk factors of cardiovascular diseases in overweight postmenopausal women: A 6-month double-blind controlled trial. *Br. J. Nutr.* **2011**, *105*, 1199–1209. [CrossRef]
85. Dewell, A.; Hollenbeck, C.B.; Bruce, B. The effects of soy-derived phytoestrogens on serum lipids and lipoproteins in moderately hypercholesterolemic postmenopausal women. *J. Clin. Endocrinol. Metab.* **2002**, *87*, 118–121. [CrossRef] [PubMed]

86. Engelbert, A.K.; Soukup, S.T.; Roth, A.; Hoffmann, N.; Graf, D.; Watzl, B.; Kulling, S.E.; Bub, A. Isoflavone supplementation in postmenopausal women does not affect leukocyte LDL receptor and scavenger receptor CD36 expression: A double-blind, randomized, placebo-controlled trial. *Mol. Nutr. Food Res.* **2016**, *60*, 2008–2019. [CrossRef] [PubMed]
87. Garrido, A.; De la Maza, M.P.; Hirsch, S.; Valladares, L. Soy isoflavones affect platelet thromboxane A2 receptor density but not plasma lipids in menopausal women. *Maturitas* **2006**, *54*, 270–276. [CrossRef] [PubMed]
88. Ho, S.C.; Chen, Y.M.; Ho, S.S.S.; Woo, J.L.F. Soy isoflavone supplementation and fasting serum glucose and lipid profile among postmenopausal Chinese women: A double-blind, randomized, placebo-controlled trial. *Menopause* **2007**, *14*, 905–912. [CrossRef]
89. Kim, J.; Lee, H.; Lee, O.; Lee, K.H.; Lee, Y.B.; Young, K.D.; Jeong, Y.H.; Choue, R. Isoflavone supplementation influenced levels of triglyceride and luteunizing hormone in Korean postmenopausal women. *Arch. Pharm. Res.* **2013**, *36*, 306–313. [CrossRef]
90. Nahas, E.A.P.; Nahas-Neto, J.; Orsatti, F.L.; Carvalho, E.P.; Oliveira, M.L.C.S.; Dias, R. Efficacy and safety of a soy isoflavone extract in postmenopausal women: A randomized, double-blind, and placebo-controlled study. *Maturitas* **2007**, *58*, 249–258. [CrossRef] [PubMed]
91. Öztürk Turhan, N.O.; Duvan, C.I.; Bolkan, F.; Onaran, Y. Effect of isoflavone on plasma nitrite/nitrate, homocysteine, and lipid levels in Turkish women in the early postmenopausal period: A randomized controlled trial. *Turk. J. Med. Sci.* **2009**, *39*, 367–375.
92. Wu, J.; Oka, J.; Tabata, I.; Higuchi, M.; Toda, T.; Fuku, N.; Ezaki, J.; Sugiyama, F.; Uchiyama, S.; Yamada, K.; et al. Effects of isoflavone and exercise on BMD and fat mass in postmenopausal Japanese women: A 1-year randomized placebo-controlled trial. *J. Bone Miner Res.* **2006**, *21*, 780–789. [CrossRef]
93. Colacurci, N.; Chiàntera, A.; Fornaro, F.; de Novellis, V.; Manzella, D.; Arciello, A.; Chiàntera, V.; Improta, L.; Paolisso, G. Effects of soy isoflavones on endothelial function in healthy postmenopausal women. *Menopause* **2005**, *12*, 299–307. [CrossRef] [PubMed]
94. Atkinson, C.; Oosthuizen, W.; Scollen, S.; Loktionov, A.; Day, N.E.; Bingham, S.A. Modest protective effects of isoflavones from a red clover derived dietary supplement on cardiovascular disease risk factors in perimenopausal women, and evidence of an interaction with ApoE genotype in 49–65 year old women. *J. Nutr.* **2004**, *134*, 1759–1764. [CrossRef]
95. Clifton-Bligh, P.B.; Nery, M.L.; Clifton-Bligh, R.J.; Visvalingam, S.; Fulcher, G.R.; Byth, K.; Baber, R. Red clover isoflavones enriched with formononetin lower serum LDL cholesterol—a randomized, double-blind, placebo-controlled study. *Eur. J. Clin. Nutr.* **2015**, *69*, 134–142. [CrossRef]
96. Hale, G.E.; Hughes, C.L.; Robboy, S.J.; Agarwal, S.K.; Bievre, M. A double-blind randomized study on the effects of red clover isoflavones on the endometrium. *Menopause* **2001**, *8*, 338–346. [CrossRef]
97. Hidalgo, L.A.; Chedraui, P.A.; Morocho, N.; Ross, S.; San Miguel, G. The effect of red clover isoflavones on menopausal symptoms, lipids and vaginal cytology in menopausal women: A randomized, double-blind, placebo-controlled study. *Gynecol. Endocrinol.* **2005**, *21*, 257–264. [CrossRef]
98. Lambert, M.N.T.; Thorup, A.C.; Hansen, E.S.S.; Jeppesen, P.B. Combined red clover isoflavones and probiotics potently reduce menopausal vasomotor symptoms. *PLoS ONE* **2017**, *12*, e0176590. [CrossRef] [PubMed]
99. Lambert, M.N.T.; Thybo, C.B.; Lykkeboe, S.; Rasmussen, L.M.; Frette, X.; Christensen, L.P.; Jeppesen, P.B. Combined bioavailable isoflavones and probiotics improve bone status and estrogen metabolism in postmenopausal osteopenic women: A randomized controlled trial. *Am. J. Clin. Nutr.* **2017**, *106*, 909. [CrossRef] [PubMed]
100. Schult, T.M.; Ensrud, K.E.; Blackwell, T.; Ettinger, B.; Wallace, R.; Tice, J.A. Effect of isoflavones on lipids and bone turnover markers in menopausal women. *Maturitas* **2004**, *48*, 209–218. [CrossRef] [PubMed]
101. Hadi, A.; Askarpour, M.; Shekoufeh, S.; Ghaedi, E.; Symonds, M.E.; Miraghajani, M. Effect of flaxseed supplementation on lipid profile: An updated systematic review and dose-response meta-analysis of sixty-two randomized controlled trials. *Pharmacol. Res.* **2020**, *152*, 104622. [CrossRef] [PubMed]
102. Yang, C.; Xia, H.; Wan, M.; Lu, Y.; Xu, D.; Yang, X.; Yang, L.; Sun, G. Comparisons of the effects of different flaxseed products consumption on lipid profiles, inflammatory cytokines and anthropometric indices in patients with dyslipidemia related diseases: Systematic review and a dose–response meta-analysis of randomized controlled trials. *Nutr. Metab.* **2021**, *18*, 91.
103. Zhan, S.; Ho, S.C. Meta-analysis of the effects of soy protein containing isoflavones on the lipid profile. *Am. J. Clin. Nutr.* **2005**, *81*, 397–408. [CrossRef] [PubMed]
104. Luís, A.; Domingues, F.; Pereira, L. Effects of red clover on perimenopausal and postmenopausal women's blood lipid profile: A meta-analysis. *Climacteric* **2018**, *21*, 446–453. [CrossRef] [PubMed]
105. Kanadys, W.; Baranska, A.; Jedrych, M.; Religioni, U.; Janiszewska, M. Effects of red clover (*Trifolium pratense*) isoflavones on the lipid profile of perimenopausal and postmenopausal women—A systematic review and meta-analysis. *Maturitas* **2020**, *132*, 7–16. [CrossRef] [PubMed]

nutrients

MDPI

Article

Screening and Isolation of Potential Anti-Inflammatory Compounds from *Saxifraga atrata* via Affinity Ultrafiltration-HPLC and Multi-Target Molecular Docking Analyses

Gang Li [1], Yan Fang [1], Yonggui Ma [2], Yangzom Dawa [2], Qilan Wang [3], Jing Gan [1] and Jun Dang [3,*]

1 Center for Mitochondria and Healthy Aging, College of Life Sciences, Yantai University, Yantai 264005, China; ligang_ytu@126.com (G.L.); fyy2008fx@163.com (Y.F.); ganjing@ytu.edu.cn (J.G.)
2 Key Laboratory of Medicinal Animal and Plant Resources of Qinghai-Tibetan Plateau in Qinghai Province, College of Life Science, Qinghai Normal University, Xining 810001, China; 2025041@qhnu.edu.cn (Y.M.); dawayangzom@126.com (Y.D.)
3 Qinghai Provincial Key Laboratory of Tibetan Medicine Research, Key Laboratory of Tibetan Medicine Research, Chinese Academy of Sciences, Northwest Institute of Plateau Biology, Xining 810001, China; wql@nwipb.cas.cn
* Correspondence: dangjun@nwipb.cas.cn

Citation: Li, G.; Fang, Y.; Ma, Y.; Dawa, Y.; Wang, Q.; Gan, J.; Dang, J. Screening and Isolation of Potential Anti-Inflammatory Compounds from *Saxifraga atrata* via Affinity Ultrafiltration-HPLC and Multi-Target Molecular Docking Analyses. *Nutrients* **2022**, *14*, 2405. https://doi.org/10.3390/nu14122405

Academic Editors: Haixia Yang and Jianjun Deng

Received: 7 May 2022
Accepted: 7 June 2022
Published: 9 June 2022

Abstract: In this study, a 100 g sample of *Saxifraga atrata* was processed to separate 1.3 g of 11-O-(4′-O-methylgalloyl)-bergenin (Fr1) after 1 cycle of MCI GEL® CHP20P medium pressure liquid chromatography using methanol/water. Subsequently, COX-2 affinity ultrafiltration coupled with reversed-phase liquid chromatography was successfully used to screen for potential COX-2 ligands in this target fraction (Fr1). After 20 reversed-phase liquid chromatography runs, 74.1 mg of >99% pure 11-O-(4′-O-methylgalloyl)-bergenin (Fr11) was obtained. In addition, the anti-inflammatory activity of 11-O-(4′-O-methylgalloyl)-bergenin was further validated through molecular docking analyses which suggested it was capable of binding strongly to ALOX15, iNOS, ERBB2, SELE, and NF-κB. As such, the AA metabolism, MAPK, and NF-κB signaling pathways were hypothesized to be the main pathways through which 11-O-(4′-O-methylgalloyl)-bergenin regulates inflammatory responses, potentially functioning by reducing pro-inflammatory cytokine production, blocking pro-inflammatory factor binding to cognate receptors and inhibiting the expression of key proteins. In summary, affinity ultrafiltration-HPLC coupling technology can rapidly screen for multi-target bioactive components and when combined with molecular docking analyses, this approach can further elucidate the pharmacological mechanisms of action for these compounds, providing valuable information to guide the further development of new multi-target drugs derived from natural products.

Keywords: affinity ultrafiltration; molecular docking; preparative isolation; *Saxifraga atrata*; 11-O-(4′-O-methylgalloyl)-bergenin

1. Introduction

Inflammatory responses are generally induced by inflammatory cytokines and associated inflammatory mediators [1,2]. Important inflammatory factors associated with pyrexia and related symptoms include interleukin-1 (IL-1), IL-6, and tumor necrosis factor (TNF), while pain-related inflammatory factors including prostaglandin E2 (PGE2) and bradykinin (BK), PGE2, and nitric oxide (NO) are all related to inflammation-associated vasodilation. Increases in vascular permeability in inflammatory settings are tied to the activity of histamine (HA), BK, and reactive oxygen species metabolites. Inflammatory factors associated with increased vascular permeability include histamine (HA), BK, and reactive oxygen species metabolites (ROMs), whereas oxygen free radicals, lysosomal enzymes, and

NO can induce tissue damage. Notably, several signal transduction pathways coordinate the onset and termination of inflammatory responses [3,4]. As such, inhibiting the release of these inflammatory cytokines and mediators has the potential to effectively mitigate tissue damage caused during the inflammatory process. At present, antibacterial and antiviral anti-inflammatory drugs are often used in clinical treatment, while the long-term use of these drugs is prone to drug resistance and related complications. Therefore, the development of therapeutically effective and less toxic anti-inflammatory drugs is essential as a means of reliably treating inflammatory diseases in the clinic. Natural products (NPs) often serve as critical precursors in the field of drug development. Tibetan medicine is one of the most comprehensive traditional medical systems in the world [5–7]. The search for potential anti-inflammatory active substances derived from traditional Tibetan medicines has thus emerged as a promising approach to the development of novel drugs to prevent or treat inflammatory diseases.

Saxifraga L. is the largest genus in the Saxifragaceae family, with around 500 species found primarily in the circumpolar and alpine regions of the Northern Hemisphere [8]. *Saxifraga atrata* (*S. atrata*) is a member of the *Saxifraga Sect. Micranthes* and grows at elevations of 3000 to 4200 m in alpine meadows or flowstone beaches [9]. In traditional Tibetan medicine, *S. atrata* flowers are used to reduce fevers and treat lung diseases [10]. However, the morphological characteristics of different *Saxifraga* L. species are so similar that it can be difficult to differentiate them from a macroscopic perspective. Moreover, as there are limited data available regarding the active compounds present in *S. atrata*, quality control and bioactivity determination efforts remain challenging. It is thus critical that a large-scale approach to isolating and purifying standard bioactive substances from *S. atrata* be established in order to better develop its medicinal value. The complex makeup of *S. atrata*, as well as its limited solubility, have impeded the isolation of adequately pure compounds from this species to date. Silica-gel column chromatography [11] and recrystallization [12] are traditional approaches to preparing, separating, and purifying NPs. However, they are all limited by low efficiency and low recovery rates [13]. In recent years, additional extraction techniques have emerged, including high-speed counter-current chromatography (HSCCC), which separates components based upon a liquid−liquid partitioning principle based on the differential partition coefficients of different substances in two phases [14,15]. The principle of HSCCC theory is based on a hydrodynamic equilibrium system. One criterion used to judge whether the system has reached equilibrium is the observation of whether the effluent liquid was stratified. If stratification is observed, the sample can be injected into the high-speed counter-current chromatography system with separation then being achieved using a spiral tube separation based on the differential partition coefficients of the different components in the two phases, thus forming a multi-stage extraction process, with the results of this separation being recorded using a data collection system and workstation. Controlling these separation effects during the instrument's operation necessitates a range of approaches, leading to drawbacks including poor separation resolution, the need to determine the partition coefficient, and complex operating procedures. High resolution and repeatability are essential for the separation and preparation of high-purity compounds [16–18]. To overcome these issues, additional separation techniques are needed to facilitate the large-scale purification of high-purity bioactive compounds derived from *S. atrata*.

Preparative high-performance liquid chromatography (prep-HPLC) has been established as an effective approach to separating a single compound from complex systems, such as biological samples and NPs. Because of its superior column efficiency, separation repeatability, online detection, and autonomous control, this technology has been extensively employed in a variety of fields [19,20]. However, to preserve the chromatographic column and simplify the subsequent separation and purification steps, prep-HPLC cannot directly separate crude extracts. Instead, target compounds must first be enriched from the crude extracts via sample pretreatment, while non-target components should be eliminated. Based on the principle of analytical liquid chromatography, which takes advantage of the fact that different substances have different partition coefficients in a system consisting of a stationary phase and a mobile phase, medium pressure liquid chromatography (MPLC) represents an effective pretreatment approach [21–23]. As such, the preferred approach to the large-scale preparation of high-purity bioactive compounds is a combination of MPLC and prep-HPLC [24,25]. Affinity ultrafiltration-HPLC, a high-throughput screening technique combining affinity ultrafiltration and HPLC, has emerged as a focus of growing interest owing to its ability to rapidly, efficiently, and sensitively screen for bioactive components derived from complex NPs. Affinity ultrafiltration generally includes four steps: affinity incubation, centrifugal ultrafiltration, ligand dissociation, and analysis and detection. Purified, non-covalent complexes must be collected following affinity ultrafiltration, after which appropriate methods such as the addition of organic reagents, ultrasonic treatment, or changing the pH to denature the drug target such that the small molecule ligands of interest dissociate from non-covalent complexes to permit subsequent analysis and detection [26,27].

Since its initial development in the mid-1970s, molecular docking has proven to be an important tool to help understand how compounds interact with their molecular targets, assisting in drug design and discovery [28]. Molecular docking is one of the most commonly used methods in the field of structure-based drug design, which focuses on the study of molecular interactions, as it can predict the appropriate target binding site with considerable accuracy, calculate the affinity of the resultant complex, determine the relative positions and orientations of the ligands and receptors, and thereby study the mechanisms governing the activity of a given agonist, inhibitor, or drug, laying the foundation for new drug design [29–31]. Molecular docking methods are invaluable in the drug research field, providing an effective tool for the discovery and optimization of lead compounds. This approach can be used to study the interaction of small molecule probes with intracellular biomolecules to identify the targets of small molecules in an organism, and to enable breakthroughs in new drug development. Based on structural biology, the structure and function of important proteins related to normal physiological processes and diseases such as cancer and inflammation can be systematically studied and analyzed to define the three-dimensional (3D) structures of protein drug targets and conduct kinetic simulation studies of drug–target protein interactions. Moreover, existing drug molecules and the active ingredients of herbal medicines can be leveraged to design new lead compounds with enhanced activity through further modification and optimization.

To the best of our knowledge, research focused on *S. atrata* to date has primarily centered on the isolation of antioxidants [20]. There has only been one report regarding the separation and purification of standard substances from this species. Briefly, using MPLC technology combining a polyamide with MCI GEL® CHP20P stationary phase, Dang et al. [32] effectively isolated and purified bergenin from *S. atrata* with a purity greater than 99%. In the present study, the MCI GEL® CHP20 stationary phase was a styrene–divinylbenzene matrix with exceptional hydrophobicity that was able to efficiently separate polar compounds [33,34]. First, the target component was enriched on a medium pressure chromatographic column with MCI GEL® CHP20 as the stationary phase. Subsequently, a specific anti-inflammatory ingredient was effectively screened from Fr1 by the method of affinity ultrafiltration-RPLC. After that, using only one step of preparative reversed-phase liquid chromatography (prep-RPLC), a high purity sample of the anti-inflammatory active compound (11-O-(4′-O-methylgalloyl)-bergenin, Fr11) was obtained from the target fraction. Ultimately, molecular docking analyses were then used to explore interactions between this bioactive compound and the inflammatory-related targets of cyclooxygenase-2 (COX-2), 5-lipoxygenase (5-LOX), arachidonic acid 15-lipoxygenase (ALOX15), p38 mitogen-activated protein kinase (p38MAPK), c-Jun NH_2-terminal kinase (JNK), extracellular signal-regulated kinase 5 (ERK5), epidermal growth factor receptor (EGFR), epidermal growth factor receptor 2 (ERBB2), toll-like receptors (TLR), nuclear factor kappa-B (NF-κB), TNF, tumor necrosis factor receptor 1A (TNFR1A), nitric oxide synthase (iNOS), IL-6, IL-23 receptor (IL-23R), IL-1β, IL-10, and e-selectin (SELE) in order to evaluate the anti-inflammatory activity of this sample. The predicted results suggest that the isolated potential ligand (11-O-(4′-O-methylgalloyl)-bergenin, Fr11) exhibits significant anti-inflammatory activity through multi-pathway and multi-target synergy, providing novel basis for the treatment of inflammatory diseases. Therefore, an integrated combination of affinity ultrafiltration-HPLC and multi-target molecular docking provides a powerful approach to the multi-dimensional mining of anti-inflammatory active components in *S. atrata*. This approach can be leveraged for the large-scale purification of anti-inflammatory isocoumarin standards from other NP extracts, thereby supporting the development of the NP-based pharmaceutical industry.

2. Materials and Methods

2.1. Apparatus and Chemicals

An MPLC workstation consisting of two NP7000 prep-HPLC pumps, an NU3000 UV–Vis detector, a 5 mL manual injector, and an LC workstation was used for this study (Hanbon Science & Technology Co., Ltd., Huaian, China). HPLC analyses were conducted using an LC-16A instrument equipped with a column thermostat and autosampler following sample degassing performed using a DGU-20A3R instrument (Shimadzu Instruments Co., Shanghai, China). A Waters QDa ESI mass spectrometer (Waters Instruments Co., Milford, MA, USA) or a Q Exactive Orbitrap instrument (Thermo Fisher Scientific Inc., Waltham, MA, USA) mass spectrometer was utilized when collecting ESI-MS spectra, while ^{1}H and ^{13}C NMR spectra were obtained with a Bruker Avance 600 MHz (Bruker, Karlsruhe, Germany) with MeOH-d_4 as the solvent.

MCI GEL® CHP20P (120 μm) separation materials were obtained from Mitsubishi Chemical Corporation (Tokyo, Japan). Two ReproSil-Pur C18 AQ columns (4.6 × 250 mm, 5 μm and 20 × 250 mm, 5 μm) were from Maisch Corporation (Munich, Germany). The Click XIon column (4.6 × 250 mm, 5 μm) was from ACCHROM Corporation (Beijing, China). Acetonitrile (ACN), preparative methanol (CH_3OH), and HPLC-grade ACN were from Kelon Chemical Reagent Factory (Chengdu, China). A Moore water purification station was used to obtain ultrapure water for HPLC from deionized water (Chongqing, China). Human cyclooxygenase 2 was from Sigma-Aldrich (St. Louis, MO, USA). PBS was purchased from Solarbio Science & Technology Co. Ltd. (Beijing, China). Ultrafiltration membranes (YM-30, 30 kDa) were from Millipore Co. Ltd (Burlington, MA, USA).

2.2. Preparation of the Target Fraction

The *S. atrata* materials used in this study were harvested from Menyuan County, Haibei Tibetan Autonomous Prefecture, Qinghai Province (3922 m, 37°38′, 09.81″ N, 101°31′, 51.77″ E) in July 2017. The plant was identified as *S. atrata* by Qingbo Gao of the Northwest Institute of Plateau Biology of the Chinese Academy of Sciences. Medicinal material specimens are kept in the Key Laboratory of Adaptation and Evolution of Plateau Biota, Chinese Academy of Sciences (No. Gao2017080).

The harvested fresh sample was allowed to air dry under cool conditions and ground to produce a powder, after which 100 g of the resultant powder were extracted twice using methanol (4.0 L of methanol per extraction for 24 h). The obtained extract (8 L) was filtered and condensed by evaporation under decreased pressure to a volume of about 500 mL, mixed with 10 g silica gel and dried at 40 °C. Then the sample was crushed and passed through a sieve to yield a 21.8 g sample. The dried silica gel mixture was loaded into a medium pressure chromatography tower (49 × 100 mm), which was linked to a medium pressure chromatography column (49 × 460 mm) containing 1.2 L of MCI GEL CHP20P stationary phase. Elution was conducted using a methanol/water system for 0–150 min (0–100% methanol) at a 57.0 mL/min flow rate, with a 254 nm detection wavelength. This same method was repeated once, yielding an 840.0 mg sample after concentrating the target fraction with a recovery of 7.1%.

2.3. Screening and Isolation of Potential Anti-Inflammatory Active Ingredients Based on Affinity Ultrafiltration-HPLC

2.3.1. Screening of the Potential Anti-Inflammatory Active Ingredient in the Target Fraction with Affinity Ultrafiltration

Initially, 1 mg of the *S. atrata* fraction Fr1 was dissolved in 500 μL of PBS (pH 7.2) to prepare a 2 mg/mL sample solution for affinity ultrafiltration screening. The test sample solution (Fr1, 100 μL) was incubated with 10 μL of COX-2 (4 U) for 45 min at 37 °C in an incubator shaker. This mixture was then transferred to a 30 kDa ultrafiltration tube, centrifuged at 10,000 rpm for 10 min and washed 3 times with 200 μL of PBS to remove unbound ligands. Next, COX-2-bound ligands were released from this complex by incubation with 200 μL of methanol (90%, *v*/*v*) for 10 min at room temperature and centrifuged at 10,000 rpm for 10 min. This process was repeated in triplicate, and the filtrate was then collected, freeze-dried, and re-dissolved with 50 μL of methanol for HPLC analysis. To prepare a negative control sample, the enzyme was denatured via incubation for 10 min in boiling water, after which the operating procedures were identical to those above.

2.3.2. Purification of Potential COX-2 Ligand via Reversed-Phase Liquid Chromatograph

To further purify the target sample, this 840.0 mg sample was dissolved in a 10.0 mL methanol/water (7:3, *v*/*v*) solution. It was then filtered through a 0.45 μm organic filter membrane to obtain an 84.0 mg/mL sample solution. The purification step was completed on a ReproSil-pur C18 AQ preparative column (20 × 250 mm, 5 μm), with 0.2% *v*/*v* formic acid in water as mobile phase A and chromatographic acetonitrile as mobile phase B. The injection volume was 0.5 mL and the sample was eluted with 10% acetonitrile isocratic elution at a 19 mL/min flow rate for 20 min, with 254 nm as the detection wavelength.

2.4. Analysis of the Purity of the Isolated Candidate COX-2 Ligand

Click XIon and ReproSil-Pur C18 AQ analytical columns were used to assess the purity of the isolated standard substance. The mobile phase A was composed of 0.2% (*v*/*v*) formic acid in water, while mobile phase B consisted of ACN. Both analytical columns were used to perform gradient elution for 30 min in 95–55% ACN and 5–50% ACN, with a flow rate of 1.0 mL/min and a 5 μL injection volume. The detection wavelength was 254 nm.

2.5. Molecular Docking Analyses

Molecular docking is a technique that simulates the interactions between small molecule ligands and receptor biomolecules based on the "Induced Fit Theory" of ligand−receptor interaction. By continuously optimizing the position of the small molecule compound and the dihedral angle of the flexible bond inside the molecule, the optimal conformation of the small molecule for interaction with the target macromolecule can be identified and its binding mode and affinity can be assessed, with the ligand exhibiting the highest affinity for a given receptor that is closest to its natural conformation; this is identified via a predictive scoring function. The 2D structure of the active compound was converted into a 3D structure using Chem3D 16.0. Then, the crystal structures of COX-2 (PDB ID: 5IKQ), 5-LOX (PDB ID: 3V99), ALOX15 (PDB ID: 2P0M), p38MAPK (PDB ID: 1A9U), JNK (PDB ID: 3DA6), ERK5 (PDB ID: 4IC7), EGFR (PDB ID: 2ITX), ERBB2 (PDB ID: 3PP0), TLR (PDB ID: 2Z7X), NF-κB (PDB ID: 3RZF), TNF (PDB ID: 2E7A), TNFR1A (PDB ID: 1ICH), iNOS (PDB ID: 2ORO), IL-6 (PDB ID: 1P9M), IL-23R (PDB ID: 3DUH), IL-1β (PDB ID: 5I1B), IL-10 (PDB ID: 2ILK), and SELE (PDB ID: 1GIT), were downloaded from the PDB (https://www.rcsb.org/ (accessed on 6 April 2022) database, and water molecules and free radicals were removed using the PyMOL software, followed by the addition of polar hydrogen atoms and Kollman charges. Finally, the format of the compounds and protein molecules were converted to the pdbqt format using AutoDock for molecular docking and the calculation of molecular binding energies. Minimization was performed using the Lamarckian genetic algorithm and the pseudo-Solis and Wets methods with default parameters. A total of 100 peptide conformations were defined based on docking score values, where conformations exhibiting the lowest binding energy were selected for model development. After the docking was completed, the interaction of the ligand with key amino acid residues in the active site of the protein was assessed based on the scoring results. It is generally believed that the lower the binding energy, the stronger the affinity and the better the ligand and receptor binding.

3. Results

3.1. Target Fraction MCI GEL® CHP20P Medium Pressure Liquid Chromatography Sample Pretreatment

The methanol extract of *S. atrata* contained large amounts of chlorophyll, which had the potential to interfere with the subsequent separation of the target compound and cause column contamination. It was thus necessary to eliminate it prior to any further chromatographic experiments. The MCI GEL® CHP20P filler used medium pressure chromatographic separation consists of a matrix of the styrene−divinylbenzene copolymer, and polymer-based filler can maintain the stability of the spherical structures and associated properties following exposure to extreme acid−base solutions and organic solvents, thus ensuring the repeatability of target compound isolation during purification [33]. This enabled appropriate process development and optimization of separation conditions to achieve high resolution and product recovery. Based on these adsorption properties, we initially loaded 21.8 g of the mixed sample to a small medium pressure column (49 × 100 mm) and connected it to a medium pressure column (49 × 460 mm) equipped with 1.2 L of MCI GEL® CHP20P for dry sample loading preparation. The separation chromatogram for this extract when using MCI GEL® CHP20P and methanol−water as an eluent is shown in Figure 1. After 1 cycle, the target component was collected, concentrated, and weighed to yield 840.0 mg of sample (Fr1, recovery 7.1%). This compound was then dissolved in a methanol/water solution (7:3 v/v, 10.0 mL, 84.0 mg/mL) and filtered using a 0.45 µM organic filter membrane to facilitate further purification.

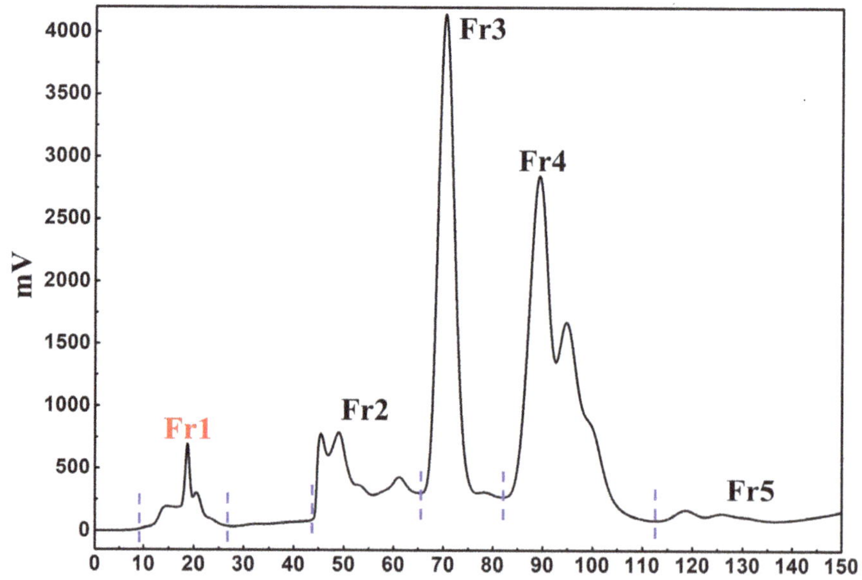

Figure 1. Separation chromatogram of *S. atrata* amorphous silica gel mixture. Conditions: mobile phase A: water and B: methanol; gradient: 0–150 min, 0–100% B; monitoring wavelength: 254 nm; flow rate: 57.0 mL/min; sample loading: 21.8 g of *S. atrata* amorphous silica gel combination; column temperature: room temperature.

3.2. Screening and Isolation of Potential Anti-Inflammatory Active Ingredient Based on Affinity Ultrafiltration-HPLC

3.2.1. Screening of the Potential Anti-Inflammatory Active Ingredient in the Target Fraction

The target fraction (Fr1) was first analyzed using a Click XIon column (4.6 × 250 mm, 5 μm). The presence of a major component (peak 1) in this fraction was evident (Figure 2A), but many impurities (red dashed lines 2, 3, and 4) were also evident when using this column, and the response value was relatively low. Hydrophilic interaction chromatography (HILIC) and RPLC are known to offer effective complementary selectivity. To improve the purity of the main component (peak 1), Fr1 was re-analyzed using a ReproSil-Pur C18 AQ column (4.6 × 250 mm, 5 μm). The analytical chromatogram, shown as the black line in Figure 2B, also revealed a main component (peak 1). Observation of the black line in Figure 2B revealed that a portion of the peaks (red dotted line 5) came out in the first 5 min, possibly due to the presence of large polar substances in this fraction. Strongly polar compounds were poorly retained during reversed-phase liquid chromatography. Some impurities (red dotted line 6) were also observed behind the main peak. A comparison of the chromatographs derived from these two columns (Figure 2A,B) revealed that on the ReproSil-Pur C18 AQ analytical column, the main peak and impurities exhibited a somewhat longer peak time interval. This was more conducive to the optimization of elution conditions, allowing for the more efficient preparation of the principal component of this fraction (peak 1 in Figure 2B). When comparing the experimental conditions used for these two analytical columns, the consumption of organic reagents was reduced when using the ReproSil-Pur C18 AQ analytical column 5–50% ACN gradient elution relative to the Click XIon analytical column 95–55% ACN gradient elution during the same 30 min elution period, making the former approach more environmentally friendly. This provided further support for the different selectivity of these two analytical columns. While polar compounds were weakly retained on the RPLC analytical column such that they were not effectively separated for analysis, the HILIC approach is often used in the separation of highly polar molecules, such that these two approaches can complement one another.

To determine whether the main component in Fr1 was the potential anti-inflammatory active component of interest, we next performed a one-step affinity ultrafiltration screening step. COX-2 is an inducible isoenzyme, while COX-2 activity under basal conditions is very limited, it can be readily upregulated under inflammatory conditions, leading to an increase in PEG2, PGE1, and PGI2 production in inflammatory tissues, thus perpetuating inflammatory responses and tissue damage [35]. Affinity ultrafiltration has unique applications in the discovery of small molecule drugs due to its high sensitivity and selectivity. The magnitude of the ability of a small molecule ligand to bind to an enzyme can be expressed by the relative binding affinity (RBA). This RBA value is obtained by calculating the peak area ratio of the compound after incubation with the active and inactivated enzymes. The peak areas of the peaks were obtained from the integration of the high-performance liquid chromatograph. Compounds with an RBA greater than 1.5 are considered to be potential ligands. The RBA value for a given compound is calculated as follows:

$$\mathrm{RBA} = \mathrm{A_S}/\mathrm{A_0}$$

where A_S represents the peak area of the sample incubated with the activated enzyme and A_0 represents the peak area of the sample incubated with the inactivated enzyme. As shown in the red and blue lines in Figure 2B, the *S. atrata* Fr1 main component exhibited a good RBA value of 1.63.

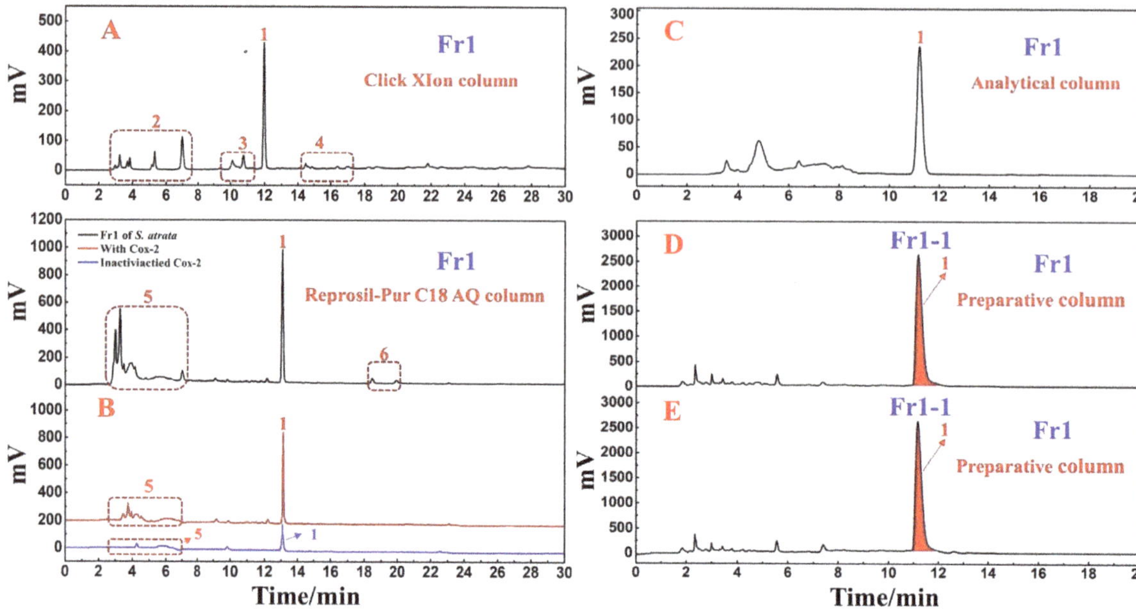

Figure 2. The chromatogram derived from the Click XIon analytical column (**A**) An affinity ultrafiltration-HPLC chromatogram for potential COX-2 ligands present in *S. atrata* target fraction Fr1 (**B**). The black, red, and blue lines, respectively, correspond to the HPLC chromatograms of *S. atrata* target fraction Fr1 without COX-2, with activated COX-2, and with inactivated COX-2. The analytical (**C**) and preparative (**D**,**E**) chromatograms of Fr1 on the ReproSil-Pur C18 AQ column. Number 1 represents the target main component in Fr1, while red dashed boxes 2, 3, 4, 5, and 6 highlight the impurities present in Fr1. Conditions: mobile phase A: 0.2% *v*/*v* formic acid in water, B: ACN; gradient: 0–30 min, 95–55% B for Click XIon analytical column (**A**), 0–30 min, 5–50% B for Reprosil-Pur C18 AQ analytical column (**B**) and 0–20 min, 10% B for Reprosil-Pur C18 AQ column (**C**–**E**); detection wavelength: 254 nm; flow rate: 1.0 mL/min (**A**–**C**) and 19.0 mL/min (**D**,**E**); injection volume: 5 μL (**A**–**C**) and 0.5 mL (**D**,**E**); column temperature: 30 °C for analysis and room temperature for preparation.

3.2.2. Purification of the Potential COX-2 Ligand with Reversed-Phase Liquid Chromatography

Based on the above analytical findings, a ReproSil-Pur C18 AQ preparative column (20 × 250 mm, 5 μm) was chosen for further preparation in order to increase the purity of the primary component (peak 1 in Figure 2B) while mitigating environmental harm to the greatest extent possible. The elution parameters were optimized before employing the ReproSil-Pur C18 AQ column for separation and preparation. Under isocratic elution conditions using ACN-0.2% *v*/*v* formic acid in water (0–20 min, 10% ACN), the main peak (peak 1 in Figure 2C) exhibited sufficient resolution. As a result, we conducted the following studies using these conditions. After linear amplification, the main compound (peak 1 in Figure 2D,E) was separated and purified on the ReproSil-Pur C18 AQ preparative column using the established conditions. The result of this purification step is shown in Figure 2D,E. Reproducible chromatographic separation is critically important in order to prepare a compound with the highest possible purity, minimizing the mixing of components during repeated injections. In this study, we recorded the retention time values of the target components after 10 repeated injections to assess the reproducibility of the process. As shown in Figure 2D,E, the system exhibited good reproducibility, and therefore could be used to enrich the target component via repeated collection. The sample was effectively dissolved in a 70% methanol/water solvent mixture (86.7 mg/mL), prepared with 10% ACN as the eluent at a flow rate of 19.0 mL/min and an injection volume of 0.5 mL. Fr11 was collected and concentrated after 20 replicate purification steps, yielding 74.1 mg of the sample with a recovery of 5.7%.

3.3. Analyses of the Purity and Structural Characterization of Potential COX-2 Ligands

To clarify the purity of the target compound following the above isolation procedures, different stationary phases with distinct polarities and separation mechanisms were used: one was a HILIC column (Click XIon column) and the other was a pure water-resistant RP-C18 column (ReproSil-Pur C18 AQ column). As shown in Figure 3A,B, the target compound was >99% purity, obtained for each column. RPLC relies on hydrophobic interactions between the hydrophobic stationary phase and the solute to achieve the efficient separation of weakly and moderately polar compounds. HILIC consists of a polar stationary phase and a polar mobile phase. The mobile phase used is similar to that employed for RPLC, with a weak eluent as the organic phase and a strong eluent as the aqueous phase, enabling this approach to achieve column efficiency and symmetrical peak shapes equivalent to those produced via RPLC. Thus, confirmation of the purity of the target compound Fr11 was achieved through both the Click XIon and ReproSil-Pur C18 AQ columns, with both strategies supporting the high purity of this target compound Fr11.

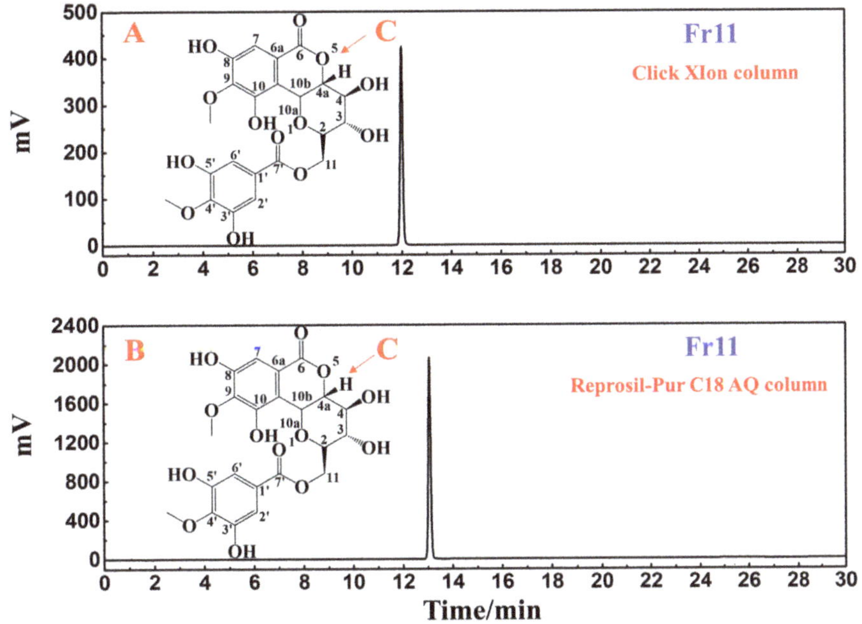

Figure 3. Analyses figures of the purity of the target compound in Fr11 were performed using Click XIon (**A**) and ReproSil-Pur C18 AQ (**B**) analytical columns. (**C**) shows the chemical structure of the isolated compound. Conditions: mobile phase A: 0.2% v/v formic acid in water and B: ACN; gradient: 0–30 min, 95–55% B for (**A**), 0–30 min, 5–50% B for (**B**); detection wavelength: 254 nm; flow rate: 1.0 mL/min; injection volume: 5 µL; column temperature: 30 °C.

The obtained ESI-MS, ^{1}H NMR, and ^{13}C NMR spectra for the target compound in Fr11 were compared with published literature to clarify the structural characteristics of this compound (Figure 3C). The structural identification of Fr11 based on these data is shown in Supplementary Figures S1–S3, with all of these data (summarized below) supporting the identity of this target compound as 11-O-(4′-O-methylgalloyl)-bergenin.

Compound Fr11 (11-O-(4′-O-methylgalloyl)-bergenin, 74.1 mg, white powder, ESI-MS m/z: 493.32 [M-H]$^{-}$, calc. for $C_{22}H_{22}O_{13}$ m/z 494.1060): ^{1}H NMR (600 MHz, MeOH-

d_4) 7.09 (2H, s, H-2′, 6′), 7.08 (1H, s, H-7), 5.03 (1H, d, J = 10.5 Hz, H-10b), 4.87 (H, dd, J = 5.2, 4.2 Hz, H-11a), 4.41 (1H, dd, J = 11.2, 6.7 Hz, H-11b), 4.10 (1H, dd, J = 10.5, 9.5 Hz, H-4a), 3.97 (1H, m, H-2), 3.90 (3H, s, C-9-OCH_3), 3.86 (3H, s, C-4′-OCH_3), 3.85 (1H, m, H-4), 3.35 (1H, s, H-3); ^{13}C NMR (151 MHz, MeOH-d_4) 167.8 (C-4′), 165.7 (C-6), 152.4 (C-8), 151.9 (C-3′, 5′), 149.4 (C-10), 142.4, (C-9) 141.5 (C-7′), 126.2 (C-1′), 119.5 (C-6a), 117.1 (C-10a), 111.2 (C-7), 110.4 (C-2′, 6′), 81.4 (C-4a), 80.6 (C-2), 75.5 (C-4), 74.5 (C-10b), 71.9 (C-3), 62.0 (C-11), 61.0 (C-9-OCH_3), 60.8 (C-4′-OCH_3). These ESI-MS, ^{1}H NMR, and ^{13}C NMR spectra thus revealed this compound to be 11-O-(4′-O-methylgalloyl)-bergenin based on a comparison with published literature sources [36].

3.4. Molecular Docking Analyses of the Interactions between 11-O-(4′-O-methylgalloyl)-bergenin and Inflammation-Related Targets

Although we successfully screened and isolated candidate COX-2 ligand from *S. atrata* via affinity ultrafiltration-HPLC, the mechanistic basis for the predicted ligand−enzyme interactions remained unclear. To further explore the role of this potential anti-inflammatory active derivative of *S. atrata* (11-O-(4′-O-methylgalloyl)-bergenin, Fr11), a molecular docking method was used to study the binding mode and binding energy of 11-O-(4′-O-methylgalloyl)-bergenin when interacting with inflammation-related targets. Inflammatory responses are shaped through interactions between various inflammatory and anti-inflammatory factors that regulate the biological effects of inflammatory cells through different signal transduction pathways, with different inflammatory cell types, cytokines, adhesion molecules, and inflammatory mediators, all cooperating to determine the characteristics of a given inflammatory response. After reviewing the relevant literature [37,38], a molecular docking analysis was carried out for common inflammatory signaling pathways including arachidonic acid (AA) metabolism, MAPK, and TLR/myeloid differentiation factor 88 (MyD88)/NF-κB pathways. Figure 4 illustrated a summary of the involved inflammatory-related mechanisms in the molecular docking analysis.

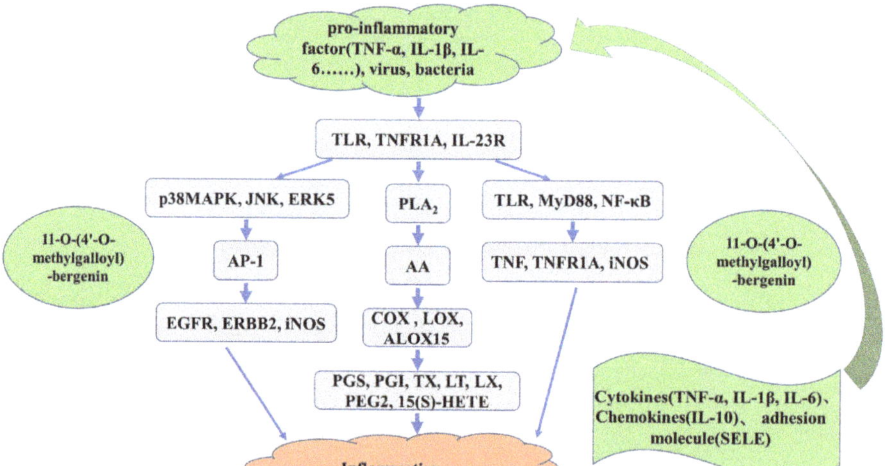

Figure 4. The schematic diagram of the involved inflammatory-related mechanisms.

3.4.1. AA Metabolic Pathway

Phospholipase A_2 (PLA_2) leases AA from the membrane phospholipid, whereupon it can be used to produce prostaglandins, prostacyclins, thromboxanes, leukotrienes, lipoxins, arachidonic acid ethanolamine, and epoxyeicosatrienoic acids via the COX and LOX pathways [35]. These metabolites affect neutrophil recruitment and infiltration, platelet aggregation, epithelial barrier function, vascular permeability and bronchoconstriction, in addition to inducing inflammatory responses. The ALOX15 protease is a heme iron-free dioxygenase that catalyzes and esterifies the peroxidation of polyunsaturated fatty acids and produces a series of bioactive lipid intermediates. Notably, ALOX15 is involved in early inflammatory responses and its metabolite 15-hydroxyeicosatetraenoic acid (15(S)-HETE) is a potent pro-inflammatory chemoattractant for neutrophils and leukocytes.

Given the above considerations, COX-2, 5-LOX, and ALOX15 were selected for molecular docking with 11-O-(4′-O-methylgalloyl)-bergenin (Fr11) to assess its potential anti-inflammatory activity. These docking results are shown in Figure 5. When docking with COX-2, 11-O-(4′-O-methylgalloyl)-bergenin (Fr11) was located in the pocket surrounded by multiple amino acids (LEU-153, TYR-131, PRO-154, GLU-466, GLN-462, HIS-39, GLY-136, CYS-47, CYS-36, PRO-157, and ASP-158), interacting through van der waals, conventional hydrogen bond, carbon hydrogen bond, and pi-alkyl acting force interactions (Figure 5A). When docking with 5-LOX, 11-O-(4′-O-methylgalloyl)-bergenin (Fr11) was located in a pocket surrounded by ASP-170, GLU-622, ARG-401, PHE-402, LEU-615, and ALA-398, interacting through conventional hydrogen bond and pi-sigma and pi-alkyl acting force interactions (Figure 5B). When docking with ALOX15, 11-O-(4′-O-methylgalloyl)-bergenin (Fr11) was located in a pocket surrounded by LEU-597, HIS-545, LEU-549, ASP-600, MET-640, ASP-550, LEU-215, and GLU-650, interacting through conventional hydrogen bond, carbon hydrogen bond, and pi-anion and pi-alkyl acting force interactions (Figure 5C). The binding of COX-2 and 5-LOX to 11-O-(4′-O-methylgalloyl)-bergenin (Fr11) was dominated by hydrogen bonding (Figure 5A,B), while for ALOX15, this binding was dominated by hydrophobic interactions (Figure 5C). As shown in Table 1, the binding energy values for interactions between 11-O-(4′-O-methylgalloyl)-bergenin (Fr11) and 5-LOX, COX-2, and ALOX15 in this study were −3.66 kcal/mol, −6.35 kcal/mol, and −9.36 kcal/mol, respectively. Overall, these results suggested that the docking of 11-O-(4′-O-methylgalloyl)-bergenin (Fr11) with 5-LOX was not stable, such that 5-LOX is unlikely to represent a target of the anti-inflammatory activity of this compound. In contrast, the docking binding energy for the interaction between COX-2 and 11-O-(4′-O-methylgalloyl)-bergenin (Fr11) of −6.35 kcal/mol largely supported the high RBA value (1.63) detected via affinity ultrafiltration-HPLC. In addition, 11-O-(4′-O-methylgalloyl)-bergenin (Fr11) exhibited a binding energy of −9.36 kcal/mol when interacting with ALOX15, binding to amino acids in 2P0M primarily through hydrophobic interactions and thus reducing the energy of the system, forming a stable structure. Thus, we can predict that 11-O-(4′-O-methylgalloyl)-bergenin (Fr11) is likely to exert its anti-inflammatory effects mainly through the COX-2 pathway by modulating AA metabolism, with such activity also being closely related to ALOX15.

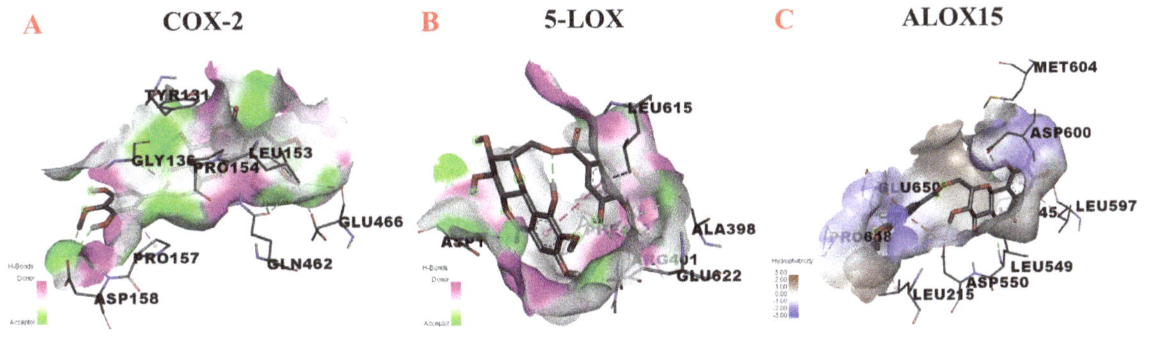

Figure 5. Molecular docking analysis of the putative binding between 11-O-(4′-O-methylgalloyl)-bergenin and COX-2, 5-LOX, and ALOX15. (**A**–**C**) correspond to the binding models for interactions between 11-O-(4′-O-methylgalloyl)-bergenin and COX-2, 5-LOX, and ALOX15, respectively. Hydrophobic or H-bond interactions are displayed as colored surface.

Table 1. Intermolecular interactions between Fr11 and COX-2, 5-LOX, and ALOX15.

Proteins	Protein Function	Binding Energy (kcal/mol)	Binding Residues	Type
COX-2	Inflammation, pain	−6.35	LEU-153	Pi-Alkyl
			TYR-131	Conventional Hydrogen Bond
			PRO-154	Pi-Alkyl
			GLU-466	Carbon Hydrogen Bond
			GLN-462	Conventional Hydrogen Bond
			HIS-39	van der Waals
			GLY-136	Carbon Hydrogen Bond
			CYS-47	van der Waals
			CYS-36	Carbon Hydrogen Bond
			PRO-157	Pi-Alkyl
			ASP-158	Conventional Hydrogen Bond
5-LOX	Inflammation, pain	−3.66	ASP-170	Conventional Hydrogen Bond
			GLU-622	Conventional Hydrogen Bond
			ARG-401	Conventional Hydrogen Bond
			PHE-402	Pi-Alkyl
			LEU-615	Conventional Hydrogen Bond
			LEU-615	Pi-Sigma
			ALA-398	Conventional Hydrogen Bond
ALOX15	Inflammation, immunity, neuroprotection	−9.36	LEU-597	Pi-Alkyl
			HIS-545	Carbon Hydrogen Bond
			HIS-545	Conventional Hydrogen Bond
			LEU-549	Conventional Hydrogen Bond
			ASP-600	Conventional Hydrogen Bond
			MET-640	Pi-Alkyl
			ASP-550	Conventional Hydrogen Bond
			ASP-550	Pi-Anion
			LEU-215	Pi-Alkyl
			GLU-650	Conventional Hydrogen Bond
			GLU-650	Pi-Anion

3.4.2. MAPK Signaling Pathway

The MAPK is an important signaling pathway that plays a critical role in mediating diverse cellular responses, leveraging a highly conserved three-level kinase cascade to transmit signals. Extracellular stimuli activate MAPK kinase kinase (MKKK) proteins, which in turn activate MAPK kinase (MKK), leading to the activation of MAPK proteins via the dual phosphorylation of tyrosine and threonine [38]. JNK, p38 MAPK, and ERK5 are the major subfamilies of MAPK. JNK, also known as stress-activated protein kinase (SAPK), is functionally similar to p38 MAPK, as both can be activated by various inflammatory cytokines and play important roles in stress responses related to conditions such as inflammation and apoptosis [39]. ERK5 mainly regulates cell growth and differentiation and is mediated by the upstream Ras/Raf signal protein, transmitting stimulatory signals to the nucleus and thereby regulating the proliferation, differentiation, and apoptosis of macrophages [40]. EGFR can bind to receptor tyrosine kinase (RTK) and when EGFR expression is enhanced, it can induce epithelial growth factor to form a complex with RTK through the Ras/Raf/MAPK signaling pathway, influencing epithelial cell proliferation and differentiation in a manner that can drive inflammatory activity. ERBB2 is a 185 kDa cell membrane receptor encoded by the proto-oncogene erbB-2. ERBB2 phosphorylation also causes ERBB2 to activate the Ras/Raf/MAPK signaling pathway [41].

In light of the above information, we next assessed potential interactions between 11-O-(4′-O-methylgalloyl)-bergenin (Fr11) and p38MAPK, JNK, ERK5, EGFR, and ERBB2 through molecular docking analyses. The results are shown in Figure 6. When docking with p38 MAPK, 11-O-(4′-O-methylgalloyl)-bergenin (Fr11) was predicted to bind to ASP-168, GLU-71, LYS-53, ILE-84, PHE-169, LEU-75, and TYR-35 in 1A9U, interacting through conventional hydrogen bond, pi-alkyl, pi-anion and pi-pi stacked acting force (Figure 6A). When docking with JNK, 11-O-(4′-O-methylgalloyl)-bergenin (Fr11) was predicted to bind to VAL-78, LYS-93, LEU-206, GLU-147, ASN-194, SER-72, SER-193, and ASN-152 in 3DA6, interacting through conventional hydrogen bond, carbon hydrogen bond, pi-sigma and pi-alkyl acting force (Figure 6B). When docking with ERK5, 11-O-(4′-O-methylgalloyl)-bergenin (Fr11) was predicted to bind to LEU-189, ASP-143, MET-140, ASN-63, GLY-62, ASP-138, GLY-67, VAL-69, and LYS-84 through van der waals, conventional hydrogen bond, pi-cation, amide-pi stacked and pi-alkyl acting force (Figure 6C). When docking with EGFR, 11-O-(4′-O-methylgalloyl)-bergenin (Fr11) was predicted to bind to ARG-841, ASN-842, THR-854, GLU-762, LEU-718, LYS-745, VAL-726, MET-766, MET-793, LEU-844, and ALA-743, interacting through conventional hydrogen bond, sulfur-x, and pi-alkyl acting force interactions (Figure 6D). When docking with ERBB2, 11-O-(4′-O-methylgalloyl)-bergenin (Fr11) was predicted to bind to ALA-751, LEU-852, LEU-726, MET-801, ARG-849, GLN-799, ASN-850, THR-862, GLY-729, LYS-753, MET-774, PHE-864, and LEU-785, interacting through conventional hydrogen bond, carbon hydrogen bond, pi-sigma, pi-pi t-shaped, and pi-alkyl acting force interactions (Figure 6E). As such, 11-O-(4′-O-methylgalloyl)-bergenin (Fr11) was able to access the binding sites of p38MAPK, JNK, ERK5, EGFR, and ERBB2, binding to these sites primarily through hydrophobic interactions (Figure 6A–E). The binding energy values calculated for the interaction between this compound and p38MAPK, JNK, ERK5, EGFR, and ERBB2 were −6.21 kcal/mol, −6.87 kcal/mol, −6.31 kcal/mol, −6.51 kcal/mol, and −8.35 kcal/mol, respectively (Table 2). As all of these values were below −5.0 kcal/mol, this was considered indicative of good binding activity. It was thus hypothesized that 11-O-(4′-O-methylgalloyl)-bergenin (Fr11) exerts its anti-inflammatory activity by reducing the release of pro-inflammatory mediators including leukotriene (LTs) and TNF-α, primarily via interactions with the p38MAPK, JNK, ERK5, EGFR and ERBB2 components of the MAPK signaling pathway, with the inhibition of ERBB2 being the most robust, suggesting that this may be an important mechanism whereby it can regulate inflammatory responses.

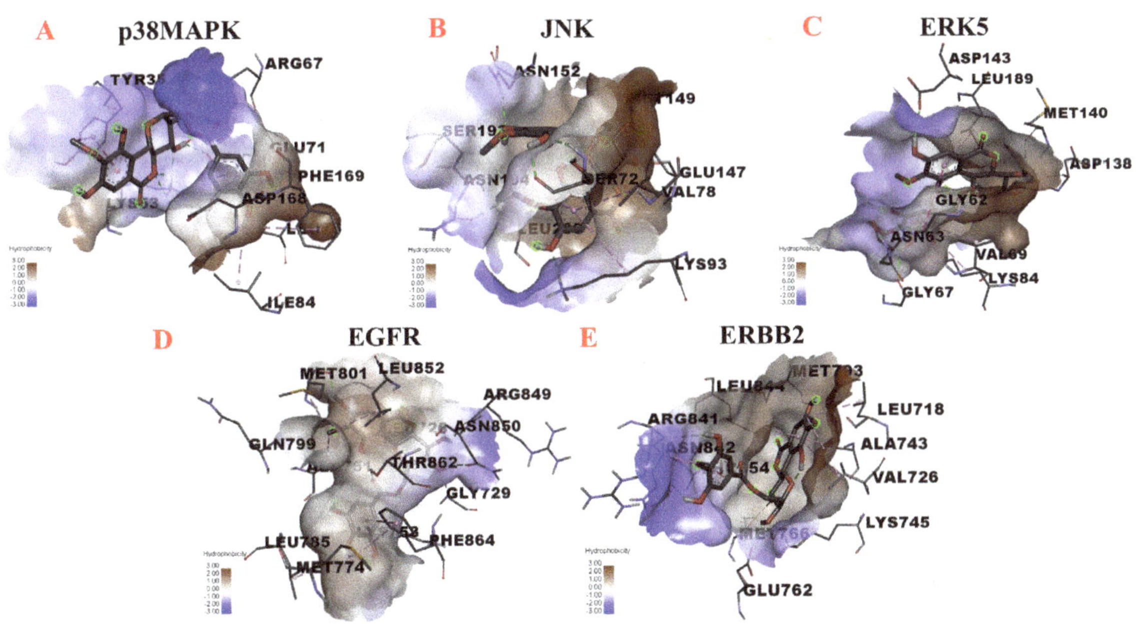

Figure 6. Molecular docking analyses of the putative binding between 11-O-(4′-O-methylgalloyl)-bergenin and p38 MAPK, JNK, ERK5, EGFR, and ERBB2. (**A–E**) correspond to the binding models for interactions between 11-O-(4′-O-methylgalloyl)-bergenin and p38 MAPK, JNK, ERK5, EGFR, and ERBB2, respectively. Hydrophobic interactions are displayed as a colored surface.

Table 2. Intermolecular interactions between Fr11 and p38 MAPK, JNK, ERK5, EGFR, and ERBB2.

Proteins	Protein Function	Binding Energy (kcal/mol)	Binding Residues	Type
p38 MAPK	Inflammation, apoptosis, proliferation, differentiation	−6.21	ASP-168 GLU-71 GLU-71 LYS-53 ILE-84 PHE-169 LEU-75 TYR-35 TYR-35	Conventional Hydrogen Bond Pi-Alkyl Pi-Anion Conventional Hydrogen Bond Pi-Pi Stacked Pi-Alkyl Pi-Alkyl Pi-Pi Stacked Pi-Alkyl
JNK	Inflammation, apoptosis, proliferation, differentiation	−6.87	VAL-78 LYS-93 LYS-93 LEU-206 LEU-206 GLU-147 ASN-194 SER-72 SER-72 SER-193 ASN-152	Pi-Sigma Conventional Hydrogen Bond Pi-Alkyl Pi-Sigma Pi-Alkyl Conventional Hydrogen Bond Carbon Hydrogen Bond Conventional Hydrogen Bond Carbon Hydrogen Bond Conventional Hydrogen Bond Conventional Hydrogen Bond
ERK5	proliferation, differentiation, development	−6.31	LEU-189 ASP-143 MET-140 ASN-63 GLY-62 ASP-138 GLY-67 VAL-69 LYS-84	Pi-Alkyl Conventional Hydrogen Bond Conventional Hydrogen Bond van der Waals Amide-Pi Stacked Conventional Hydrogen Bond Conventional Hydrogen Bond Pi-Alkyl Pi-Cation
EGFR	Inflammation, proliferation, differentiation	−6.51	ARG-841 ARG-841 ASN-842 THR-854 GLU-762 LEU-718 LYS-745 VAL-726 MET-766 MET-793 LEU-844 ALA-743	Conventional Hydrogen Bond Pi-Alkyl Conventional Hydrogen Bond Conventional Hydrogen Bond Conventional Hydrogen Bond Pi-Alkyl Conventional Hydrogen Bond Pi-Alkyl Sulfur-X Conventional Hydrogen Bond Pi-Alkyl Pi-Alkyl
ERBB2	Inflammation, proliferation	−8.35	ALA-751 LEU-852 LEU-726 MET-801 MET-801 ARG-849 GLN-799 ASN-850 THR-862 GLY-729 LYS-753 MET-774 PHE-864 PHE-864 LEU-785	Pi-Alkyl Pi-Sigma Pi-Sigma Conventional Hydrogen Bond Pi-Alkyl Conventional Hydrogen Bond Carbon Hydrogen Bond Conventional Hydrogen Bond Conventional Hydrogen Bond Carbon Hydrogen Bond Pi-Alkyl Pi-Alkyl Pi-Alkyl Pi-Pi T-Shaped Pi-Alkyl

3.4.3. TLR/MyD88/NF-κB Pathway

The TLR/MyD88/NF-κB pathway is an important signaling pathway that regulates inflammatory responses. TLR proteins are innate immune receptors, with TLR2 and TLR4 playing particularly central roles in inflammatory responses. MyD88 is a key molecule in the TLR signaling pathway and is involved in the initiation of inflammation through its role in the transduction of upstream signal information. TLR2 and TLR4 can bind to MyD88 after engaging their cognate ligands, thereby activating the downstream transcription factor NF-κB and driving the upregulation of inflammatory mediators, including IL-1β, IL-6, TNF-α, and other inflammatory cytokines, resulting in an inflammatory cascade response [42]. As a core transcriptional regulator of inflammation, NF-κB is rapidly activated in response to pathogens and in the context of immune system activation, while the inhibition of the NF-κB pathway can effectively treat many inflammatory diseases [43]. Blocking the NF-κB signaling pathway can inhibit the high expression of iNOS and reduce the levels of it and inflammatory mediators including NO, IL-6, and TNF-α, thereby exerting anti-inflammatory effects [44]. The TNF signaling pathway also plays a role in the induction of systemic inflammatory responses and the acute phase response. TNF can trigger the activation of many pathways, including the NF-κB and MAPK pathways, and can be separated into two structurally distinct isoforms (TNF-α and TNF-β). Macrophages secrete large quantities of TNF-α after foaming when activated, in turn promoting NF-κB pathway activation, downregulating the expression of iNOS downstream of NF-κB and thereby attenuating the local inflammatory response [45]. TNFR1A is a type I TNFR that mediates inflammatory responses mainly through NF-κB signaling, the induction of apoptosis, and the promotion of IL-6 secretion [46]. There are three different isoforms of iNOS, including endothelial nitric oxide synthase (eNOS), which is expressed under normal physiological conditions, neuronal nitric oxide synthase (nNOS) and iNOS, which is induced following organismal injury where upon it can promote the production of large amounts of NO. Excessive NO production can result in acute or chronic inflammation and given its role as the most important mediator of NO production, iNOS is an important regulator of the overall inflammatory response [47].

Given the above, we assessed potential interactions between 11-O-(4′-O-methylgalloyl)-bergenin (Fr11) and TLR, NF-κB, TNF, TNFR1A, and iNOS via a molecular docking approach. When docking with TLR, 11-O-(4′-O-methylgalloyl) bergenin (Fr11) was predicted to bind to a multiple amino acid pocket surrounded (ASN-103, LYS-104, HIS-78, ASN-79, ARG-80, GLN-54, TYR-56, and LYS-33) by conventional hydrogen bond, carbon hydrogen bond, amide-pi stacked, and pi-alkyl acting force interactions (Figure 7A). When docking with NF-κB, 11-O-(4′-O-methylgalloyl)-bergenin (Fr11) was predicted to bind to a pocket surrounded by LEU-21, CYS-99, VAL-29, THR-23, MET-65, GLU-97, ILE-165, LYS-44, MET-96, and GLU-61 via conventional hydrogen bond, carbon hydrogen bond, pi-sigma, pi-sulfur, and pi-alkyl acting force interactions (Figure 7B). When docking with TNF, 11-O-(4′-O-methylgalloyl)-bergenin (Fr11) was predicted to bind to a pocket surrounded by PRO-100, GLN-102, SER-99, ARG-103, PRO-106, ASN-112, and GLY-68 through conventional hydrogen bond, carbon hydrogen bond, and pi-cation acting force interactions (Figure 7C). When docking with TNFR1A, 11-O-(4′-O-methylgalloyl)-bergenin (Fr11) was predicted to bind to a pocket surrounded by GLY-364, ASP-357, GLU-360, LYS-343, LEU-359, MET-374, TRP-342, and ALA-370 through conventional hydrogen bond, carbon hydrogen bond, pi-anion, pi-pi stacked and pi-alkyl acting force interactions (Figure 7D). When docking with iNOS, 11-O-(4′-O-methylgalloyl)-bergenin (Fr11) was predicted to bind to a pocket surrounded by MET-368, ILE-195, GLY-365, GLN- 199, ASN-364, TRP-366, CYS-194, TYR-483, TRP-188, ALA-191, MET-349, PRO-192, and TYR-485 through van der waals, conventional hydrogen bond, pi-cation, amide-pi stacked and pi-alkyl acting force interactions (Figure 7E). As shown in Table 3, the respective binding energy values for interactions between 11-O-(4′-O-methylgalloyl)-bergenin (Fr11) and iNOS and NF-κB were −9.30 kcal/mol, −7.33 kcal/mol, respectively, with both being less than −7.0 kcal/mol, consistent with a strong binding interaction. Moreover, the binding energy values for interactions between 11-O-(4′-O-methylgalloyl)-bergenin and TNF, TNFR1A, and TLR were −5.62 kcal/mol, −5.87 kcal/mol, and −5.42 kcal/mol, respectively, with these values being below −5.0 kcal/mol, consistent with good bonding stability. Given these results and the associated signaling pathways, these data suggest that 11-O-(4′-O-methylgalloyl)-bergenin (Fr11) may regulate inflammatory responses primarily by reducing the production of pro-inflammatory cytokines and inhibiting the expression of key proteins that initiate inflammatory responses.

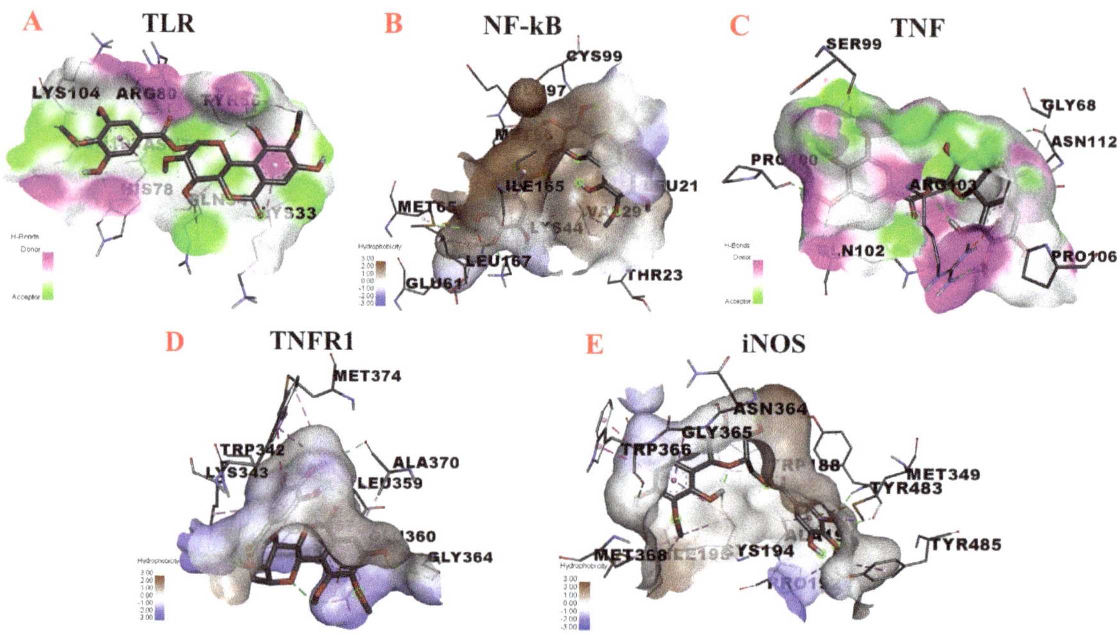

Figure 7. Molecular docking analyses of the putative binding between 11-O-(4′-O-methylgalloyl)-bergenin and TLR, NF-κB, TNF, TNFR1A, and iNOS. (**A**–**E**) correspond to the binding models for interactions between 11-O-(4′-O-methylgalloyl)-bergenin and TLR, NF-κB, TNF, TNFR1A, and iNOS, respectively. Hydrophobic or H-bond interactions are displayed as a colored surface.

Table 3. Intermolecular interactions between Fr11 and TLR, NF-κB, TNF, TNFR1A, and iNOS.

Proteins	Protein Function	Binding Energy (kcal/mol)	Binding Residues	Type
TLR	Inflammation, immunity, survival, proliferation	−5.42	ASN-103	Carbon Hydrogen Bond
			LYS-104	Carbon Hydrogen Bond
			HIS-78	Conventional Hydrogen Bond
			ASN-79	Amide-Pi Stacked
			ARG-80	Conventional Hydrogen Bond
			GLN-54	Conventional Hydrogen Bond
			TYR-56	Conventional Hydrogen Bond
			LYS-33	Pi-Alkyl

Table 3. *Cont.*

Proteins	Protein Function	Binding Energy (kcal/mol)	Binding Residues	Type
NF-κB	Inflammation, immunity	−7.33	LEU-21	Conventional Hydrogen Bond
			LEU-21	Pi-Sigma
			CYS-99	Conventional Hydrogen Bond
			VAL-29	Pi-Alkyl
			THR-23	Carbon Hydrogen Bond
			MET-65	Conventional Hydrogen Bond
			MET-65	Pi-Alkyl
			MET-65	Pi-Sulfur
			GLU-97	Conventional Hydrogen Bond
			ILE-165	Pi-Alkyl
			LYS-44	Conventional Hydrogen Bond
			LYS-44	Pi-Alkyl
			MET-96	Pi-Alkyl
			GLU-61	Carbon Hydrogen Bond
TNF	Fever, proliferation, differentiation, inflammation, cytotoxicity	−5.62	PRO-100	Conventional Hydrogen Bond
			GLN-102	Conventional Hydrogen Bond
			SER-99	Conventional Hydrogen Bond
			SER-99	Carbon Hydrogen Bond
			ARG-103	Conventional Hydrogen Bond
			ARG-103	Pi-Cation
			PRO-106	Carbon Hydrogen Bond
			ASN-112	Conventional Hydrogen Bond
			GLY-68	Carbon Hydrogen Bond
TNFR1A	Inflammation, apoptosis	−5.87	GLY-364	Conventional Hydrogen Bond
			ASP-357	Conventional Hydrogen Bond
			GLU-360	Conventional Hydrogen Bond
			GLU-360	Pi-Anion
			LYS-343	Conventional Hydrogen Bond
			LYS-343	Pi-Alkyl
			LEU-359	Pi-Alkyl
			MET-374	Pi-Alkyl
			TRP-342	Pi-Alkyl
			TRP-342	Pi-Pi Stacked
			ALA-370	Carbon Hydrogen Bond
iNOS	Inflammation, apoptosis	−9.30	MET-368	Pi-Alkyl
			ILE-195	Pi-Alkyl
			GLY-365	Pi-Sigma
			GLN-199	Conventional Hydrogen Bond
			ASN-364	Conventional Hydrogen Bond
			TRP-366	Conventional Hydrogen Bond
			TRP-366	Pi-Sulfur
			CYS-194	Conventional Hydrogen Bond
			CYS-194	Pi-Alkyl
			TYR-483	Conventional Hydrogen Bond
			TRP-188	Carbon Hydrogen Bond
			ALA-191	Pi-Sigma
			MET-349	Conventional Hydrogen Bond
			PRO-192	Pi-Alkyl
			TYR-485	Pi-Alkyl

3.4.4. Inflammation-Associated Factors and Adhesion Molecule

Interleukins (ILs) are an important class of inflammatory factors, with at least 38 ILs having been identified that play important roles in the regulation of the maturation, activation, and proliferation of immune cells. The regulation of the levels of specific IL family proteins is a key indicator used to evaluate the anti-inflammatory efficacy of many drugs. Here, we selected four representative inflammatory factors for molecular docking analyses exploring the mechanistic basis for potential anti-inflammatory activity [48]. IL-6 belongs to a family of glycoproteins that vary in molecular weight size from 26-30 kDa due to cell-specific post-translational modifications. IL-6 participates in immune regulation, hematopoietic, and inflammatory processes. Macrophages can secrete IL-6 in response to pathogen-associated molecular patterns (PAMPs), which induce an intracellular signaling cascade response, which leads to the production of inflammatory cytokines. IL-23R is a receptor subunit specific for IL-23, which acts as a pro-inflammatory factor that induces macrophages to produce LTs and TNF-α influencing tissue-specific autoimmune inflammation. IL-1β is a cytokine that plays an important role in regulating, inflammatory responses, and mediating the activation, proliferation, and differentiation of T cells and B cells. IL-1β is thought to be associated with macrophage foam cell formation. IL-10 can be produced and secreted by a variety of immune cells such as macrophages and B cells whereupon it functions as an important anti-inflammatory factor that can negatively regulate immune negative responses and maintain inflammatory homeostasis. SELE is a cell surface glycoprotein that mediates cell—cell and cell—extracellular matrix adhesion. SELE upregulation is a key component of the initiation of the inflammatory response. During inflammatory responses, leukocytes can extravasate from the plasma through adhesion to vascular endothelial cells, in turn contributing to tissue edema and further inflammation [49].

Given the above, inflammatory factors (IL-6, IL-23R, IL-1β and IL-10) and an adhesion molecule (SELE) were selected in this study to further explore the potential anti-inflammatory targets of 11-O-(4′-O-methylgalloyl)-bergenin (Fr11), as shown in Figure 8. When docking with IL-6, 11-O-(4′-O-methylgalloyl)-bergenin (Fr11) was predicted to bind to PHE-136, ARG-154, ASN-135, PRO-157, GLU-133, THR-130, GLY-127, and GLU-129 in 1P9M via conventional hydrogen bond, carbon hydrogen bond, pi-anion, pi-alkyl and pi-sigma acting force (Figure 8A). When docking with IL-23R, 11-O-(4′-O-methylgalloyl)-bergenin (Fr11) was predicted to bind to LYS-104, SER-203, PHE-106, LEU-107, TYR-201, ASN-200, THR-202, TRP-90, and MET-189 in 3DUH via conventional hydrogen bond, carbon hydrogen bond, pi-donor hydrogen bond, pi-pi t-shaped and pi-alkyl acting force (Figure 8B). When docking with IL-1β, 11-O-(4′-O-methylgalloyl)-bergenin (Fr11) was predicted to bind to GLU-64, LYS-63, VAL-40, MET-20, GLN-38, VAL-41, LYS-65, GLU-37, GLN-39, and MET-36 in 5I1B via conventional hydrogen bond, carbon hydrogen bond and pi-alkyl acting force (Figure 8C). When interacting with IL-10, 11-O-(4′-O-methylgalloyl)-bergenin (Fr11) was predicted to bind to PHE-143, VAL-124, GLU-142, ALA-139, and LYS-138 sites via conventional hydrogen bond, carbon hydrogen bond, pi-alkyl, pi-sigma and amide-pi stacked acting force (Figure 8D). When docking with SELE, 11-O-(4′-O-methylgalloyl)-bergenin (Fr11) was predicted to bind to CYS-325, THR-327, ASN-269, ALA-326, ASP-272, LEU-273, LYS-270, GLY-45, ARG-178, GLU-43, THR-48, VAL-179, SER-47, THR-177, and LYS-51, interacting through conventional hydrogen bond, pi-sigma, pi-lone pair, sulfur-x and pi-alkyl acting force interactions (Figure 8E). Binding between this compound and IL-6, IL-23R, and SELE was dominated by hydrogen bonding (Figure 8A,B,E), whereas its interactions with IL-1β and were dominated by hydrophobic interactions (Figure 8C,D). As shown in Table 4, the binding energy values for interactions between 11-O-(4′-O-methylgalloyl)-bergenin (Fr11) and IL-6, IL-23R, IL-1β, and IL-10 were −5.62 kcal/mol, −6.20 kcal/mol, −5.52 kcal/mol and −4.02 kcal/mol, respectively, whereas the binding energy for SELE was −7.93 kcal/mol. Lower binding energy values generally correspond to more stable interactions, with values below −5.0 kcal/mol and −7.0 kcal/mol, corresponding to good and strong binding activity, respectively. As such, we speculate that 11-O-(4′-O-methylgalloyl)-bergenin (Fr11) may exert its anti-inflammatory activity primarily via inhibiting the production of pro-inflammatory cytokines and the expression of the adhesion molecule, rather than by promoting the release of anti-inflammatory cytokines.

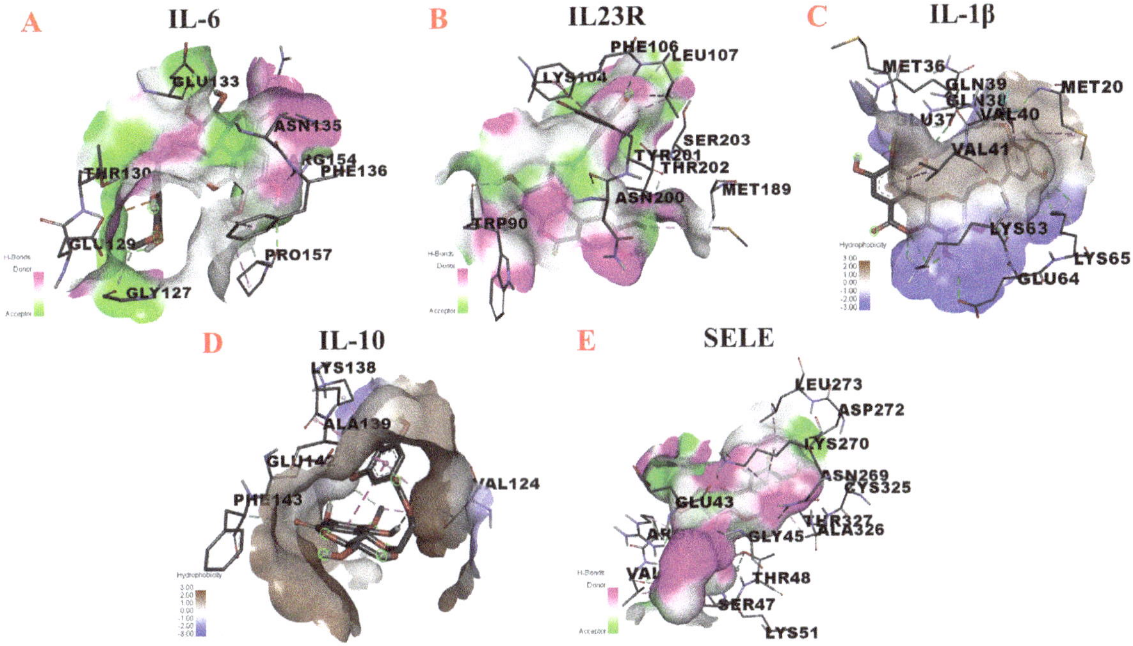

Figure 8. Molecular docking analysis of the putative binding between 11-O-(4′-O-methylgalloyl)-bergenin and IL-6, IL-23R, IL-1β, IL-10, and SELE. (**A–E**) correspond to the binding models for interactions between 11-O-(4′-O-methylgalloyl)-bergenin and IL-6, IL-23R, IL-1β, IL-10, and SELE, respectively. Hydrophobic or H-bond interactions are displayed as a colored surface.

Table 4. Intermolecular interactions between Fr11 and IL-6, IL-23R, IL-1β, and IL-10.

Proteins	Protein Function	Binding Energy (kcal/mol)	Binding Residues	Type
IL-6	Inflammation, immunity, hematopoiesis	−5.62	PHE-136 PHE-136 ARG-154 ASN-135 PRO-157 PRO-157 GLU-133 THR-130 GLY-127 GLU-129	Conventional Hydrogen Bond Pi-Alkyl Pi-Sigma Conventional Hydrogen Bond Conventional Hydrogen Bond Pi-Alkyl Conventional Hydrogen Bond Carbon Hydrogen Bond Carbon Hydrogen Bond Conventional Hydrogen Bond Pi-Anion
IL-23R	Inflammation, immunity	−6.20	LYS-104 SER-203 PHE-106 LEU-107 LEU-107 LEU-107 TYR-201 ASN-200 THR-202 TRP-90 MET-189	Conventional Hydrogen Bond Carbon Hydrogen Bond Pi-Pi T-Shaped Conventional Hydrogen Bond Pi-Alkyl Pi-Lone Pair Carbon Hydrogen Bond Conventional Hydrogen Bond Carbon Hydrogen Bond Conventional Hydrogen Bond Pi-Alkyl

Table 4. *Cont.*

Proteins	Protein Function	Binding Energy (kcal/mol)	Binding Residues	Type
IL-1β	Inflammation, immunity, proliferation, differentiation	−5.52	GLU-64	Conventional Hydrogen Bond
			LYS-63	Conventional Hydrogen Bond
			VAL-40	Pi-Alkyl
			MET-20	Conventional Hydrogen Bond
			MET-20	Pi-Alkyl
			GLN-38	Conventional Hydrogen Bond
			VAL-41	Pi-Alkyl
			LYS-65	Conventional Hydrogen Bond
			GLU-37	Conventional Hydrogen Bond
			GLN-39	Conventional Hydrogen Bond
			MET-36	Carbon Hydrogen Bond
IL-10	Inflammation, immunity, apoptosis	−4.02	PHE-143	Carbon Hydrogen Bond
			VAL-124	Pi-Alkyl
			GLU-142	Conventional Hydrogen Bond
			ALA-139	Pi-Sigma
			LYS-138	Amide-Pi Stacked
			LYS-138	Pi-Alkyl
SELE	Inflammation	−7.93	CYS-325	Sulfur-X
			THR-327	Pi-Sigma
			ASN-269	Conventional Hydrogen Bond
			ALA-326	Pi-Alkyl
			ASP-272	Conventional Hydrogen Bond
			ASP-272	Pi-Alkyl
			LEU-273	Pi-Alkyl
			LYS-270	Conventional Hydrogen Bond
			LYS-270	Pi-Sigma
			GLY-45	Conventional Hydrogen Bond
			ARG-178	Conventional Hydrogen Bond
			ARG-178	Pi-Lone Pair
			GLU-43	Conventional Hydrogen Bond
			THR-48	Conventional Hydrogen Bond
			VAL-179	Conventional Hydrogen Bond
			SER-47	Conventional Hydrogen Bond
			THR-177	Pi-Lone Pair
			LYS-51	Conventional Hydrogen Bond

The inflammatory response is a complex process that involves multiple genes and signaling pathways. By combining the above analysis results with target and pathway analyses, we can predict that 11-O-(4′-O-methylgalloyl)-bergenin (Fr11), a bioactive component of *S. atrata* isolated via affinity ultrafiltration-HPLC, may exert anti-inflammatory effects through four mechanisms: blocking the binding of pro-inflammatory factors to their cognate receptors, inhibiting the expression of key proteins that initiate the inflammatory response, reducing the production of pro-inflammatory cytokines, and regulating cell proliferation so as to indirectly regulate the inflammatory response. As such, this study provides a convenient means of exploring the mechanisms of interaction between inflammation-related targets and *S. atrata*-derived ligands, guiding the future development of anti-inflammatory active components from *S. atrata*.

4. Discussion

Uncertainty regarding the main components of NPs is a problem that is commonly encountered when attempting to establish quality control standards for these products. As such, the ability to isolate high-purity standard compounds from complex NPs is critical to the development of the NPs industry as a whole. NPs are also important precursors for the design of novel anti-inflammatory drugs. Inflammation is a physiological response to adverse stimuli and associated damage, and can cause cellular degeneration, necrosis, and abnormal metabolic activity [50].

Nonsteroidal anti-inflammatory drugs (NSAIDs) are a broad class of anti-inflammatory agents. Since aspirin was first synthesized in 1898, more than 100 types of NSAIDs have been marketed under thousands of brand names, including aspirin, acetaminophen, indomethacin, ibuprofen, and rofecoxib [51]. They are widely used in clinical practice for the treatment of rheumatoid arthritis, fevers, and pain. However, prolonged NSAID use can result in the emergence of drug resistance and associated complications [52]. There is thus an urgent need to screen for novel anti-inflammatory drugs. Molecular docking approaches offer significant technical advantages as a means of evaluating medicinal compounds and their putative pharmacological targets, offering a means of more accurately screening and clarifying potential pharmacodynamic and pharmacological mechanisms of action for drugs of interest.

Accordingly, in this study, a one-step MCI GEL® CHP20P MPLC approach was successfully performed to pretreat *S. atrata* samples, after which a one-step affinity ultrafiltration-RPLC strategy was used to screen for potential COX-2 ligand in the target fraction Fr1. Subsequent preparation RPLC was then used to isolate 11-O-(4′-O-methylgalloyl)-bergenin (Fr11), a potent anti-inflammatory active ingredient that was >99% pure. These results thus demonstrate the feasibility of specifically isolating COX-2 ligands with potential anti-inflammatory activity from *S. atrata*. To explore the binding ability of 11-O-(4′-O-methylgalloyl)-bergenin (Fr11) when interacting with inflammation-related targets, molecular docking analyses were conducted that revealed it to bind to the following compounds, which are ranked in descending order based on predicted binding energy values: ALOX15, iNOS, ERBB2, SELE, NF-κB, JNK, EGFR, COX-2, ERK5, p38MAPK, IL-23R, TNFR1A, TNF, IL-6, IL-1β, TLR, IL-10, and 5-LOX.

This study is the first to our knowledge to use a molecular docking approach to report on the potential anti-inflammatory mechanism of action of the *S. atrata*-derived active ingredient 11-O-(4′-O-methylgalloyl)-bergenin (Fr11), revealing this compound exhibits a high degree of binding activity towards ALOX15, iNOS, ERBB2, SELE, and NF-κB. Combined target and pathway analyses thus suggested that 11-O-(4′-O-methylgalloyl)-bergenin (Fr11) may primarily regulate inflammatory responses through the targeting of the AA metabolism, MAPK, and NF-κB signaling pathways, exerting its anti-inflammatory activity through four primary mechanisms: blocking the binding between pro-inflammatory factors and their cognate receptors, inhibiting the expression of key proteins that initiate the inflammatory response, reducing the production of pro-inflammatory cytokines, and regulating cellular proliferation so as to indirectly regulate the inflammatory response. A key advantage of this study is that it provides a sound theoretical basis for further research and development focused on the anti-inflammatory effects of isocoumarins such as 11-O-(4′-O-methylgalloyl)-bergenin (Fr11), revealing novel directions for research exploring the multi-component and multi-target modes of action of traditional Chinese medicinal compounds.

It is important to note that molecular docking techniques are constantly being updated, that some data are uncertain, and that the predicted results need to be validated by future cellular, animal-based, and clinical trials. These results suggest that the prioritization of mechanistic studies exploring pathways associated with ALOX15, iNOS, ERBB2, and SELE may of particular relevance for efforts to clarify the molecular basis for the activity of 11-O-(4′-O-methylgalloyl)-bergenin. For example, flow cytometry can be utilized as a means of assessing reactive oxygen species production and mitochondrial transmembrane potential, whereas western blotting can be used to evaluate the expression of proteins of interest. PCR can be used to assess mitochondrial copy numbers, thus offering comprehensive insight regarding the mechanism of action for this *S. atrata*-derived compound as an anti-inflammatory mediator. This approach will provide a robust experimental foundation for the design of new anti-inflammatory drugs.

5. Conclusions

The medicinal efficacy of NPs is primarily attributable to the complex chemical characteristics of their secondary metabolites, which can synergistically interact with multiple drug targets or multiple biochemical pathways to exert their overall efficacy. In this study, an MCI GEL® CHP20P medium pressure column with a methanol−water mobile phase was used for the effective enrichment of the methanol extract of *S. atrata* and the target fraction (Fr1) was successfully prepared. By comparing the separation effects of Click XIon and ReproSil-Pur C18 AQ analytical columns for this target fraction (Fr1), a ReproSil-Pur C18 AQ preparative column was selected for subsequent separation and purification. COX-2 affinity ultrafiltration coupled with reversed-phase liquid chromatography was successfully used to screen for candidate COX-2 ligands in this target fraction (Fr1). The preparative isolation of this potential COX-2 ligand was then completed via isocratic elution in 10% ACN, and an HPLC purity analysis conducted using HILIC and RP-18 columns confirmed that the sample (74.1 mg) collected in 20 replicate cycles was 11-O-(4′-O-methylgalloyl)-bergenin and that it was >99% pure. To further validate the anti-inflammatory effect of the thus putative COX-2 ligand, molecular docking techniques were used to predict its anti-inflammatory targets. This comprehensive analytical approach revealed that can readily bind to ALOX15, iNOS, ERBB2, and SELE, providing a theoretical justification for future justification assays. Overall, these results and future validation efforts will provide a basis for the discovery of additional drugs derived from *S. atrata* and other medicinal plants that can be used to treat various inflammatory diseases based upon correlations between bioactive compounds and their putative biological targets.

Supplementary Materials: The following supporting information can be downloaded at: https://www.mdpi.com/article/10.3390/nu14122405/s1, Figure S1: Schematic of the study design; Figure S2: Actual micro gel chp20p medium pressure liquid chromatography system; Figure S3: Schematic diagram of the principle of affinity ultrafiltration-HPLC; Figure S4: ESI mass spectrum of 11-O-(4′-O-methylgalloyl)-bergenin; Figure S5: ^{1}H NMR spectrum (600 MHz) of 11-O-(4′-O-methylgalloyl)-bergenin (in MeOH-d_4); Figure S6: ^{13}C NMR spectrum (151 MHz) of 11-O-(4′-O-methylgalloyl)-bergenin (in MeOH-d_4).

Author Contributions: Conceptualization, methodology, J.D. and Y.F.; writing—original draft, Y.F. and G.L.; writing—review and editing, Y.F. and G.L.; formal analysis, Y.D.; investigation, Y.M.; data curation, Q.W.; visualization, J.G.; supervision, project administration, and funding acquisition, J.D. and G.L. All authors have read and agreed to the published version of the manuscript.

Funding: This research was funded by the Young Scholars in Western China, Chinese Academy of Sciences (2022), the Kunlun Talent and High-end Innovation and Entrepreneurship Talents (2021), Key Laboratory of Medicinal Animal and Plant Resources of Qinghai-Tibetan Plateau in Qinghai Province (2020-ZJ-Y04), the Joint Research Project of Three-River Headwaters National Park, Chinese Academy of Sciences and the People's Government of Qinghai Province (LHZX-2021-02), Yantai Science and Technology Innovation Development Project (2021XDHZ078) and the Shandong Provincial Science Foundation (ZR2020MH380).

Institutional Review Board Statement: Not applicable.

Informed Consent Statement: Not applicable.

Data Availability Statement: All data generated or analyzed during this study are included in this published article (and its Supplementary Materials file).

Conflicts of Interest: The authors have stated that they have no known conflicting financial interests or personal ties that may have influenced the work presented in this publication.

References

1. Kany, S.; Vollrath, J.T.; Relja, B. Cytokines in Inflammatory Disease. *Int. J. Mol. Sci.* **2019**, *20*, 6008. [CrossRef] [PubMed]
2. Borsini, A.; Zunszain, P.A.; Thuret, S.; Pariante, C.M. The role of inflammatory cytokines as key modulators of neurogenesis. *Trends Neurosci.* **2015**, *38*, 145–157. [CrossRef] [PubMed]
3. Kim, J.H.; Yi, Y.S.; Kim, M.Y.; Cho, J.Y. Role of ginsenosides, the main active components of Panax ginseng, in inflammatory responses and diseases. *J. Ginseng. Res.* **2017**, *41*, 435–443. [CrossRef] [PubMed]
4. Tahamtan, A.; Teymoori-Rad, M.; Nakstad, B.; Salimi, V. Anti-Inflammatory MicroRNAs and Their Potential for Inflammatory Diseases Treatment. *Front. Immunol.* **2018**, *9*, 1377. [CrossRef] [PubMed]
5. Shen, X.F.; Zeng, Y.; Li, J.C.; Tang, C.; Zhang, Y.; Meng, X.L. The anti-arthritic activity of total glycosides from Pterocephalus hookeri, a traditional Tibetan herbal medicine. *Pharm. Biol.* **2017**, *55*, 560–570. [CrossRef] [PubMed]
6. Li, Y.; Li, F.; Zheng, T.T.; Shi, L.; Zhang, Z.G.; Niu, T.M.; Wang, Q.Y.; Zhao, D.S.; Li, W.; Zhao, P. Lamiophlomis herba: A comprehensive overview of its chemical constituents, pharmacology, clinical applications, and quality control. *Biomed. Pharmacother.* **2021**, *144*, 112299. [CrossRef]
7. Zhang, M.R.; Jiang, K.; Yang, J.L.; Shi, Y.P. Flavonoids as key bioactive components of *Oxytropis falcata* bunge, a traditional anti-inflammatory and analgesic Tibetan medicine. *Nat. Prod. Res.* **2020**, *34*, 3335–3352. [CrossRef]
8. Gerschwitz-Eidt, M.A.; Kadereit, J.W. Species composition of Saxifraga sect. Saxifraga subsect. Arachnoideae (Saxifragaceae) based on DNA sequence evidence. *Willdenowia* **2020**, *50*, 225–233. [CrossRef]
9. Durovic, S.Z.; Kuzmanovic, N.; Tomovic, G.; Niketic, M. Nomenclatural notes and typification of the names related to *Silene saxifraga* group (Caryophyllaceae). *Phytotaxa* **2018**, *344*, 201–214. [CrossRef]
10. Yang, Y.; He, T.; Lu, S.; Huang, R.; Wang, Z. *Records of Tibetan Medicine*; Qinghai Publishing House: Qinghai, China, 1991; pp. 450–451.
11. Zhao, X.M.; Li, R.J.; Zhou, C.; Zhang, J.; He, C.H.; Zheng, Y.T.; Wu, W.D.; Zhang, H.B. Separation and purification of deoxynivalenol (DON) mycotoxin from wheat culture using a simple two-step silica gel column chromatography. *J. Integr. Agric.* **2016**, *15*, 694–701. [CrossRef]
12. Huang, M.X.; Pineau, A.; Bouaziz, O.; Vu, T.D. Recrystallization induced plasticity in austenite and ferrite. *Mater. Sci. Eng. A* **2012**, *541*, 196–198. [CrossRef]
13. Jandera, P.; Janas, P. Recent advances in stationary phases and understanding of retention in hydrophilic interaction chromatography. A review. *Anal. Chim. Acta* **2017**, *967*, 12–32. [CrossRef] [PubMed]
14. Yang, Y.G.; Ju, Z.C.; Yang, Y.B.; Zhang, Y.H.; Yang, L.; Wang, Z.T. Phytochemical analysis of Panax species: A review. *J. Ginseng. Res.* **2021**, *45*, 1–21. [CrossRef]
15. Gong, Y.; Huang, X.Y.; Pei, D.; Duan, W.D.; Zhang, X.; Sun, X.; Di, D.L. The applicability of high-speed counter current chromatography to the separation of natural antioxidants. *J. Chromatogr. A* **2020**, *1623*, 461150. [CrossRef] [PubMed]
16. Zhao, X.; Zhang, S.S.; Zhang, X.K.; He, F.; Duan, C.Q. An effective method for the semi-preparative isolation of high-purity anthocyanin monomers from grape pomace. *Food Chem.* **2020**, *310*, 125830. [CrossRef]
17. Sun, G.Y.; Abuduaini, M.; Adili, G.; Zhao, Y.X.; Aisa, H.A. Dual-tautomerism separation method based on asymmetric transformation: Gram-scale preparation of high-purity punicalagin from pomegranate peel wastes. *J. Chromatogr. A* **2021**, *1651*, 462281. [CrossRef]
18. Fan, C.; Li, N.; Cao, X.L.; Wen, L.J. Ionic liquid-modified countercurrent chromatographic isolation of high-purity delphinidin-3-rutinoside from eggplant peel. *J. Food Sci.* **2020**, *85*, 1132–1139. [CrossRef]
19. Wang, Z.; Xie, T.T.; Yan, X.; Xue, S.; Chen, J.W.; Wu, Z.; Qiu, Y.K. Gradual gradient two-dimensional preparative liquid chromatography system for preparative separation of complex natural products. *Chromatographia* **2019**, *82*, 543–552. [CrossRef]
20. Dawa, Y.Z.; Du, Y.R.; Wang, Q.; Chen, C.B.; Zou, D.L.; Qi, D.S.; Ma, J.B.; Dang, J. Targeted isolation of 1,1-diphenyl-2-picrylhydrazyl inhibitors from *Saxifraga atrata* using medium- and high- pressure liquid chromatography combined with online high performance liquid chromatography-1,1-diphenyl-2-picrylhydrazyl detection. *J. Chromatogr. A* **2021**, *1635*, 461690. [CrossRef]
21. Ma, Z.H.; Wei, X.Y.; Zhou, M.Y.; Liu, G.H.; Liu, F.J.; Liu, Z.Q.; Yu, X.Y.; Zong, Z.M. Isolation and purification of carbazole contained in anthracene slag by extraction combined with medium pressure liquid chromatography. *Chin. J. Chem. Eng.* **2019**, *27*, 2925–2929. [CrossRef]
22. Dang, J.; Wang, Q.; Wang, Q.L.; Yuan, C.; Li, G.; Ji, T.F. Preparative isolation of antioxidative gallic acid derivatives from *Saxifraga tangutica* using a class separation method based on medium-pressure liquid chromatography and reversed-phase liquid chromatography. *J. Sep. Sci.* **2021**, *44*, 3734–3746. [CrossRef] [PubMed]
23. Lee, D.Y.; Choi, S.I.; Han, S.H.; Lee, Y.J.; Choi, J.G.; Lee, Y.S.; Choi, J.H.; Lee, S.E.; Kim, G.S. Potential of Pseudoshikonin I Isolated from Lithospermi Radix as Inhibitors of MMPs in IL-1 beta-Induced SW1353 Cells. *Int. J. Mol. Sci.* **2016**, *17*, 1350. [CrossRef] [PubMed]
24. Dang, J.; Ma, J.B.; Dawa, Y.Z.; Liu, C.; Ji, T.F.; Wang, Q.L. Preparative separation of 1,1-diphenyl-2-picrylhydrazyl inhibitors originating from *Saxifraga sinomontana* employing medium-pressure liquid chromatography in combination with reversed-phase liquid chromatography. *Rsc. Adv.* **2021**, *11*, 38739–38749. [CrossRef] [PubMed]

25. Pan, G.Q.; Shen, J.W.; Ma, Y.H.; He, Y.F.; Bao, Y.; Li, R.R.; Wang, S.S.; Wang, Q.; Lin, P.C.; Dang, J. Preparative separation of isoquinoline alkaloids from *Corydalis impatiens* using a middle-pressure chromatogram isolated gel column coupled with two-dimensional liquid chromatography. *J. Sep. Sci.* **2019**, *42*, 3182–3190. [CrossRef] [PubMed]
26. Fan, M.X.; Chen, G.L.; Sun, B.Q.; Wu, J.L.; Li, N.; Sarker, S.D.; Nahar, L.; Guo, M.Q. Screening for natural inhibitors of human topoisomerases from medicinal plants with bio-affinity ultrafiltration and LC-MS. *Phytochem. Rev.* **2020**, *19*, 1231–1261. [CrossRef]
27. Zhang, H.; Chen, G.L.; Yang, J.P.; Yang, C.L.; Guo, M.Q. Screening and characterisation of potential antioxidant, hypoglycemic and hypolipidemic components revealed in *Portulaca oleracea* via multi-target affinity ultrafiltration LC-MS and molecular docking. *Phytochem. Anal.* **2022**, *33*, 272–285. [CrossRef]
28. Pinzi, L.; Rastelli, G. Molecular Docking: Shifting Paradigms in Drug Discovery. *Int. J. Mol. Sci.* **2019**, *20*, 4331. [CrossRef]
29. Salman, M.M.; Al-Obaidi, Z.; Kitchen, P.; Loreto, A.; Bill, R.M.; Wade-Martins, R. Advances in Applying Computer-Aided Drug Design for Neurodegenerative Diseases. *Int. J. Mol. Sci.* **2021**, *22*, 4688. [CrossRef]
30. Cheng, F.J.; Huynh, T.K.; Yang, C.S.; Hu, D.W.; Shen, Y.C.; Tu, C.Y.; Wu, Y.C.; Tang, C.H.; Huang, W.C.; Chen, Y.; et al. Hesperidin Is a Potential Inhibitor against SARS-CoV-2 Infection. *Nutrients* **2021**, *13*, 2800. [CrossRef]
31. Cinelli, M.A.; Do, H.T.; Miley, G.P.; Silverman, R.B. Inducible nitric oxide synthase: Regulation, structure, and inhibition. *Med. Res. Rev.* **2020**, *40*, 158–189. [CrossRef]
32. Dang, J.; Ma, J.B.; Du, Y.R.; Dawa, Y.; Wang, Q.; Chen, C.B.; Wang, Q.L.; Tao, Y.D.; Ji, T.F. Large-scale preparative isolation of bergenin standard substance from *Saxifraga atrata* using polyamide coupled with MCI GEL®CHP20P as stationary phases in medium pressure chromatography. *J. Chromatogr. B* **2021**, *1170*, 122617. [CrossRef] [PubMed]
33. Garcia-Galan, C.; Barbosa, O.; Hernandez, K.; dos Santos, J.C.S.; Rodrigues, R.C.; Fernandez-Lafuente, R. Evaluation of Styrene-Divinylbenzene Beads as a Support to Immobilize Lipases. *Molecules* **2014**, *19*, 7629–7645. [CrossRef] [PubMed]
34. Yang, X.; Lv, Y.P. Purification, characterization, and DNA damage protection of active components from tartary buckwheat (Fagopyrum tataricum) hull. *Food Sci. Biotechnol.* **2015**, *24*, 1959–1966. [CrossRef]
35. Zielinska, D.; Zielinski, H.; Laparra-Llopis, J.M.; Szawara-Nowak, D.; Honke, J.; Gimenez-Bastida, J.A. Caffeic Acid Modulates Processes Associated with Intestinal Inflammation. *Nutrients* **2021**, *13*, 554. [CrossRef]
36. Lee, Y.Y.; Jang, D.S.; Jin, J.L.; Yun-Choi, H.S. Anti-platelet aggregating and anti-oxidative activities of 11-O-(4′-O-methylgalloyl)-bergenin, a new compound isolated from Crassula cv. 'Himaturi'. *Planta Med.* **2005**, *71*, 776–777. [CrossRef]
37. Zarrin, A.A.; Bao, K.; Lupardus, P.; Vucic, D. Kinase inhibition in autoimmunity and inflammation. *Nat. Rev. Drug Discov.* **2021**, *20*, 39–63. [CrossRef]
38. Bai, J.R.; Zhang, Y.S.; Tang, C.; Hou, Y.; Ai, X.P.; Chen, X.R.; Zhang, Y.; Wang, X.B.; Meng, X.L. Gallic acid: Pharmacological activities and molecular mechanisms involved in inflammation-related diseases. *Biomed. Pharmacother.* **2021**, *133*, 110985. [CrossRef]
39. Chen, J.; Ye, C.; Wan, C.; Li, G.; Peng, L.C.; Peng, Y.Y.; Fang, R.D. The Roles of c-Jun N-Terminal Kinase (JNK) in Infectious Diseases. *Int. J. Mol. Sci.* **2021**, *22*, 9640. [CrossRef]
40. Lu, N.; Malemud, C.J. Extracellular Signal-Regulated Kinase: A Regulator of Cell Growth, Inflammation, Chondrocyte and Bone Cell Receptor-Mediated Gene Expression. *Int. J. Mol. Sci.* **2019**, *20*, 3792. [CrossRef]
41. Maisel, S.A.; Broka, D.; Atwell, B.; Bunch, T.; Kupp, R.; Singh, S.K.; Mehta, S.; Schroeder, J. Stapled EGFR peptide reduces inflammatory breast cancer and inhibits additional HER-driven models of cancer. *J. Transl. Med.* **2019**, *17*, 201. [CrossRef]
42. Akdis, M.; Burgler, S.; Crameri, R.; Eiwegger, T.; Fujita, H.; Gomez, E.; Klunker, S.; Meyer, N.; O'Mahony, L.; Palomares, O.; et al. Interleukins, from 1 to 37, and interferon-gamma: Receptors, functions, and roles in diseases. *J. Allergy Clin. Immunol.* **2011**, *127*, 701–721. [CrossRef] [PubMed]
43. Grancieri, M.; Martino, H.S.D.; de Mejia, E.G. Protein Digests and Pure Peptides from Chia Seed Prevented Adipogenesis and Inflammation by Inhibiting PPAR gamma and NF-kappa B Pathways in 3T3L-1 Adipocytes. *Nutrients* **2021**, *13*, 176. [CrossRef] [PubMed]
44. Stachon, T.; Latta, L.; Kolev, K.; Seitz, B.; Langenbucher, A.; Szentmáry, N. Increased NF-κB and iNOS Expression in Keratoconus Keratocytes—Hints for an Inflammatory Component? *Klin. Monbl. Augenheilkd.* **2021**, *238*, 1010–1017. [CrossRef] [PubMed]
45. Lohberger, B.; Bernhart, E.; Stuendl, N.; Glaenzer, D.; Leithner, A.; Rinner, B.; Bauer, R.; Kretschmer, N. Periplocin mediates TRAIL-induced apoptosis and cell cycle arrest in human myxofibrosarcoma cells via the ERK/p38/JNK pathway. *Phytomedicine* **2020**, *76*, 153262. [CrossRef] [PubMed]
46. Liu, A.R.; Ramakrishnan, P. Regulation of Nuclear Factor-kappaB Function by O-GlcNAcylation in Inflammation and Cancer. *Front. Cell. Dev. Biol.* **2021**, *9*, 751761. [CrossRef] [PubMed]
47. Wang, B.; Wu, L.J.; Chen, J.; Dong, L.L.; Chen, C.; Wen, Z.; Hu, J.; Fleming, I.; Wang, D.W. Metabolism pathways of arachidonic acids: Mechanisms and potential therapeutic targets. *Signal Transduct. Target. Ther.* **2021**, *26*, 94. [CrossRef]
48. Du, B.; Lin, C.Y.; Bian, Z.X.; Xu, B.J. An insight into anti-inflammatory effects of fungal beta-glucans. *Trends Food Sci. Technol.* **2015**, *41*, 49–59. [CrossRef]
49. Huang, X.F.; Cheng, W.B.; Jiang, Y.; Liu, Q.; Liu, X.H.; Xu, W.F.; Huang, H.T. A network pharmacology-based strategy for predicting anti-inflammatory targets of ephedra in treating asthma. *Int. Immunopharmacol.* **2020**, *83*, 106423. [CrossRef]
50. Li, Y.; Huang, H.; Liu, B.; Zhang, Y.; Pan, X.; Yu, X.Y.; Shen, Z.; Song, Y.H. Inflammasomes as therapeutic targets in human diseases. *Signal Transduct. Target. Ther.* **2021**, *6*, 247. [CrossRef]

51. Doña, I.; Pérez-Sánchez, N.; Eguiluz-Gracia, I.; Muñoz-Cano, R.; Bartra, J.; Torres, M.J.; Cornejo-García, J.A. Progress in understanding hypersensitivity reactions to nonsteroidal anti-inflammatory drugs. *Allergy* **2020**, *75*, 561–575. [CrossRef]
52. Rastogi, A.; Tiwari, M.K.; Ghangrekar, M.M. A review on environmental occurrence, toxicity and microbial degradation of Non-Steroidal Anti-Inflammatory Drugs (NSAIDs). *J. Environ. Manag.* **2021**, *300*, 113694. [CrossRef] [PubMed]

nutrients

MDPI

Article

Ginsenoside Rh4 Suppresses Metastasis of Gastric Cancer via SIX1-Dependent TGF-β/Smad2/3 Signaling Pathway

Hongbo Jiang [1,2], Pei Ma [1,2], Zhiguang Duan [1,2], Yannan Liu [1,2], Shihong Shen [1,2], Yu Mi [1,2,*] and Daidi Fan [1,2,*]

1 Shaanxi Key Laboratory of Degradable Biomedical Materials, Shaanxi R&D Center of Biomaterials, Fermentation Engineering, School of Chemical Engineering, Northwest University, Taibai North Road 229, Xi'an 710069, China; jianghongbo@stumail.nwu.edu.cn (H.J.); mapei@nwu.edu.cn (P.M.); duanzhiguang@nwu.edu.cn (Z.D.); liuyannan@nwu.edu.cn (Y.L.); shenshihong@nwu.edu.cn (S.S.)

2 Biotech and Biomed Research Institute, Northwest University, Taibai North Road 229, Xi'an 710069, China

* Correspondence: mi_yu@nwu.edu.cn (Y.M.); fandaidi@nwu.edu.cn (D.F.); Tel.: +86-29-88305118 (Y.M. & D.F.)

Abstract: Gastric cancer (GC) is the leading causes of cancer-related death worldwide. Surgery remains the cornerstone of gastric cancer treatment, and new strategies with adjuvant chemotherapy are currently gaining more and more acceptance. Ginsenoside Rh4 has excellent antitumor activity. Conversely, the mechanisms involved in treatment of GC are not completely understood. In this study, we certified that Rh4 showed strong anti-GC efficiency in vitro and in vivo. MTT and colony formation assays were performed to exhibit that Rh4 significantly inhibited cellular proliferation and colony formation. Results from the wound healing assay, transwell assays, and Western blotting indicated that Rh4 restrained GC cell migration and invasion by reversing epithelial–mesenchymal transition (EMT). Further validation by proteomic screening, co-treatment with disitertide, and SIX1 signal silencing revealed that SIX1, a target of Rh4, induced EMT by activating the TGF-β/Smad2/3 signaling pathway. In summary, our discoveries demonstrated the essential basis of the anti-GC metastatic effects of Rh4 via suppressing the SIX1–TGF-β/Smad2/3 signaling axis, which delivers a new idea for the clinical treatment of GC.

Keywords: ginsenoside Rh4; SIX1; TGF-β/Smad2/3 pathway; gastric cancer

Citation: Jiang, H.; Ma, P.; Duan, Z.; Liu, Y.; Shen, S.; Mi, Y.; Fan, D. Ginsenoside Rh4 Suppresses Metastasis of Gastric Cancer via SIX1-Dependent TGF-β/Smad2/3 Signaling Pathway. *Nutrients* **2022**, *14*, 1564. https://doi.org/10.3390/nu14081564

Academic Editor: Anna Tresserra-Rimbau

Received: 13 March 2022
Accepted: 5 April 2022
Published: 9 April 2022

1. Introduction

Gastric cancer (GC) remains a non-negligible cancer worldwide, ranking fifth for incidence and fourth for mortality globally. As reported by *CA: A Cancer Journal for Clinicians*, there were over one million new cases in 2020 with an estimated 769,000 deaths caused by GC [1]. Despite the great advancement in the treatment and diagnosis of GC in recently years, the morbidity and mortality of GC have remained uncontrolled. Metastasis and resistance to chemotherapeutic agents are major problems in GC treatment [2]. Therefore, it is necessary to deeply understand the mechanisms of GC metastasis and identify therapeutic agents to achieve precise treatment of patients and prolong their survival time.

Epithelial–mesenchymal transition (EMT) plays a key role in cellular processes associated with cancer metastasis and progression, in which tumor cells lose epithelial features, accompanied by reduced cell junctions and acquisition of invasive and metastatic capabilities, as well as stem-cell phenotypes [3,4]. GC genesis, progression, and metastasis are closely related to EMT. The levels of transforming growth factor-β1 (TGF-β1), Snail1 (Snail, a key regulator in EMT), and Vimentin (an important component protein of cytoskeleton) are all upregulated in patients with dysplasia or early GC, while the level of E-cadherin is decreased [5]. When aberrant EMT is activated, cadherin switches from E-cadherin to N-cadherin, which increases cell motility and invasiveness in GC progression [6]. Zhang et al. demonstrated that CCR7 upregulates Snail expression to induce EMT, leading to GC progression, migration, and invasion [7]. Accumulating evidence has confirmed TGF-β1 to be a key growth factor which induces EMT of various tumors [8–10]. It has been reported that

TGF-β1-induced EMT is dependent on the downstream effector, Smad. TGF-β receptor II (TGF-βRII) phosphorylates TGF-β receptor I (TGF-βRI) in response to TGF-β1, which in turn activates Smads. These activated receptor-regulated Smads (R-Smads) form complexes that then act as transcriptional regulators ectopically into the nucleus to regulate the expression of EMT-related proteins such as epithelial cadherin (E-cadherin) and Snail [11,12]. Furthermore, a study by Irina et al. revealed that the TGF-β1-induced cell invasion and EMT phenotype in lung cancer cells are Smad2/3-dependent [13].

Sineoculis homeobox homolog 1 (SIX1) is a member of the *SIX* gene superfamily located on human chromosome 14 [14], and it is accepted as a transcription factor involved in the regulation of embryonic development [15]. Among the recently studied molecules reportedly linked to EMT is SIX1 [16]. As recorded in multiple studies, SIX1 can be highly expressed in different tumors, including liver cancer, colorectal cancer, gastric cancer, and breast cancer, thus affecting their clinical prognosis [14,17] and signal transduction [18], as well as promoting tumor progression. Furthermore, SIX1-induced upregulation of TGF-β1 is critical for propagation of TGF-β/Smad2/3 signaling, induction of EMT persistence, and metastasis [19,20]. Farabaugh et al. showed that SIX1 induces EMT by activating the TGF-β signaling pathway and interacting with its pro-metastatic function, thereby causing cancerous epithelial cells to become cancerous mesenchymal cells [20,21]. Although some studies in other cancers indicated that SIX1 regulates EMT processes through the TGF-β/Smad2/3 signaling pathway in papillary thyroid cancer [22], there is no direct evidence that SIX1 induces EMT via the TGF-β/Smad2/3 signaling pathway in GC. Thus, it is necessary to investigate the effects of SIX1 and TGF-β/Smad2/3 on EMT in GC.

Ginseng is a medicinal plant widely used in traditional and modern medicine, and its active ingredients are its secondary metabolites, ginsenosides, which have excellent pharmacological effects in clinical applications, such as for the treatment of cardiovascular diseases [23], as well as antitumor [24,25], antidiabetic [26], anti-inflammatory [27–29], anti-obesity [30,31], and neuroprotective effects [32], and immunity enhancement [33]. Ginsenoside Rh4 is known to be a tetracyclic triterpenoid saponin composed of a glucosin and a triterpene sapogenin [34]. Earlier studies showed that ginsenosides affect EMT through inhibiting TGF-β1 expression [35], activating PI3K/Akt [36], and downregulating NF-κB [37]. However, there are no reports about the anti-metastasis effect of Rh4 on GC, and its mechanism remains unknown. In our study, the results showed that Rh4 could target SIX1 to inhibit the TGF-β/Smad2/3 signaling pathway and effectively suppressed EMT metastasis in GC cells according to the wound healing assay, transwell assay, proteome sequencing, and Western blot.

2. Materials and Methods

2.1. Materials and Chemicals

Ginsenoside Rh4 (Supplementary Figure S1A, purity $\geq$ 99%) was purchased from Puruifa Technology Development Co., Ltd. (Chengdu, China). Methylthiazoletetrazolium (MTT) and dimethyl sulfoxide (DMSO) were obtained from Aladdin Biotechnology (Shanghai, China). Fetal bovine serum (FBS), streptomycin, and penicillin were purchased from Gibco (Grand Island, NY, USA). Dulbecco's modified Eagle medium (DMEM) was purchased from HyClone (Logan, UT, USA). The 24-well Transwell chambers were obtained from Corning (New York, NY, USA). Lipofectamine 2000 and TRIzol Reagent were supplied by Invitrogen (Carlsbad, CA, USA). SYBR Green Master Mixture was obtained from Takara (Otsu, Japan). Primary antibodies against E-cadherin, Vimentin, N-cadherin, Snail1, SIX1, and TGF-β were purchased from Proteintech Group Inc. (Chicago, IL, USA). The primary antibodies against Smad3, P-Smad3, and β-actin were supplied by Abcam (Cambridge UK). Goat anti-mouse IgG and goat anti-rabbit IgG were obtained from Abbkine (Wuhan, China). Oxaliplatin was obtained from Qilu Pharmaceutical Co., Ltd. (Shandong, China).

2.2. Cell Culture

Human gastric carcinoma cell lines (HGC-27, BGC-823) were supplied by the American Type Culture Collection (ATCC, VA, USA). Cells were cultured in DMEM with 10% FBS and 1% penicillin/streptomycin and maintained in an atmosphere of 5% CO_2 at 37 °C.

2.3. Cell Proliferation Assay

HGC-27 and BGC-823 (1×10^5 cells/well) were seeded in 96-well plates. After adherence, different concentrations of Rh4 (0, 20, 40, 60, 80, 100, and 120 μM) were injected into each well for 24 h. Then, 50 μL of MTT (5 mg/mL) and 150 μL of DMSO were added into the plates sequentially. Absorbance was measured at 490 nm using a microplate reader (Power Wave XS2, Bio-Tek Instruments Inc., Winooski, VT, USA).

2.4. Colony Formation Assay

HGC-27 and BGC-823 (500 cells/well) were seeded evenly in six-well plates and incubated overnight. Then, cells were cocultured with Rh4 (0, 40, 80, and 120 μM) for 48 h. Finally, cells were fixed, stained, and counted when visible colonies formed. The independent experiment was repeated three times.

2.5. AO/EB Staining

HGC-27 and BGC-823 (2×10^5 cells/well) were cultured evenly in six-well plates and incubated overnight. After adherence, cells treated with Rh4 (0, 40, 80, and 120 μM) for 24 h. Then, each well was added with an acridine orange/ethidium bromide (AO/EB) mixture for 30 to 45 min, before photographing the results. Three separate experiments were conducted.

2.6. Wound Healing Assay

GC cell lines (HGC-27, BGC-823) clones were grown to confluency. A linear wound was made by scraping a non-opening Pasteur pipette across the confluent cell layer. Cells were washed twice to remove detached cells and debris, treated with serum-free DMEM medium containing different concentrations of Rh4 (0, 20 μM, and 40 μM). Then, the sizes of the wounds at the same location were observed and measured at the indicated times (0 h, 12 h, 24 h) by ImageJ software. The independent experiment was repeated three times.

2.7. Transwell Assay

HGC-27 and BGC-823 ($5 \times 10^3/2 \times 10^4$ cells per well) were seeded without or with Matrigel coating (1 mg/mL, BD Matrigel ™) in the top chamber of the Transwell. Then, 600 μL of DMEM with 10% FBS was added to the Transwell lower chamber. The migrates/invasive cells were fixed, stained, and counted by an inverted microscope (200× magnification), 24 h later. At least three independent experiments were conducted [38,39].

2.8. Western Blot

The total protein of HGC-27 and BGC-823 was collected and extracted using lysis buffer. Cell extracts were measured by the BCA kit (Beyotime, Shanghai, China), and equal quantities of proteins (15 μg) were separated by 10% SDS-PAGE. The isolated proteins were transferred onto PVDF membranes by Trans-Blot Turbo (BioRad, Hercules, CA, USA). PVDF membranes were sequentially incubated with different primary antibodies, eluted with buffer, incubated with HRP-conjugated secondary antibodies, eluted with buffer, and imaged by a Gel Image system (Tanon5200, Shanghai, China). Data were expressed as the mean ± SD of at least three independent experiments.

2.9. Quantitative Real-Time Polymerase Chain Reaction (qRT-PCR)

Total RNA was derived from HGC-27 and BGC-823, followed by reverse transcription to obtain cDNA. The relative mRNA levels were calculated using the $2^{-\Delta\Delta CT}$ method. The expression of β-actin was used as an internal control. β-actin: 5′–CTGTCCCTGTATGCCTCT-

-3′ (forward), 5′–ATGTCACGCACGATTTCC–3′ (reverse); E-cadherin: 5′–GGAACTATGAA AAGTGGGCTTG–3′ (forward), 5′–GGCATCAGGCTCCACAGT–3′ (reverse); N-cadherin: 5′–GGTGGAGGAGAAGAAGACCAG–3′ (forward), 5′–GGCATCAGGCTCCACAGT–3′ (reverse); Vimentin: 5′–GACGCCATCAACACCGAGTT–3′ (forward), 5′–CTTTGTCGTTGG TTAGCTGGT–3′ (reverse); Snail1: 5′–CTTCCAGCAGCCCTACGAC–3′ (forward); 5′–CGGT GGGGTTGAGGATCT–3′ (reverse). Detail of qRT-PCR method was described in the supplementary method 1.1.

2.10. Proteomics Analysis

Cell lysis and digestion: The HGC-27 cellular pellet was resuspended in boiling lysis buffer and boiled for 10 min before further lysis by sonication with a microtip probe. The protein concentration was determined using a BCA kit.

iTRAQ labeling of peptides and strong cation exchange (SCX) chromatography: Aliquots of 100 μg of peptide from each grade were labeled using the iTRAQ assay. The labeled peptides were incubated for 2 h at room temperature, then mixed in equal proportions, and dried by vacuum centrifugation. Further separation of iTRAQ-labeled peptides was achieved using high performance liquid chromatography.

LC–MS/MS analysis: All samples were evaluated on an Easy Nano Liquid Chromatography apparatus (nLC) coupled to a Q-Exactive HF-X MS (Thermo Fisher Scientific, Waltham, MA, USA) 20.

Data processing and bioinformatics analysis of proteins: The mass spectra were analyzed with MaxQuant software version 1.5.8.3 (Max Planck Institute of Biochemistry, Munich, Bavaria, Germany). Bioinformatics analysis was performed using the Gene Ontology (GO) database to annotate the differentially expressed proteins. Unsupervised hierarchical clustering analysis, PCA, and volcano plot analysis were performed in Perseus. The Kyoto Encyclopedia of Genes and Genomes (KEGG) database provided the analysis of the pathways of differentially expressed proteins. Enrichment analyses for GO and KEGG pathways were carried out to reveal the enriched pathways of significantly altered proteins and to identify their functions. Detail of proteomics anailsis was decribed in the supplementary method 1.2.

2.11. SiRNA Transfection

The specific siRNA of SIX1 (sense: 5′–GUCAGCAACUGGUUUAAGATT–3′ and antisense: 5′–UCUUAAACCAGUUGCUGACTT–3′) was provided by GenePharma (Shanghai, China). HGC-27 and BGC-823 were transfected for 48 h with Lipofectamine ™ 2000 (Carlsbad, CA, USA) and 100 pM siRNA duplexes in six-well plates, and then treated with or with Rh4 (40 μM) for follow-up research.

2.12. Tail Vein Injection Model

This experiment was conducted with permission from the Animal Ethics Committee of Northwest University (NWU-AWC-20210310M) and complied with the requirements of the Law of the People's Republic of China on Laboratory Animals. Male BALB/c nude mice (5 weeks old, weighing about 21 g) were acquired from GemPharmatech (Jiangsu, China). Lung metastasis was studied by washing 2×10^5 HGC-27 in serum-free DMEM and injecting it intravenously into mice ($n = 40$). The successfully modeled mice were randomly divided into the following six groups ($n = 10$, intraperitoneal injection): normal group, normal + Rh4 group (100 mg per kg daily), control group (model mice), low-dose Rh4 group (50 mg per kg daily), high-dose Rh4 group (100 mg per kg daily), and cisplatin group (10 mg per kg per 3 days) [40]. All mice were euthanized by continuous injection for 3 weeks, and critical organs were isolated and weighed. Then, some organs were paraffin-embedded, while some were stored at −80 °C. Bouin's solution was used to fix and stain some lungs, whose metastases were distinguished as white colonies.

2.13. Histopathology and Immunohistochemistry Assay (IHC)

The major organs (heart, lung, spleen, kidney, and liver) of nude mice were isolated, fixed in 4% paraformaldehyde, weighed, and paraffin-embedded. For H&E, sectioning and staining were applied. For IHC, paraffin was sectioned, pretreated, incubated with primary antibody, incubated with secondary antibody, stained, and photographed under an optical microscope (Nikon, Japan).

2.14. Immunofluorescence Assay (IF)

HGC-27 and BGC-823 were adhered to the slides and cocultured with different concentrations of Rh4 for 24 h. Subsequently, cells were fixed, blocked in 5% bovine serum albumin, incubated with primary antibody, incubated with secondary antibody, and photographed for analysis by Olympus confocal microscopy (Tokyo, Japan).

2.15. Statistical Analysis

Statistical analysis was performed by SPSS 20.0 (Chicago, IL, USA) and Prism 9 (San Diego, CA, USA) software. Differences between groups were derived from one-way analysis of variance (ANOVA). Results are from at least three separate experiments and are expressed as the mean $\pm$ standard deviation (SD). A p-value <0.05 reveals statistically significant differences.

3. Results

3.1. Ginsenoside Rh4 Inhibited GC Cell Growth In Vitro

To investigate the antiproliferative activity of Rh4 on GC cells, HGC-27 and BGC-823 were incubated with different concentrations of Rh4 for 24 h and subjected to MTT assay. The results demonstrated that Rh4 significantly reduced the vigor of HGC-27 and BGC-823 in a dose- and time-dependent manner (Supplementary Figure S1B,C). The IC_{50} values for Rh4-treated HGC-27 and BGC-823 were 83.15 μM and 92.38 μM, respectively. Moreover, compared to the control, the analysis of colony formation assay revealed that Rh4 suppressed the formation of HGC-27 and BGC-823 colonies (Supplementary Figure S1D). Meanwhile, AO/EB staining experiments proved that Rh4 significantly reduced the viability of HGC-27 and BGC-823 in a dose-dependent manner (Supplementary Figure S1E). In brief, these results confirmed that Rh4 has a strong antitumor effect on gastric cells in vitro.

3.2. Ginsenoside Rh4 Restrains Migration and Invasion of GC Cells

To test the antimigration and anti-invasion functions of Rh4 on GC cells, wound healing assays and transwell assays were conducted. HGC-27 and BGC-823 were incubated with Rh4 at concentrations of 20 and 40 μM, respectively. The wound healing assay showed that the wound healing percentages decreased evidently in a dose-dependent manner (Figure 1A, * $p < 0.05$, *** $p < 0.001$). Furthermore, the transwell assay for migration proved that the cell metastasis ability of the Rh4 treatment group showed a dose-dependent reduction of 32.44–81.4% in HGC-27 and 30.97–47.24% in BGC-823 (Figure 1B, ** $p < 0.05$, *** $p < 0.001$), indicating the inhibitory effect of Rh4 on migration. A Matrigel invasion assay was carried out to evaluate the ability of HGC-27 and BGC-823 to invade through the Matrigel and membrane barrier of the transwell. The results revealed that the treatment of Rh4 could effectively attenuate HGC-27 and BGC-823 invasion in a dose-dependent manner (Figure 1C, *** $p < 0.001$). These results demonstrated that the migration and invasion ability were limited in a dose-dependent manner after treatment with Rh4 in HGC-27 and BGC-823.

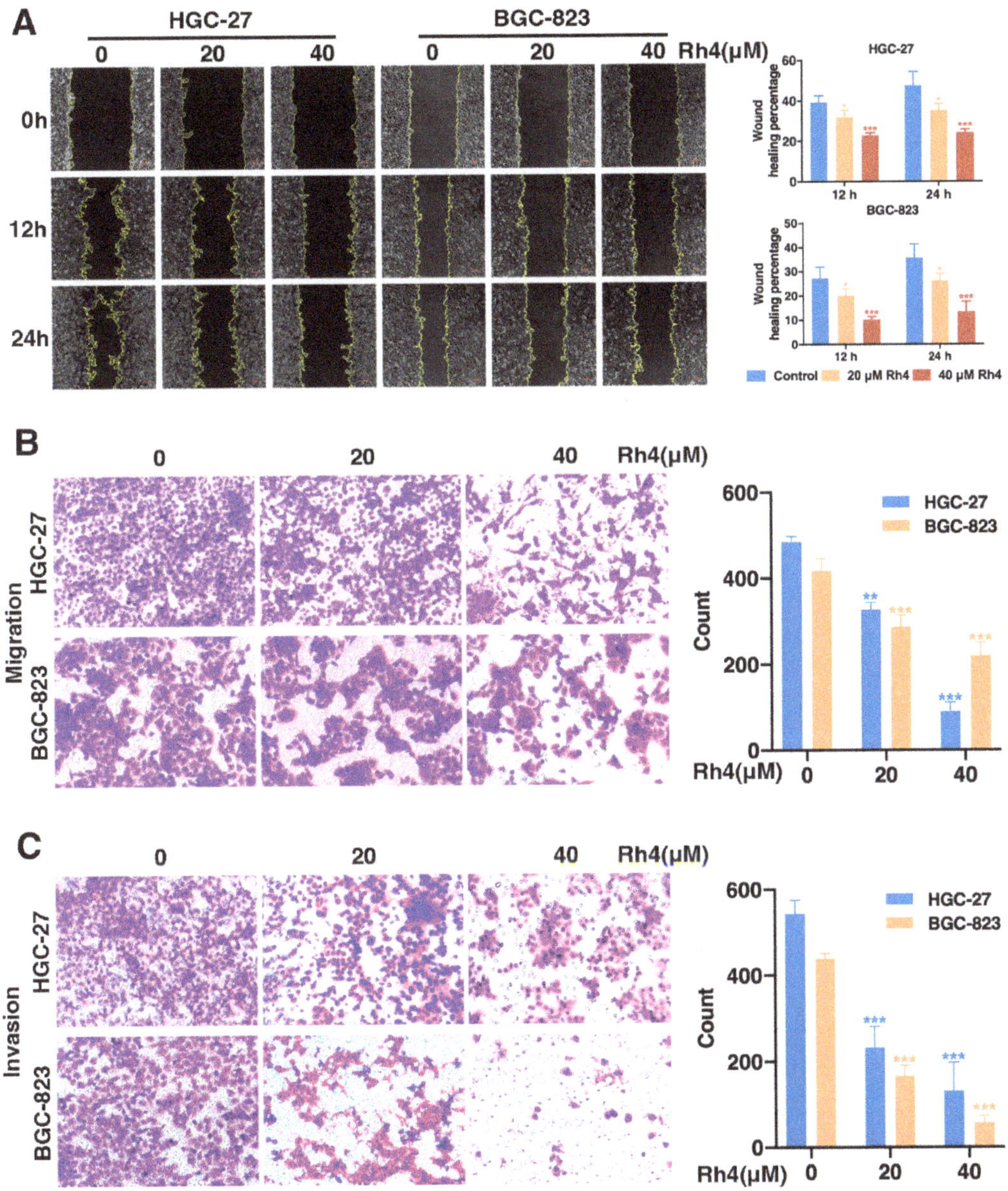

Figure 1. Ginsenoside Rh4 inhibits GC metastasis in vitro. Wound healing assays (**A**) and transwell migration (**B**) and invasion (**C**) assays were performed to investigate the migration and invasion ability alteration of BC cells with the treatment of Rh4. Data were processed using ImageJ software (NH, Bethesda, MD, USA). Quantification charts are listed on the right. Statistics are exhibited as the mean ± SD of triplicate independent experiments; * $p < 0.05$, ** $p < 0.01$, *** $p < 0.001$.

3.3. *Ginsenoside Rh4 Inhibited GC Metastasis In Vivo, and Causes Low Toxicity and Side-Effects*

A tail vein injection model was built to verify the impact of ginsenoside Rh4 on GC tumor metastasis in vivo [41]. Compared to the control, the number of lung nodules treated with Rh4 and oxaliplatin (drug commonly used in adjuvant therapy for gastric cancer) showed an obvious reduction (Supplementary Figure S2A,B). These results suggest that Rh4 suppressed tumor growth and metastasis in vivo.

Furthermore, the body weight of mice treated with low and high doses of Rh4 were not significantly different from the control, while the oxaliplatin group showed a dramatic decrease, indicating no systemic toxicity after Rh4 administration (Supplementary Figure S2C). The H&E staining of organ tissues showed that the heart, spleen, and lung damage were visible in the oxaliplatin group, compared with no effect in the Rh4 group (Supplementary Figure S2D). Additionally, there were no statistically significant differences in organ indicators between each group, except for a significantly lower splenic index in the oxaliplatin group (Table 1). In summary, these results demonstrated that Rh4 causes no toxic side-effects in GC treatment.

Table 1. The organ index of mice.

Organ index	Control	Normal	Normal + H-Rh4 (100 mg/kg)	L-Rh4 (50 mg/kg)	H-Rh4 (100 mg/kg)	Oxaliplatin (10 mg/kg)
Heart	0.52 ± 0.03	0.49 ± 0.02	0.50 ± 0.04	0.49 ± 0.07	0.54 ± 0.03	0.53 ± 0.05
Liver	5.46 ± 0.25	5.43 ± 0.46	5.39 ± 0.16	5.88 ± 0.62	5.22 ± 0.18	4.79 ± 0.32
Spleen	1.45 ± 0.23	1.44 ± 0.18	1.47 ± 0.15	1.50 ± 0.19 ##	1.41 ± 0.28 ##	0.39 ± 0.12 **
Lung	0.78 ± 0.16	0.72 ± 0.09	0.73 ± 0.15	0.71 ± 0.07	0.72 ± 0.13	0.71 ± 0.15
Kidney	1.44 ± 0.16	1.47 + 0.11	1.46 ± 0.09	1.49 ± 0.20	1.45 ± 0.16	1.47 ± 0.16

Statistics are exhibited as the mean ± SD (n = 5 in each group); ** $p < 0.01$ compared with the control group, ## $p < 0.01$ compared with the disitertide-treated group.

3.4. Ginsenoside Rh4 Reverses EMT In Vitro and In Vivo

To evaluate whether ginsenoside Rh4 inhibits EMT in HGC-27 and BGC-823, the expression of EMT-related proteins and mRNA was studied. Rh4 attenuated the expression of N-cadherin and Snail and enhanced that of E-cadherin (epithelial marker) in a dose-dependent manner in HGC-27 and BGC-823 (Figure 2A). Additionally, qRT-PCR analysis showed that Rh4 treatment (20 and 40 µM) upregulated the transcription level of E-cadherin and downregulated the transcription levels of N-cadherin, Vimentin, and Snail in HGC-27 and BGC-823 (Figure 2B). Moreover, changes in the protein content of HGC-27 and BGC-823 were more clearly assessed by immunofluorescence (IF), and it was observed that Rh4 decreased E-cadherin levels and increased N-cadherin levels (Figure 2C). Overall, it was demonstrated that ginsenoside Rh4 specifically interfered with the expression of N-cadherin, E-cadherin, and other biomarkers to reverse the EMT process.

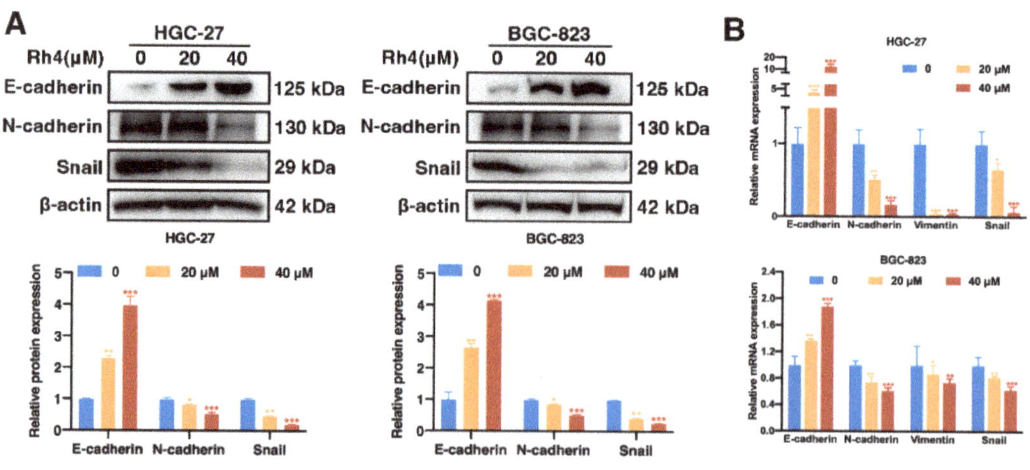

Figure 2. *Cont.*

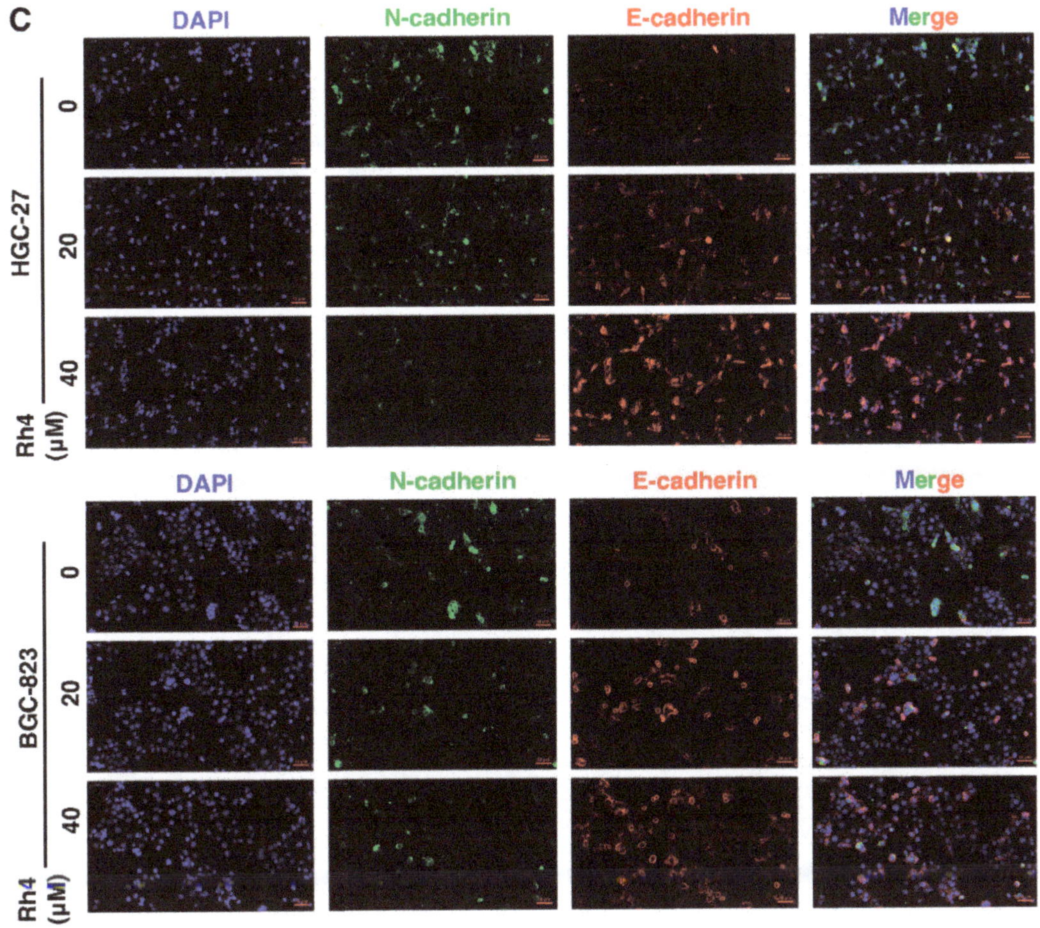

Figure 2. Ginsenoside Rh4 reverses EMT in vitro. (**A**) Western blot. Quantification plots are shown below. (**B**) qRT-PCR assay. β-actin was used as an endogenous reference. (**C**) IF assay. E-cadherin (red) and N-cadherin (green) are presented. DAPI (blue) was used to mark the cell nuclei. Scale bars = 50 μm. Statistics are exhibited as the mean ± SD of triplicate independent experiments; * $p < 0.05$, ** $p < 0.01$, *** $p < 0.001$.

Lung tissue samples of the HGC-27 tail vein injection model were analyzed by Western blot (Figure 3A), IF (Figure 3B), and immunohistochemical analysis (Figure 3C). The results showed consistent trends in EMT-related proteins in vivo and in vitro.

3.5. SIX1 Is Involved in EMT Suppression by Rh4

To further investigate how Rh4 inhibits EMT in GC cells, proteomic analysis was performed using quantitative mass spectrometry (MS) of the proteome (Figure 4A). Three biological replicates were collected per condition. In the dataset, 7609 proteins were quantified. In order to assess whether the response to Rh4 is obvious at the proteomic level, principal component analysis (PCA) was performed by the sample (Figure 4B). This analysis showed that the different sample proteomes were separated into two distinct groups on component 1, reflecting their response to Rh4. Next, differentially regulated proteins were identified between the control and Rh4-treated cell groups by performing a microarray significance analysis (SAM) statistical test. A total of 218 regulated proteins were found, with 98 in the Rh4-treated cell group. Significantly more proteins (120) were downregulated (Figure 4C). To determine how Rh4 inhibits GC metastasis, enrichment analysis of the KEGG signaling pathway was performed (Figure 4E). It was found that the TGF-β signaling pathway was the most significantly enriched. At the same time, the upstream protein SIX1 of the TGF-β signaling pathway was regulated by Rh4 (Figure 4C,D). TCGA dataset showed that SIX1 expression levels in normal tissues were higher than in tumor tissues (p = 0.0018) (Figure 4F). Kaplan–Meier survival analysis showed that patients with high SIX1 expression had a shorter overall survival period (Figure 4G). These results suggest that ginsenoside Rh4 may suppress EMT of GC cells via mediating the TGF-β/Smad2/3 signaling pathway by targeting SIX1.

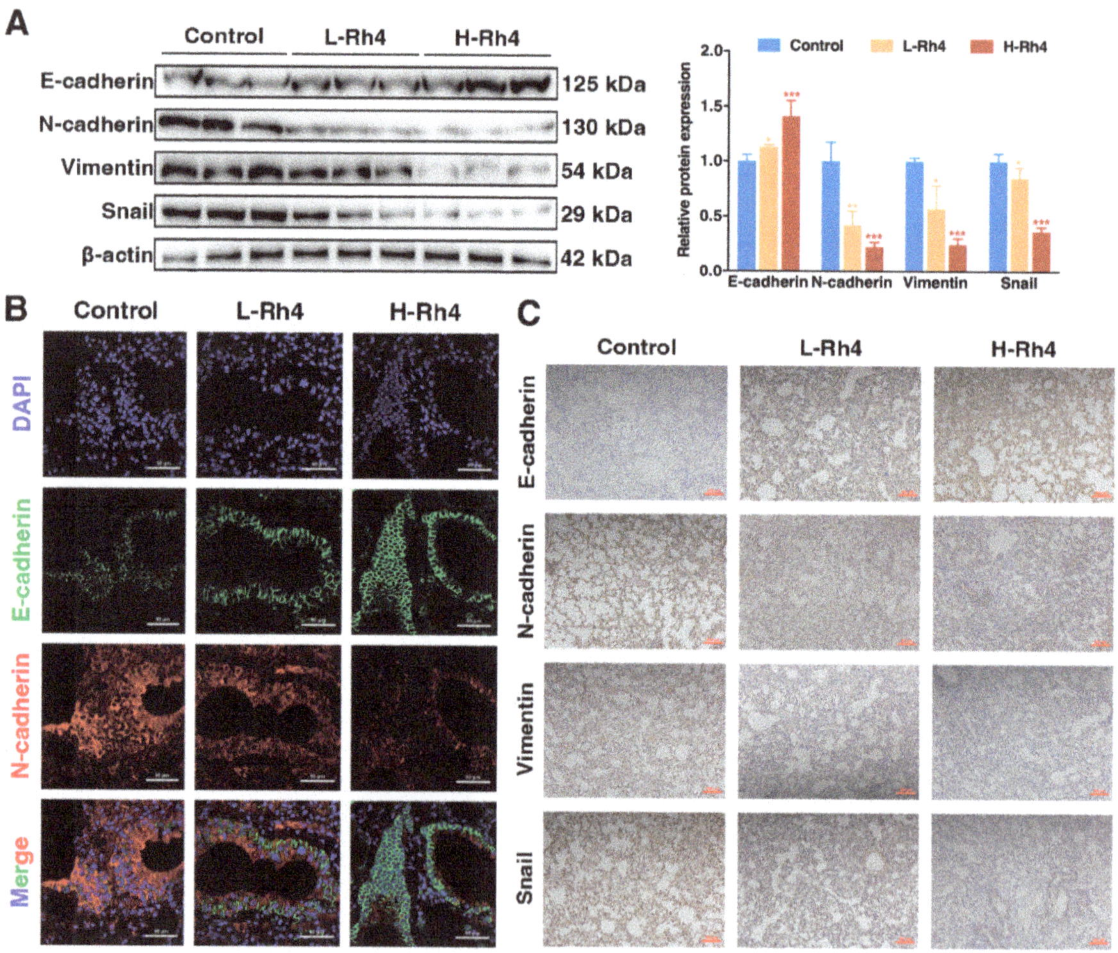

Figure 3. Ginsenoside Rh4 reverses the EMT procedure of GC in vivo. (**A**) Western blot showing increased E-cadherin and decreased N-cadherin, Vimentin, and Snail in tumor tissues under Rh4 treatment. β-Actin was used as an endogenous reference. Quantification plots are shown on the right. (**B**) IF assay. The downward adjustment of E-cadherin and upward adjustment of N-cadherin were reversed by Rh4 in vivo. Scale bars = 50 μm. (**C**) IHC assays were performed to assess changes in E-cadherin, N-cadherin, Vimentin, and Snail. Scale bars = 200 μm. Statistics are exhibited as the mean ± SD of triplicate independent experiments; * $p < 0.05$, ** $p < 0.01$, *** $p < 0.001$.

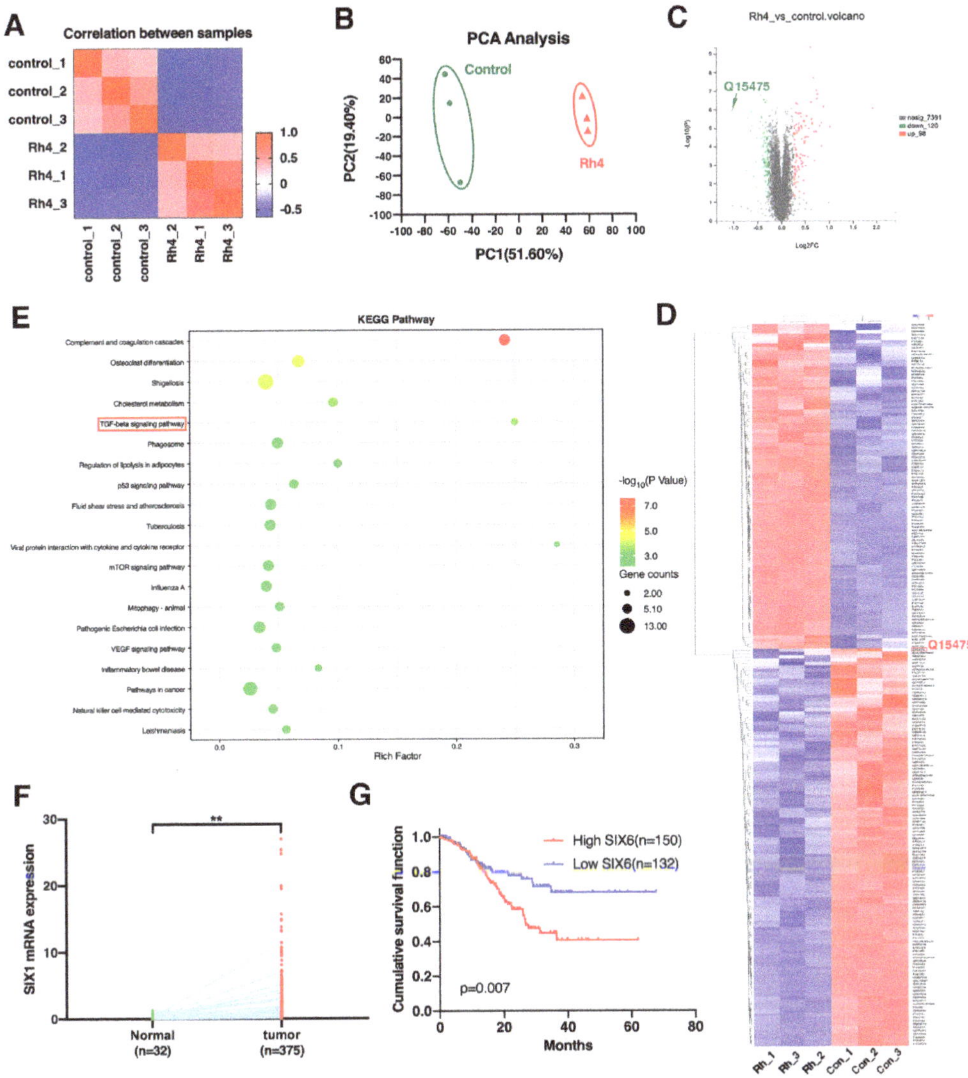

Figure 4. Proteomics and clinical data indicating that ginsenoside Rh4 targeted SIX1, suppressing EMT in GC. (**A**) Correlation heatmap of samples. (**B**) Principal component analysis (PCA) of significantly regulated proteins in the HGC-27 groups with or without Rh4 treatment (n = 3 biologically independent experiments). (**C**)Volcano plot analysis for the Rh4 treatment/control comparisons. Proteins regulated on the proteome level are marked in red (upregulated) and green (downregulated). Q15475 represents the SIX1 protein. (**D**) Heatmap hierarchical clustering from differentially expressed proteins (n = 218) identified across all sample groups. Q15475 represents the SIX1 protein. (**E**) Statistics of KEGG pathway enrichment. (**F**) TCGA analysis of SIX1 expression levels in GC tissues and paracancerous tissues (** $p < 0.01$). (**G**) Survival curves synthesized using Kaplan–Meier method.

3.6. Ginsenoside Rh4 Reverses EMT through SIX1–TGF-β/Smad2/3 Signaling Axis In Vitro and In Vivo

To verify the mechanism of action of Rh4 and whether SIX1 is a key target for regulation and antitumor activity, the related protein expression under the combined treatment of

Rh4 and TGF-β1 inhibitor (disitertide) was evaluated. Expression of TGF-β1 and P-Smad3 was restrained in the disitertide group compared to the disitertide-negative control, and levels of these TGF-β/Smad2/3 pathway-related proteins were further downregulated by Rh4 (40 μM). Complementarily, coincubation with disitertide and Rh4 enhanced the effect of Rh4 treatment alone on E-cadherin, N-cadherin, Vimentin, and Snail. (Figure 5A). The disitertide group had no effect on the expression of SIX1, while the siRNA transfection assay of SIX1 displayed that Rh4 restrained SIX1 expression specifically (Figure 5B), proving that SIX1 is the target protein of Rh4. The outcomes demonstrated that ginsenoside Rh4 reverses the EMT of GC cells via regulating the TGF-β/Smad2/3 signaling pathway by targeting SIX1.

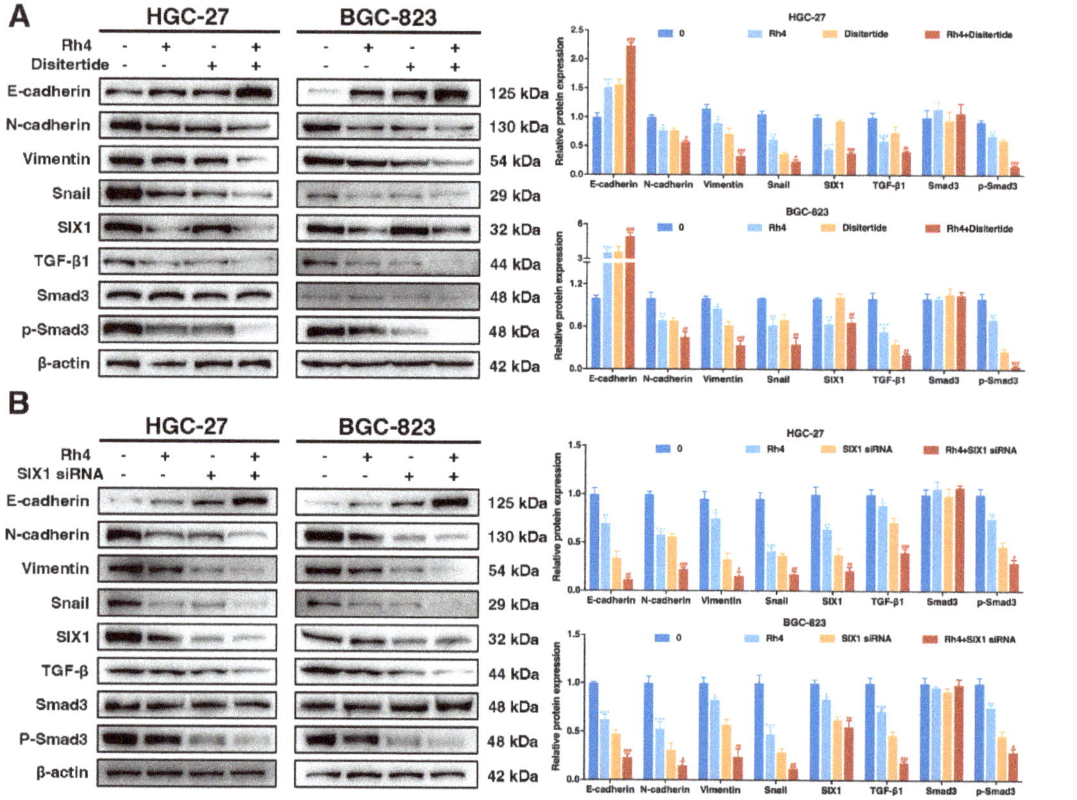

Figure 5. SIX1 is critical for Rh4-induced TGF-β/Smad2/3 signaling pathway and EMT process inhibition in GC. (**A**) Western blot showing the effects of EMT biomarkers and TGF-β/Smad2/3 signaling protein expression in GC cells (HGC-27 and BGC-823) after incubation with disitertide and Rh4. (**B**) Western blot showing effects of EMT biomarkers and TGF-β/Smad2/3 signaling protein expression in GC cells (HGC-27 and BGC-823) after incubation with SIX1-siRNA and Rh4. β-Actin was used as an endogenous reference. Quantification plots are shown on the right. Statistics are exhibited as the mean ± SD of triplicate independent experiments; * $p < 0.05$, ** $p < 0.01$, and *** $p < 0.001$ compared with the control group; # $p < 0.05$, ## $p < 0.01$, and ### $p < 0.001$ compared with the disitertide-treated group.

To further assess whether Rh4 inhibited the metastasis of GC cells in vivo, lung tissue samples of the HGC-27 tail vein injection model were analyzed by Western blot, IF, and IHC assays. The lung tissue treated with Rh4 showed weak TGF-β/Smad2/3 signaling pathway signals, as evidenced by Western blot (Figure 6A), IHC (Figure 6B), and IF assays

(Figure 6C), consistent with the results in vitro. This means that the metastasis of GC cells in vivo can be regulated by treatment with Rh4.

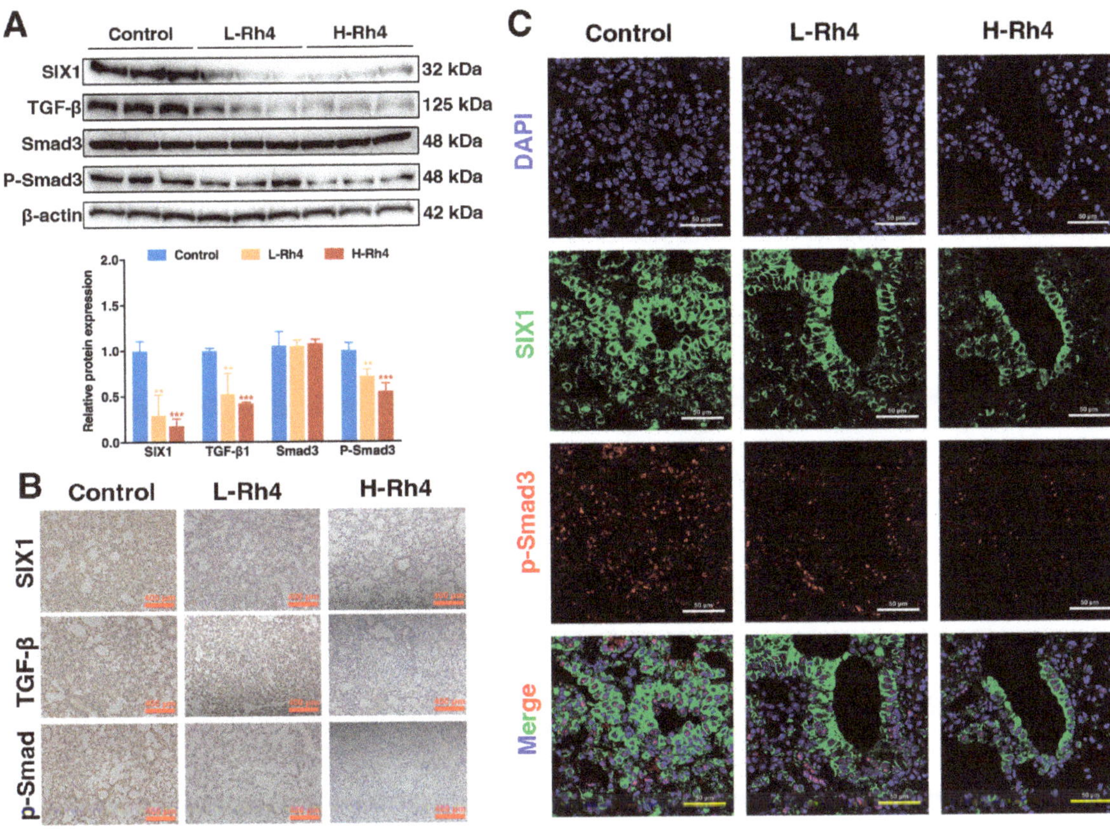

Figure 6. Ginsenoside Rh4 attenuated TGF-β/Smad2/3 signaling pathway of GC in vivo. (**A**) Western blot showing expression of SIX1, TGF-β, Smad3, and P-Smad3 in tumor tissues under Rh4 treatment. (**B**) IHC assay performed to assess changes in SIX1, TGF-β, and P-Smad3. Scale bars = 200 μm. (**C**) IF assay. Upregulation of SIX1 and P-Smad3 is reversed by Rh4 in vivo. Scale bars = 50 μm. Statistics are exhibited as the mean ± SD of triplicate independent experiments; ** $p < 0.01$, *** $p < 0.001$.

4. Discussion

Ginsenoside Rh4, as the natural product of ginseng, has been confirmed to have immunomodulatory, anticancer, anti-inflammatory, and antidiabetic properties [42–44], demonstrating its essential role in the treatment and prevention of ranges of diseases. However, the anti-GC effect of Rh4 and the underlying molecular mechanism are largely unknown. In this research, we corroborated that ginsenoside Rh4 markedly suppressed growth, proliferation, EMT, and invasion by blocking the activation of the SIX1–TGF-β/Smad2/3 signaling axis in vitro and in vivo (Figure 7), which preliminarily revealed the anti-metastasis effects of Rh4. In vitro, according to the MTT assay, colony formation assays, and AO/EB staining, Rh4 obviously decreased the survival ratios to 20% and restrained colony formation at the dose of 120 μM. In vivo, the number of tumor nodules was 51.22% and 92.68% for 50 mg/kg and 100 mg/kg Rh4 intraperitoneal injections, respectively, thus confirming the outstanding efficacy of Rh4. Simultaneously, unlike oxaliplatin, Rh4 did not cause weight loss in tumor-bearing nude mice. In addition, H&E staining showed that the treatment with Rh4 did not negatively affect the normal structure and function of heart, lung, liver, spleen, and kidney, indicating that Rh4 has little toxicity toward the organism during GC treatment. As a result, the strong anti-GC effect and low toxicity of Rh4 suggest that it can be used as an effective drug for GC treatment.

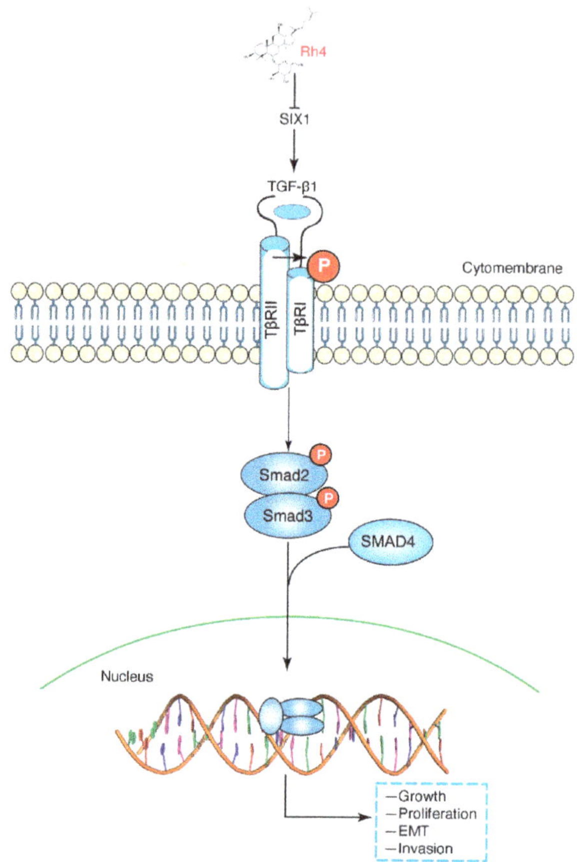

Figure 7. Schematic representation of the proposed mechanism of Rh4 in GC cells. Rh4 targets SIX1 to inhibit TGF-β/Smad2/3 signaling, preventing the metastasis of GC.

EMT is an essential procedure in tumor metastasis, which hardly occurs in normal cells. Under the influence of various factors such as the tumor microenvironment, tumor epithelial cells lose their epithelial phenotype (e.g., epithelial polarity and cell junctions), promoting a shift to a mesenchymal phenotype, which makes the cells more invasive and migratory to other cells [45]. Studies have shown that tumor metastasis is inseparable from the continuous EMT process [46], which is often associated with EMT-related protein changes. In the EMT procedure, the most important change among these biomarkers is the decrease in E-cadherin, which keeps cells tightly adherent, and the increase in N-cadherin, a characteristic protein of mesenchymal cells, as well as Vimentin, an essential structural protein of the cell, whose expression in mesenchymal cells is higher than that in epithelial cells. In addition, expression of EMT-promoting transcription factor Snail suppresses E-cadherin transcription. From our latest research, results from transwell assays and cell scratch tests demonstrated that coincubation with Rh4 significantly inhibited the migration and invasive ability of GC cells. In vivo and in vitro, it was obtained from Western blot, qRT-PCR, and IHC analysis that Rh4 decreased the level of N-cadherin while it enhanced that of E-cadherin, suggesting that the EMT of GC cells was visibly suppressed. As obtained by Western blot, qRT-PCR, and immunohistochemical analysis, Rh4 attenuated N-cadherin expression and increased E-cadherin expression, indicating that the EMT of GC cells was significantly inhibited in vivo and in vitro.

To our knowledge, TGFβ receptor type I (TβRI) and type II (TβRII) bind io TGF-β ligands, phosphorylating Smad2/3, which plays an important role in proliferation, apoptosis, and differentiation [47,48]. The TGF-β/Smad2/3 signaling pathway was suppressed by silencing Trim59 in bladder cancer [49]. Moreover, nobiletin blocked TGFβ1/Smad3 signaling to inhibit EMT in human non-small-cell lung cancer [50]. Similarly, we confirmed that TGF-β/Smad2/3 signaling was refrained with Rh4 treatment. Western blot and immunohistochemical analysis of cellular proteins and tumor tissue proteins concluded that Rh4 significantly inhibited the expression of TGF-β1, P-Smad3, and Snail and blocked the EMT process in GC cells.

SIX1 is an essential developmental transcription factor that not only has a promotive effect on embryonic muscle, but its overexpression also promotes tumorigenesis [51]. SIX1 has been reported to promote the proliferation and metastasis of a variety of cancer cells such as breast cancer, hepatocellular carcinoma, and GC cells. According to recent studies, the SIX1-induced upregulation of TGF-β1 is critical for the propagation of TGF-β/Smad2/3 signaling, induction of EMT persistence, and metastasis [19,20]. Research from Farabaugh et al. revealed that Eya2 activates the TGF-β signaling pathway by interacting with the pro-metastatic function of SIX1 to induce EMT, thereby causing cancer epithelial cells to become cancer stem cells [20]. In our study, proteomic screening proved that Rh4 targeted SIX1 via the TGF-β/Smad2/3 signaling axis to suppress EMT. Subsequently, cotreatment with disitertide and the SIX1-siRNA transfection assay confirmed this conclusion. In conclusion, our discoveries demonstrated that Rh4 suppressed the metastasis of GC via inhibiting the SIX1–TGF-β/Smad2/3 signaling axis.

5. Conclusions

In conclusion, our studies confirmed that Rh4 displayed significant anti-GC effects via inhibiting the TGF-β/Smad2/3 signaling pathway, and they revealed the key role of SIX1 in this process. In metastasis and invasion models, Rh4 significantly inhibited the metastatic and invasive ability of GC cells in a dose-dependent manner in vitro. In the GC tail vein injection model, Rh4 showed significant antitumor metastatic effects with few negative effects in vivo. Proteomics combined with inhibitor experiments and SIX1-siRNA transfection assay showed that Rh4 could inhibit the TGF-β/Smad2/3 signaling pathway via binding to SIX1. Our study not only offered new insights into Rh4 as a possible anticancer agent, but also highlighted SIX1 as a potential molecular target for the regulation of Rh4 and TGF-β/Smad2/3 signaling pathways.

Supplementary Materials: The following supporting information can be downloaded at: https://www.mdpi.com/article/10.3390/nu14081564/s1, Figure S1: Ginsenoside Rh4 inhibits GC cells growth in vitro, Figure S2: Ginsenoside Rh4 inhibits GC metastasis in vivo, and causes low toxicity and side effects.

Author Contributions: Conceptualization, H.J. and Y.M.; methodology, H.J., P.M., Z.D., Y.L. and Y.M.; software, H.J.; validation, H.J. and Y.M.; formal analysis, H.J.; investigation, H.J. and P.M.; resources, S.S., Y.M. and D.F.; data curation, H.J., Y.L. and Y.M.; writing—original draft preparation, H.J.; writing—review and editing, H.J. and Y.M.; visualization, H.J.; supervision, Y.M.; project administration, Y.M. and D.F.; Funding acquisition, S.S., Y.M. and D.F. All authors have read and agreed to the published version of the manuscript.

Funding: This research was funded by the National Key R&D Program of China, Grant No. 2021YFC2101000, by the National Natural Sciences Foundation of China, Grant Nos. 21776227 and 22108224, and by the Natural Science Foundation of Shaanxi Province, China, Grant No. 2020JQ-570.

Institutional Review Board Statement: The animal study protocol was approved by the Animal Ethics Committee of Northwest University (NWU-AWC-20210310M).

Informed Consent Statement: Not applicable.

Data Availability Statement: The data presented in this study are available within the text and figures.

Conflicts of Interest: The authors declare no conflict of interest.

References

1. Sung, H.; Ferlay, J.; Siegel, R.L.; Laversanne, M.; Soerjomataram, I.; Jemal, A.; Bray, F. Global Cancer Statistics 2020: GLOBOCAN Estimates of Incidence and Mortality Worldwide for 36 Cancers in 185 Countries. *CA Cancer J. Clin.* **2021**, *71*, 209–249. [CrossRef] [PubMed]
2. Miller, K.D.; Siegel, R.L.; Lin, C.C.; Mariotto, A.B.; Kramer, J.L.; Rowland, J.H.; Stein, K.D.; Alteri, R.; Jemal, A. Cancer treatment and survivorship statistics, 2016. *CA Cancer J. Clin.* **2016**, *66*, 271–289. [CrossRef] [PubMed]
3. Yilmaz, M.; Christofori, G. EMT, the cytoskeleton, and cancer cell invasion. *Cancer Metastasis Rev.* **2009**, *28*, 15–33. [CrossRef] [PubMed]
4. Choi, C.Y.; Lim, S.-C.; Lee, T.-B.; Han, S.I. Molecular Basis of Resveratrol-Induced Resensitization of Acquired Drug-Resistant Cancer Cells. *Nutrients* **2022**, *14*, 699. [CrossRef]
5. Huang, L.; Wu, R.-L.; Xu, A.M. Epithelial-mesenchymal transition in gastric cancer. *Am. J. Transl. Res.* **2015**, *7*, 2141–2158.
6. Peng, Z.; Wang, C.-X.; Fang, E.-H.; Wang, G.-B.; Tong, Q. Role of epithelial-mesenchymal transition in gastric cancer initiation and progression. *World. J. Gastroenterol.* **2014**, *20*, 5403–5410. [CrossRef]
7. Zhang, J.; Zhou, Y.; Yang, Y. CCR7 pathway induces epithelial–mesenchymal transition through up-regulation of Snail signaling in gastric cancer. *Med. Oncol.* **2015**, *32*, 17. [CrossRef]
8. Borthwick, L.A.; Gardner, A.; De Soyza, A.; Mann, D.A.; Fisher, A.J. Transforming growth factor-β1 (TGF-β1) driven epithelial to mesenchymal transition (EMT) is accentuated by tumour necrosis factor α (TNFα) via crosstalk between the SMAD and NF-κB pathways. *Cancer Microenviron.* **2012**, *5*, 45–57. [CrossRef]
9. Wu, Y.; Tang, S.-J.; Sun, G.; Sun, K. CXCR7 mediates TGFβ1-promoted EMT and tumor-initiating features in lung cancer. *Oncogene* **2016**, *35*, 2123–2132. [CrossRef]
10. Petanidis, S.; Kioseoglou, E.; Domvri, K.; Zarogoulidis, P.; Carthy, J.M.; Anestakis, D.; Moustakas, A.; Salifoglou, A. In vitro and ex vivo vanadium antitumor activity in (TGF-β)-induced EMT. Synergistic activity with carboplatin and correlation with tumor metastasis in cancer patients. *Int. J. Biochem. Cell Biol.* **2016**, *74*, 121–134. [CrossRef]
11. Pang, L.; Li, Q.; Wei, C.; Zou, H.; Li, S.; Cao, W.; He, J.; Zhou, Y.; Ju, X.; Lan, J. TGF-β1/Smad signaling pathway regulates epithelial-to-mesenchymal transition in esophageal squamous cell carcinoma: In vitro and clinical analyses of cell lines and nomadic Kazakh patients from northwest Xinjiang, China. *PLoS ONE* **2014**, *9*, e112300. [CrossRef]
12. Ko, H.; So, Y.; Jeon, H.; Jeong, M.-H.; Choi, H.-K.; Ryu, S.-H.; Lee, S.-W.; Yoon, H.-G.; Choi, K.-C. TGF-β1-induced epithelial–mesenchymal transition and acetylation of Smad2 and Smad3 are negatively regulated by EGCG in human A549 lung cancer cells. *Cancer Lett.* **2013**, *335*, 205–213. [CrossRef] [PubMed]
13. Kolosova, I.; Nethery, D.; Kern, J.A. Role of Smad2/3 and p38 MAP kinase in TGF-β1-induced epithelial–mesenchymal transition of pulmonary epithelial cells. *J. Cell. Physiol.* **2011**, *226*, 1248–1254. [CrossRef] [PubMed]
14. Emadi-Baygi, M.; Nikpour, P.; Emadi-Andani, E. SIX1 overexpression in diffuse-type and grade III gastric tumors: Features that are associated with poor prognosis. *Adv. Biomed. Res.* **2015**, *4*, 139. [CrossRef]
15. Liu, Y.; Nandi, S.; Martel, A.; Antoun, A.; Ioshikhes, I.; Blais, A. Discovery, optimization and validation of an optimal DNA-binding sequence for the Six1 homeodomain transcription factor. *Nucleic Acids Res.* **2012**, *40*, 8227–8239. [CrossRef]
16. Zeng, J.; Wei, M.; Shi, R.; Cai, C.; Liu, X.; Li, T.; Ma, W. MiR-204-5p/Six1 feedback loop promotes epithelial–mesenchymal transition in breast cancer. *Tumor Biol.* **2016**, *37*, 2729–2735. [CrossRef]
17. Xu, H.-X.; Wu, K.-J.; Tian, Y.-J.; Liu, Q.; Han, N.; He, X.-L.; Yuan, X.; Wu, G.S.; Wu, K.-M. Expression profile of SIX family members correlates with clinic-pathological features and prognosis of breast cancer: A systematic review and meta-analysis. *Medicine* **2016**, *95*, e4085. [CrossRef]
18. Nishimura, T.; Tamaoki, M.; Komatsuzaki, R.; Oue, N.; Taniguchi, H.; Komatsu, M.; Aoyagi, K.; Minashi, K.; Chiwaki, F.; Shinohara, H. SIX 1 maintains tumor basal cells via transforming growth factor-β pathway and associates with poor prognosis in esophageal cancer. *Cancer Sci.* **2017**, *108*, 216–225. [CrossRef]
19. Smith, A.L.; Iwanaga, R.; Drasin, D.J.; Micalizzi, D.S.; Vartuli, R.L.; Tan, A.-C.; Ford, H.L. The miR-106b-25 cluster targets Smad7, activates TGF-β signaling, and induces EMT and tumor initiating cell characteristics downstream of Six1 in human breast cancer. *Oncogene* **2012**, *31*, 5162–5171. [CrossRef]
20. Micalizzi, D.S.; Wang, C.-A.; Farabaugh, S.M.; Schiemann, W.P.; Ford, H.L. Homeoprotein Six1 increases TGF-β type I receptor and converts TGF-β signaling from suppressive to supportive for tumor growth. *Cancer Res.* **2010**, *70*, 10371–10380. [CrossRef]
21. Iwanaga, R.; Wang, C.-A.; Micalizzi, D.S.; Harrell, J.C.; Jedlicka, P.; Sartorius, C.A.; Kabos, P.; Farabaugh, S.M.; Bradford, A.P.; Ford, H.L. Expression of Six1 in luminal breast cancers predicts poor prognosis and promotes increases in tumor initiating cells by activation of extracellular signal-regulated kinase and transforming growth factor-beta signaling pathways. *Breast Cancer Res.* **2012**, *14*, R100. [CrossRef] [PubMed]
22. Min, W.-P.; Wei, X.-F. Silencing SIX1 inhibits epithelial mesenchymal transition through regulating TGF-β/Smad2/3 signaling pathway in papillary thyroid carcinoma. *Auris Nasus Larynx* **2021**, *48*, 487–495. [CrossRef] [PubMed]

23. ompson Coon, J.; Ernst, E. Panax ginseng: A systematic review of adverse effects and drug interactions. *Focus Altern. Complement. Erapies* **2002**, *7*, 111–112. [CrossRef]
24. Li, B.; Zhao, J.; Wang, C.-Z.; Searle, J.; He, T.-C.; Yuan, C.-S.; Du, W. Ginsenoside Rh2 induces apoptosis and paraptosis-like cell death in colorectal cancer cells through activation of p53. *Cancer Lett.* **2011**, *301*, 185–192. [CrossRef]
25. Liu, Y.; Fan, D. The Preparation of Ginsenoside Rg5, Its Antitumor Activity against Breast Cancer Cells and Its Targeting of PI3K. *Nutrients* **2020**, *12*, 246. [CrossRef]
26. Kang, K.S.; Ham, J.; Kim, Y.-J.; Park, J.H.; Cho, E.-J.; Yamabe, N. Heat-processed Panax ginseng and diabetic renal damage: Active components and action mechanism. *J. Ginseng Res.* **2013**, *37*, 379. [CrossRef]
27. Yayeh, T.; Jung, K.-H.; Jeong, H.Y.; Park, J.-H.; Song, Y.-B.; Kwak, Y.-S.; Kang, H.-S.; Cho, J.Y.; Oh, J.-W.; Kim, S.-K. Korean red ginseng saponin fraction downregulates proinflammatory mediators in LPS stimulated RAW264. 7 cells and protects mice against endotoxic shock. *J. Ginseng Res.* **2012**, *36*, 263. [CrossRef]
28. Yi, Y.-S. Roles of ginsenosides in inflammasome activation. *J. Ginseng Res.* **2019**, *43*, 172–178. [CrossRef]
29. Wang, F.; Park, J.-S.; Ma, Y.; Ma, H.; Lee, Y.-J.; Lee, G.-R.; Yoo, H.-S.; Hong, J.-T.; Roh, Y.-S. Ginseng Saponin Enriched in Rh1 and Rg2 Ameliorates Nonalcoholic Fatty Liver Disease by Inhibiting Inflammasome Activation. *Nutrients* **2021**, *13*, 856. [CrossRef]
30. Fan, X.; Zhang, C.; Niu, S.; Fan, B.; Gu, D.; Jiang, K.; Li, R.; Li, S. Ginsenoside Rg1 attenuates hepatic insulin resistance induced by high-fat and high-sugar by inhibiting inflammation. *Eur. J. Pharmacol.* **2019**, *854*, 247–255. [CrossRef]
31. Kim, K.; Nam, K.H.; Yi, S.A.; Park, J.W.; Han, J.-W.; Lee, J. Ginsenoside Rg3 Induces Browning of 3T3-L1 Adipocytes by Activating AMPK Signaling. *Nutrients* **2020**, *12*, 427. [CrossRef] [PubMed]
32. Ying, Y.; Zhang, Y.-l.; Ma, C.-j.; Li, M.-q.; Tang, C.-y.; Yang, Y.-f.; Zeng, J.-h.; Huang, X.-y.; Yi, J.; Wang, X.-m. Neuroprotective effects of ginsenoside Rg1 against hyperphosphorylated tau-induced diabetic retinal neurodegeneration via activation of IRS-1/Akt/GSK3β signaling. *J. Agric. Food Chem.* **2019**, *67*, 8348–8360. [CrossRef]
33. Kim, M.-Y.; Cho, J.Y. 20S-dihydroprotopanaxadiol, a ginsenoside derivative, boosts innate immune responses of monocytes and macrophages. *J. Ginseng Res.* **2013**, *37*, 293. [CrossRef] [PubMed]
34. Baek, N.-I.; Kim, D.S.; Lee, Y.H.; Park, J.D.; Lee, C.B.; Kim, S.I. Ginsenoside Rh4, a genuine dammarane glycoside from Korean red ginseng. *Planta Medica* **1996**, *62*, 86–87. [CrossRef] [PubMed]
35. Wang, P.; Du, X.; Xiong, M.; Cui, J.; Yang, Q.; Wang, W.; Chen, Y.; Zhang, T. Ginsenoside Rd attenuates breast cancer metastasis implicating derepressing microRNA-18a-regulated Smad2 expression. *Sci. Rep.* **2016**, *6*, 33709. [CrossRef]
36. Zhang, K.; Li, Y. Effects of ginsenoside compound K combined with cisplatin on the proliferation, apoptosis and epithelial mesenchymal transition in MCF-7 cells of human breast cancer. *Pharm. Biol.* **2016**, *54*, 561–568. [CrossRef]
37. Li, L.; Wang, Y.; Qi, B.; Yuan, D.; Dong, S.; Guo, D.; Zhang, C.; Yu, M. Suppression of PMA-induced tumor cell invasion and migration by ginsenoside Rg1 via the inhibition of NF-κB-dependent MMP-9 expression. *Oncol. Rep.* **2014**, *32*, 1779–1786. [CrossRef]
38. Kuang, J.; Li, L.; Guo, L.; Su, Y.; Wang, Y.; Xu, Y.; Wang, X.; Meng, S.; Lei, L.; Xu, L. RNF8 promotes epithelial-mesenchymal transition of breast cancer cells. *J. Exp. Clin. Cancer Res.* **2016**, *35*, 88. [CrossRef]
39. Wang, L.; Shi, Y.; Ju, P.; Liu, R.; Yeo, S.P.; Xia, Y.; Owlanj, H.; Feng, Z. Silencing of diphthamide synthesis 3 (Dph3) reduces metastasis of murine melanoma. *PLoS ONE* **2012**, *7*, e49988. [CrossRef]
40. Yu, L.; Gu, C.; Zhong, D.; Shi, L.; Kong, Y.; Zhou, Z.; Liu, S. Induction of autophagy counteracts the anticancer effect of cisplatin in human esophageal cancer cells with acquired drug resistance. *Cancer Lett.* **2014**, *355*, 34–45. [CrossRef]
41. Yang, L.; Qiu, J.; Xiao, Y.; Hu, X.; Liu, Q.; Chen, L.; Huang, W.; Li, X.; Li, L.; Zhang, J. AP-2β inhibits hepatocellular carcinoma invasion and metastasis through Slug and Snail to suppress epithelial-mesenchymal transition. *Theranostics* **2018**, *8*, 3707. [CrossRef] [PubMed]
42. Baek, S.-H.; Shin, B.-k.; Kim, N.J.; Chang, S.-Y.; Park, J.H. Protective effect of ginsenosides Rk3 and Rh4 on cisplatin-induced acute kidney injury in vitro and in vivo. *J. Ginseng Res.* **2017**, *41*, 233–239. [CrossRef] [PubMed]
43. Park, S.A.; Kim, E.H.; Na, H.K.; Surh, Y.J. KG-135 Inhibits COX-2 Expression by Blocking the Activation of JNK and AP-1 in Phorbol Ester–Stimulated Human Breast Epithelial Cells. *Ann. N. Y. Acad. Sci.* **2007**, *1095*, 545–553. [CrossRef] [PubMed]
44. Deng, J.; Liu, Y.; Duan, Z.; Zhu, C.; Hui, J.; Mi, Y.; Ma, P.; Ma, X.; Fan, D.; Yang, H. Protopanaxadiol and protopanaxatriol-type saponins ameliorate glucose and lipid metabolism in type 2 diabetes mellitus in high-fat diet/streptozocin-induced mice. *Front. Pharmacol.* **2017**, *8*, 506. [CrossRef]
45. Majc, B.; Sever, T.; Zarić, M.; Breznik, B.; Turk, B.; Lah, T.T. Epithelial-to-mesenchymal transition as the driver of changing carcinoma and glioblastoma microenvironment. *Biochim. Biophys. Acta (BBA)-Mol. Cell Res.* **2020**, *1867*, 118782. [CrossRef]
46. Yang, J.; Mani, S.A.; Donaher, J.L.; Ramaswamy, S.; Itzykson, R.A.; Come, C.; Savagner, P.; Gitelman, I.; Richardson, A.; Weinberg, R.A. Twist, a master regulator of morphogenesis, plays an essential role in tumor metastasis. *Cell* **2004**, *117*, 927–939. [CrossRef]
47. Kim, J.; Kong, J.; Chang, H.; Kim, H.; Kim, A. EGF induces epithelial-mesenchymal transition through phospho-Smad2/3-Snail signaling pathway in breast cancer cells. *Oncotarget* **2016**, *7*, 85021. [CrossRef]
48. Lu, M.; Liu, Z.; Li, B.; Wang, G.; Li, D.; Zhu, Y. The high expression of long non-coding RNA PANDAR indicates a poor prognosis for colorectal cancer and promotes metastasis by EMT pathway. *J. Cancer Res. Clin. Oncol.* **2017**, *143*, 71–81. [CrossRef]
49. Chen, W.; Zhao, K.; Miao, C.; Xu, A.; Zhang, J.; Zhu, J.; Su, S.; Wang, Z. Silencing Trim59 inhibits invasion/migration and epithelial-to-mesenchymal transition via TGF-β/Smad2/3 signaling pathway in bladder cancer cells. *OncoTargets Ther.* **2017**, *10*, 1503. [CrossRef]

50. Da, C.; Liu, Y.; Zhan, Y.; Liu, K.; Wang, R. Nobiletin inhibits epithelial-mesenchymal transition of human non-small cell lung cancer cells by antagonizing the TGF-β1/Smad3 signaling pathway. *Oncol. Rep.* **2016**, *35*, 2767–2774. [CrossRef]
51. Wu, W.; Ren, Z.; Li, P.; Yu, D.; Chen, J.; Huang, R.; Liu, H. Six1: A critical transcription factor in tumorigenesis. *Int. J. Cancer* **2015**, *136*, 1245–1253. [CrossRef] [PubMed]

nutrients

MDPI

Systematic Review

Effect of Lycopene Intake on the Fasting Blood Glucose Level: A Systematic Review with Meta-Analysis

Takuro Inoue [1,2,*], Kazutaka Yoshida [3], Erika Sasaki [3], Koichi Aizawa [4] and Hiroharu Kamioka [2]

1 Agricultural and Bio Resource Development Department, KAGOME Co., Ltd., 17 Nishitomiyama, Nasushiobara 329-2762, Japan
2 Department of Ecological Symbiotic Science, Tokyo University of Agriculture, 1-1-1 Sakuragaoka, Setagaya-ku, Tokyo 156-8502, Japan
3 Diet & Well-Being Research Department, KAGOME Co., Ltd., 17 Nishitomiyama, Nasushiobara 329-2762, Japan
4 Health & Wellness Business Department, KAGOME Co., Ltd., Nihonbashi-Hamacho F-Tower, 3-21-1 Nihonbashi-Hamacho, Chuo-ku, Tokyo 103-8461, Japan
* Correspondence: takuro_inoue@kagome.co.jp; Tel.: +81-287-36-2935

Abstract: Lycopene is a lipophilic unsaturated carotenoid exhibiting a strong singlet oxygen-quenching ability. Herein, we investigated the effect of lycopene intake on the fasting blood glucose (FBG) level by conducting a systematic review and meta-analyses. We searched 15 databases (from the earliest date to June 2022 for PubMed or to August or September 2018 for the other databases) and included human interventional studies that assessed the effects of oral lycopene intake on FBG levels of participants $\geq$ 18 years of age. Three authors independently selected applicable studies and then assessed the study quality. Data were pooled as standardized mean difference (SMD) and analyzed by the random-effects model. Heterogeneity was assessed by I^2 statistics. A meta-analysis including 11 trial arms (n = 750) revealed a tendency towards a significant decrease in FBG level with not-important heterogeneity [SMD = −0.15 (95% CI: −0.31, 0.00), p = 0.05, I^2 = 9%]. Subgroup meta-analysis including two studies (n = 152) in type 2 diabetes patients revealed significantly decreased FBG levels with not-important heterogeneity [SMD = −0.37 (95% CI: −0.69, −0.05), p = 0.02, I^2 = 0%]. Most studies meeting the eligibility criteria had a moderate risk of bias. The funnel plot for FBG suggested an absence of publication bias. In conclusion, this systematic review and meta-analyses suggested that lycopene intake exerted an FBG-decreasing effect.

Keywords: lycopene; fasting blood glucose; diabetes mellitus; systematic review; meta-analysis

Citation: Inoue, T.; Yoshida, K.; Sasaki, E.; Aizawa, K.; Kamioka, H. Effect of Lycopene Intake on the Fasting Blood Glucose Level: A Systematic Review with Meta-Analysis. *Nutrients* **2023**, *15*, 122. https://doi.org/10.3390/nu15010122

Academic Editors: Haixia Yang and Jianjun Deng

Received: 23 November 2022
Revised: 16 December 2022
Accepted: 23 December 2022
Published: 27 December 2022

1. Introduction

Type 2 diabetes (T2D) is a chronic metabolic disease characterized by high blood glucose levels, causing serious damage to the cardiovascular, renal, respiratory, as well as other systems [1]. The global diabetes prevalence is currently rising and has been estimated to be 10.9% (700 million people) by 2045 [2], while the global health expenditure for diabetes is expected to reach USD 776 billion in 2045 [3]. Therefore, preventing the initiation and progression of T2D is a critical global issue.

Glycemic control is one of the most important approaches to treating T2D [4], and the cornerstone of T2D treatment is a healthy lifestyle, which includes the adoption of a healthy diet, increased physical activity, maintenance of healthy body weight, and a smoking cessation plan [3]. Th glycemic index (GI) introduced in 1981 [5] and glycemic load (GL) based on GI [6] are well-known indices to estimate the postprandial blood glucose level rise, and some systematic reviews reported the usefulness of low GI diets and/or low GL diets for diabetes mellitus patients. Ojo et al. reported that low GI diets were more effective in controlling FBG and Hemoglobin A1c (HbA1c) than higher GI diets in T2D patients [7]. Chiavaroli et al. reported that low GI and/or GL diets reduced FBG and HbA1c in comparison with higher GI and/or GL diets in type 1 and type 2 diabetes patients [8]. Recently, the intake of antioxidant-rich foods is also recommended as part of the lifestyle [9] since oxidative stress is considered a major characteristic of the pathogenesis and development of T2D [10]. The total antioxidant capacity of the diet was suggested to play a role in reducing the risk of T2D in middle-aged women [11], and fasting blood glucose (FBG) levels were found to be significantly lower in T2D patients with a better oxidative balance score [12]. Some systematic reviews have also demonstrated that the intake of fruits and/or vegetables is inversely associated with the risk of T2D [13–15]. Since fruits and vegetables are rich in vitamins, flavonoids, and carotenoids, these antioxidants can be expected to play an important role in controlling the glycemic condition and/or providing a defense against T2D by reducing oxidative stress.

Lycopene is a lipophilic unsaturated carotenoid found in red-colored fruits and vegetables, including tomatoes, watermelon, red grapefruit, papaya, apricot, and guava. It exhibits a strong singlet oxygen-quenching ability, which is twice as high as that of beta-carotene and 100 times higher than that of alpha-tocopherol as a physical quenching rate [16]. Lycopene has been reported to exert beneficial effects in preventing many diseases, for example, cancer [17], cardiovascular diseases [18], diabetes mellitus [19], skin diseases [20], bone diseases [21], etc. Regarding antidiabetic effects, a higher dietary lycopene intake has been observed in non-T2D men compared to T2D men [22]. Increased plasma or serum lycopene levels have been reported to be associated with lower risks of T2D [23] and also better glycemic control (lower FBG) in T2D patients [24,25]. Recently, a review article summarized the lycopene effects on glycemic control in T2D. However, it was a narrative review, and a comprehensive literature search was not yet performed [19].

Several systematic reviews were conducted trying to evaluate the effect of tomato and/or its components on the FBG level. One systematic review with meta-analysis reported no significant difference in the FBG level between the tomato intervention and control groups [26]. Regarding lycopene, two systematic reviews did not address the effects on the FBG level because of inconsistency [27] or data unavailability of the included studies [28]. In these systematic reviews, possible limitations include the fact that only two to four electronic bibliographic databases were used for the literature search [26–28], and eligible studies were restricted to English or other Germanic/Romanic languages [27,28]. Therefore, there is a need for a more exhaustive literature search to find studies listed in other databases and/or reported in languages other than English and Germanic/Romanic. In this study, we performed a systematic review with meta-analysis to summarize the evidence relative to the effect of lycopene intake on the FBG level that was collected in human interventional trials, using more bibliographic databases and without restricting the study eligibility criteria by language.

2. Materials and Methods

2.1. Protocol and Registration

This systematic review and meta-analyses were conducted with the research question "Does oral lycopene intake improve FBG level, one of the most important biomarkers of diabetes mellitus, in participants $\geq$ 18 years of age?" and reported in accordance with the PRISMA 2009 statement [29]. The protocol was registered with PROSPERO, the International Prospective Register of Systematic Reviews, before starting the review (Registration number CRD42018104595).

2.2. Literature Search

We searched the following 15 databases: PubMed (MEDLINE), Web of Science (Core Collection, FSTA, Derwent Innovations Index, Medline, Zoological Record, BIOSIS Citation Index, Current Chemical Reactions, Data Citation Index, Current Contents Connect, Index Chemicus), Cochrane Library, SciFinder, Global Index Medicus, Western Pacific Region Index Medicus, CINAHL, Reaxys, Ichushi-Web, JDream III (JMEDPlus), AGRIS, University Hospital Medical Information Network-Clinical Trials Registry, International Clinical Trials

Registry Platform, ClinicalTrials.gov, and PROSPERO. We searched PubMed (MEDLINE) from the earliest date to June 2022 and the other databases from each earliest date to August or September 2018. The database search strategy is presented in Supplementary Table S1. We also searched the reference lists of the included relevant papers and the latest reviews.

2.3. Study Selection

Studies were selected similar to a previous report [30]. Based on the research question, we set the selection criteria as follows: (A) participants were $\geq$18 years of age; (B) intervention was the oral intake of test foods containing lycopene; (C) control was the oral intake of test foods not containing lycopene, oral intake of test foods containing lower levels of lycopene than intervention, or nothing; (D) outcome was FBG level; and (E) study design was a randomized controlled parallel trial (RCT-P), quasi-RCT-P, non-RCT-P, randomized controlled crossover trial (RCT-C), quasi-RCT-C, or non-RCT-C. For the studies retrieved from the literature search, three authors (K.Y., E.S., and K.A.) independently reviewed the titles and abstracts of them to identify studies that potentially met the selection criteria and reviewed the full text of selected studies to assess their eligibility. If there was any uncertainty or disagreement about eligibility, it was discussed with another author (T.I.) and resolved. Proceedings, grey literature, and unpublished studies were excluded. Eligibility was not restricted by language.

2.4. Data Extraction

We extracted data from the included studies similar to a previous report [30] for quality assessment and evidence synthesis, using a standardized, pre-piloted form. Extracted data included: citation, author, title, objective, setting, trial registration identifier, participant characteristics, intervention conditions, control conditions, outcomes, study design, randomization, blinding (participant, care provider, and outcome assessor), number of randomized participants, number of analyzed participants, results, conclusion, adverse events, cost of intervention, and funding. We extracted the mean and standard deviation (SD) values for FBG before and after the intervention. We also extracted the mean difference and SD between values before and after the intervention. When the SD values of the mean difference were not reported, we calculated them using the formula: square root $[(SD_{before})^2 + (SD_{after})^2 - 2R \times SD_{before} \times SD_{after}]$, assuming a correlation coefficient $R = 0.5$ [31]. The unit of FBG level was represented in mg/dL; if the values were originally published in mmol/L, they were converted to mg/dL by multiplying a factor of 18. Three authors (K.Y., E.S., and K.A.) independently extracted data, and any discrepancies were discussed with another author (T.I.) and resolved. If necessary, missing data were requested from the study authors via e-mail.

2.5. Quality Assessment

Three authors (K.Y., E.S., and K.A.) independently assessed the risk of bias in the reviewed studies, similar to a previous report [30], using a modified checklist of the Cochrane Handbook [32]. Briefly, the checklist included 13 items as follows: (A) randomization; (B) concealment of allocation; (C) blinding of participants; (D) blinding of care providers; (E) blinding of outcome assessors; (F) rate of drop-out; (G) intention-to-treat analysis; (H) selective outcome reporting; (I) similarity of baseline; (J) co-intervention; (K) compliance; (L) outcome assessment timing; and (M) other potential bias source. We scored each item as "there is no risk of bias" (+), "there is a risk of bias", or "unclear" (−). Based on the total number of (−), we evaluated each study as follows: 0–3, low risk of bias; 4–8, moderate risk of bias; 9–13, high risk of bias. If there were any uncertainties and disagreements on the risk of bias, they were discussed with another author (T.I.) and resolved.

2.6. Statistical Analysis

We conducted meta-analyses similar to a previous report [30] using Review Manager (RevMan Version 5.3 for Windows, The Cochrane Collaboration, Copenhagen, Denmark).

We used the mean difference and its SD values to evaluate the intervention effect. To include studies with more than two intervention groups in meta-analyses, we combined relevant intervention groups using a standard formula [31] to create single pair-wise comparisons. To compare effect sizes across studies, we used the standardized mean differences (SMDs) with 95% CI as a summary statistic. The random-effects model [33] was used to calculate the pooled SMDs, and a two-sided p-value < 0.05 was considered statistically significant. Heterogeneity was evaluated in the Forest plot [34] according to the I^2 statistics defined as follows: 0–40%, not-important; 30–60%, moderate; 50–90%, substantial; and 75–100%, considerable [31]. We also evaluated the inconsistency of evidence according to the I^2 statistics. We evaluated the publication bias by visual inspection of a funnel plot.

2.7. Subgroup Analysis

To investigate the factors that influenced the effect of lycopene on FBG and potential sources of heterogeneity, we planned in advance to conduct the subgroup analyses on the following viewpoints: (A) study design (focused on RCT-P); (B) types of test foods (supplement type or others); (C) length of the intervention period (shorter period or longer period); (D) lycopene level in test foods (lower level or higher level); and (E) participants' characteristics (healthy or others). Additionally, the following subgroup analysis was conducted post hoc: separating studies by participants' characteristics (diabetes mellitus participants and others).

3. Results

3.1. Search Results

The results of the study selection process are described in Figure 1. The literature searches (database search and additional sources search) yielded 3818 records, including duplicates, of which 15 studies met the eligibility criteria and were qualitatively assessed for risk of bias (Table 1) [35–49]. Ten of these 15 studies were included in the meta-analysis (Table 2) [36,37,39,41,42,44–47,49], and five studies were excluded due to either no post-intervention data in the control group or no data available (Table 3) [35,38,40,43,48]. Of the 15 studies, 13 were reported in English, and the others were reported either in Chinese (n = 1) [42] or Russian (n = 1) [36].

Table 1. Quality assessment of the selected studies.

Selected Studies	Sources of Risk of Bias *													Total Number of "−"
	(A)	(B)	(C)	(D)	(E)	(F)	(G)	(H)	(I)	(J)	(K)	(L)	(M)	
Upritchard 2000 [35]	+	+	−	−	+	−	−	−	−	+	−	+	+	7
Olfer'ev 2004 [36]	−	−	+	−	+	−	−	−	+	+	−	−	+	8
Engelhard 2006 [37]	−	−	+	−	+	+	−	−	+	+	−	−	+	7
Neyestani 2007 [38]	−	−	+	+	−	+	+	−	+	+	−	−	+	6
Devaraj 2008 [39]	−	−	+	−	+	−	−	−	+	+	−	−	+	8
Kim 2011 [40]	−	−	+	−	+	−	−	−	+	+	−	−	+	8
Thies 2012 [41]	−	−	−	−	+	−	−	+	+	+	−	−	+	8
Zeng 2013 [42]	−	−	−	−	+	−	−	−	+	+	−	+	+	8
Samaras 2014 [43]	−	−	−	−	+	+	+	−	−	+	−	−	−	9
Tsitsimpikou 2014 [44]	−	−	−	−	+	+	+	−	+	+	−	−	+	7
Deplanque 2016 [45]	−	+	+	+	+	+	−	−	+	+	−	+	−	5
Chernyshova 2019 [46]	−	−	−	−	+	+	+	−	−	+	−	−	+	8
Nishimura 2019 [47]	+	+	+	+	+	−	−	+	+	+	+	+	+	2
Wiese 2019 [48]	−	−	+	−	+	+	+	−	−	+	−	−	+	7
Takagi 2020 [49]	−	−	+	−	+	−	−	−	−	+	−	+	+	8

+, "there is no risk of bias"; −, "there is a risk of bias" or "unclear". * Sources of risk of bias corresponded to the following criteria: (A) randomization; (B) concealment of allocation; (C) blinding of participants; (D) blinding of care providers; (E) blinding of outcome assessors; (F) rate of drop-out; (G) intention-to-treat analysis; (H) selective outcome reporting; (I) similarity of baseline; (J) co-intervention; (K) compliance; (L) outcome assessment timing, and (M) other potential bias source. A larger number for "−" indicates a higher risk of bias.

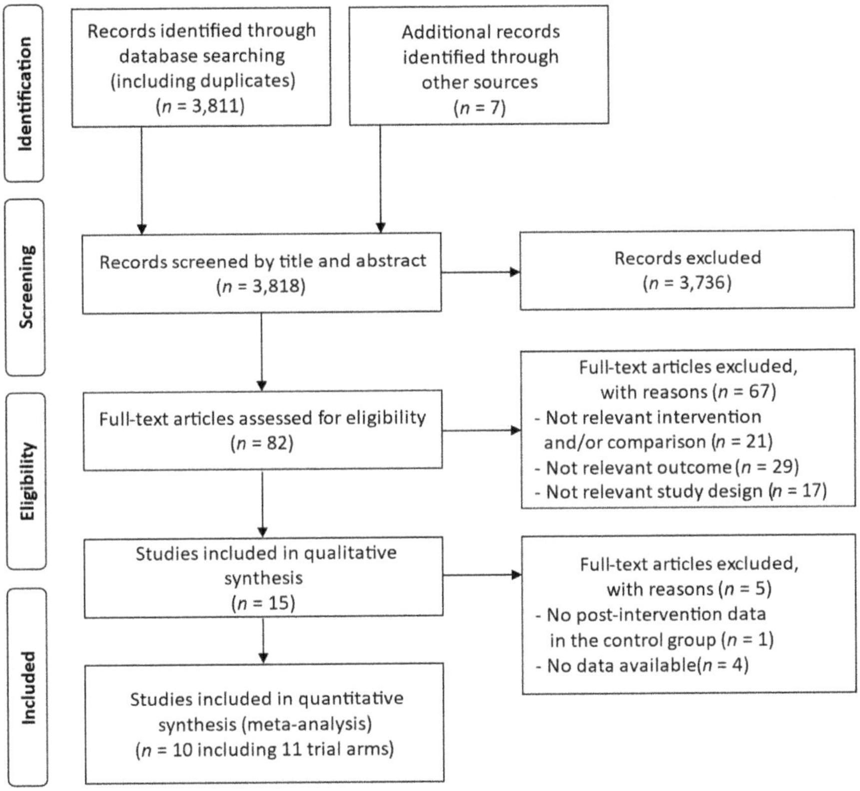

Figure 1. Flow diagram of the study selection process.

Table 2. Characteristics of the included studies.

Sample Size, Sex	Participant, Age (Years)	Sample Size, Sex	Intervention/Control	Lycopene Dosage Per Day	Duration (Intake Period)	Outcome (Blood Biomarkers of Glucose Metabolism)	Study Design
Olfer'ev 2004 [36], Russia	Type 2 diabetic postmenopausal women, mean age 66.4	I: 20 (all F) C: 15 (all F)	I: Tomato extract capsule (3/day) C: Placebo capsule (3/day)	I: 30 mg C: 0 mg	12 weeks	FBG	RCT-P
Engelhard 2006 [37], Israel	Grade-1 hypertensive subjects, age range 30–73	I: 31 (13 F/18 M) C: 31 (13 F/18 M)	I: Tomato extract capsule (1/day) C: Placebo capsule (1/day)	I: 15 mg C: 0 mg	8 weeks	FBG	non-RCT-C
Devaraj 2008 [39], USA	Healthy subjects, age range ≥40	I1: 21 (17 F/4 M) I2: 17 (13 F/4 M) I3: 21 (14 F/7 M) C: 18 (14 F/4 M)	I1: Lycopene capsule (1/day) I2: Lycopene capsule (1/day) I3: Lycopene capsule (1/day) C: Placebo capsule (1/day)	I1: 6.5 mg I2: 15 mg I3: 30 mg C: 0 mg	8 weeks	FBG	RCT-P
Thies 2012 [41], UK	Moderate overweight subjects, age range 40–65	I1: 68 (40 F/28 M) I2: 81 (46 F/35 M) C: 76 (46 F/30 M)	I1: Low-tomato diet and tomato extract capsule (1/day) I2: High-tomato diet C: Low-tomato diet	I1: 10 mg I2: 32–50 mg C: 0.3 mg	12 weeks	FBG Insulin HOMA-IR QUICKI	RCT-P
Zeng 2013 [42], China	Type 2 diabetic patients, age range ≥60	I: 58 C: 59	I: Lycopene capsule (4/day) C: Placebo capsule (4/day)	I: 30 mg C: 0 mg	6 months	FBG PBG HbA1c	RCT-P
Tsitsimpikou 2014 [44], Greece	Metabolic syndrome subjects, mean age 54.9	I: 15 (2 F/13 M) C: 12 (1 F/11 M)	I: Tomato juice C: None	I: NA C: 0 mg	2 months	FBG Insulin FIRI	non-RCT-P
Deplanque 2016 [45], France	Healthy subjects, mean age 34.9	I: 75 C: 70	I: Tomato extract capsule (1/day) C: Placebo capsule (1/day)	I: 15 mg C: 0 mg	2 weeks	FBG	RCT-P
Chernyshova 2019 [46], Russia	Healthy subjects, mean age 33.4	I: 10 (5 F/5 M) C: 10 (5 F/5 M)	I: Lycopene-enriched ice cream (50 g/day) C: Ice cream (50 g/day)	I: 7 mg C: 0 mg	4 weeks	FBG	RCT-C
Nishimura 2019 [47], Japan	Healthy subjects, age range 30–70	I: 49 C: 49	I: Semidried high-lycopene tomato (50 g/day) C: Semidried lycopene-free tomato (50 g/day)	I: 22.0–27.8 mg C: 0 mg	12 weeks	FBG HbA1c HOMA-IR	RCT-P
Takagi 2020 [49], Japan	Obese men, age range 40–65	I1: 7 (all M) I2: 5 (all M) C1: 7 (all M) C2: 5 (all M)	I1: Carrot and kale juice (high lycopene + high lutein) (200 mL/day) I2: Carrot and cabbage juice (high lycopene + low lutein) (200 mL/day) C1: Carrot and kale juice (low lycopene + high lutein) (200 mL/day) C2: Carrot and cabbage juice (low lycopene + low lutein) (200 mL/day)	I1: 7.56 mg I2: 8.6 mg C1: 0 mg C2: 0 mg	8 weeks	FBG	RCT-P

I, intervention group; C, control group; F, female; M, male; FBG, fasting blood glucose; HOMA-IR, homeostasis model assessment-insulin resistance; QUICKI, Quantitative Insulin-Sensitivity Check Index; PBG, postprandial blood glucose; HbA1c, Hemoglobin A1c; FIRI, Fasting Insulin Resistance Index; RCT-P, randomized controlled parallel trial; RCT-C, randomized controlled crossover trial; NA, not available.

Table 3. Characteristics of the excluded studies.

Sample Size, Sex	Participant, Age (Years)	Sample Size, Sex	Intervention/Control	Lycopene Dosage Per Day	Duration (Intake Period)	Outcome (Blood Biomarkers of Glucose Metabolism)	Study Design	Reason for Exclusion
Upritchard 2000 [35], New Zealand	Type 2 diabetic patients, mean age 59	I1: 15 (5 F/10 M) I2: 12 (6 F/6 M) I3: 12 (6 F/6 M) C: 13 (3 F/10 M)	I1: Tomato juice (500 mL/day) I2: Vitamin E (800 U/day) I3: Vitamin C (500 mg/day) C: Placebo capsule (1/day)	I1: NA I2: 0 mg I3: 0 mg C: 0 mg	4 weeks	FBG HbA1c	RCT-P	No data available
Neyestani 2007 [38], Iran	Type 2 diabetic patients, mean age 54	I: 16 (9 F/7 M) C: 19 (10 F/9 M)	I: Lycopene supplement C: Placebo supplement	I: 10 mg C: 0 mg	8 weeks	FBG HbA1c	non-RCT-P	No data available
Kim 2011 [40], Korea	Healthy subjects, mean age 34.3	I1: 41 (all M) I2: 37 (all M) C: 38 (all M)	I1: Tomato extract capsule (1/day) I2: Tomato extract capsule (1/day) C: Placebo capsule (1/day)	I1: 6 mg I2: 15 mg C: 0 mg	8 weeks	FBG	RCT-P	No data available
Samaras 2014 [43], Greece	Ultra-marathon runners, mean age 44.9	I1: 15 (2 F/13 M) I2: 16 (2 F/14 M) C: 12 (all M)	I1: Tomato juice I2: Protein bar C: Carbohydrate supplementation beverage	I1: NA I2: NA C: NA	2 months	FBG	non-RCT-P	No post-intervention data in the control group
Wiese 2019 [48], Russia	Moderate obese subjects, mean age 55	I1: 6 (3 F/3 M) I2: 6 (3 F/3 M) C1: 6 (3 F/3 M) C2: 6 (3 F/3 M)	I1: Lycopene-enriched dark chocolate (10 g/day) I2: Lycopene capsule (1/day) C1: Dark chocolate (10 g/day) C2: Lycopene capsule (1/day)	I1: 7 mg I2: 30 mg C1: 0 mg C2: 7 mg	1 month	FBG	RCT-P	No data available

-For abbreviations, see Table 2.

3.2. *Study Characteristics*

The study characteristics for the 15 studies that met the eligibility criteria are described in Tables 2 and 3. Study locations included Russia (n = 3), Greece (n = 2), Japan (n = 2), Israel (n = 1), USA (n = 1), UK (n = 1), China (n = 1), France (n = 1), New Zealand (n = 1), Iran (n = 1), and Korea (n = 1). Ten studies used an RCT-P [35,36,39–42,45,47–49], one used an RCT-C [46], three used a non-RCT-P [38,43,44], and one used a non-RCT-C design [37]. In six studies, participants were healthy subjects [39,40,45–47] and ultra-marathon runners [43]. In the other studies, participants were moderately overweight [41], obese [48,49], or had a metabolic syndrome [44], grade-1 hypertension [37], or type 2 diabetes [35,36,38,42]. For test foods, nine studies used either tomato extract capsules or lycopene supplements or synthetic lycopene capsules [36–42,45,48], three studies used tomato juice [35,43,44], one study used semi-dried tomatoes [47], one study used lycopene-enriched ice cream [46], one study used lycopene-enriched dark chocolate [48], one study used a tomato-rich diet [41], and one study used carrot and kale juice and carrot and cabbage juice [49]. The dosages of lycopene ranged from 6 to 50 mg/day, and intake periods ranged from 2 weeks to 6 months.

3.3. *Quality Assessment of the Studies*

Of the 15 studies considered, one study [47] was assessed as having a low risk of bias, while 13 studies [35–42,44–46,48,49] were assessed as having a moderate risk of bias, and one study [43] was assessed as having a high risk of bias (Table 1). In most studies, randomization, concealment of allocation, blinding of care provider, intention-to-treat analysis, selective outcome reporting, compliance, and outcome assessment timing were not reported in detail (Table 1).

3.4. *Meta-Analysis*

The meta-analysis, which included 10 studies (11 trial arms) with a total of 750 participants, revealed a tendency towards a significant decrease in FBG in the lycopene group compared with the control group [SMD = −0.15 (95% CI: −0.31, 0.00), p = 0.05], and heterogeneity was not important (I^2 = 9%) (Figure 2a).

Some of the subgroup meta-analyses revealed a significant decrease in FBG in the lycopene group compared to the control group. The RCT-P study design, which included 7 studies (8 trial arms) with a total of 641 participants, revealed a significantly decreased FBG with not-important heterogeneity [SMD = −0.21 (95% CI: −0.37, −0.06), p = 0.008, I^2 = 0%] (Figure 2b). The other study designs, which included 3 studies with a total of 109 participants, revealed no significant change in FBG with not-important heterogeneity [SMD = 0.20 (95% CI: −0.18, 0.58), p = 0.30, I^2 = 0%] (Figure 2c). The T2D participants, which included 2 studies with a total of 152 T2D participants, exhibited a significantly decreased FBG with not-important heterogeneity [SMD = −0.37 (95% CI: −0.69, −0.05), p = 0.02, I^2 = 0%] (Figure 2d). Participants other than T2D, which included 8 studies (9 trial arms) with a total of 598 participants, showed no significant change in FBG with not-important heterogeneity [SMD = −0.10 (95% CI: −0.27, 0.08), p = 0.28, I^2 = 8%] (Figure 2e). The other subgroup meta-analyses revealed no significant changes in FBG in the lycopene group compared with the control group (Supplementary Figure S1). Subgroup meta-analysis for lycopene levels in test foods could not be conducted due to a lack of lycopene dose information.

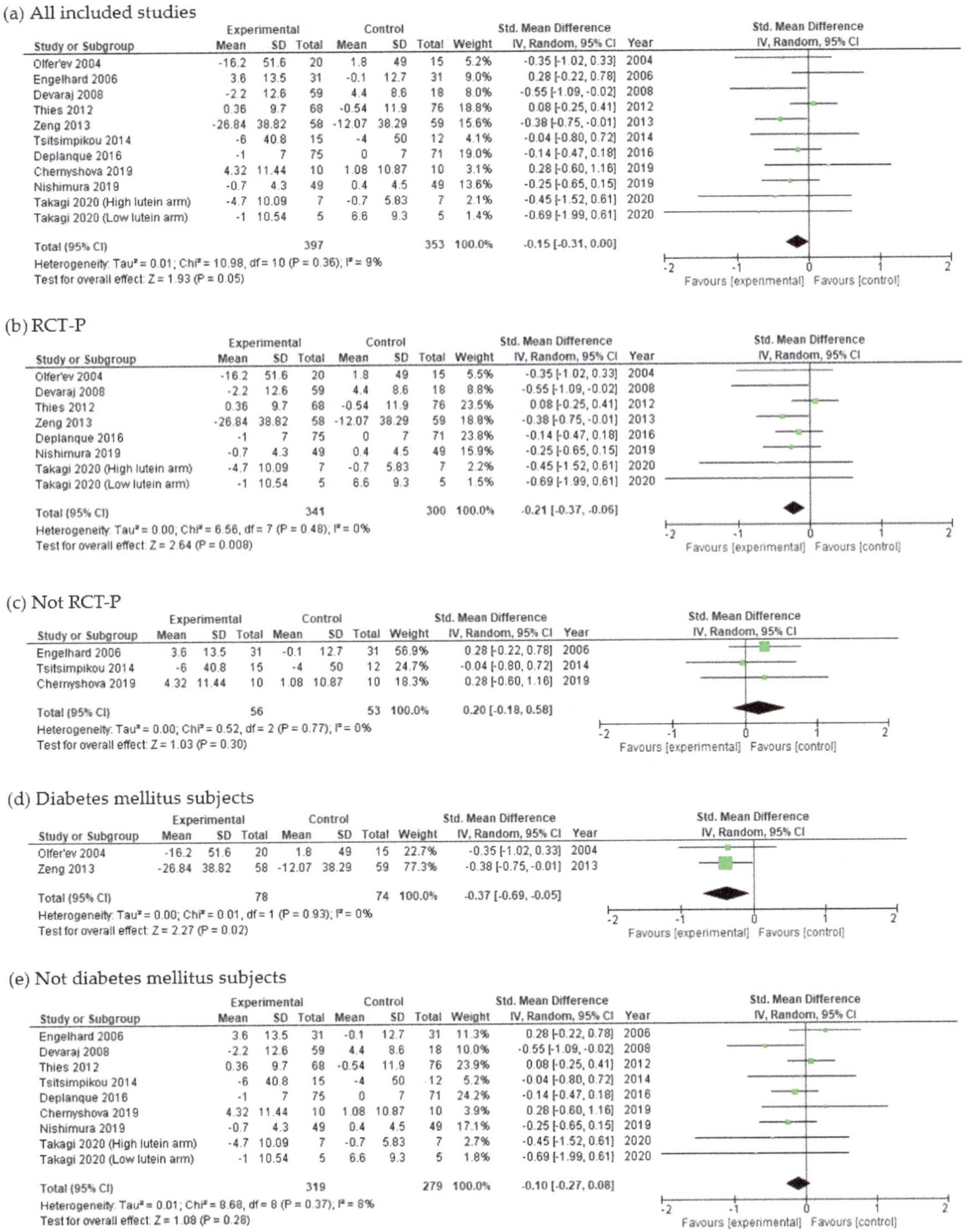

Figure 2. Meta-analysis and subgroup meta-analyses of the effects of lycopene on fasting blood

glucose (FBG): (**a**) all included studies (n = 11 trial arms), (**b**) RCT-P (n = 8 trial arms), (**c**) not RCT-P (n = 3 studies), (**d**) diabetes mellitus subjects (n = 2 studies), and (**e**) not diabetes mellitus subjects (n = 9 trial arms). The green squares represent the standardized mean difference in each study. The black diamonds represent the pooled effects in each meta-analysis. RCT-P, randomized controlled parallel trial; Std., standardized; SD, standard deviation; IV, inverse variance; CI, confidence interval [36,37,39,41,42,44–47,49].

3.5. Publication Bias

The funnel plot for FBG suggested an absence of publication bias (Figure 3). In subgroup analyses, assessments of publication bias were not meaningful because too few studies were included.

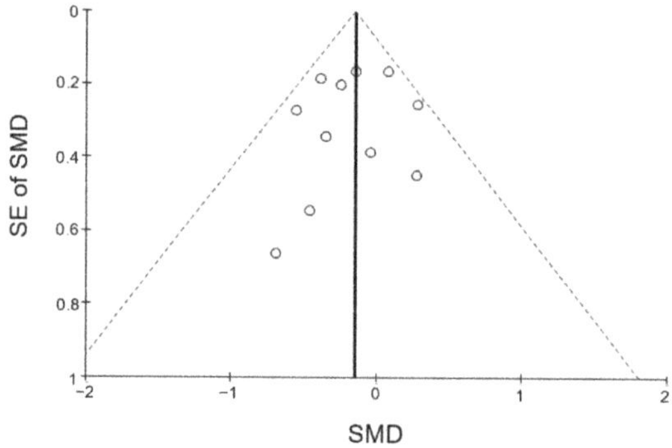

Figure 3. Funnel plot of studies included in the meta-analysis on fasting blood glucose (n = 11 trial arms). The vertical solid line represents the pooled effect size, and the dashed lines represent the 95% confidence interval. SMD, standardized mean difference; SE, standard error.

4. Discussion

4.1. Effects of Lycopene on FBG

The meta-analysis with all included studies revealed a tendency towards a significant decrease in the FBG level, and the subgroup meta-analysis restricted to T2D patients suggested a significant decrease in the FBG level in the lycopene group compared with that in the control group. To the best of our knowledge, this is the first systematic review supporting the indication that lycopene improves the FBG level in T2D patients.

Our literature search and study selection using 15 bibliographic databases without restricting the study eligibility criteria by language was extended to studies in Chinese and Russian, with qualities that were similar to those of studies in English. A Chinese study in T2D patients reported that lycopene intake (30 mg/day for 6 months) significantly improved the FBG level compared to the pre-intake level [42]. A Russian study in T2D patients disclosed no significant effect of lycopene intake (30 mg/day for 12 weeks) on the FBG level, although the decrease of the FBG level in the lycopene group was larger than that in the control group [36]. Both studies were not included in the previous systematic reviews [26–28]. Therefore, this systematic review could provide novel insights by including those studies. In this study, we excluded one study [50] included in the previous systematic review [26] because the control intervention did not meet our study selection criteria (comparison between polyphenol-enriched tomato juice and standard tomato juice). Although the difference in the included studies among these systematic reviews might be due to the differences in the literature search strategy and detailed study eligibility criteria, the methodological quality of each systematic review should be assessed using critical appraisal instruments, such as AMSTAR2 [51].

In this systematic review, several pre-set subgroup meta-analyses and one post hoc subgroup meta-analysis were conducted, and two subgroup meta-analyses restricted to RCT-P and T2D patients revealed a significantly decreased FBG with not-important heterogeneities. A randomized controlled trial is generally the highest quality study design and the gold standard in interventional clinical trials [52], while non-randomized controlled trials might have a potentially higher risk of bias. In this study, the subgroup meta-analysis restricted to RCT-P (excluding not-RCT-P) disclosed a significantly decreased FBG. Although the meta-analysis with all included studies revealed only a tendency towards a significant decrease in the FBG level, a significant decrease might be confirmed by reporting more high-quality primary studies.

For studies included in the meta-analyses, the average FBG level in the pre-intake period in T2D patients was much higher than that in not-T2D patients (161.7 mg/dL vs. 93.4 mg/dL), which was assumed to be the reason why the effect of lycopene was exerted only in T2D patients. In food studies, it is often difficult to identify appropriate placebo foods when using complex test foods. However, the studies in T2D patients included in the meta-analyses used supplement-type test foods such as tomato extract capsules [36] and lycopene capsules [42]. Therefore, the effects of ingredients other than lycopene can be expected to be small, although they could not be completely ruled out. The effects of medications should be considered when patients participate in interventional clinical studies. In T2D patients studies included in the meta-analyses [36,42], the T2D participants had received medications. However, the effects of medications seemed to be small because the drugs prescribed prior to the study were continued throughout the study in both the intervention and the control group [36], or there was no significant difference in medication status between the two groups [42].

Therefore, lycopene intake can be expected to have a positive effect on FBG levels even under medication, although limited to individuals with high FBG levels. However, two studies [35,38] with T2D participants were excluded from the meta-analyses due to the data unavailability for this study, and additional primary studies are needed to clarify the FBG-improving effect of lycopene.

4.2. Effect Size and Possible Mechanisms

In both two studies with T2D patients, the dosage of lycopene was 30 mg/day, and a meta-analysis restricted to these studies showed that the SMD for FBG was −0.37 (95% CI: −0.69, −0.05), that is, the mean difference for FBG was −15.25 (95% CI: −28.15, −2.35) mg/dL, which was about 9.4% of the FBG level at the pre-intake point. Gao et al. conducted a cross-sectional study to determine whether increased carotenoids intake was associated with a reduced risk of gestational diabetes mellitus and reported an inverse association between lycopene intake and FBG; each 1 mg increase in lycopene intake was associated with a 0.09 mg/dL decrease in FBG [53]. However, it is difficult to simply compare these results due to differences in participants and study designs. Some foods other than tomatoes have been reported to have FBG-improving effects in T2D patients. Shabani et al. reported that garlic intake improved FBG levels, and its effect size was −10.90 mg/dL (95% CI: −16.40, −5.40) in a systematic review [54]. Suksomboon et al. reported that *Aloe vera* intake also improved FBG levels, and its effect size was −21.06 mg/dL (95% CI: −42.3, 0.00) in a systematic review [55]. Therefore, lycopene intake can be expected to have an effect similar to those of garlic and *Aloe vera* in improving FBG levels and to provide better FBG control in combination with these foods. In addition, if the FBG-improving effect size of each food and their combination could be clearly shown, it would be an easy-to-understand guideline for T2D patients to try them in their diet and further expected to be reflected in the standards of diet treatment in diabetes.

Previous studies have suggested some possible mechanisms by which lycopene affects FBG levels. Hashimoto et al. examined the effect of lycopene on glucose tolerance in normal rats and found that a lycopene-rich tomato intake improved glucose tolerance via an increase in plasma leptin levels that enhanced insulin sensitivity [56]. Some reports using diabetic model rats indicated the importance of the antioxidative effect of lycopene. Yin et al. reported that lycopene intervention decreased the FBG level in T2D model rats, and lycopene might improve glucose metabolism by reducing oxidized low-density lipoprotein cholesterol [57]. Zheng et al. described that lycopene intervention decreased the FBG level in a dose-dependent manner in T2D rats and concluded that lycopene protected against diabetic progression and prevented further complications of diabetic rats through ameliorating oxidative stress and inflammation as well as improving the systemic antioxidative capacity [58]. It is possible that lycopene affects FBG levels through multiple pathways, and further evidence, especially in human studies, needs to be accumulated.

4.3. Strengths and Limitations

The strengths of this study include an extensive literature search that used many databases not restricted by language. Although the latest additional literature search was conducted only in PubMed (MEDLINE), a basic search in 15 databases and additional search enabled us to find studies reported in languages other than English.

This study is not without limitations. Most included studies had a moderate risk of bias. Since any missing data could not be obtained from the study authors, we excluded studies with no data available and did not consider them in the meta-analyses. Although we imputed partly missing data according to the Cochrane Handbook [31], this procedure might have created a risk of bias. Since the protocol for this study was set in August 2018, this systematic review corresponded only partially to the latest PRISMA 2020 [59] and PRISMA-S [60] guidelines. In addition, the definite criteria to assess the strength of evidence were not set in advance or assessed in this study. The strength of evidence should be assessed using an appropriate instrument, for example, the GRADE approach [61]. In this study, we focused only on the FBG levels. However, other blood biomarkers (for example, insulin, homeostasis model of risk assessment-insulin resistance (HOMA-IR), Hemoglobin A1c and C peptide, etc.) should be evaluated to better understand the effects of lycopene on glucose metabolism.

5. Conclusions

This systematic review and meta-analysis adopted a more exhaustive literature search than the previous systematic reviews, that is, using 15 databases without restricting the study eligibility criteria by language, and highlighted an FBG-decreasing effect of lycopene intake, especially in T2D patients. In order to clarify this effect, additional clinical trials in T2D patients are needed, not only to evaluate the effect of lycopene on FBG but also on other glucose metabolism markers.

Supplementary Materials: The following supporting information can be downloaded at: https://www.mdpi.com/article/10.3390/nu15010122/s1, Table S1: Search strategy for electronic bibliographic databases; Figure S1: Subgroup meta-analyses of the effects of lycopene on fasting blood glucose.

Author Contributions: Conceptualization, T.I., K.Y., E.S., K.A. and H.K.; methodology, T.I. and H.K.; formal analysis, T.I.; investigation, T.I., K.Y., E.S. and K.A.; resources, T.I., K.Y., E.S. and K.A.; data curation, T.I., K.Y., E.S. and K.A.; writing—original draft preparation, T.I.; writing—review and editing, T.I., K.Y., E.S., K.A. and H.K.; visualization, T.I., K.Y., E.S. and K.A.; supervision, H.K.; project administration, T.I.; funding acquisition, T.I., K.Y., E.S. and K.A. All authors have read and agreed to the published version of the manuscript.

Funding: This research was funded by KAGOME Co., Ltd. and received no external funding.

Institutional Review Board Statement: Not applicable.

Informed Consent Statement: Not applicable.

Data Availability Statement: The protocol of this study was registered with PROSPERO (Registration number CRD42018104595). More detailed information and the datasets generated and/or analyzed in this systematic review are available from the corresponding author upon reasonable request.

Acknowledgments: We profoundly acknowledge Mari Makishi for her literature search.

Conflicts of Interest: T.I., K.Y., E.S., and K.A. are employees of KAGOME Co., Ltd. and hold stocks of the company. H.K. was previously compensated for supervising other systematic reviews of KAGOME Co., Ltd. submitted as scientific bases for the system of Food with Function Claims (FFC) in Japan. However, regarding this study, which is not for the submission of the FFC, H.K. did not receive any reward from KAGOME Co., Ltd.

References

1. Bassi-Dibai, D.; Santos-de-Araújo, A.D.; Dibai-Filho, A.V.; de Azevedo, L.F.S.; Goulart, C.D.L.; Luz, G.C.P.; Burke, P.R.; Garcia-Araújo, A.S.; Borghi-Silva, A. Rehabilitation of individuals with diabetes mellitus: Focus on diabetic myopathy. *Front. Endocrinol.* **2022**, *13*, 869921. [CrossRef] [PubMed]
2. Saeedi, P.; Petersohn, I.; Salpea, P.; Malanda, B.; Karuranga, S.; Unwin, N.; Colagiuri, S.; Guariguata, L.; Motala, A.A.; Ogurtsova, K.; et al. Global and regional diabetes prevalence estimates for 2019 and projections for 2030 and 2045: Results from the International Diabetes Federation Diabetes Atlas, 9th edition. *Diabetes Res. Clin. Pract.* **2019**, *157*, 107843. [CrossRef] [PubMed]
3. Karuranga, S.; Fernandes, J.R.; Huang, Y.; Malanda, B. (Eds.) *IDF Diabetes Atlas*, 8th ed.; International Diabetes Federation: Brussels, Belgium, 2017. Available online: https://diabetesatlas.org/atlas/eighth-edition/ (accessed on 12 November 2022).
4. Turner, R.C.; Cull, C.A.; Frighi, V.; Holman, R.R. Glycemic control with diet, sulfonylurea, metformin, or insulin in patients with type 2 diabetes mellitus: Progressive requirement for multiple therapies (UKPDS 49). UK Prospective Diabetes Study (UKPDS) Group. *JAMA* **1999**, *281*, 2005–2012. [CrossRef] [PubMed]
5. Jenkins, D.J.; Wolever, T.M.; Taylor, R.H.; Barker, H.; Fielden, H.; Baldwin, J.M.; Bowling, A.C.; Newman, H.C.; Jenkins, A.L.; Goff, D.V. Glycemic index of foods: A physiological basis for carbohydrate exchange. *Am. J. Clin. Nutr.* **1981**, *34*, 362–366. [CrossRef]
6. Salmerón, J.; Ascherio, A.; Rimm, E.B.; Colditz, G.A.; Spiegelman, D.; Jenkins, D.J.; Stampfer, M.J.; Wing, A.L.; Willett, W.C. Dietary fiber, glycemic load, and risk of NIDDM in men. *Diabetes Care* **1997**, *20*, 545–550. [CrossRef]
7. Ojo, O.; Ojo, O.O.; Adebowale, F.; Wang, X. The effect of dietary glycaemic index on glycaemia in patients with type 2 diabetes: A systematic review and meta-analysis of randomized controlled trials. *Nutrients* **2018**, *10*, 373. [CrossRef]
8. Chiavaroli, L.; Lee, D.; Ahmed, A.; Cheung, A.; Khan, T.A.; Mejia, S.B.; Mirrahimi, A.; Jenkins, D.J.A.; Livesey, G.; Wolever, T.M.S.; et al. Effect of low glycaemic index or load dietary patterns on glycaemic control and cardiometabolic risk factors in diabetes: Systematic review and meta-analysis of randomised controlled trials. *BMJ* **2021**, *374*, n1651. [CrossRef]
9. Abdali, D.; Samson, S.E.; Grover, A.K. How effective are antioxidant supplements in obesity and diabetes? *Med. Princ. Pract.* **2015**, *24*, 201–215. [CrossRef]
10. Rehman, K.; Akash, M.S.H. Mechanism of generation of oxidative stress and pathophysiology of type 2 diabetes mellitus: How are they interlinked? *J. Cell. Biochem.* **2017**, *118*, 3577–3585. [CrossRef]
11. Mancini, F.R.; Affret, A.; Dow, C.; Balkau, B.; Bonnet, F.; Boutron-Ruault, M.C.; Fagherazzi, G. Dietary antioxidant capacity and risk of type 2 diabetes in the large prospective E3N-EPIC cohort. *Diabetologia* **2018**, *61*, 308–316. [CrossRef]
12. Golmohammadi, M.; Ayremlou, P.; Zarrin, R. Higher oxidative balance score is associated with better glycemic control among Iranian adults with type-2 diabetes. *Int. J. Vitam. Nutr. Res.* **2021**, *91*, 31–39. [CrossRef] [PubMed]
13. Li, M.; Fan, Y.; Zhang, X.; Hou, W.; Tang, Z. Fruit and vegetable intake and risk of type 2 diabetes mellitus: Meta-analysis of prospective cohort studies. *BMJ Open* **2014**, *4*, e005497. [CrossRef] [PubMed]
14. Wang, P.; Fang, J.; Gao, Z.; Zhang, C.; Xie, S. Higher intake of fruits, vegetables or their fiber reduces the risk of type 2 diabetes: A meta-analysis. *J. Diabetes Investig.* **2016**, *7*, 56–69. [CrossRef] [PubMed]
15. Schwingshackl, L.; Hoffmann, G.; Lampousi, A.M.; Knüppel, S.; Iqbal, K.; Schwedhelm, C.; Bechthold, A.; Schlesinger, S.; Boeing, H. Food groups and risk of type 2 diabetes mellitus: A systematic review and meta-analysis of prospective studies. *Eur. J. Epidemiol.* **2017**, *32*, 363–375. [CrossRef] [PubMed]
16. Di Mascio, P.; Kaiser, S.; Sies, H. Lycopene as the most efficient biological carotenoid singlet oxygen quencher. *Arch. Biochem. Biophys.* **1989**, *274*, 532–538. [CrossRef] [PubMed]
17. Kapała, A.; Szlendak, M.; Motacka, E. The anti-cancer activity of lycopene: A systematic review of human and animal studies. *Nutrients* **2022**, *14*, 5152. [CrossRef] [PubMed]
18. Przybylska, S.; Tokarczyk, G. Lycopene in the prevention of cardiovascular diseases. *Int. J. Mol. Sci.* **2022**, *23*, 1957. [CrossRef] [PubMed]
19. Leh, H.E.; Lee, L.K. Lycopene: A potent antioxidant for the amelioration of type II diabetes mellitus. *Molecules* **2022**, *27*, 2335. [CrossRef]
20. Stahl, W.; Heinrich, U.; Aust, O.; Tronnier, H.; Sies, H. Lycopene-rich products and dietary photoprotection. *Photochem. Photobiol. Sci.* **2006**, *5*, 238–242. [CrossRef]
21. Costa-Rodrigues, J.; Fernandes, M.H.; Pinho, O.; Monteiro, P.R.R. Modulation of human osteoclastogenesis and osteoblastogenesis by lycopene. *J. Nutr. Biochem.* **2018**, *57*, 26–34. [CrossRef]
22. Quansah, D.Y.; Ha, K.; Jun, S.; Kim, S.A.; Shin, S.; Wie, G.A.; Joung, H. Associations of dietary antioxidants and risk of type 2 diabetes: Data from the 2007-2012 Korea national health and nutrition examination survey. *Molecules* **2017**, *22*, 1664. [CrossRef]
23. Jiang, Y.; Sun, Z.; Tong, W.; Yang, K.; Guo, K.; Liu, G.; Pan, A. Dietary intake and circulating concentrations of carotenoids and risk of type 2 diabetes: A dose-response meta-analysis of prospective observational studies. *Adv. Nutr.* **2021**, *12*, 1723–1733. [CrossRef]

24. Leh, H.E.; Sopian, M.M.; Bakar, M.H.A.; Lee, L.K. The role of lycopene for the amelioration of glycaemic status and peripheral antioxidant capacity among the type II diabetes mellitus patients: A case–control study. *Ann. Med.* **2021**, *53*, 1059–1065. [CrossRef]
25. Mummidi, S.; Farook, V.S.; Reddivari, L.; Hernandez-Ruiz, J.; Diaz-Badillo, A.; Fowler, S.P.; Resendez, R.G.; Akhtar, F.; Lehman, D.M.; Jenkinson, C.P.; et al. Serum carotenoids and pediatric metabolic index predict insulin sensitivity in Mexican American children. *Sci. Rep.* **2021**, *11*, 871. [CrossRef]
26. Li, H.; Chen, A.; Zhao, L.; Bhagavathula, A.S.; Amirthalingam, P.; Rahmani, J.; Salehisahlabadi, A.; Abdulazeem, H.M.; Adebayo, O.; Yin, X. Effect of tomato consumption on fasting blood glucose and lipid profiles: A systematic review and meta-analysis of randomized controlled trials. *Phytother. Res.* **2020**, *34*, 1956–1965. [CrossRef]
27. Tierney, A.C.; Rumble, C.E.; Billings, L.M.; George, E.S. Effect of dietary and supplemental lycopene on cardiovascular risk factors: A systematic review and meta-analysis. *Adv. Nutr.* **2020**, *11*, 1453–1488. [CrossRef]
28. Valero, M.A.; Vidal, A.; Burgos, R.; Calvo, F.L.; Martínez, C.; Luengo, L.M.; Cuerda, C. Meta-analysis on the role of lycopene in type 2 diabetes mellitus. *Nutr. Hosp.* **2011**, *26*, 1236–1241. [CrossRef]
29. Moher, D.; Liberati, A.; Tetzlaff, J.; Altman, D.G.; PRISMA Group. Preferred reporting items for systematic reviews and meta-analyses: The PRISMA Statement. *PLoS Med.* **2009**, *6*, e1000097. [CrossRef]
30. Inoue, T.; Yoshida, K.; Sasaki, E.; Aizawa, K.; Kamioka, H. Effects of lycopene intake on HDL-cholesterol and triglyceride levels: A systematic review with meta-analysis. *J. Food Sci.* **2021**, *86*, 3285–3302. [CrossRef]
31. Higgins, J.P.T.; Green, S. (Eds.) *Cochrane Handbook for Systematic Reviews of Interventions, Version 5.1.0*; The Cochrane Collaboration: London, UK, 2021. Available online: https://training.cochrane.org/handbook/archive/v5.1/ (accessed on 12 November 2022).
32. Furlan, A.D.; Malmivaara, A.; Chou, R.; Maher, C.G.; Deyo, R.A.; Schoene, M.; Bronfort, G.; van Tulder, M.W.; Editorial Board of the Cochrane Back, Neck Group. 2015 updated method guideline for systematic reviews in the cochrane back and neck group. *Spine* **2015**, *40*, 1660–1673. [CrossRef]
33. DerSimonian, R.; Laird, N. Meta-analysis in clinical trials. *Control. Clin. Trials* **1986**, *7*, 177–188. [CrossRef] [PubMed]
34. Higgins, J.P.T.; Thompson, S.G. Quantifying heterogeneity in a meta-analysis. *Stat. Med.* **2002**, *21*, 1539–1558. [CrossRef] [PubMed]
35. Upritchard, J.E.; Sutherland, W.H.F.; Mann, J.I. Effect of supplementation with tomato juice, vitamin E, and vitamin C on LDL oxidation and products of inflammatory activity in type 2 diabetes. *Diabetes Care* **2000**, *23*, 733–738. [CrossRef]
36. Olfer'ev, A.M.; Il'ina, M.V.; Berzak, N.V.; Stetsenko, A.V.; Olfer'ev, M.A.; Chudakova, I.A.; Kapitanov, A.B.; Shamarin, V.M. Effect of lycopene on blood lipoproteids in women with type 2 diabetes mellitus in postmenopause. *Vopr. Pitan.* **2004**, *73*, 19–23. [PubMed]
37. Engelhard, Y.N.; Gazer, B.; Paran, E. Natural antioxidants from tomato extract reduce blood pressure in patients with grade-1 hypertension: A double-blind, placebo-controlled pilot study. *Am. Heart J.* **2006**, *151*, 100. [CrossRef] [PubMed]
38. Neyestani, T.R.; Shariat-Zadeh, N.; Gharavi, A.; Kalayi, A.; Khalaji, H. The opposite associations of lycopene and body fat mass with humoral immunity in type 2 diabetes mellitus: A possible role in atherogenesis. *Iran. J. Allergy Asthma Immunol.* **2007**, *6*, 79–87.
39. Devaraj, S.; Mathur, S.; Basu, A.; Aung, H.H.; Vasu, V.T.; Meyers, S.; Jialal, I. A dose-response study on the effects of purified lycopene supplementation on biomarkers of oxidative stress. *J. Am. Coll. Nutr.* **2008**, *27*, 267–273. [CrossRef]
40. Kim, J.Y.; Paik, J.K.; Kim, O.Y.; Park, H.W.; Lee, J.H.; Jang, Y.; Lee, J.H. Effects of lycopene supplementation on oxidative stress and markers of endothelial function in healthy men. *Atherosclerosis* **2011**, *215*, 189–195. [CrossRef]
41. Thies, F.; Masson, L.F.; Rudd, A.; Vaughan, N.; Tsang, C.; Brittenden, J.; Simpson, W.G.; Duthie, S.; Horgan, G.W.; Duthie, G. Effect of a tomato-rich diet on markers of cardiovascular disease risk in moderately overweight, disease-free, middle-aged adults: A randomized controlled trial. *Am. J. Clin. Nutr.* **2012**, *95*, 1013–1022. [CrossRef]
42. Zeng, Y.; Zhao, H.; Li, J.; Liu, Y.; Wang, T.; Mu, G. Effects of lycopene on cytokines and carotid intima-media thickness in type 2 diabetic patients. *Acta Nutr. Sin.* **2013**, *35*, 268–272.
43. Samaras, A.; Tsarouhas, K.; Paschalidis, E.; Giamouzis, G.; Triposkiadis, F.; Tsitsimpikou, C.; Becker, A.T.; Goutzourelas, N.; Kouretas, D. Effect of a special carbohydrate-protein bar and tomato juice supplementation on oxidative stress markers and vascular endothelial dynamics in ultra-marathon runners. *Food Chem. Toxicol.* **2014**, *69*, 231–236. [CrossRef] [PubMed]
44. Tsitsimpikou, C.; Tsarouhas, K.; Kioukia-Fougia, N.; Skondra, C.; Fragkiadaki, P.; Papalexis, P.; Stamatopoulos, P.; Kaplanis, I.; Hayes, A.W.; Tsatsakis, A.; et al. Dietary supplementation with tomato-juice in patients with metabolic syndrome: A suggestion to alleviate detrimental clinical factors. *Food Chem. Toxicol.* **2014**, *74*, 9–13. [CrossRef] [PubMed]
45. Deplanque, X.; Muscente-Paque, D.; Chappuis, E. Proprietary tomato extract improves metabolic response to high-fat meal in healthy normal weight subjects. *Food Nutr. Res.* **2016**, *60*, 32537. [CrossRef] [PubMed]
46. Chernyshova, M.P.; Pristenskiy, D.V.; Lozbiakova, M.V.; Chalyk, N.E.; Bandaletova, T.Y.; Petyaev, I.M. Systemic and skin-targeting beneficial effects of lycopene-enriched ice cream: A pilot study. *J. Dairy Sci.* **2019**, *102*, 14–25. [CrossRef]
47. Nishimura, M.; Tominaga, N.; Ishikawa-Takano, Y.; Maeda-Yamamoto, M.; Nishihira, J. Effect of 12-week daily intake of the high-lycopene tomato (*Solanum Lycopersicum*), a variety named "PR-7", on lipid metabolism: A randomized, double-blind, placebo-controlled, parallel-group study. *Nutrients* **2019**, *11*, 1177. [CrossRef]
48. Wiese, M.; Bashmakov, Y.; Chalyk, N.; Nielsen, D.S.; Krych, Ł.; Kot, W.; Klochkov, V.; Pristensky, D.; Bandaletova, T.; Chernyshova, M.; et al. Prebiotic effect of lycopene and dark chocolate on gut microbiome with systemic changes in liver metabolism, skeletal muscles and skin in moderately obese persons. *BioMed Res. Int.* **2019**, *2019*, 4625279. [CrossRef]

49. Takagi, T.; Hayashi, R.; Nakai, Y.; Okada, S.; Miyashita, R.; Yamada, M.; Mihara, Y.; Mizushima, K.; Morita, M.; Uchiyama, K.; et al. Dietary intake of carotenoid-rich vegetables reduces visceral adiposity in obese Japanese men—A randomized, double-blind trial. *Nutrients* **2020**, *12*, 2342. [CrossRef]
50. Michaličková, D.; Belović, M.; Ilić, N.; Kotur-Stevuljević, J.; Slanař, O.; Šobajić, S. Comparison of polyphenol-enriched tomato juice and standard tomato juice for cardiovascular benefits in subjects with stage 1 hypertension: A randomized controlled study. *Plant Foods Hum. Nutr.* **2019**, *74*, 122–127. [CrossRef]
51. Shea, B.J.; Reeves, B.C.; Wells, G.; Thuku, M.; Hamel, C.; Moran, J.; Moher, D.; Tugwell, P.; Welch, V.; Kristjansson, E.; et al. AMSTAR 2: A critical appraisal tool for systematic reviews that include randomised or non-randomised studies of healthcare interventions, or both. *BMJ* **2017**, *358*, j4008. [CrossRef]
52. Kapoor, M.C. Types of studies and research design. *Indian J. Anaesth.* **2016**, *60*, 626–630. [CrossRef]
53. Gao, Q.; Zhong, C.; Zhou, X.; Chen, R.; Xiong, T.; Hong, M.; Li, Q.; Kong, M.; Han, W.; Sun, G.; et al. The association between intake of dietary lycopene and other carotenoids and gestational diabetes mellitus risk during mid-trimester: A cross-sectional study. *Br. J. Nutr.* **2019**, *121*, 1405–1412. [CrossRef]
54. Shabani, E.; Sayemiri, K.; Mohammadpour, M. The effect of garlic on lipid profile and glucose parameters in diabetic patients: A systematic review and meta-analysis. *Prim. Care Diabetes* **2019**, *13*, 28–42. [CrossRef] [PubMed]
55. Suksomboon, N.; Poolsup, N.; Punthanitisarn, S. Effect of *Aloe vera* on glycaemic control in prediabetes and type 2 diabetes: A systematic review and meta-analysis. *J. Clin. Pharm. Ther.* **2016**, *41*, 180–188. [CrossRef] [PubMed]
56. Hashimoto, N.; Tominaga, N.; Wakagi, M.; Ishikawa-Takano, Y. Consumption of lycopene-rich tomatoes improved glucose homeostasis in rats via an increase in leptin levels. *J. Nat. Med.* **2020**, *74*, 252–256. [CrossRef] [PubMed]
57. Yin, Y.; Zheng, Z.; Jiang, Z. Effects of lycopene on metabolism of glycolipid in type 2 diabetic rats. *Biomed. Pharmacother.* **2019**, *109*, 2070–2077. [CrossRef] [PubMed]
58. Zheng, Z.; Yin, Y.; Lu, R.; Jiang, Z. Lycopene ameliorated oxidative stress and inflammation in type 2 diabetic rats. *J. Food Sci.* **2019**, *84*, 1194–1200. [CrossRef]
59. Page, M.J.; McKenzie, J.E.; Bossuyt, P.M.; Boutron, I.; Hoffmann, T.C.; Mulrow, C.D.; Shamseer, L.; Tetzlaff, J.M.; Akl, E.A.; Brennan, S.E.; et al. The PRISMA 2020 statement: An updated guideline for reporting systematic reviews. *BMJ* **2021**, *372*, n71. [CrossRef]
60. Rethlefsen, M.L.; Kirtley, S.; Waffenschmidt, S.; Ayala, A.P.; Moher, D.; Page, M.J.; Koffel, J.B.; PRISMA-S Group. PRISMA-S: An extension to the PRISMA statement for reporting literature searches in systematic reviews. *Syst. Rev.* **2021**, *10*, 39. [CrossRef]
61. Guyatt, G.; Oxman, Q.; Vist, G.; Kunz, R.; Falck-Ytter, Y.; Alonso-Coello, P.; Schünemann, H.; GRADE Working Group. GRADE: An emerging consensus on rating quality of evidence and strength of recommendations. *BMJ* **2008**, *336*, 924–926. [CrossRef]

MDPI
St. Alban-Anlage 66
4052 Basel
Switzerland
www.mdpi.com

Nutrients Editorial Office
E-mail: nutrients@mdpi.com
www.mdpi.com/journal/nutrients

www.ingramcontent.com/pod-product-compliance
Lightning Source LLC
Chambersburg PA
CBHW040807250926
51850CB00007BA/1107
* 9 7 8 3 7 2 5 8 1 2 4 5 5 *